Flower Development

Methods and Protocols

Edited by

José Luis Riechmann

Center for Research in Agricultural Genomics and Institució Catalana de Recerca i Estudis Avançats, Barcelona, Spain

Frank Wellmer

Trinity College Dublin, Ireland

Humana Press

Editors
José Luis Riechmann
Center for Research in Agricultural Genomics and
Institució Catalana de Recerca i Estudis
Avançats, Barcelona, Spain

Frank Wellmer
Trinity College Dublin
Ireland

ISSN 1064-3745 ISSN 1940-6029 (electronic)
ISBN 978-1-4614-9407-2 ISBN 978-1-4614-9408-9 (eBook)
DOI 10.1007/978-1-4614-9408-9
Springer New York Heidelberg Dordrecht London

Library of Congress Control Number: 2013956367

Printed on acid-free paper

Humana Press is a brand of Springer
Springer is part of Springer Science+Business Media (www.springer.com)

Methods in Molecular Biology™

Series Editor
John M. Walker
School of Life Sciences
University of Hertfordshire
Hatfield, Hertfordshire, AL10 9AB, UK

For further volumes:
http://www.springer.com/series/7651

Preface

Over the past 25 years, detailed insights into the genetic and molecular mechanisms that control flower development in different angiosperm species have been obtained and many key regulators of flower morphogenesis have been identified. As in all fields of biology, these advances have been made possible largely through technological advances, such as the introduction of methods of molecular genetics, genomics, and systems biology into plant research.

To facilitate further progress in the field of flower development, the present book provides a collection of protocols for many of the experimental approaches that are currently used to study the formation of flowers, from genetic methods and phenotypic analyses to genome-wide experiments, modeling, and system-wide approaches. In addition, several introductory chapters provide an overview of the last 25 years of molecular and genetic studies of flower development, and a current perspective of the field, highlighting open questions and future directions. Methods chapters are organized in five major sections: genetic and phenotypic analyses; microscopy and histology; experimental systems; specific molecular biology methods that are frequently used in flower development studies; and genomics and systems biology. Each chapter contains a brief introduction, step-by-step methods, a list of necessary materials, and a Notes section with tips on troubleshooting. Comprehensive and up-to-date, we hope that this book on flower development will become an essential guide for plant developmental biologists, from the more novice to the experienced researcher, and for those considering venturing into the field.

Barcelona, Spain — *José Luis Riechmann*
Dublin, Ireland — *Frank Wellmer*

Contents

Contributors

ELENA R. ALVAREZ-BUYLLA • *Laboratorio de Genética Molecular, Desarrollo y Evolución de Plantas, Instituto de Ecología, Universidad Nacional Autónoma de México, Ciudad Universitaria, México D.F., Mexico; Centro de Ciencias de la Complejidad, Universidad Nacional Autónoma de México, Ciudad Universitaria, México D.F., Mexico; Department of Plant and Microbial Biology, College of Natural Resources, University of California, Berkeley, Berkeley, CA, USA*

EUGENIO AZPEITIA • *Laboratorio de Genética Molecular, Desarrollo y Evolución de Plantas, Instituto de Ecología, Universidad Nacional Autónoma de México, Ciudad Universitaria, México D.F., Mexico; Centro de Ciencias de la Complejidad, Universidad Nacional Autónoma de México, Ciudad Universitaria, México D.F., Mexico*

VICENTE BALANZÀ • *Instituto de Biología Molecular y Celular de Plantas, Consejo Superior de Investigaciones Científicas-Universidad Politécnica de Valencia, Valencia, Spain*

PATRICIA BALLESTER • *Instituto de Biología Molecular y Celular de Plantas, Consejo Superior de Investigaciones Científicas-Universidad Politécnica de Valencia, Valencia, Spain*

MOHAMMED BENDAHMANE • *Laboratory Reproduction et Dévelopement des Plantes, Institut National de la Recherche Agronomique, Centre National de la Recherche Scientifique, Université Lyon 1, Ecole Normale Supérieure, Lyon, France*

JOHN L. BOWMAN • *School of Biological Sciences, Monash University, Clayton, Australia*

MARIANA BUSTAMANTE • *Center for Research in Agricultural Genomics CSIC-IRTA-UAB-UB, Universidad Autónoma de Barcelona, Barcelona, Spain*

ORIOL CASAGRAN • *Center for Research in Agricultural Genomics CSIC-IRTA-UAB-UB, Universidad Autónoma de Barcelona, Barcelona, Spain*

BARRY CAUSIER • *Centre for Plant Sciences, School of Biology, University of Leeds, Leeds, UK*

FANG CHANG • *State Key Laboratory of Genetic Engineering, Institute of Plant Biology, School of Life Sciences, Fudan University, Shanghai, China*

XUEMEI CHEN • *Department of Botany and Plant Sciences, Institute of Integrative Genome Biology, University of California, Riverside, CA, USA; Howard Hughes Medical Institute, University of California, Riverside, CA, USA*

ZHIHAO CHENG • *State Key Laboratory of Genetic Engineering, Institute of Plant Biology, School of Life Sciences, Fudan University, Shanghai, China*

MONICA COLOMBO • *Instituto de Biología Molecular y Celular de Plantas, Consejo Superior de Investigaciones Científicas-Universidad Politécnica de Valencia, Valencia, Spain*

BRENDAN DAVIES • *Centre for Plant Sciences, School of Biology, University of Leeds, Leeds, UK*

JOSÉ DAVILA-VELDERRAIN • *Laboratorio de Genética Molecular, Desarrollo y Evolución de Plantas, Instituto de Ecología, Universidad Nacional Autónoma de México, Ciudad Universitaria, México D.F., Mexico; Centro de Ciencias de la Complejidad, Universidad Nacional Autónoma de México, Ciudad Universitaria, México D.F., Mexico*

THANH THERESA DINH • *Department of Botany and Plant Sciences, Institute of Integrative Genome Biology, University of California, Riverside, CA, USA; NSF ChemGen IGERT program, University of California, Riverside, CA, USA*

Balaji Enugutti • *Entwicklungsbiologie der Pflanzen, Wissenschaftszentrum Weihenstephan, Technische Universität München, Freising, Germany*

Cristina Ferrándiz • *Instituto de Biología Molecular y Celular de Plantas, Consejo Superior de Investigaciones Científicas-Universidad Politécnica de Valencia, Valencia, Spain*

Thilia Ferrier • *Center for Research in Agricultural Genomics CSIC-IRTA-UAB-UB, Universidad Autónoma de Barcelona, Barcelona, Spain*

Jennifer C. Fletcher • *Plant Gene Expression Center, USDA-ARS/UC, Berkeley, CA, USA; Department of Plant and Microbial Biology, University of California at Berkeley, Berkeley, CA, USA*

Chloé Fourquin • *Instituto de Biología Molecular y Celular de Plantas, Consejo Superior de Investigaciones Científicas-Universidad Politécnica de Valencia, Valencia, Spain*

Robert G. Franks • *Department of Plant and Microbial Biology, North Carolina State University, Raleigh, NC, USA*

Eng-Seng Gan • *Temasek Life Sciences Laboratory, National University of Singapore, Singapore, Singapore; Department of Biological Sciences, Faculty of Science, National University of Singapore, Singapore, Singapore*

Emmanuelle Graciet • *Smurfit Institute of Genetics, Trinity College, Dublin, Ireland*

Veronica Gregis • *Dipartimento di Scienze Biomolecolari e Biotecnologie, Universita degli Studi di Milano, Milan, Italy*

Ueli Grossniklaus • *Institute of Plant Biology and Zürich-Basel Plant Science Center, University of Zürich, Zürich, Switzerland*

Jun He • *State Key Laboratory of Plant Genomics, Institute of Genetics and Developmental Biology, Chinese Academy of Sciences, Beijing, China*

Marcus G. Heisler • *Developmental Biology Unit, European Molecular Biology Laboratory, Heidelberg, Germany; School of Biological Sciences, University of Sydney, Sydney, Australia*

Hiro-Yuki Hirano • *Department of Biological Sciences, Graduate School of Science, The University of Tokyo, Bunkyo-ku, Tokyo, Japan*

Toshiro Ito • *Temasek Life Sciences Laboratory, National University of Singapore, Singapore, Singapore; Department of Biological Sciences, Faculty of Science, National University of Singapore, Singapore, Singapore*

Thomas P. Jack • *Department of Biological Sciences, Dartmouth College, Hanover, NH, USA*

Yuling Jiao • *State Key Laboratory of Plant Genomics, Institute of Genetics and Developmental Biology, Chinese Academy of Sciences, Beijing, China*

Jian Jin • *Center for Research in Agricultural Genomics CSIC-IRTA-UAB-UB, Universidad Autónoma de Barcelona, Barcelona, Spain*

Yue Jin • *State Key Laboratory of Genetic Engineering, Institute of Plant Biology, School of Life Sciences, Fudan University, Shanghai, China*

Martin M. Kater • *Dipartimento di Scienze Biomolecolari e Biotecnologie, Universita degli Studi di Milano, Milan, Italy*

Beth A. Krizek • *Department of Biological Sciences, University of South Carolina, Columbia, SC, USA*

Shaofang Li • *Department of Botany and Plant Sciences, Institute of Integrative Genome Biology, University of California, Riverside, CA, USA*

Xigang Liu • *Department of Botany and Plant Sciences, Institute of Integrative Genome Biology, University of California, Riverside, CA, USA*

Jan U. Lohmann • *Department of Stem Cell Biology, Centre for Organismal Studies, University of Heidelberg, Heidelberg, Germany*

PINGLI LU • *Department of Biology, The Pennsylvania State University, University Park, PA, USA; The Huck Institutes of the Life Sciences, The Pennsylvania State University, University Park, PA, USA*

ELIZABETH LUSCHER • *Department of Botany and Plant Sciences, Institute of Integrative Genome Biology, University of California, Riverside, CA, USA*

HONG MA • *State Key Laboratory of Genetic Engineering, Institute of Plant Biology, School of Life Sciences, Fudan University, Shanghai, China; Center for Evolutionary Biology, School of Life Sciences, Fudan University, Shanghai, China*

IRENE MARTÍNEZ-FERNÁNDEZ • *Instituto de Biología Molecular y Celular de Plantas, Consejo Superior de Investigaciones Científicas-Universidad Politécnica de Valencia, Valencia, Spain*

JOSÉ TOMÁS MATUS • *Center for Research in Agricultural Genomics CSIC-IRTA-UAB-UB, Universidad Autónoma de Barcelona, Barcelona, Spain*

ANNA MEDZIHRADSZKY • *Department of Stem Cell Biology, Centre for Organismal Studies, University of Heidelberg, Heidelberg, Germany*

ELLIOT M. MEYEROWITZ • *Division of Biology, California Institute of Technology, Pasadena, CA, USA*

MONA M. MONFARED • *Plant Gene Expression Center, USDA-ARS/UC, Berkeley, CA, USA; Department of Plant and Microbial Biology, University of California at Berkeley, Berkeley, CA, USA*

JANAKI S. MUDUNKOTHGE • *Department of Biological Sciences, University of South Carolina, Columbia, SC, USA*

TANIA NOLAN • *Sigma-Aldrich, Haverhill, UK*

DIARMUID SEOSAMH Ó'MAOILÉIDIGH • *Smurfit Institute of Genetics, Trinity College Dublin, Dublin, Ireland*

CAROLYN OHNO • *Developmental Biology Unit, European Molecular Biology Laboratory, Heidelberg, Germany*

NATHANAËL PRUNET • *Department of Biological Sciences, Dartmouth College, Hanover, NH, USA*

G. VENUGOPALA REDDY • *Department of Botany and Plant Sciences, Center for Plant Cell Biology, Institute of Integrative Genome Biology, University of California, Riverside, CA, USA*

JOSÉ LUIS RIECHMANN • *Center for Research in Agricultural Genomics CSIC-IRTA-UAB-UB, Universidad Autónoma de Barcelona, Barcelona, Spain; Institució Catalana de Recerca i Estudis Avançats, Barcelona, Spain*

KAY SCHNEITZ • *Entwicklungsbiologie der Pflanzen, Wissenschaftszentrum Weihenstephan, Technische Universität München, Freising, Germany*

PAMELA S. SOLTIS • *Florida Museum of Natural History, University of Florida, Gainesville, FL, USA*

DOUGLAS E. SOLTIS • *Florida Museum of Natural History, University of Florida, Gainesville, FL, USA; Department of Biology, University of Florida, Gainesville, FL, USA*

JUDIT SZÉCSI • *Laboratory Reproduction et Développement des Plantes, Institut National de la Recherche Agronomique, Centre National de la Recherche Scientifique, Université Lyon 1, Ecole Normale Supérieure, Lyon, France*

WAKANA TANAKA • *Department of Biological Sciences, Graduate School of Science, The University of Tokyo, Bunkyo-ku, Tokyo, Japan*

Ljudmilla Timofejeva • *Department of Plant Genetics, Institute of Experimental Biology at the Estonian Agricultural University, Harku, Harju maakond, Estonia*
Taiyo Toriba • *Department of Biological Sciences, Graduate School of Science, The University of Tokyo, Bunkyo-ku, Tokyo, Japan*
Carlos Villarreal • *Instituto de Física, Universidad Nacional Autónoma de México, Ciudad Universitaria, México D.F., Mexico; Centro de Ciencias de la Complejidad, Universidad Nacional Autónoma de México, Ciudad Universitaria, México D.F., Mexico*
Ying Wang • *State Key Laboratory of Plant Genomics, Institute of Genetics and Developmental Biology, Chinese Academy of Sciences, Beijing, China*
Yingxiang Wang • *State Key Laboratory of Genetic Engineering, Institute of Plant Biology, School of Life Sciences, Fudan University, Shanghai, China*
Frank Wellmer • *Smurfit Institute of Genetics, Trinity College Dublin, Dublin, Ireland*
Barbara Wippermann • *Laboratory Reproduction et Développement des Plantes, Institut National de la Recherche Agronomique, Centre National de la Recherche Scientifique, Université Lyon 1, Ecole Normale Supérieure, Lyon, France*
So Youn Won • *Department of Botany and Plant Sciences, Institute of Integrative Genome Biology, University of California, Riverside, CA, USA*
Samuel E. Wuest • *Institute of Evolutionary Biology and Environmental Studies & Zürich-Basel Plant Science Center, University of Zurich, Zürich, Switzerland; Institute of Plant Biology and Zürich-Basel Plant Science Center, University of Zürich, Zürich, Switzerland*
Yifeng Xu • *Temasek Life Sciences Laboratory, National University of Singapore, Singapore, Singapore*
Zaibao Zhang • *State Key Laboratory of Genetic Engineering, Institute of Plant Biology, School of Life Sciences, Fudan University, Shanghai, China*

Part I

Review and Overview Chapters

Chapter 1

Flower Development in *Arabidopsis*: There Is More to It Than Learning Your ABCs

Nathanaël Prunet and Thomas P. Jack

Abstract

The field of Arabidopsis flower development began in the early 1980s with the initial description of several mutants including *apetala1*, *apetala2*, and *agamous* that altered floral organ identity (Koornneef and van der Veen, Theor Appl Genet 58:257–263, 1980; Koornneef et al., J Hered 74:265–272, 1983). By the end of the 1980s, these mutants were receiving more focused attention to determine precisely how they affected flower development (Komaki et al., Development 104:195–203, 1988; Bowman et al., Plant Cell 1:37–52, 1989). In the last quarter century, impressive progress has been made in characterizing the gene products and molecular mechanisms that control the key events in flower development. In this review, we briefly summarize the highlights of work from the past 25 years but focus on advances in the field in the last several years.

Key words Flower development, *Arabidopsis*, Meristem, Transcription factor, Transcriptional network, Floral determinacy, Boundary, Founder cells, Homeotic genes

1 Anatomy and Development of Arabidopsis Flowers

Arabidopsis thaliana flowers form on the flanks of the shoot apical meristem after the vegetative- to -reproductive transition. The flower consists of four organ types that develop in distinct floral whorls: four green sepals arise at precise positions in the first whorl, four white petals in the second, six pollen-producing stamens in the third, and two fused carpels in the fourth. During the initial stages of flower development (until stage 2 [5]), the flower primordium consists of a small cluster of morphologically undifferentiated cells. As the primordium enlarges, floral organ anlagen become morphologically distinct. The sepal primordia become visible at stage 3, the stamen primordia at stage 4, and carpel and petal primordia at stage 5. From stage 6–8, the floral organ primordia enlarge and begin to differentiate. Megasporogenesis and microsporogenesis begin at stage 7, and the pollen matures shortly before anthesis at stage 13.

José Luis Riechmann and Frank Wellmer (eds.), *Flower Development: Methods and Protocols*, Methods in Molecular Biology, vol. 1110, DOI 10.1007/978-1-4614-9408-9_1,

Unlike the shoot apical meristem, which is indeterminate, the floral meristem is determinate. The shoot meristem continues to produce flowers throughout the life of the plant while the floral meristem leads to a stereotypic number of floral organs in each of the four whorls.

The Arabidopsis flower is arranged in three clonally distinct cell layers, the epidermal L1 layer, the underlying subepidermal L2 layer, and the internal cells that derive from the L3.

2 20 Years of ABCs: From a Genetic to a Molecular Model

Floral organ identity has been extensively studied in Arabidopsis for more than 20 years. The floral homeotic mutants *agamous (ag), pistillata (pi), apetala1 (ap1), apetala2 (ap2), and apetala3 (ap3)* were among the first Arabidopsis mutants to be characterized in detail [4]. The floral homeotic mutants have dramatic phenotypes that involve homeotic transformation of floral organ identity. Initially, mutants with organ identity defects in adjacent whorls of the flower were grouped into three classes: A class, B class, and C class. A class mutants such as *ap1* and *ap2* exhibit organ identity defects primarily in first whorl sepals and second whorl petals. The B class mutants *ap3* and *pi* exhibit homeotic transformations of petals to sepals in whorl 2 and stamens to carpels in whorl 3. In the C class mutant *ag*, stamens develop as petals, and carpels are replaced by a whole new flower bud, starting with a whorl of sepals. In addition, *ag* mutants exhibit a loss of floral determinacy; *ag* mutant flowers can consist of more than 100 floral organs, all of which are sepals and petals [4, 6]. Initially, the ABC model was proposed based on genetic experiments, by careful examination of the phenotypes of single, double, and triple mutants. There are two major tenets of the ABC model [7]. First, the model postulates that different combinations of A, B, and C activity specify unique organ identity: A class alone specifies sepals, A + B petals, B + C stamens, and C alone carpels (Fig. 1a). The second major tenet of the ABC model is that A class and C class are mutually repressive; in the absence of A activity C activity is present throughout the flower and vice versa.

Throughout the 1990s, the five key ABC genes were cloned (*AP1* [8], *AP2* [9], *PI* [10], *AP3* [11], and *AG* [12]) and the major tenets of the ABC model were confirmed by molecular experiments. Key among these was the demonstration that the gene products (i.e., RNA and protein) for *AG*, *AP3,* and *PI* accumulate in the region of the flower that exhibits defects in mutants. For example, *ag* mutants exhibit defects in stamens and carpels, and it is in these regions where *AG* RNA accumulates [12].

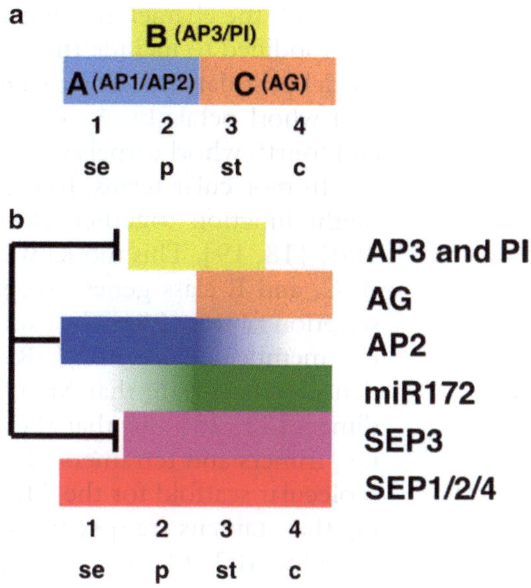

Fig. 1 "Classic" and "revised" *ABC* model of flower development. (**a**) The "classic" ABC model. Three activities in the flower, *A*, *B*, and *C*, are active in adjacent whorls of the flower: *A* in whorls 1 and 2, *B* in whorls 2 and 3, and *C* in whorls 3 and 4. Combinatorial action of these activities specifies organ identity: *A* alone specifies sepals, *A*+ *B* specifies petals, *B*+ *C* specifies stamens, and *C* alone specifies carpels. As originally proposed, the key *A* class genes in Arabidopsis are *AP1* and *AP2*, the *B* class genes are *AP3* and *PI*, and the *C* class gene is *AG*. (**b**) The "revised" *ABC* model includes the *E* class genes *SEP1*, *SEP2*, *SEP3*, and *SEP4*. This model also contains miR172, a negative regulator of *AP2*. Note that the activity of *AP2* is not strictly confined to whorls 1 and 2. Similarly, the activity of miR172 is not strictly confined to whorls 3 and 4. Also shown are the effects of *AP2* as a negative regulator of *SEP3* and *AP3* in whorl 1

A second critical test of the ABC model was to ectopically express the B and C class genes. Consistent with the predictions of the ABC model, ectopic expression of *AG* results in flowers that resemble strong *ap2* mutants [13, 14], while ectopic expression of both *AP3* and *PI* results in flowers that convert sepals to petals and carpels to stamens, demonstrating that B class genes are not only necessary for proper development of petals and stamens but also sufficient to direct petal and stamen development when ectopically expressed in the flower [15].

In the early 2000s, the ABC model was expanded to include a fourth class of genes, the E class genes. The four E class genes in Arabidopsis, the closely related *SEPALLATA* (*SEP*) *1/2/3/4* genes, function redundantly to specify floral organ identity [16, 17]. Single- and double-*sep* mutants exhibit only subtle phenotypic defects, but in the *sep1/2/3/4* quadruple mutant, all four floral organs develop as leaf-like structures [17], a phenotype similar to the *abc* (*ap2 pi ag*) triple mutant [6].

With the characterization of the E class genes, the ABC model was modified to include the E class genes (the ABCE model). This model postulates that first whorl sepals are specified by A+E, second whorl petals by A+B+E, third whorl stamens by B+C+E, and fourth whorl carpels by C+E.

In molecular terms, to explain how the ABCE gene products might function together, the "quartet" model was proposed in 2001 [18, 19]. This model was based on the fact that all of the A, B, C, and E class genes except one encode MADS domain transcription factors [20]. The exception is the A class gene AP2, which is a member of the AP2/ERF family [21]. Multiple lines of evidence demonstrate that MADS domain proteins bind to DNA as dimers [22, 23] and that they can form higher order complexes, i.e., trimers and tetramers [24–27]. The quartet model provides a molecular scaffold for the ABCE model, for example, by postulating that stamens are specified by a tetramer of MADS proteins that includes AP3, PI, AG, and SEP3. However, demonstration that these protein complexes actually exist in plants was slow to materialize, but this has finally been demonstrated [28] (*see* below in Sect. 3 for a more detailed description).

The last decade has been spent working out key molecular details and probing the aspects of the ABCE model that have proven difficult to reconcile. Areas of recent progress include (1) defining the precise nature of MADS protein complexes bound to DNA and providing evidence for MADS protein complexes *in planta*, (2) the role/function of A class genes such as *AP2*, (3) accelerating efforts to define the regulatory networks controlled by floral control genes, (4) control of floral organ positioning, and (5) emerging details of floral meristem determinacy. The remainder of the review focuses on describing recent progress in these five areas.

3 MADS Domain Proteins Form Higher Order Complexes *In Planta*

Initially, evidence that B, C, and E MADS domain proteins form tetramers was supported by several rather circumstantial lines of evidence, the first of which is yeast three and four hybrid experiments that demonstrated that AP3, PI, AG, and SEP3 could form protein complexes in yeast [24–26]. However, these yeast experiments did not firmly establish that the complex is a tetramer. The second line of circumstantial evidence comes from in vivo experiments, the most dramatic of which is the demonstration that simultaneous expression of AP3, PI, SEP3, and AG could convert cauline leaves into organs that resemble stamens [26, 29]. The third is EMSA/gel shift experiments that indicated that a complex larger than a dimer could bind to a consensus MADS binding sequence called a CArG box [24].

Recent experiments have shown definitively that MADS tetramers can bind to a DNA fragment that contains two CArG box sequences [28, 30, 31]. These experiments demonstrate that MADS tetramers consist of a dimer of dimers. Binding of the two MADS dimers to the two CArG sites is cooperative and results in bending of the DNA between the two binding sites, presumably allowing a single tetramer to simultaneously contact two binding sequences [31]. These in vitro experiments have included mixtures of four proteins predicted by the quartet model to form a complex. For example, adding a combination of AP3, PI, SEP3, and AG results in the formation of discrete quaternary complexes, one consisting of two SEP3/AG heterodimers and a second consisting of one SEP3/AG heterodimer and one AP3/PI heterodimer. Again, these experiments provide strong support for the quartet model [28, 30].

Exciting recent progress has demonstrated that MADS protein complexes exist in vivo. To identify MADS complexes *in planta*, transgenic plants were constructed that contained fusions of AP1, AG, AP3, and SEP3 to the green fluorescent protein (GFP). Immunoprecipitation from floral extracts using anti-GFP antibodies followed by LC-MS identified in vivo interaction partners. As predicted by the quartet model, AP3 interacts with PI, SEP3, AP1, and AG. AP1 interacts with AP3, PI, SEP1, and SEP3, while AG interacts with AP3, PI, and SEP3 [28]. Thus, the strongest interactions observed are precisely those predicted by the quartet model, providing the strongest evidence to date that MADS proteins are present in higher order complexes in plant cells.

Smaczniak et al. also provide evidence that, in the presence of a combination of MADS proteins, different MADS complexes can form depending on the relative concentration of the monomers. In the presence of AG, SEP3, AP3, and PI, Smaczniak et al. clearly show that reducing the concentration of AP3 and PI leads to tetramers that consist of only AG and SEP3. The fact that changes in the concentration of MADS monomers can shift the types of tetramers that form offers an explanation for some old observations that involve changing the levels of the MADS genes. For example, *35S:AP3* flowers exhibit partial transformations of fourth whorl carpels to stamens, likely due to AP3 (from ectopic expression) dimerizing with PI (transiently expressed in whorl 4) and the AP3/PI dimers displacing AG/SEP heterodimers from the AG/SEP–AG/SEP tetramers. When the level of B class gene products is increased in *35S:AP3 35S:PI* double transgenics, the carpel-to-stamen transformations are more complete, likely due to a more complete displacement of AG/SEP by AP3/PI. Conversely, increased AG levels in whorl 3 of flowers expressing *AG*'s activator *WUSCHEL* (*WUS*) under the control of the *AP3* promoter (*pAP3:WUS*) trigger a partial transformation of stamens into carpels, likely due to AG/SEP dimers displacing AP3/PI dimers from AG/SEP–AP3/PI tetramers [32].

Non-MADS proteins were also identified as interaction partners of plant ABC MADS proteins. SEP3 was found to be a component of a complex of approximately 670 kD, much larger than the predicted size of a MADS tetramer [28]. Proteins present in various MADS complexes include a number of nucleosome remodeling factors, including the histone demethylase RELATIVE OF EARLY FLOWERING 6 (REF6), the SWI/SNF ATPases BRAHMA (BRM) and SPLAYED (SYD), the CHD remodeler PICKLE (PKL), as well as the ISWI-type remodelers CHROMATIN REMODELING 4 (CHR4), CHR11, and CHR17. The transcriptional corepressors SEUSS and LEUNIG-HOMOLOG (LUH) were also identified as components of MADS protein complexes. A number of non-MADS proteins know to be involved in floral meristem specification were also isolated including the KNOTTED-like homeodomain proteins BELLRINGER (BLR), KNOTTED-LIKE 3 (KNAT3), and BELL-LIKE HOMEODOMAIN 1 (BLH1). These KNOTTED-like proteins interact most strongly with AP1, which is consistent with the role of AP1 in floral meristem identity and of the previously described function of the homeodomain partner proteins in maintenance of the shoot apical meristem. Other proteins known to function in early flower development such as AUXIN RESPONSE FACTOR 2 (ARF2) and SQUAMOSA PROMOTER BINDING PROTEIN LIKE 8 (SPL8) were also shown to be components of AP1 protein complexes [28].

In summary, the recent progress in confirming the molecular predictions of the quartet model has been impressive. These recent advances begin to shed light on several long-standing questions. First, it has long been unclear how functional specificity of MADS proteins was achieved since all MADS proteins recognize the CArG box consensus sequence. Perhaps the non-sequence-specific DNA-binding proteins that are present in MADS protein complexes contribute to functional specificity. Second, it has been unclear how MADS proteins bring about transcriptional activation since some MADS proteins like SEP3 and AP1 possess transcriptional activation domains while others like AP3, PI, and AG do not. It has long been speculated that since MADS proteins form tetramers, then perhaps if one subunit of the tetramer contains a protein with an activation domain (like SEP3 or AP1) then the entire tetramer might function to activate transcription. But the demonstration that MADS proteins are present in a complex with a wide array of chromatin remodelers, co-activators, and corepressors suggests that other mechanisms of transcriptional activation/repression may be operative.

4 The Role of AP2 and A Class Genes

The A class genes have never fit neatly into the ABC paradigm. In the classical ABC model, A class genes were postulated to have two activities. The first is the specification of the organ identity of first whorl sepals and second whorl petals. The second activity is repression of C class activity in whorls one and two. Initially based on floral phenotypes, *AP1* and *AP2* were postulated to be A class genes. However, closer examination of the *ap1* phenotype reveals aspects of the phenotype that do not fit the paradigm. First, even in null *ap1* mutants, petals occasionally develop [33], and development of petals is enhanced in certain double-mutant combinations, e.g., *ap1 ag* [33] or *35S:SEP3 ap1* [34], demonstrating that *AP1* is not strictly required for petal development. In addition, from the *ap1* mutant phenotype, there was no reason to think that *AP1* functions to repress C class genes in whorls 1 and 2 since the first and second whorls of *ap1* mutants do not exhibit characteristics of carpels and stamens. Instead, whorl 1 of *ap1* mutants is bract-like, with secondary flowers often forming in the axils of these bracts, while the second whorl of *ap1* mutants is most often missing. Unlike the B class genes, which function only in floral organ identity, both *AP1* and *AP2* also play a key role in floral meristem identity [33, 35].

At first glance, *AP2* appears to both specify organ identity and repress C class in whorls 1 and 2. In contrast to *ap1*, strong *ap2* mutants never produce petals, suggesting that *AP2* is necessary for the specification of petal identity [6]. Similarly, whorls 1 and 2 clearly develop as reproductive organs in *ap2* mutants, and *AG* is ectopically expressed providing strong support that *AP2* functions to repress *AG* in whorls 1 and 2 [36]. But close examination of strong *ap2* mutants like *ap2-2* reveals phenotypic defects in whorls three and four, i.e., stamens are missing in whorl 3 and carpels are often unfused in whorl 4 suggesting that *AP2* activity is not confined to the outer two whorls [6]. The second bit of information that did not fit the prediction is the spatial expression of *AP2* mRNA. Unlike *AP3*, *PI*, and *AG*, whose mRNA accumulates in the region of the flower that exhibits defects in mutants, *AP2* mRNA accumulates outside of the perianth [9]. Although it has been known for almost 20 years that *AP2* RNA accumulates in all four floral whorls [9, 37], a recent paper [38] provides new detail and demonstrates that the *AP2* expression pattern is more complex; specifically, *AP2* mRNA is expressed more strongly in whorls 1, 2, and 3 compared to whorl 4. Nonetheless, the fact that *AP2* is localized to stamens raised some key questions, e.g., why does *AP2* not repress C class activity if *AP2* and *AG* mRNA expression overlap in whorl 3?

The discovery that miR172 targeted *AP2* in 2003 [37, 39] offered a potential explanation for why *AP2* and A class activity is strongest in the perianth. *AP2* as well as several other *AP2*-like genes contain a binding site for miR172. Evidence that miR172 functions to downregulate *AP2* comes from ectopic experiments; expression of miR172 throughout the flower results in flowers that resemble strong *ap2* mutants [37]. Conversely, constitutive expression of a wild-type version of *AP2* has little or no effect, but ectopic expression of miRNA-resistant forms of *AP2* (i.e., *AP2m*) under various promoters triggers strong *ag*-like defects in whorls 3 and 4, including stamen-to-petal transformations and a loss of flower meristem termination, associated with a decrease in *AG* expression ([37, 38, 40]; the control of floral meristem termination is described below in Sect. 9). Evidence to date suggests that miR172 can regulate *AP2* both by targeting its mRNA for degradation and by repressing translation [37, 41, 42].

Interestingly, ectopic expression of *AP2m* also triggers phenotypes that are not simply explained by the effect of *AP2* on *AG*. *35S:AP2m* and *pAP2-AP2m* flowers have only petals or stamens inside of whorl 2, due to expanded expression of *AP3* and *PI*, and resemble *ag superman* (*sup*) flowers more than *ag* flowers. However, this phenotype was suggested to be an indirect effect of *AP2* expression in the center of the flower, [40] (for more detail on the role of *SUP*, *see* below in Sect. 8). Also the effect of ectopic *AP2* on floral stem cells is partly independent of *AG*. The flower meristem of *35S:AP2m* and *pAP2-AP2m* plants is massively enlarged, probably due to the antagonistic effect of *AP2* on the CLAVATA (CLV) pathway [40, 43]. Similarly, *pAP3:AP2m* lines sometimes exhibit a strong over-proliferation of organs within the third whorl, a phenotype that is associated with the ectopic formation of *WUS* ectopic foci [38].

One key question is whether miR172 is responsible for confining *AP2* activity to the perianth. If miR172 expression were confined to the reproductive organs, it would provide an explanation for why AP2 (and A class) activity is strongest in sepals and petals. Recent meticulous examination of the miR172 expression pattern demonstrates that miR172 accumulates most strongly in developing stamens and carpels, with some expression in petals as well [38]. Thus, miR172 and *AP2* RNAs overlap in developing stamens and to a lesser degree in developing petals. These data rule out simple models where, for example, the domain of *AP2* activity is strictly defined by miR172. Instead, more complex models that depend on the relative levels of *AP2*, miR172, and *AG* must be considered.

One model of how A and C activities relate and that is consistent with present data is as follows (Fig. 1b). The classic ABC model postulates mutual antagonism between A and C classes, meaning that there is no overlap in activity. Extending this, the assumption had been that there would be no overlap in expression of the key A class (*AP2*) and C class (*AG*) genes. In the revised model, there is no strict antagonism between A and C activity or expression. The key observation that is not explainable by the classic ABC model is the simultaneous accumulation of RNAs for *AP2*, *AG*, and miR172 in developing third whorl stamens. Thus, rather than strict antagonism, it appears that the relative levels/activity of *AP2*, *AG*, and miR172 appear to be critical for the specification of floral organ identity. Several key experiments highlight this delicate balance. Interfering with miR172 function in whorl 3 either by destroying the miR172-binding site in *AP2* (*pAP3:AP2m*) or by downregulating miR172 accumulation with a microRNA target mimic (*pAP3:MIM172*) results in increased *AP2* activity in whorl 3, leading to decreased *AG* activity and subsequent conversion of stamens to petals [38]. Similarly, some of the *35S:AP2m* transgenics exhibit flowers that only produce petals inside of whorl 2, presumably due to a total loss of *AG* expression [37, 40]. However, other *35S:AP2m* lines as well as *pAP2-AP2m* transgenics produce only stamens inside of whorl 2, a phenotype associated with a loss of *AG* expression only in the center of the flower [40], suggesting that *AG* expression has a different sensitivity to *AP2* in different subdomains [40]. By contrast, high-level expression of miR172, e.g., *35S:miR172*, results in flowers that resemble *ap2* mutants, presumably due to miR172 downregulating *AP2* throughout the flower and concomitant increased expression of *AG* throughout the flower [37]. These experiments suggest that miR172 functions to downregulate the level of *AP2* activity, but the level of downregulation is dependent on the relative levels of *AP2* and in turn *AG*. As miR172 downregulates *AP2*, it allows for a higher expression of *AG*, which also affects *AP2* expression. In whorl 3, the level of miR172 is high, and *AP2* activity remains below the level required both to direct petal identity and to repress *AG;* thus, stamens develop. By contrast, in whorl 2, the level of miR172 is lower, *AP2* activity is at a level sufficient to repress *AG*, and thus petals develop.

In summary, one of the two major tenets of the ABC model, namely, that A and C class activities are mutually repressive, is an oversimplification. The reality is that the picture is more complex (Fig. 1b).

5 A Biochemical Role for AP2

A molecular role for AP2 has recently been uncovered through its interaction with the TOPLESS (TPL) protein. *tpl* mutants have a dramatic embryo phenotype where the shoot meristem and cotyledons are converted into a second root [44]. *TPL* encodes a protein that is a member of the Groucho/Tup1 family of transcriptional corepressors. TPL has no sequence-specific DNA binding activity; instead, TPL carries out its function by interacting with a protein domain called the EAR motif, present in a subset of transcriptional repressors [45–47].

The AP2 protein contains an EAR motif and has been shown to interact with TPL in yeast two-hybrid assays [48, 49]. A suggestion that this interaction might be biologically relevant comes from the analysis of weak *tpl* mutants, which exhibit weak sepal-to-petal transformations suggesting that B class genes are active in whorl 1. The sepal-to-petal transformations are enhanced in *tpl* mutants that are heterozygous for *ap2-2* (e.g., *ap2-2/AP2*). Demonstration that *AP2* functions as a transcriptional repressor comes from fusions of TPL to the DNA-binding domain of AP2; these TPL-AP2(DB) fusions rescue the floral defects of *ap2-2* mutants [48]. This result is somewhat at odds with a study that looked at AP2 transcriptional targets and showed that AP2 directly activates as well as represses target genes [50] but could be reconciled by postulating that the transcriptional repression activity of AP2 is functionally more important than the transcriptional activation activity. AP2 binds directly to the large regulatory second intron of *AG* [50–53] and recruits both TPL and the histone deacetylase HDA19, likely explaining the repression of *AG* (C class) by *AP2* (A class).

Surprisingly, *AP2* plays a role in repressing B and E class genes in whorl 1. In weak *tpl* mutants, *AP3, PI,* and *SEP3* are misexpressed in whorl 1. AP2, TPL, and HDA19 bind to regulatory sequences in the *AP3* and *SEP3* promoters, and binding of this repression complex is necessary for repression of both *AP3* and *SEP3* in whorl 1 [48].

These results add additional detail to the ABC model. Not only does *AP2* function to repress *AG* (C class), as predicted by the ABC model, but AP2 also functions to repress *AP3* (B class) and *SEP3* (E class) in whorl 1. The repression of B class genes is particularly surprising considering that *AP2*, as an A class genes, is postulated to direct petal identity, but *AP2* clearly represses a gene (*AP3*) that is absolutely required to get petals. It is also difficult to explain why *AP2* does not repress *AP3* and *SEP3* in whorl 2, since *AP2,* as well as its partner proteins TPL and HDA19, is clearly expressed in whorl 2. Perhaps miR172 modulates *AP2* expression in whorl 2 to a level that makes *AP2* unable to repress *AP3* and *SEP3.* Alternatively, another factor such as *UNUSUAL FLORAL ORGANS* [54–57] could function to overcome AP2/TPL/HDA19 repression in whorl 2. Future experiments will shed light on these details.

6 Deciphering Regulatory Networks of Floral Transcription Factors

Plant developmental biologists have long been interested in defining the genes that are controlled by "selector" genes like *LFY*, *AP1*, *AP2*, *AP3/PI*, *AG*, and *SEP3.* Early efforts to systematically define target genes were laborious and led to the identification of only a very small subset of target genes (e.g., [58]). However, the last decade has seen the development of techniques that revolutionized the ability to define transcriptional regulatory networks. The availability of microarrays allowed gene expression changes to be determined on a genome-wide scale but could not distinguish between direct and indirect transcriptional targets. Techniques such as ChIP-chip and ChIP-seq allow the identification of all genomic sites bound by a particular transcription factor. Combining these very rich data sets has allowed detailed gene regulatory networks to begin to be assembled.

To date, ChIP-chip or ChIP-seq experiments have been carried out for LFY [59], AP1 [60], SEP3 [61], AP2 [50], and AP3/PI [62]. A few general lessons can be learned from these studies. First, all of these key floral factors bind to thousands of genomic sites, about 1,500 for LFY and AP3/PI, 2,000 for AP1, 2,200 for AP2, and 4,200 for SEP3. It is worth keeping in mind that the definition of what comprises a "target gene" varies in different studies so, the numbers of targets cannot be compared with a high degree of precision. For LFY, AP1, SEP3, and AP3/PI, slightly less than half of the bound sites exhibit expression changes in the presence/absence of LFY, SEP3, AP1, and AP3/PI, though the level of gene expression change is quite small in some cases. For AP2, less than a quarter of bound genes exhibit expression changes [50]. As expected for a heterodimeric transcription factor, a high percentage of sites bound by AP3 were also bound by PI [62]. Sixty-four percent of the genes bound by AP1 are also bound by SEP3 [60], a result consistent with SEP3 being a key component of MADS protein complexes. Similarly, there was a high degree of correlation of binding sites for AP3/PI, SEP3, AP1, and LFY, consistent with these genes regulating common targets in flower development [62].

Many different types of genes are controlled by these transcription factors. The most predominant classes of targets are (1) other transcription factors, (2) genes associated with cell growth and cell proliferation, and (3) components of signaling pathways, including hormone signaling. Other targets are genes that encode structural proteins or function in metabolism (e.g., lipid biosynthesis [61]). These key floral organ identity proteins regulate genes at multiple levels; some genes are directly regulated, while others exhibit indirect regulation. The floral organ identity proteins AP3/PI, SEP3, AP1, and AP2 are expressed persistently throughout development (LFY expression ceases earlier in flower development, at stage 7) and function in a wide range of cell and tissue types and activate different targets at different developmental stages, from floral organogenesis to cell type specification. For example, AG binds to *KNUCKLES* in whorl 4 primordia beginning at stage 3 [63], to *SPOROCYTELESS/NOZZLE* beginning at stage 6 in developing microspores of the stamen [64], and to the jasmonate biosynthetic gene *DAD1* after stage 10 in stamen filaments [65].

The temporal dynamics of AP3/PI function reveal that AP3/PI control the expression of almost 1,000 genes at very early floral stages (stages 3–4), but less than 60 genes at mid stages of flower development (stage 6), and close to 500 genes at late stages of flower development [62]. In addition, the specific targets controlled by AP3/PI at early stages are largely non-overlapping with those controlled at late stages; less than 3 % of the genes controlled by AP3/PI at stages 3–4 exhibit expression changes at late floral stages. This elegant recent study provides the best evidence to date for the emerging paradigm that the floral organ identity genes regulate distinct sets of genes at different times in development.

For all of the floral transcription factors studied in detail to date, it is clear that there is considerable cross talk between pathways, with extensive positive and negative feedback loops, and a large degree of regulatory redundancy. LFY, AP1, AP2, AP3/PI, AG, and SEP3 also function as both transcriptional activators and repressors, though AP1 (particularly at early stages of flower development) and AP2 function mostly to repress transcription, while SEP3 functions more often as a transcriptional activator. The precise biochemical function of these transcription factors likely depends on interactions with co-activator or corepressor proteins.

The transcriptional network controlled by *LFY*, *SEP3*, *AP1*, *AP3/PI*, and *AP2* is in sharp contrast to the network controlled by *WUSCHEL* (*WUS*), a key gene in the specification of the stem cell niche in the shoot and flower. WUS binds only to slightly more than 100 sites in the genome, and less than ten of these genes exhibit gene expression changes in the presence/absence of WUS [66]. Although WUS appears to control a smaller transcriptional network, the magnitude of the differences in the WUS regulatory network compared to the other networks is difficult to quantify since the methods used to define target genes in these studies are variable.

7 Floral Organ Positioning and Growth

The process of floral organ growth begins with the specification of a small group of "founder" cells or "initial" cells that subsequently give rise to each of the floral organs [67]. The position where each of the floral organs is formed is precisely specified. As development proceeds, these founder cells proliferate first to form floral organ primordia, and later organ outgrowth occurs to produce each organ. Although much is known about how floral organ identity is controlled, much less is known about founder cell specification and subsequent outgrowth of the floral organ primordia. Organ outgrowth and organ identity specification are two separable processes during flower development, as demonstrated by a large number of experiments conducted over the last two decades (e.g., [68–71]). One recent elegant example comes from experiments that utilized inducible artificial miRNAs designed to target AP3 and PI [62]. A single short induction of the amiRNA targeted to *AP3* leads to a transient reduction of *AP3* RNA and protein resulting in differential phenotypic effects in stamens and petals. For petals, downregulation of *AP3* between stages 3 and 6 did not affect organ identity, but downregulation between stages 7 and 9 resulted in petal-to-sepal homeotic transformations. This clearly demonstrates that organ identity in petals is not determined prior to stage 7 and that formation of organ primordia is separable from specification of organ identity.

Classical fate mapping and cell lineage experiments demonstrate that each floral organ derives from a small number of cells: eight cells for each medial sepal, two cells for each petal, four cells for each stamen, and eight cells for carpels [69, 72]. But how these cells attain their fate in a morphologically undifferentiated floral meristem is an open question that has been difficult to address due to the lack of molecular markers that label the founder cells.

Clearly, auxin is a key player in organ outgrowth as mutants defective in auxin production, transport, and signaling exhibit defects in floral organ formation. *PINFORMED1* encodes a protein that functions in polar auxin transport. *pin1* mutants develop a pin-shaped inflorescence that fails to develop flowers or floral organs [73–75]. Flower formation can be restored to *pin1* mutants by exogenous application of auxin [74, 75], demonstrating that auxin is required for lateral organ formation on the flanks of the shoot apical meristem. Mutants with defects in auxin biosynthesis such as the *yucca* [76] and *taa/tar* [77] lack most floral organs, and mutants deficient in auxin signaling such as the *auxin response factor* (*arf*) mutants also exhibit defects in floral organ formation [78, 79]. These phenotypes suggest a role for auxin in the initiation of floral organs.

It is clear from detailed studies on the shoot apical meristem that formation of an auxin maximum, brought about by the action of the PIN1 polar auxin transporter, precedes the morphological formation of the lateral organ primordia such as leaves and flower buds [80]. A similar mechanism appears to be operative in specification of each of the floral organ primordia during stages 2–4. Since auxin levels cannot be directly measured in living plants, transcriptional readouts of the auxin response such as *DR5:GFP* serve as a proxy. Close examination of the *DR5:GFP* expression pattern during early flower development via live imaging [80, 81] reveals that *DR5:GFP* is expressed in the periphery of the inflorescence meristem, at the sites of incipient flower initiation. However, *DR5:GFP* expression is not detected in stage 1 flower buds, but it recommences at stage 2, marking positions where the four sepals will develop [81]. At stage 3, *DR5:GFP* marks the locations where the two lateral stamens will develop. By stage 4, it marks the locations of the four medial stamens and incipient petal primordia. *DR5:GFP* is probably also expressed at stage 5 at the site of incipient carpel primordia, but this has not been described due to the fact that sepal primordia cover the flower bud at this stage and hinder the detection of GFP in the underlying tissues. A recent study of the *petal loss* (*ptl*) mutant confirms the role of auxin maxima in the initiation of petal primordia [82]. In *ptl* flowers, petals are missing or abnormally oriented. This correlates with defects in *DR5:GFP* expression, which is often absent or distorted at the expected sites of petal initiation.

One unresolved question is whether formation of the auxin maxima is the earliest marker of the sites of floral organ primordia initiation. Recent evidence suggests that initial expression of *DR5:GFP* at the positions of the floral organ primordia may occur after expression of *DORNRÖSCHENLIKE* (*DRNL*), a gene that encodes an AP2/ERF transcription factor. A *pDRNL:erGFP* reporter similarly marks the location of the incipient floral primordia, but prior to the formation of the auxin maxima, at least as measured by *DR5:GFP* [81] (Fig. 2c). Although null *drnl* alleles have yet to be described, *drnl* mutants exhibit defects in the development of all four types of floral organs, particularly in stamens, which are sometimes missing or arrest as small filamentous primordia [83] (Fig. 2b). This phenotype, associated with the *DRNL* expression pattern, suggests a role for *DRNL* in specifying the sites of organ initiation [81].

Although the *DR5:GFP* maxima in developing floral organs appears after *DRNL* is activated, this is not definitive evidence that *DRNL* activation is independent of auxin. First, reorientation of the PIN1 auxin efflux transporter precedes the formation of the auxin maxima, and a clear comparison of the temporal dynamics of *pPIN1-PIN1:GFP* and *pDRNL:erGFP* has not been performed. Second, there is a temporal lag in the expression of *DR5:GFP* in response to auxin, which is estimated to be 1–2 h [84], which complicates the analysis. Third, it cannot be ruled out that a non-canonical auxin response pathway [85] is involved in floral organ founder cell specification.

The present model for floral organ outgrowth postulates that the PIN1 auxin efflux proteins reorient auxin flow resulting in an auxin concentration maximum at the site of the incipient floral organ. This auxin maximum then activates genes that lead to organ outgrowth and differentiation. If *DRNL* expression precedes the auxin maximum, then *DRNL* could be either necessary for the auxin maximum to form, i.e., act upstream of auxin, similar to the AP2 gene *PLETHORA* in the root [86], or expressed and function in floral organ founder cells completely independently of auxin, perhaps by functioning to make groups of founder cells "competent" to respond to an auxin signal, similar to the requirement for the mobile transcription factor *TMO7* in proper hypophysis specification [85]. A third possibility is that *DRNL* functions downstream of the auxin signal. The presence of several auxin response elements in the *DRNL* regulatory region [83] made this an appealing model,

Fig. 2 pDRNL:erGFP expression pattern in early-stage flowers pDRNL:erGFP is not detected in the center of the shoot apical meristem (marked with *asterisk*) but is detected in stage 1 flowers. During stage 2, pDRNL:erGFP marks the position of the four sepal primordia. During stage 3, pDRNL:erGFP in a ring marks the future position of the medial stamens; clusters of expression marking the position of the petal primordia extend from the ring

but experiments performed to date have failed to exhibit changes in the *DRNL* expression pattern in early floral stages when the auxin response elements are mutated [81].

It is presently unclear what happens to founder cell specification in mutants that alter floral organ number, e.g., *ap2-2* mutants have fewer floral organs, but it is not known if the reduction in floral organ number is due to a failure to properly specify founder cells. By contrast, *superman* mutants exhibit an increase in organ number; presumably, the additional organs derive from supernumerary founder cell populations, but this has yet to be experimentally demonstrated.

8 Boundary Formation

As mentioned above in Sect. 7, floral organs are specified in stereotypic positions in the flower primordia. Between each floral organ (i.e., intrawhorl), and between the whorls of floral organs (i.e., interwhorl), are regions of reduced cell growth/proliferation that form boundaries [87]. Clearly, these two events, floral organ primordia specification and boundary specification, need to be coordinated. Similar boundaries form between the shoot meristem and the lateral organs that develop on the flanks of the shoot meristem. It is critical that there be coordination between specification of floral organ identity and boundary specification. For example, proper expression of *SUPERMAN* (*SUP*), which is involved in the specification of the boundary between stamens and carpels [88, 89], is dependent on the B class genes *AP3* and *PI* and the C class gene *AG* [90].

The key genes that specify boundaries in both the shoot apical meristem and developing flowers are the *CUP SHAPED COTYLEDON* (*CUC*) genes. The three *CUC* genes, *CUC1*, *CUC2*, and *CUC3*, function in shoot meristem maintenance, lateral organ separation, and floral organ separation [91, 92]. The *CUC* genes encode proteins that are members of the NAC transcription factor family. Loss of *CUC* gene activity results in fused floral organs or an increase in floral organ number [93], suggesting that *CUC* genes function at lateral organ boundaries by suppressing cell growth between lateral organs [87, 94]. *CUC1* and *CUC2* mRNAs are targeted for degradation by the miR164 family of miRNAs [95, 96]. A mutation in the *miR164c* gene, called *early extra petals1* (*eep1*), results in extra petals on the first flowers on the inflorescence and smaller, more widely spaced petals [93], two phenotypes that could be explained by the specification of more and larger boundaries, respectively. Similarly, ectopic expression of constitutively active miRNA-resistant forms of *CUC1* and *CUC2* results in suppression of floral organ growth and smaller floral organs [95–97]. Based on these phenotypes, the *CUC* genes are postulated to be important for formation of lateral organ boundaries by suppressing cell growth between lateral organs [94]. However, somewhat surprisingly, a dominant, gain-of-function allele of *CUC2* with a single point mutation in the miR164-binding site disrupting miRNA targeting results in larger floral organs [98]. Although at odds with the phenotype of transgenic lines expressing miR164-resistant version of *CUC1*, this result suggests that CUC may not simply function to suppress cell proliferation.

Several additional genes also function in the flower in interwhorl and intrawhorl boundary regions. For example, the *PTL* gene functions to specify the intrawhorl boundary in whorl 1. In *ptl* mutants, sepals are fused and petal number is reduced [99].

PTL is strongly expressed between sepal primordia early in flower development and at the margins of developing sepals later on [100, 101]. *PTL* represses proliferation in the intersepal boundary regions: careful examination of *ptl* mutant flowers revealed that the intersepal zones are enlarged compared to wild type [101]. Accordingly, ectopic expression of *PTL* in the flower results in strong growth suppression as evidenced by missing or dramatically reduced floral organs [100]. The sepal fusion phenotype of *ptl* mutant flowers may thus result from increased cell proliferation in the intersepal zones. Alternatively, sepals may fuse later during the development of *ptl* flowers due to the lack of *PTL* activity in the sepal margins. Sepal fusions are enhanced in double mutants between *ptl* and *cuc1* or *cuc2* [100, 101], suggesting that *PTL* and the *CUC* genes control intrasepal boundaries via different mechanisms. *PTL* expression is not detectable in petal primordia, suggesting that *PTL* acts non-cell autonomously to promote petal formation [100]. Interestingly, the size of the intrasepal boundaries correlates with petal number; larger boundaries result in more petals while smaller boundaries (or fused sepals) result in fewer petals. This suggests that there is a mobile signal that moves from the intrasepal areas to the second whorl to control petal number. Indeed, *PTL* was recently shown to make auxin available in the intersepal regions for the formation of an auxin maxima in the neighboring petal founder cells [82].

The *RABBIT EARS* (*RBE*) gene encodes a C_2H_2-type zinc finger protein with an EAR repressor domain that specifies both the intrasepal boundary and the intrawhorl boundary between the third and fourth whorls. The *rbe* mutant exhibits a reduction in petal growth and number, an increase in sepal size and number, as well as occasional fusions between adjacent sepals [102, 103]. A close examination of the *rbe* phenotype also reveals the presence of staminoid petals and stamens in whorl 2 [102]. With the exception of the change in organ identity in the second whorl, the *rbe* phenotype resembles that of the *cuc1 cuc2* double mutant [91]. However, *RBE* is expressed from late stage 2 to stage 6 in petal primordia, that is, in a domain very different from where the *CUC* genes are expressed [102, 103]. Two studies shed light on the origin of the *rbe* phenotype and support a role of *RBE* in both intra- and interwhorl boundaries. *RBE* was recently shown to directly repress miR164c, resulting in an overall increase in miR164 levels and a decrease in *CUC1* and *CUC2* expression [104]. Accordingly, both the *eep1/mir164c-1* mutation and a miR164-resistant version of *CUC1* expressed under the *CUC1* promoter (*pCUC1-CUC1m*) suppress the sepal fusion and loss of petal phenotypes of the *rbe* mutant, suggesting that *RBE* function is partly mediated by the modulation of the miR164/CUC module. However, neither *eep1/mir164c-1* nor *pCUC1-CUC1m* rescues the increase in sepal number or the change in organ identity in the second whorl of *rbe* flowers [104]. Indeed, *RBE* also functions to exclude *AG* from the second whorl, thus maintaining the boundary between whorls 2 and 3 [102]. *AG* is ectopically expressed to varying degrees in the second whorl organs of *rbe* flowers. This ectopic expression of *AG* can account for all the phenotypes of *rbe*, as *ag* is fully epistatic to *rbe* with regard to organ identity, number, and fusion in whorls 1 and 2 [102]. Supporting this hypothesis, a petal-to-stamen transformation and reduction in petal number similar to that seen in *rbe* flowers have been described in other mutants with ectopic *AG* expression in the perianth, such as *ap2* or *leunig* (*lug*) [4, 6, 105]. It is less clear how *AG* may affect the intrasepal boundary in the first whorl of *rbe* mutant flowers.

SUP maintains the boundary between whorls 3 and 4 [88, 89, 106]. *SUP* encodes a C_2H_2-type zinc finger protein that is closely related to RBE. *sup* mutant flowers exhibit an increase in stamen number, usually at the expense of carpels, which are reduced, staminoid, or missing. This phenotype is associated with the expansion of B class genes *AP3* and *PI* towards the center of the flower but does not result from a simple homeotic transformation of carpels into stamens, as the overall number of reproductive organs is significantly increased compared to the wild type [10, 88, 89]. Two hypotheses have been proposed to explain how *SUP* controls the stamen/carpel boundary. *SUP* was initially hypothesized to function in the fourth whorl and to block *AP3* and *PI* expression from spreading ectopically in the center of the flower [88, 89]. Several lines of evidence lend support to this "ectopic" model. First, class B mutants are mostly epistatic to *sup* mutants, suggesting that *AP3* and *PI* mediate *SUP* function. Second, ectopic expression of *AP3*, alone or together with *PI*, under the control of the *35S* promoter (*35S:AP3* and *35S:AP3/PI*), phenocopies *sup*, suggesting that expanded expression of B class gene in the center of the flower is sufficient to explain the *sup* phenotype [15, 107]. It is worth noting that this model implies an indirect role for *SUP* in promoting stem cell termination by opposing B class genes expression in whorl 4 (for further discussion about the influence of class B genes on stem cell termination, *see* Sect. 9 below; about the potential role of *SUP* in this process, *see* [108]). However, the realization that *SUP* is expressed in the inner part of whorl 3 at the boundary with whorl 4, rather than in whorl 4, led to the proposition of an alternate, "proliferation" model, which hypothesizes that *SUP* functions to repress proliferation in a cell-autonomous fashion [106]. According to this model, the production of extra stamens and the expansion of B class gene expression towards the center of the flower bud result from an over-proliferation of *SUP/AP3/PI*-expressing cells from the inner part of whorl 3 rather than from a homeotic transformation of whorl 4 cells into whorl 3-type cells. The "proliferation" model is further supported by the observation that overexpressing *SUP* under the control of the *35S* promoter (*35S:SUP*) results in dwarfed plants due to reduced proliferation rates throughout development [109].

However, ectopic expression of *SUP* throughout the flower under the control of sequences from the *AP1* promoter (*pAP1:SUP*) causes both a decrease in overall floral organ number (which likely results from proliferation defects) and homeotic transformations of petals into sepals and stamens into carpels, associated with a downregulation of *AP3* and its activator *UNUSUAL FLORAL ORGANS* (*UFO*) [110]. The "proliferation" model also implies that cell proliferation is strongly reduced in the fourth whorl of *sup* flowers, but this is not supported by the fact that cells on both sides of the boundary between whorl 3 and 4 proliferate at approximately the same rate [87]. Thus, there is insufficient evidence to discriminate between the "ectopic" and the "proliferation" models, partly due to the insufficient resolution of the techniques that have been used so far to address this question. It is worth noting that both models are not exclusive. Further analyses will be necessary to fully understand how *SUP* functions to maintain the boundary between whorls 3 and 4.

Many other genes could be included in this discussion about the establishment and maintenance of inter- and intrawhorl boundaries in the flower. However, with the exception of the *CUC* and *SUP* genes, which are expressed in boundary domains and which only have a boundary function, most of the genes involved in the development of boundaries cannot be strictly considered to be boundary genes, as they are expressed in much broader domains and/or contribute to other functions. The proper expression pattern of ABC genes, for instance, is required for the establishment of proper interwhorl boundaries and relies on complex regulation. The mutual antagonism between A and C genes and the opposite influence of *AP2* on *AP3* expression in different parts of the flower have been described above in Sect. 4. The exclusion of *AG* from the perianth depends on various factors such as *LUG* or *CURLY LEAF* (*CLF*; [105, 111], while the activation of B class genes in whorls 2 and 3 relies on *UFO* [56]. Even *RBE*, whose main function is to control intra- and interwhorl boundaries, is expressed in petal primordia rather than in boundary regions [102, 103]. Conversely, *HANABA TARANU* (*HAN*) is expressed specifically in the intra- and interwhorl boundaries [112]. Accordingly, *HAN* affects boundary formation and organ positioning but also controls the proper expression of *WUS*, showing that its function is not limited to boundaries. The complexity of the control of boundary formation and its interplay with organ positioning and identity probably reflects the existence of intricate communication networks between organs and boundaries.

9 Regulation of Stem Cells in the Flower

Stem cell maintenance in the floral meristem is disrupted in stage 6 flower buds when carpel primordia arise. This developmental event, known as flower determinacy or floral meristem termination, is crucial for the fertility of the flower, as mutants with defects in stem cell termination set many fewer, if any, seeds, either because they lack proper carpels (e.g., strong *ag* mutants [4, 6]) or due to the formation of extra floral organs within the gynoecium, which interferes with the development of ovules (e.g., *clavata* [113] and *knuckles* (*knu*) [114] mutants [63, 115–118]).

As described above in Sect. 4, the function of *AG* is not restricted to promoting stamen and carpel identity. *AG* is also the main switch for the arrest of stem cell maintenance in the developing flower. Strong *ag* mutants (e.g., *ag-1*) exhibit a homeotic transformation of stamens into petals, but the fourth whorl is not simply transformed into sepals: it is replaced by a new flower bud, which in turn develops into an abnormal flower, thus producing a fully indeterminate flower-within-a-flower pattern [4, 6]. Weaker *ag* mutants (e.g., *ag-4*) have fewer defects in floral organ identity but still exhibit indeterminate flowers [119].

A loss or delay of floral meristem termination has been observed in various other mutants and transgenic lines [15, 37, 40, 107, 115–118, 120–131]. However, unlike *AG*, none of these other genes are strictly required for stem cell termination in the flower, as indicated by the fact that floral meristem termination is fully disrupted only in multiple mutants and transgenic lines but not in single mutants. With the exception of *CRABS CLAW* (*CRC*) and *KNU*, which act downstream of *AG* [63, 132], most of the genes involved in the control of stem cell termination act upstream of *AG* and participate in the complex, multi-level regulation of *AG*. The *CLV*, *ULTRAPETALA1* (*ULT1*), and *PERIANTHIA* (*PAN*) genes, for instance, activate the transcription of *AG* in a subdomain in the center of the flower, directly in the case of *ULT1* and *PAN* [115, 121, 122, 127, 133]. *AG* pre-mRNA undergoes splicing, a process that depends on *HUA1* and *HUA2* and *HUA ENHANCER2* (*HEN2*) and *HEN4* [134]. Finally, the AG protein needs to interact with its SEP partners in order to carry out its determinacy function [16, 17, 135].

Numerous studies have contributed to demonstrating the importance of various microRNAs—and of the miR172 family in particular—in the control of flower meristem termination. The miR172 family is composed of three unique miRNA sequences produced from five independent loci, *MIR172a–e*. Mutants with reduced miR172 levels, such as *squint* (*sqn*) or *powerdress* (*pwr*), as well as a mutant for *MIR172d*, *mir172d-1*, exhibit a potential for indeterminacy in various genetic backgrounds [128, 131, 136]. The weakness of their indeterminacy phenotype as single mutants is probably due to sufficient residual levels of miR172 or to partial redundancy with other *MIR172* loci. miR172 promotes stem cell termination by downregulating its target *AP2*, which in turn represses *AG*; while overexpression of a wild-type version of *AP2*

(*35S:AP2*) results in only subtle phenotypic defects, overexpression of a miR172-resistant version of *AP2* (*35S:AP2m1/3*) results in a total loss of determinacy, with a completely indeterminate meristem at the center of the flower that continuously produces stamens or petals [37, 40]. Other microRNAs have been shown to be involved in the control of determinacy, but their role in this process is not as clear as that for miR172. For example, a mutant in *miR160a* simultaneously develops over-determinate and indeterminate flowers [137]. The amount of miR165/166 and of its HD-ZIP III targets also influences flower meristem termination. Gain-of-function, miR165/166-resistant alleles of *PHABULOSA* (*PHB*) and *PHAVOLUTA* (*PHV*) trigger flower indeterminacy in some mutant backgrounds but so does overexpression of miR165/166 [124]. However, increased levels of miR166 in the *jabba1-D* mutant, and the subsequent decrease in *PHB*, *PHV*, and *CORONA* (*CNA*) expression, result in a reduction in carpel number [138]. By contrast, the simultaneous loss of function of *PHB*, *PHV*, and *CNA* triggers an increase in carpel number and the occasional production of extra organs within the gynoecium [139]. The precise mechanism by which miR160 and miR165/166 influence stem cell termination remains unclear, but it is worth noting that contrary to miR172, they may act independently of *AG*.

Consistent with the role of various microRNAs in flower determinacy, several genes involved in microRNA biosynthesis (e.g., *HEN1* and *DICER LIKE1/CARPEL FACTORY* [*DCL1/CAF*]) or in microRNA-induced mRNA degradation (e.g., *ARGONAUTE1* [140] and *10*) have been shown to contribute to stem cell termination [123, 124, 141, 142]. Accumulation of miR172 is strongly reduced in *hen1* and *dcl1/caf* mutants, and accordingly, the AP2 protein over-accumulates to similar levels as in lines overexpressing a microRNA-resistant version of *AP2* [37]. However, the indeterminacy phenotype of *hen1* and *dcl1/caf* is much weaker than that of the *35S:AP2m1/3* lines [37, 40, 123, 141]. This may be due to a decrease in the levels of other microRNAs in *hen1* and *dcl1/caf*, some of which may have an opposite effect to that of miR172 on flower determinacy. Indeed, the miR169 family dampens the expression of the orthologs of *AG* in snapdragon (*Antirrhinum majus*) and petunia (*Petunia hybrida*) and may have a similar role in Arabidopsis [143]. A decrease in miR169 levels may thus counterbalance the decrease in miR172 in *hen1* and *dcl1/caf* and account for their moderate indeterminacy phenotype. Similar reasons may explain the weak indeterminacy phenotype of *ago1* and *ago10*.

Flower meristem termination is associated with the female developmental program. It relies on a genetic pathway centered on *AG*, which is the master gene for both male and female programs. However, *AG* is required in the fourth whorl but not in the third to disrupt stem cell maintenance, as *pAP3:AG ag-3* flowers are fully indeterminate [144]. Indeed, B class genes, which promote the male program together with *AG*, oppose stem cell termination when ectopically expressed in the center of the flower. Overexpression of *AP3* under the control of the *35S* promoter, alone or together with *PI*, as well as overexpression of *UFO*, an activator of B class genes, trigger a carpel-to-stamen transformation but also significantly delay floral meristem termination, and the fourth whorl sometimes is replaced by several whorls of stamens [15, 56, 107]. Conversely, *ap3* and *pi* mutant flowers are over-determinate, with a strong reduction in floral organ number in whorls 3 and 4 [4, 6]. One possible explanation for how class B genes oppose stem cell termination is that the relative dose of AP3/PI and AG proteins determines the type of quartets formed by AG [108]. An increase in the level of AP3/PI in the fourth whorl may trigger a switch from AG/SEP-AG/SEP quartets, which are responsible for AG's fourth whorl functions, to AG/SEP-AP3/PI, which are responsible for AG's third whorl function. Such a switch from AG/SEP-AG/SEP to AG/SEP-AP3/PI quartets following a decrease in the dose of AG has recently been described in vitro using EMSA experiments [28].

The link between flower meristem termination and the female developmental program is further strengthened by the fact that several other genes involved in carpel development also contribute to the arrest of stem cell maintenance in the flower. *CRC* and *SPATULA* (*SPT*) encode transcription factors that control carpel growth, polarity, and fusion and act redundantly to promote flower determinacy [120]. *crc* and *spt* single mutant flowers are determinate, but the *crc spt* double mutants are slightly indeterminate. Mutation of *CRC* also triggers a strong loss of flower meristem termination in various genetic backgrounds including *ag-1/AG*, *jaiba* (*jab*), *ult1*, and *sqn* [120, 128, 130]. *CRC* is a direct target of *AG* [145], and *SPT* probably also functions downstream of *AG*. However, how both genes regulate stem cell termination remains unclear.

Despite the strong link between flower meristem termination and carpel development, several pieces of evidence suggest that these two programs may be separable. First, stem cell termination requires higher doses of *AG* than carpel identity [146, 147]. Second, various mutant combinations such as *sqn ult1* or *hua1 hua2 ago10* have indeterminate flowers with normal carpels [124, 128]. Third, *AG* is required in a subdomain of whorl 4, at the center of the flower meristem, which includes the stem cells and the organizing center, to promote stem cell termination. Various mutants or transgenic lines with indeterminate flowers specifically lack *AG* expression in this subdomain [40, 115, 121, 122, 128]. Indeed, *AG* initiates a specific program in this subdomain (*see* below). Despite these arguments supporting a separation between stem cell termination and carpel development, it remains unclear whether *AG* expression in this subdomain of whorl 4 is sufficient to disrupt stem cell maintenance. The minor role of genes involved in carpel polarity such as *CRC*, *SPT*, *PHB*, *PHV*, and *CNA* in the control of determinacy suggests that carpel primordia may signal back to the meristem to arrest stem cell proliferation [120, 124].

AG disrupts stem cell maintenance in the flower by turning off the expression of *WUS* [32, 148]. This repression of *WUS* by *AG* has long been thought to be indirect, because (1) *WUS* mRNA becomes undetectable at stage 6, 2 days after the onset of *AG* expression at stage 3 [12, 149], and (2) *AG* is required in both the L3 (where *WUS* is expressed) and L2 (where *WUS* is not expressed) to terminate the flower meristem [150]. Two elegant studies combining genetics and ChIP experiments have recently demonstrated that *AG* represses *WUS* both directly and indirectly and provided insights into how the timing of this repression is achieved [126, 129].

AG binds to two independent sites at the *WUS* locus and directly represses its transcription, as indicated by the fact that induction of a *35S:AG-GR ag-1* line triggers a decrease in *WUS* expression even in the presence of cycloheximide [126]. This direct repression is mediated by the PcG genes *CURLY LEAF* (*CLF*) and *TERMINAL FLOWER2/LIKE HETEROCHROMATIN PROTEIN1* (*TFL2/LHP1*). A mutation in *CLF* abolishes the decrease in *WUS* expression following the induction of the *35S:AG-GR ag-1* line, while H3K27me3 histone repressive marks and TFL2/LHP1 occupancy at the *WUS* locus are strongly reduced in an *ag-1* background [126]. *AG* starts directly repressing *WUS* shortly after the onset of *AG* expression, as *WUS* expression peaks at stage 3 and then steadily decreases as *AG* expression increases, before *WUS* suddenly turns off at stage 6.

In parallel, *AG* represses *WUS* indirectly through *KNUCKLES* (*KNU*), a C_2H_2-type zinc finger protein with an EAR repression domain [63]. *KNU* is expressed in a small domain at the center of the flower bud from stage 6 onwards; thus, KNU activation coincides precisely with *WUS* downregulation [118]. Genetic analyses suggest that *KNU* acts downstream of *AG* and upstream of *WUS* [63]. Accordingly, *KNU* is necessary for the repression of *WUS*, which remains expressed beyond stage 6 in *knu* mutant flowers, and makes them indeterminate, with extra organs developing inside the gynoecium. *KNU* is also sufficient to trigger stem cell termination independently of *AG* when overexpressed under the control of the 35S promoter [63]. *KNU* expression starts at stage 6, 2 days after the onset of *AG* expression at stage 3, and is dependent on *AG*. A similar 2-day delay between the induction of *AG* and the initiation of *KNU* expression is also observed in a *35S:AG-GR ag-1* line. Using an *ap1 cal 35S:AP1-GR* background,

Sun et al. showed that (1) AG binds to the *KNU* locus at floral stage 3, (2) the *KNU* locus is covered with H3K27me3 histone repressive marks until 2 days after flowering induction, and (3) *AG* is required for the removal of these histone repressive marks, thus allowing for *KNU* to be transcribed [63]. The expression of *KNU* is immediately followed by the downregulation of *WUS* expression, suggesting that KNU may directly repress *WUS*. Thus, the delay between the binding of AG to the *KNU* locus and its derepression sets the timing for stem cell termination.

Several pieces of data show that both the direct and indirect repression of *WUS* by *AG* are necessary to consistently terminate stem cell maintenance in the flower. The indeterminacy phenotype of ***knu*** flowers clearly indicates that the direct repression of *WUS* by *AG* alone is insufficient to trigger flower meristem termination [63, 118]. However, ***knu*** flowers are less indeterminate than those of strong ***ag*** mutants [118, 126], suggesting that the direct repression of *WUS* by AG contributes to stem cell termination. Indeed, Liu et al. elegantly showed that expression of a *WUS* reporter lacking the two CArG boxes at one of the *AG*-binding sites is maintained beyond stage 6, after the flower meristem undergoes stem cell termination [126], thus demonstrating that direct repression of *WUS* by AG is required for KNU to consistently turn *WUS* off in a timely fashion.

Prolonged expression of the *WUS* reporter with mutated CArG box sequences beyond stage 6 also shows that cells of the organizing center are not incorporated into developing carpels after stem cell termination [126]. Instead, they retain their uniqueness as a separate subpopulation of cells in the carpel and possibly could retain some meristematic potential. This raises the possibility that repression of stem cell maintenance needs to be maintained beyond stage 6. In this regard, it is worth noting that *KNU* remains expressed at the base of the gynoecium after stage 6 [63, 118] and may therefore block cells of the organizing center from re-initiating stem cells at later stages of flower development.

10 Conclusion

Tremendous progress has been made in our understanding of the mechanisms underlying flower development over the last two decades, since the proposition of the ABC model. The floral homeotic genes have been identified, and the molecular mechanisms through which they perform their function have been characterized. Recent advances led to a precise understanding of some core processes of flower development, in particular how floral organ identity is specified and how stem cell maintenance is disrupted in the center of the flower once all floral organ primordia have been initiated. However, our knowledge of the processes underlying the spatial patterning of the flower remains very incomplete. The specification of organ founder cells precedes that of organ identity, but the number and position of floral organs as well as the number of founder cells involved differ in each whorl. Although auxin and *DRNL* have been shown to be involved in this process, how they establish such a complex pattern is unclear. Contrary to the shoot apical meristem, the flower meristem generates a compact structure with a diversity of organs, and the interplay between auxin and its transporter PIN1 seems insufficient to account for the correct spatial positioning of floral organs. This process is also intrinsically linked to the establishment of boundaries between floral organs. Several genes involved in boundary formation have been identified, but their precise function remains the subject of debate and thus requires further investigation. The recent development of cutting-edge techniques such as RNA-seq, ChIP-seq, and live confocal imaging, combined with computational modeling, now provides the flower development community with powerful tools that should lead to continued important insights.

References

1. Koornneef M, van der Veen JH (1980) Induction and analysis of gibberellin sensitive mutants in *Arabidopsis thaliana* (L.) Heynh. Theor Appl Genet 58:257–263
2. Koornneef M, van Elden J, Hanhart CJ, Stam P, Braaksma FJ, Feenstra WJ (1983) Linkage map of *Arabidopsis thaliana*. J Hered 74:265–272
3. Komaki MK, Okada K, Nishino E, Shimura Y (1988) Isolation and characterization of novel mutants of *Arabidopsis thaliana* defective in flower development. Development 104:195–203
4. Bowman JL, Smyth DR, Meyerowitz EM (1989) Genes directing flower development in Arabidopsis. Plant Cell 1:37–52
5. Smyth DR, Bowman JL, Meyerowitz EM (1990) Early flower development in Arabidopsis. Plant Cell 2:755–767
6. Bowman JL, Smyth DR, Meyerowitz EM (1991) Genetic interactions among floral homeotic genes of Arabidopsis. Development 112:1–20
7. Coen ES, Meyerowitz EM (1991) The war of the whorls: genetic interactions controlling flower development. Nature 353:31–37
8. Mandel MA, Gustafson-Brown C, Savidge B, Yanofsky MF (1992) Molecular characterization of the Arabidopsis floral homeotic gene *APETALA1*. Nature 360:273–277
9. Jofuku KD, den Boer BGW, Van Montagu M, Okamuro JK (1994) Control of Arabidopsis flower and seed development by the homeotic gene *APETALA2*. Plant Cell 6:1211–1225
10. Goto K, Meyerowitz EM (1994) Function and regulation of the Arabidopsis floral homeotic gene *PISTILLATA*. Genes Dev 8:1548–1560
11. Jack T, Brockman LL, Meyerowitz EM (1992) The homeotic gene *APETALA3* of *Arabidopsis thaliana* encodes a MADS box and is expressed in petals and stamens. Cell 68:683–697
12. Yanofsky MF, Ma H, Bowman JL, Drews GN, Feldmann KA, Meyerowitz EM (1990) The protein encoded by the Arabidopsis homeotic gene *AGAMOUS* resembles transcription factors. Nature 346:35–39
13. Mizukami Y, Ma H (1992) Ectopic expression of the floral homeotic gene *AGAMOUS* in transgenic Arabidopsis plants alters floral organ identity. Cell 71:119–131
14. Mandel MA, Bowman JL, Kempin SA, Ma H, Meyerowitz EM, Yanofsky MF (1992) Manipulation of flower structure in transgenic tobacco. Cell 71:133–143
15. Krizek BA, Meyerowitz EM (1996) The Arabidopsis homeotic genes *APETALA3* and *PISTILLATA* are sufficient to provide the B

class organ identity function. Development 122:11–22
16. Pelaz S, Ditta GS, Baumann E, Wisman E, Yanofsky MF (2000) B and C floral organ identity functions require *SEPALLATA* MADS-box genes. Nature 405:200–203
17. Ditta G, Pinyopich A, Robles P, Pelaz S, Yanofsky MF (2004) The *SEP4* gene of *Arabidopsis thaliana* functions in floral organ and meristem identity. Curr Biol 14:1935–1940
18. Theissen G, Saedler H (2001) Floral quartets. Nature 409:469–471
19. Theissen G (2001) Development of floral organ identity; stories from the MADS house. Curr Opin Plant Biol 4:75–85
20. Riechmann JL, Meyerowitz EM (1997) MADS domain proteins in plant development. Biol Chem 378:1079–1101
21. Riechmann JL, Meyerowitz EM (1998) The AP2/EBEBP family of plant transcription factors. Biol Chem 379:633–646
22. Riechmann JL, Krizek BA, Meyerowitz EM (1996) Dimerization specificity of Arabidopsis MADS domain homeotic proteins APETALA1, APETALA3, PISTILLATA, and AGAMOUS. Proc Natl Acad Sci U S A 93:4793–4798
23. Riechmann JL, Wang M, Meyerowitz EM (1996) DNA-binding properties of Arabidopsis MADS domain homeotic proteins APETALA1, APETALA3, PISTILLATA and AGAMOUS. Nucleic Acids Res 24:3134–3141
24. Egea-Cortines M, Saedler H, Sommer H (1999) Ternary complex formation between the MADS-box proteins SQUAMOSA, DEFICIENS, and GLOBOSA is involved in the control of floral architecture in *Antirrhinum majus.* EMBO J 18:5370–5379
25. Fan H-Y, Hu Y, Tudor M, Ma H (1997) Specific interactions between K domains of AG and AGLs, members of the MADS domain family of DNA binding proteins. Plant J 12:999–1010
26. Honma T, Goto K (2001) Complexes of MADS-box proteins are sufficient to convert leaves into floral organs. Nature 409: 525–529
27. Immink RG, Tonaco IA, de Folter S, Shchennikova A, van Dijk AD, Busscher-Lange J, Borst JW, Angenent GC (2009) SEPALLATA3: the 'glue' for MADS box transcription factor complex formation. Genome Biol 10:R24
28. Smaczniak C, Immink RG, Muino JM, Blanvillain R, Busscher M, Busscher-Lange J, Dinh QD, Liu S, Westphal AH, Boeren S, Parcy F, Xu L, Carles CC, Angenent GC, Kaufmann K (2012) Characterization of MADS-domain transcription factor complexes in Arabidopsis flower development. Proc Natl Acad Sci U S A 109:1560–1565
29. Pelaz S, Tapia-Lopez R, Alvarez-Buylla ER, Yanofsky MF (2001) Conversion of leaves into petals in Arabidopsis. Curr Biol 11:182–184
30. Melzer R, Theissen G (2009) Reconstitution of 'floral quartets' in vitro involving class B and class E floral homeotic proteins. Nucleic Acids Res 37:2723–2736
31. Melzer R, Verelst W, Theissen G (2009) The class E floral homeotic protein SEPALLATA3 is sufficient to loop DNA in 'floral quartet'-like complexes in vitro. Nucleic Acids Res 37:144–157
32. Lohmann JU, Hong RL, Hobe M, Busch MA, Parcy F, Simon R, Weigel D (2001) A molecular link between stem cell regulation and floral patterning in Arabidopsis. Cell 105:793–803
33. Bowman JL, Alvarez J, Weigel D, Meyerowitz EM, Smyth DR (1993) Control of flower development in *Arabidopsis thaliana* by *APETALA1* and interacting genes. Development 119:721–743
34. Castillejo C, Romera-Branchat M, Pelaz S (2005) A new role of the Arabidopsis *SEPALLATA3* gene revealed by its constitutive expression. Plant J 43:586–596
35. Irish VF, Sussex IM (1990) Function of the *apetala-1* gene during Arabidopsis floral development. Plant Cell 2:741–753
36. Drews GN, Bowman JL, Meyerowitz EM (1991) Negative regulation of the Arabidopsis homeotic gene *AGAMOUS* by the *APETALA2* product. Cell 65:991–1002
37. Chen X (2004) A microRNA as a translational repressor of *APETALA2* in Arabidopsis flower development. Science 303:2022–2025
38. Wollmann H, Mica E, Todesco M, Long JA, Weigel D (2010) On reconciling the interactions between *APETALA2*, *miR172* and *AGAMOUS* with the ABC model of flower development. Development 137:3633–3642
39. Aukerman MJ, Sakai H (2003) Regulation of flowering time and floral organ identity by a microRNA and its *APETALA2*-like target genes. Plant Cell 15:2730–2741
40. Zhao L, Kim Y-J, Dinh TT, Chen X (2007) *miR172* regulates stem cell fate and defines the inner boundary of *APETALA3* and *PISTILLATA* expression domain in Arabidopsis floral meristems. Plant J 51: 840–849
41. Schwab R, Palatnik JF, Riester M, Schommer C, Schmid M, Weigel D (2005) Specific

effects of microRNAs on the plant transcriptome. Dev Cell 8:517–527
42. Jung J-H, Seo J-H, Reyes JL, Yun J, Chua NH, Park C-M (2007) The *GIGANTEA-regulated* microRNA172 mediates photoperiodic flowering independent of *CONSTANS* in Arabidopsis. Plant Cell 19:2736–2748
43. Wurschum T, Gross-Hardt R, Laux T (2006) *APETALA2* regulates the stem cell niche in the Arabidopsis shoot meristem. Plant Cell 18:295–307
44. Long JA, Woody S, Poethig S, Meyerowitz EM, Barton MK (2002) Transformation of shoots into roots in Arabidopsis embryos mutant at the *TOPLESS* locus. Development 129:2797–2806
45. Long JA, Ohno C, Smith ZR, Meyerowitz EM (2006) *TOPLESS* regulates apical embryonic fate in Arabidopsis. Science 312: 1520–1523
46. Szemenyei H, Hannon M, Long JA (2008) *TOPLESS* mediates auxin-dependent transcriptional repression during Arabidopsis embryogenesis. Science 319:1384–1386
47. Smith ZR, Long JA (2010) Control of Arabidopsis apical-basal embryo polarity by antagonistic transcription factors. Nature 464:423–426
48. Krogan NT, Hogan K, Long JA (2012) APETALA2 recruits the co-repressor TOPLESS and histone deacetylase HDA19 to negatively regulate multiple floral organ identity genes in Arabidopsis. Development 139:4189–4190
49. Causier B, Ashworth M, Guo W, Davies B (2012) The TOPLESS interactome: a framework for gene repression in Arabidopsis. Plant Physiol 158:423–438
50. Yant L, Mathieu J, Dinh TT, Ott F, Lanz C, Wollmann H, Chen X, Schmid M (2010) Orchestration of the floral transition and floral development in Arabidopsis by the bifunctional transcription factor APETALA2. Plant Cell 22:2156–2170
51. Bomblies K, Dagenais N, Weigel D (1999) Redundant enhancers mediate transcriptional repression of *AGAMOUS* by *APETALA2*. Dev Biol 216:260–264
52. Deyholos M, Sieburth L (2000) Separable whorl-specific expression and negative regulation by enhancer elements within the *AGAMOUS* second intron. Plant Cell 12:1799–1810
53. Dinh TT, Girke T, Liu X, Yant L, Schmid M, Chen X (2012) The floral homeotic protein APETALA2 recognizes and acts through an AT-rich sequence element. Development 139:1978–1986
54. Wilkinson MD, Haughn GW (1995) *UNUSUAL FLORAL ORGANS* controls meristem identity and organ primordia fate in Arabidopsis. Plant Cell 7:1485–1499
55. Levin JZ, Meyerowitz EM (1995) *UFO*: an Arabidopsis gene involved in both floral meristem and floral organ development. Plant Cell 7:529–548
56. Lee I, Wolfe DS, Weigel D (1997) *A LEAFY* co-regulator encoded by *UNUSUAL FLORAL ORGANS*. Curr Biol 7:95–104
57. Chae E, Tan QK, Hill TA, Irish VF (2008) An Arabidopsis F-box protein acts as a transcriptional co-factor to regulate floral development. Development 135:1235–1245
58. Sablowski RWM, Meyerowitz EM (1998) A homolog of *NO APICAL MERISTEM* is an immediate target of the floral homeotic genes *APETALA3/PISTILLATA*. Cell 92:93–103
59. Winter CM, Austin RS, Blanvillain-Baufume S, Reback MA, Monniaux M, Wu MF, Sang Y, Yamaguchi A, Yamaguchi N, Parker JE, Parcy F, Jensen ST, Li H, Wagner D (2011) LEAFY target genes reveal floral regulatory logic, cis motifs, and a link to biotic stimulus response. Dev Cell 20:430–443
60. Kaufmann K, Wellmer F, Muino JM, Ferrier T, Wuest SE, Kumar V, Serrano-Mislata A, Madueno F, Krajewski P, Meyerowitz EM, Angenent GC, Riechmann JL (2010) Orchestration of floral initiation by *APETALA1*. Science 328:85–89
61. Kaufmann K, Muino JM, Jauregui R, Airoldi CA, Smaczniak C, Krajewski P, Angenent GC (2009) Target genes of the MADS transcription factor SEPALLATA3: integration of developmental and hormonal pathways in the Arabidopsis flower. PLoS Biol 7:e1000090
62. Wuest SE, O'Maoileidigh DS, Rae L, Kwasniewska K, Raganelli A, Hanczaryk K, Lohan AJ, Loftus B, Graciet E, Wellmer F (2012) Molecular basis for the specification of floral organs by *APETALA3* and *PISTILLATA*. Proc Natl Acad Sci U S A 33:13452–13457
63. Sun B, Xu Y, Ng KH, Ito T (2009) A timing mechanism for stem cell maintenance and differentiation in the Arabidopsis floral meristem. Genes Dev 23:1791–1804
64. Ito T, Wellmer F, Yu H, Das P, Ito N, Alves-Ferreira M, Riechmann JL, Meyerowitz EM (2004) The homeotic protein AGAMOUS controls microsporogenesis by regulation of *SPOROCYTELESS*. Nature 430:356–360
65. Ito T, Ng KH, Lim TS, Yu H, Meyerowitz EM (2007) The homeotic protein AGAMOUS controls late stamen development by regulating a jasmonate biosynthetic gene in Arabidopsis. Plant Cell 19:3516–3529
66. Busch W, Miotk A, Ariel FD, Zhao Z, Forner J, Daum G, Suzaki T, Schuster C, Schultheiss SJ,

Leibfried A, Haubeiss S, Ha N, Chan RL, Lohmann JU (2010) Transcriptional control of a plant stem cell niche. Dev Cell 18:849–861
67. Chandler JW (2011) Founder cell specification. Trends Plant Sci 16:607–613
68. Hill JP, Lord EM (1989) Floral development in *Arabidopsis thaliana*: a comparison of the wild-type and the homeotic *pistillata* mutant. Can J Bot 67:2922–2936
69. Bossinger G, Smyth DR (1996) Initiation patterns of flower and floral organ development in *Arabidopsis thaliana*. Development 122:1093–1102
70. Crone W, Lord EM (1994) Floral initiation and development in wild-type *Arabidopsis thaliana* (Brassicaceae) and in the organ identity mutants *apetala2-1* and *agamous-1*. Can J Bot 72:384–401
71. Jenik PD, Irish VF (2000) Regulation of cell proliferation patterns by homeotic genes during Arabidopsis floral development. Development 127:1267–1276
72. Furner IJ, Pumphrey J (1993) Cell fate in the inflorescence meristem and floral buttress of *Arabidopsis thaliana*. Plant J 4:917–931
73. Okada K, Ueda J, Komaki MK, Bell CJ, Shimura Y (1991) Requirement of the auxin polar transport system in early stages of Arabidopsis floral bud formation. Plant Cell 3:677–684
74. Reinhardt D, Mandel T, Kuhlemeier C (2000) Auxin regulates the initiation and radial position of plant lateral organs. Plant Cell 12:507–518
75. Reinhardt D, Pesce ER, Stieger P, Mandel T, Baltensperger K, Bennett M, Traas J, Friml J, Kuhlemeier C (2003) Regulation of phyllotaxis by polar auxin transport. Nature 426:255–260
76. Cheng Y, Dai X, Zhao Y (2006) Auxin biosynthesis by the YUCCA flavin monooxygenases controls the formation of floral organs and vascular tissues in Arabidopsis. Genes Dev 20:1790–1799
77. Stepanova AN, Robertson-Hoyt J, Yun J, Benavente LM, Xie DY, Dolezal K, Schlereth A, Jurgens G, Alonso JM (2008) TAA1-mediated auxin biosynthesis is essential for hormone crosstalk and plant development. Cell 133:177–191
78. Nagpal P, Ellis CM, Weber H, Ploense SE, Barkawi LS, Guilfoyle TJ, Hagen G, Alonso JM, Cohen JD, Farmer EE, Ecker JR, Reed JW (2005) Auxin response factors *ARF6* and *ARF8* promote jasmonic acid production and flower maturation. Development 132: 4107–4118
79. Tabata R, Ikezaki M, Fujibe T, Aida M, Tian CE, Ueno Y, Yamamoto KT, Machida Y, Nakamura K, Ishiguro S (2010) Arabidopsis *auxin response factor6* and *8* regulate jasmonic acid biosynthesis and floral organ development via repression of class 1 KNOX genes. Plant Cell Physiol 51:164–175
80. Heisler MG, Ohno C, Das P, Sieber P, Reddy GV, Long JA, Meyerowitz EM (2005) Patterns of auxin transport and gene expression during primordium development revealed by live imaging of the Arabidopsis inflorescence meristem. Curr Biol 15:1899–1911
81. Chandler JW, Jacobs B, Cole M, Comelli P, Werr W (2011) *DORNROSCHEN-LIKE* expression marks Arabidopsis floral organ founder cells and precedes auxin response maxima. Plant Mol Biol 76:171–185
82. Lampugnani ER, Kilinc A, Smyth DR (2012) Auxin controls petal initiation in Arabidopsis. Development 140:185–194
83. Nag A, Yang Y, Jack T (2007) *DORNRÖSCHEN-LIKE*, and AP2 gene necessary for stamen emergence in Arabidopsis. Plant Mol Biol 65:219–232
84. Ottenschlager I, Wolff P, Wolverton C, Bhalerao RP, Sandberg G, Ishikawa H, Evans M, Palme K (2003) Gravity-regulated differential auxin transport from columella to lateral root cap cells. Proc Natl Acad Sci U S A 100:2987–2991
85. Schlereth A, Moller B, Liu W, Kientz M, Flipse J, Rademacher EH, Schmid M, Jurgens G, Weijers D (2010) *MONOPTEROS* controls embryonic root initiation by regulating a mobile transcription factor. Nature 464:913–916
86. Aida M, Beis D, Heidstra R, Willemsen V, Blilou I, Galinha C, Nussaume L, Noh YS, Amasino R, Scheres B (2004) The PLETHORA genes mediate patterning of the Arabidopsis root stem cell niche. Cell 119:109–120
87. Breuil-Broyer S, Morel P, de Almeida-Engler J, Coustham V, Negrutiu I, Trehin C (2004) High-resolution boundary analysis during *Arabidopsis thaliana* flower development. Plant J 38:182–192
88. Schultz EA, Pickett FB, Haughn GW (1991) The *FLO10* gene product regulates the expression domain of homeotic genes *AP3* and *PI* in Arabidopsis flowers. Plant Cell 3:1221–1237
89. Bowman JL, Sakai H, Jack T, Weigel D, Mayer U, Meyerowitz EM (1992) *SUPERMAN*, a regulator of floral homeotic genes in *Arabidopsis*. Development 114:599–615
90. Sakai H, Krizek BA, Jacobsen SE, Meyerowitz EM (2000) Regulation of *SUP* expression

identifies multiple regulators involved in Arabidopsis floral meristem development. Plant Cell 12:1607–1618
91. Aida M, Ishida T, Fukaki H, Fujusawa H, Tasaka M (1997) Genes involved in organ separation in Arabidopsis: an analysis of the *cup-shaped cotyledon* mutant. Plant Cell 9:841–857
92. Vroemen CW, Mordhorst AP, Albrecht C, Kwaaitaal MA, de Vries SC (2003) The *CUP-SHAPED COTYLEDON3* gene is required for boundary and shoot meristem formation in Arabidopsis. Plant Cell 15:1563–1577
93. Baker CC, Sieber P, Wellmer F, Meyerowitz EM (2005) The *early extra petals1* mutant uncovers a role for microRNA *miR164c* in regulating petal number in Arabidopsis. Curr Biol 15:303–315
94. Aida M, Tasaka M (2006) Morphogenesis and patterning at the organ boundaries in the higher plant shoot apex. Plant Mol Biol 60:915–928
95. Laufs P, Peaucelle A, Morin H, Traas J (2004) MicroRNA regulation of the *CUC* genes is required for boundary size control in Arabidopsis meristems. Development 131:4311–4322
96. Mallory AC, Dugas DV, Bartel DP, Bartel B (2004) Micro-RNA regulation of NAC-domain targets is required for proper formation and separation of adjacent embryonic, vegetative, and floral organs. Curr Biol 14:1035–1046
97. Sieber P, Wellmer F, Gheyselinck J, Riechmann JL, Meyerowitz EM (2007) Redundancy and specialization among plant microRNAs: role of the MIR164 family in developmental robustness. Development 134:1051–1060
98. Larue CT, Wen J, Walker JC (2009) A microRNA-transcription factor module regulates lateral organ size and patterning in Arabidopsis. Plant J 58:450–463
99. Griffith ME, da Silva Conceicao A, Smyth DR (1999) *PETAL LOSS* gene regulates initiation and orientation of second whorl organs in the Arabidopsis flower. Development 126:5635–5644
100. Brewer PB, Howles PA, Dorian K, Griffith ME, Ishida T, Kaplan-Levy RN, Kilinc A, Smyth DR (2004) *PETAL LOSS*, a trihelix transcription factor gene, regulates perianth architecture in the Arabidopsis flower. Development 131:4035–4045
101. Lampugnani ER, Kilinc A, Smyth DR (2012) *PETAL LOSS* is a boundary gene that inhibits growth between developing sepals in Arabidopsis thaliana. Plant J 71:724–735
102. Krizek BA, Lewis MW, Fletcher JC (2006) *RABBIT EARS* is a second-whorl repressor of *AGAMOUS* that maintains spatial boundaries in Arabidopsis flowers. Plant J 45:369–383
103. Takeda S, Matsumoto N, Okada K (2004) *RABBIT EARS*, encoding a SUPERMAN-like zinc finger protein, regulates petal development in *Arabidopsis thaliana*. Development 131:425–434
104. Huang T, Lopez-Giraldez F, Townsend JP, Irish VF (2012) *RBE* controls *microRNA164* expression to effect floral organogenesis. Development 139:2161–2169
105. Liu Z, Meyerowitz EM (1995) *LEUNIG* regulates *AGAMOUS* expression in Arabidopsis flowers. Development 121:975–991
106. Sakai H, Medrano LJ, Meyerowitz EM (1995) Role of *SUPERMAN* in maintaining Arabidopsis floral whorl boundaries. Nature 378:199–203
107. Jack T, Fox GL, Meyerowitz EM (1994) Arabidopsis homeotic gene *APETALA3* ectopic expression: transcriptional and post-transcriptional regulation determine floral organ identity. Cell 76:703–716
108. Prunet N, Morel P, Negrutiu I, Trehin C (2009) Time to stop: flower meristem termination. Plant Physiol 150:1764–1772
109. Hiratsu K, Ohta M, Matsui K, Ohme-Takagi M (2002) The SUPERMAN protein is an active repressor whose carboxy-terminal repression domain is required for the development of normal flowers. FEBS Lett 514:351–354
110. Yun JY, Weigel D, Lee I (2002) Ectopic expression of *SUPERMAN* suppresses development of petals and stamens. Plant Cell Physiol 43:52–57
111. Goodrich J, Puangsomlee P, Martin M, Long D, Meyerowitz EM, Coupland G (1997) A polycomb-group gene regulates homeotic gene expression in Arabidopsis. Nature 386:44–51
112. Zhao Y, Medrano L, Ohashi K, Fletcher JC, Yu H, Sakai H, Meyerowitz EM (2004) HANABA TARANU is a GATA transcription factor that regulates shoot apical meristem and flower development in Arabidopsis. Plant Cell 16:2586–2600
113. Brand U, Fletcher JC, Hobe M, Meyerowitz EM, Simon R (2000) Dependence of stem cell fate in Arabidopsis on a feedback loop regulated by *CLV3* activity. Science 289:617–619
114. Jaber E, Thiele K, Kindzierski V, Loderer C, Rybak K, Jurgens G, Mayer U, Sollner R, Wanner G, Assaad FF (2010) A putative TRAPPII tethering factor is required for cell plate assembly during cytokinesis in Arabidopsis. New Phytol 187:751–763
115. Clark SE, Running M, Meyerowitz EM (1993) *CLAVATA1*, a regulator of meristem

and flower development in *Arabidopsis*. Development 119:397–418

116. Clark SE, Running MP, Meyerowitz EM (1995) *CLAVATA3* is a specific regulator of shoot and floral meristem development affecting the same processes as *CLAVATA1*. Development 121:2057–2067
117. Kayes JM, Clark SE (1998) *CLAVATA2*, a regulator of meristem and organ development in Arabidopsis. Development 125:3843–3851
118. Payne T, Johnson SD, Koltunow AM (2004) *KNUCKLES* (*KNU*) encodes a C2H2 zinc-finger protein that regulates development of basal pattern elements of the Arabidopsis gynoecium. Development 131:3737–3749
119. Sieburth LE, Running MP, Meyerowitz EM (1995) Genetic separation of third and fourth whorl functions of *AGAMOUS*. Plant Cell 7:1249–1258
120. Alvarez J, Smyth DR (1999) *CRABS CLAW* and *SPATULA*, two Arabidopsis genes that control carpel development in parallel with *AGAMOUS*. Development 126:2356–2375
121. Das P, Ito T, Wellmer F, Vernoux T, Dedieu A, Traas J, Meyerowitz EM (2009) Floral stem cell termination involves the direct regulation of *AGAMOUS* by *PERIANTHIA*. Development 136:1605–1611
122. Fletcher JC (2001) The *ULTRAPETALA* gene controls shoot and floral meristem size in Arabidopsis. Development 128:1323–1333
123. Jacobsen SE, Running MP, Meyerowitz EM (1999) Disruption of an RNA helicase/RNAse III gene in Arabidopsis causes unregulated cell division in floral meristems. Development 126:5231–5243
124. Ji L, Liu X, Yan J, Wang W, Yumul RE, Kim YJ, Dinh TT, Liu J, Cui X, Zheng B, Agarwal M, Liu C, Cao X, Tang G, Chen X (2011) *ARGONAUTE10* and *ARGONAUTE1* regulate the termination of floral stem cells through two microRNAs in Arabidopsis. PLoS Genet 7:e1001358
125. Liu X, Huang J, Wang Y, Khanna K, Xie Z, Owen HA, Zhao D (2010) The role of floral organs in carpels, an Arabidopsis loss-of-function mutation in *MicroRNA160a*, in organogenesis and the mechanism regulating its expression. Plant J 416–428
126. Liu X, Kim YJ, Muller R, Yumul RE, Liu C, Pan Y, Cao X, Goodrich J, Chen X (2011) *AGAMOUS* terminates floral stem cell maintenance in Arabidopsis by directly repressing *WUSCHEL* through recruitment of polycomb group proteins. Plant Cell 23:3654–3670
127. Maier AT, Stehling-Sun S, Wollmann H, Demar M, Hong RL, Haubeiss S, Weigel D, Lohmann JU (2009) Dual roles of the bZIP transcription factor PERIANTHIA in the control of floral architecture and homeotic gene expression. Development 136:1613–1620
128. Prunet N, Morel P, Thierry AM, Eshed Y, Bowman JL, Negrutiu I, Trehin C (2008) *REBELOTE*, *SQUINT*, and *ULTRAPETALA1* function redundantly in the temporal regulation of floral meristem termination in *Arabidopsis thaliana*. Plant Cell 20:901–919
129. Sun B, Ito T (2010) Floral stem cells: from dynamic balance towards termination. Biochem Soc Trans 38:613–616
130. Zuniga-Mayo VM, Marsch-Martinez N, de Folter S (2012) JAIBA, a class-II HD-ZIP transcription factor involved in the regulation of meristematic activity, and important for correct gynoecium and fruit development in Arabidopsis. Plant J 71:314–326
131. Yumul RE, Kim YJ, Liu X, Wang R, Xiao L, Chen X (2013) *POWERDRESS* and diversified expression of the *MIR172* gene family bolster the floral stem cell network. PLoS Genet 9(1):e1003218
132. Bowman JL, Smyth DR (1999) *CRABS CLAW*, a gene that regulates carpel and nectary development in Arabidopsis, encodes a novel protein with zinc finger and helix-loop-helix domains. Development 126:2387–2396
133. Carles CC, Fletcher JC (2009) The SAND domain protein ULTRAPETALA1 acts as a trithorax group factor to regulate cell fate in plants. Genes Dev 23:2723–2728
134. Cheng Y, Kato N, Wang W, Li J, Chen X (2003) Two RNA binding proteins, HEN4 and HUA1, act in the processing of *AGAMOUS* pre-mRNA in *Arabidopsis thaliana*. Dev Cell 4:53–66
135. Goto K, Kyozuka J, Bowman JL (2001) Turning floral organs into leaves, leaves into floral organs. Curr Opin Genet Dev 11:449–456
136. Smith MR, Willmann MR, Wu G, Berardini TZ, Moller B, Weijers D, Poethig RS (2009) Cyclophilin 40 is required for microRNA activity in Arabidopsis. Proc Natl Acad Sci U S A 106:5424–5429
137. Liu C, Chen H, Er HL, Soo HM, Kumar PP, Han JH, Liou YC, Yu H (2008) Direct interaction of *AGL24* and *SOC1* integrates flowering signals in Arabidopsis. Development 135:1481–1491
138. Williams L, Grigg SP, Xie M, Christensen C, Fletcher JC (2005) Regulation of Arabidopsis shoot apical meristem and lateral organ formation by microRNA miR166g and its AtHD-ZIP target genes. Development 132:3657–3668

139. Prigge MJ, Otsuga D, Alonso JM, Ecker JR, Drews GN, Clark SE (2005) Class III homeodomain-leucine zipper gene family members have overlapping, antagonistic, and distinct roles in Arabidopsis development. Plant Cell 17:61–76
140. Bohmert K, Camus I, Bellini C, Bouchez D, Caboche M, Benning C (1998) *AGO1* defines a novel locus of Arabidopsis controlling leaf development. EMBO J 17:170–180
141. Chen X, Liu J, Jia D (2002) *HEN1* functions pleiotropically in Arabidopsis development and acts in C function in the flower. Development 129:1085–1094
142. Park W, Li J, Song R, Messing J, Chen X (2002) CARPEL FACTORY, a Dicer homolog, and HEN1, a novel protein, act in microRNA metabolism in *Arabidopsis thaliana*. Curr Biol 12:1484–1495
143. Cartolano M, Castillo R, Efremova N, Kuckenberg M, Zethof J, Gerats T, Schwartz-Sommer Z (2007) A conserved microRNA module exerts homeotic control over *Petunia hybrida* and *Antirrhinum majus* floral organ identity. Nat Genet 39:901–905
144. Jack T, Sieburth L, Meyerowitz EM (1997) Targeted misexpression of *AGAMOUS* in whorl 2 of Arabidopsis flowers. Plant J 11:825–839
145. Gomez-Mena C, de Folter S, Costa MM, Angenent GC, Sablowski R (2005) Transcriptional program controlled by the floral homeotic gene *AGAMOUS* during early organogenesis. Development 132:429–438
146. Chuang C-F, Meyerowitz EM (2000) Specific and heritable genetic interference by double-stranded RNA in *Arabidopsis thaliana*. Proc Natl Acad Sci U S A 97:4985–4990
147. Mizukami Y, Ma H (1995) Separation of *AG* function in floral meristem determinacy from that in reproductive organ identity by expressing antisense *AG* RNA. Plant Mol Biol 28:767–784
148. Lenhard M, Bohnert A, Jürgens G, Laux T (2001) Termination of stem cell maintenance in Arabidopsis floral meristems by interactions between *WUSCHEL* and *AGAMOUS*. Cell 105:805–814
149. Mayer KFX, Schoof H, Haecker A, Lenhard M, Jürgens G (1998) Role of *WUSCHEL* in regulating stem cell fate in the Arabidopsis shoot meristem. Cell 95:805–815
150. Sieburth LE, Drews GN, Meyerowitz EM (1998) Non-autonomy of *AGAMOUS* function in flower development: use of a Cre/loxP method for mosaic analysis in Arabidopsis. Development 125:4303–4312

Chapter 2

Flower Development in the Asterid Lineage

Barry Causier and Brendan Davies

Abstract

A complete understanding of the genetic control of flower development requires a comparative approach, involving species from across the angiosperm lineage. Using the accessible model plant *Arabidopsis thaliana* many of the genetic pathways that control development of the reproductive growth phase have been delineated. Research in other species has added to this knowledge base, revealing that, despite the myriad of floral forms found in nature, the genetic blueprint of flower development is largely conserved. However, these same studies have also highlighted differences in the way flowering is controlled in evolutionarily diverse species. Here, we review flower development in the eudicot asterid lineage, a group of plants that diverged from the rosid family, which includes *Arabidopsis*, 120 million years ago. Work on model species such as *Antirrhinum majus*, *Petunia hybrida*, and *Gerbera hybrida* has prompted a reexamination of textbook models of flower development; revealed novel mechanisms controlling floral gene expression; provided a means to trace evolution of key regulatory genes; and stimulated discussion about genetic redundancy and the fate of duplicated genes.

Key words Flower development, Asterids, Floral meristem, Floral organ identity, (A)BC model

1 Introduction

In nature, plants are bombarded by a myriad of internal and external signals that control every aspect of their development. The point at which a plant flowers, or transits from vegetative growth to reproductive growth, is strictly controlled to ensure that environmental conditions are suitable for production of the next generation. Flowers, like all postembryonic plant organs, develop from dedicated pools of stem cells maintained in meristems. The transition from seedling to flower production involves a progression from vegetative growth through inflorescence growth to flowering, with the meristems at each stage having distinct properties and producing specific lateral organs characteristic of the growth stage (Fig. 1). The vegetative meristem is an indeterminate structure that gives rise to leaves or other specialized organs, including axillary meristems. The inflorescence meristem can be either determinate (producing a

José Luis Riechmann and Frank Wellmer (eds.), *Flower Development: Methods and Protocols*, Methods in Molecular Biology, vol. 1110, DOI 10.1007/978-1-4614-9408-9_2, © Springer Science+Business Media New York 2014

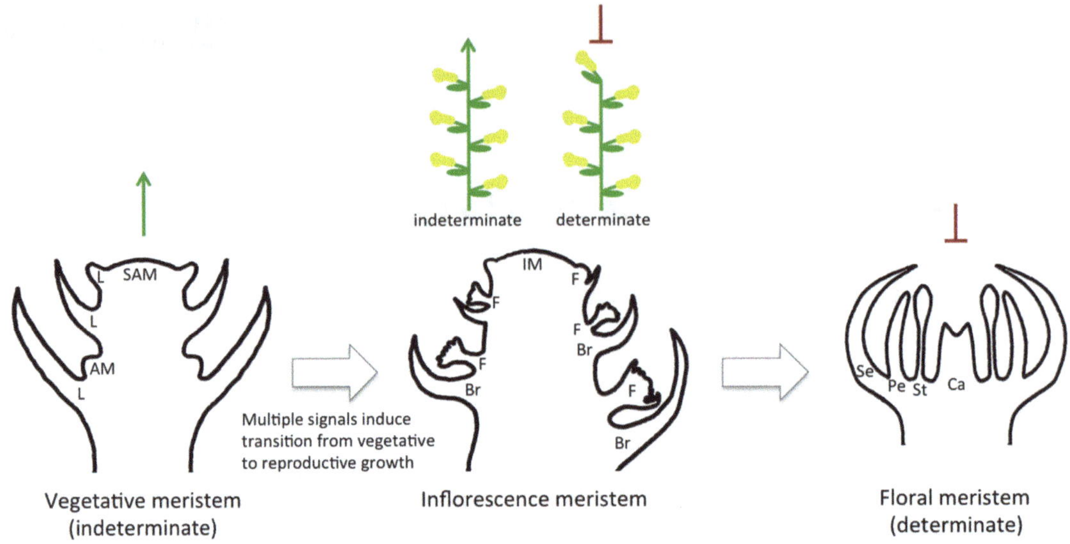

Fig. 1 Structure of vegetative, inflorescence, and floral meristems. The vegetative meristem (SAM; *left*) is an indeterminate structure (represented by the *green arrow*), which produces structures such as the leaf (L) and axillary meristems (AM) on its flanks. Under the appropriate environmental conditions, the vegetative meristem converts to an inflorescence meristem (IM; *center*). This can be either indeterminate (as shown in the diagram *above left*) or determinate, where development halts with the production of a terminal flower (*above right*, where the *barred line* represents termination of development). The inflorescence meristem produces leaves/bracts (Br) and floral merlstems (F) on its flanks. The floral meristem (*right*) is a determinate structure (represented by the *barred line*) that produces all of the organs of the flower: sepals (Se), petals (Pt), stamens (St), and carpels (Ca), with which floral meristem development terminates

fixed number of organs before initiating development of a terminal flower) or indeterminate and produces leaves, bracts, and floral meristems. The organs that comprise the flower, sepals, petals, stamens and carpels, are all derived from the floral meristem, which is usually a determinate structure (Fig. 1). Distinct genetic networks establish the identity of the vegetative, inflorescence, and floral meristems and, consequently, the development of the organs that they produce. In the model eudicot *Arabidopsis*, some of these networks have been well characterized and represent the paradigm to which flowering and other developmental processes are compared in other plants. However, as we discuss here, the broad diversity of floral form means that regulatory networks in *Arabidopsis* must be considered along with those of other plant species to allow a complete understanding of the development and evolution of flowers. In this chapter we restrict our discussions to flower development in the eudicot asterid lineage, which diverged from the rosid family that includes *Arabidopsis*, approximately 120 million years ago [1].

2 Establishing the Floral Meristem

In *Arabidopsis*, numerous factors control the switch from inflorescence meristem identity to floral meristem identity [2]. Central to this transition is *LEAFY* (*LFY*), which regulates a whole host of additional floral meristem identity genes, including the related MADS-box genes *APETALA1* (*AP1*) and *CAULIFLOWER* (*CAL*), the *SEPALLATA* (*SEP1-4*) MADS-box genes, and the *LATE MERISTEM IDENTITY* (*LMI1* and *LMI2*) genes [2–7]. A complex relationship exists between these factors, including mutual regulation, which ensure appropriate relative expression levels. In particular, following activation of *AP1* and *CAL* by LFY, these factors act together with others to reinforce expression of *LFY*, preventing the floral meristem reverting to an inflorescence meristem [2]. Loss-of-function mutations in the homologues of *LFY* or *AP1* result in the partial or the complete lack of flowering with flowers being replaced by inflorescence shoots in all species examined (reviewed in [8]). However, differences in the severity of the phenotypes have been reported, suggesting that the conditions that determine the floral state are less stringent in *Arabidopsis* than in other dicots. For example, while *lfy* mutants produce aberrant flowers [9], mutations in the *Antirrhinum FLORICAULA* (*FLO*) and *Petunia ABERRANT LEAF AND FLOWER* (*ALF*) homologues of *LFY* result in plants that never flower [10, 11]. Instead, inflorescence meristem identity is maintained and bracts develop in place of flowers. Similarly, *ap1* mutants regularly flower [12], while silencing of the *Antirrhinum* (*SQUAMOSA* (*SQUA*)) and *Petunia* (*PETUNIA FLOWERING GENE* (*PFG*)) *AP1*-like genes results in plants that rarely or never flower, respectively [13, 14]. It is unknown whether other *AP1*-like genes act redundantly with *SQUA* in *Antirrhinum*, as is the case in *Arabidopsis*, but in *Petunia* several *AP1*-class genes have been identified. Knockout alleles of *PFG* or *FLORAL BINDING PROTEIN 26* (*FBP26*) do not exhibit phenotypic abnormalities, while *pfg fbp26* double mutants have very subtle floral defects [15]. In contrast, however, transgenic cosuppression of *PFG* results in plants that remain in the vegetative phase of growth, suggesting that *PFG* participates in reproductive growth [14]. *PFG* cosuppression also results in down-regulation of *FBP26*, but the severity of the observed phenotype indicates that the expression of other *AP1*-like genes could also be affected in these lines, suggesting that the *Petunia AP1* genes function redundantly in floral meristem identity [16].

The F-box protein UNUSUAL FLORAL ORGANS (UFO) also plays a role, together with LFY, in conferring floral meristem identity in *Arabidopsis*. However, *ufo* mutants show weak floral meristem defects [17, 18], and the role of these F-box proteins in floral meristem identity was only established when *UFO* homologues in other species were characterized (reviewed in [19]).

While the *Antirrhinum fimbriata* (*fim*) mutant shows some floral reversion to inflorescence identity [20], mutations in the *Petunia DOUBLE TOP* (*DOT*) and tomato *ANATHA* (*AN*) *UFO* orthologues are more severe, with floral identity almost completely lost [21–23]. The DOT protein physically interacts with ALF, the *Petunia* LFY, resulting in ALF activation [23].

Other *Arabidopsis* genes with orthologues in studied dicot species have also been implicated in floral meristem identity. These include *APETALA2* (*AP2*), which encodes a transcription factor, and the *SEPALLATA* (*SEP*) genes, which encode multiply redundant MADS-domain transcription factors. Defects in the flowers of the *Arabidopsis ap2* mutant, including conversion of sepals to leaf- or bract-like structures, and the production of secondary flowers in the second whorl of primary flowers suggest that *AP2* specifies floral meristem identity (reviewed in [24]). In *Antirrhinum*, two *AP2*-like genes *LIPLESS1* (*LIP1*) and *LIP2* function redundantly to direct aspects of petal development. In addition, first whorl organs of the *lip1 lip2* double mutant are bract-like, suggesting an incomplete transition from inflorescence to floral meristem identity [25]. *AP2*-like genes have been identified in other eudicots, although their function in these species has yet to be established. For example, mutation of the *Petunia AP2* gene *PhAP2A* shows no floral defects, suggesting that either it acts redundantly with other AP2 factors, as shown in *Antirrhinum*, or *Petunia* might differ from other model eudicots in the way that floral meristem identity is specified [16, 26]. In *Arabidopsis*, complete loss of *SEP* gene activity, in the quadruple *sep1 sep2 sep3 sep4* mutant, which produces indeterminate "flowers" composed of just leaves, reveals a role for the *SEPs* in floral meristem identity [27]. Furthermore, the role of *SEPs* in establishing a floral context in which later acting organ identity genes can function was elegantly demonstrated in transgenic *Arabidopsis* ectopically expressing organ identity genes together with *SEP3*. These plants showed conversion of leaves to floral organ structures [28, 29]. Analysis of *SEP* homologues in other eudicots shows that the role of this class of genes is conserved, although some differences exist (reviewed in [19]). The function of the *SEP* genes in floral development is covered in greater detail in a later section.

In summary, a stepwise activation of a range of genes, mainly encoding transcription factors, is required to establish and maintain floral meristem identity. These genes are conserved in other species, and whereas their regulatory networks appear to be broadly similar, differences often manifest themselves in the severity of the observed phenotypes. While *Arabidopsis* has served as a model for elucidating the regulatory networks, the roles of several of the key genes that set the floral context only became apparent when diverse species were examined. Comparative studies are therefore important for dissecting regulatory circuits as well as for understanding the evolution of gene networks.

3 Floral Organ Identity

Once the floral meristem is initiated, floral organ primordia differentiate on its flanks (Fig. 1). In the model dicots *Arabidopsis* and *Antirrhinum*, the first floral organs to differentiate are the sepals, which are leaflike structures that enclose and protect the other floral organs as they initiate and mature. The petals form internally to the sepals, followed by the reproductive organs—a ring of male stamens surrounding the central female gynoecium (formed from two fused carpels). Over 20 years ago a model was proposed, based on the study of floral homeotic mutants of *Arabidopsis* and *Antirrhinum*, which posited that three overlapping genetic functions (A, B, and C) were sufficient to specify all the floral organ types in their appropriate positions [30]. Briefly, A-function genes alone specify sepals in the outer whorl of the flower, A plus B specify petals in the second whorl, B plus C direct stamen development in the third whorl, and C alone terminates the floral meristem by specifying development of the female organs at the center of the flower. In addition, the model predicted antagonism between the A- and C-function genes to establish the correct domains of expression.

In the intervening years, as the flowers of more species have been studied and their floral organ identity genes characterized, it has become clear that the ABC model is no longer the all-encompassing model it once promised to be. No recessive mutant conforming to the expected A-function phenotype has been reported in any species other than *Arabidopsis*. To overcome the embarrassing lack of a clear A-function role in specifying perianth identity in any studied species, it has been necessary to modify the ABC model in line with an older model of flower development [31]. The modified (A)BC model [24, 32] states that two genetic functions (B and C) act alone or in combination to specify petals (B alone), stamens (B plus C), and the gynoecium (C alone), within a floral context established by what has been termed the (A)-function. The (A)-function thus explicitly acts to define floral meristem identity and activates and regulates B- and C-function gene expression.

3.1 The (A)-Function: Floral Organ Production, in the Right Place at the Right Time

As a determinant of floral meristem identity, the (A)-function includes floral meristem identity genes that belong to the *LFY*, *AP1*, *AP2*, and *SEP* families [24, 32, 33]. In the (A)BC model, sepals represent the ground-state organ of the floral meristem, and it is not until expression of the B- and C-function genes is initiated that these default organs take on petal, stamen, and gynoecium identities. The (A)-function also includes factors that activate B- and C-function gene expression and set the boundaries of their expression through a balance between activation and repression (Fig. 2).

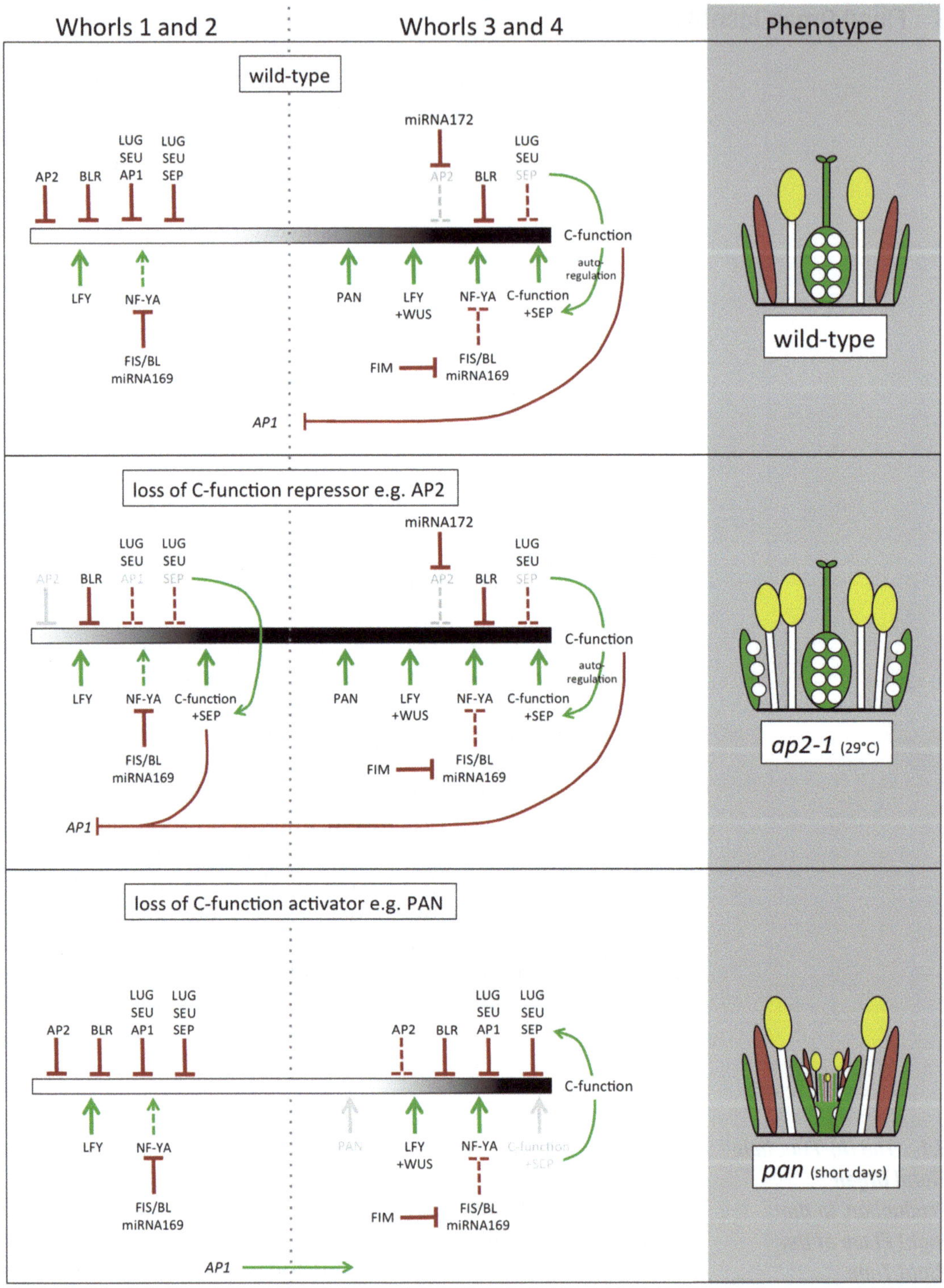

Fig. 2 The balance model of C-function expression domain control. In wild-type plants (*top panel*), C-function expression is actively excluded from floral whorls 1 and 2 by the action of various repressor proteins. Although activators of C-gene expression also function in the outer whorls, their activity is overwhelmed by that of the repressors. In *Arabidopsis*, these repressors include BLR, AP2, AP1, and SEP in complexes with the LUG–SEU corepressor and many others (*see* text). In *Antirrhinum*, and other asterid species, AP2-related factors do not

For example, in *Arabidopsis*, LFY and SEP3 activate B- and C-class genes, while the interacting MADS-domain factors AP1-AGL24 and AP1-SVP, which recruit the LEUNIG-SEUSS (LUG–SEU) corepressor complex, inhibit their early expression [34]. In addition, expression of the B-function is excluded from the *Arabidopsis* fourth whorl by the action of the *SUPERMAN* (*SUP*) gene [35]. Although counterparts of these B- and C-function regulators are conserved in other species, it is yet to be established for many of them whether they regulate B- and C-class genes in a similar way. As discussed above, the role of *LFY*, *AP1*, *SEP*, and *UFO* gene families in the establishment of the floral meristem is conserved among dicots. *UFO*, like its *Antirrhinum* orthologue *FIM*, also promotes transcription of organ identity genes, together with other factors including *LFY* [17, 18, 36]. *SUP*, and its *Antirrhinum* counterpart *OCTANDRA* (*OCT*), encodes a C2H2-type zinc finger-containing transcription factor that establishes the boundaries between whorls 3 and 4 of the flower by excluding the B-function from the fourth

Fig. 2 (continued) appear to regulate the C-function. However, STY, the *Antirrhinum* orthologue of LUG, is a negative regulator of C-function gene expression. *Antirrhinum* and *Petunia* also prevent expression of the C-function in the perianth via a microRNA (FIS/BL miRNA169) that targets an NF-YA activator of the C-function. In the inner whorls of the wild-type flower, the activity of the C-function repressors is diminished. This is achieved in a number of ways: *AP2* expression is inhibited by miRNA172 in whorls 3 and 4 of *Arabidopsis* flowers. In addition, SEP proteins are recruited into complexes with the C-function protein, disrupting formation of the SEP–LUG–SEU repression complex. The AP1–LUG–SEU repression complex also fails to form due to repression of *AP1* by the C-function. Additional activators of the C-function are upregulated at the center of the flower. In *Antirrhinum* and *Petunia*, this includes the predicted NF-YA activator that is induced due to silencing of *FIS/BL* by factors including FIM. Finally, C-function expression is maintained at the center of the flower by an autoregulatory loop. Thus, a fine balance exists between activators and repressors of the C-function to control its expression domain. Disruption to this balance by loss of either a repressor or an activator of C-function expression has consequences on both the C-function expression domain and floral morphology. Loss of a repressor (*middle panel*) results in a shift in favor of C-function activation in the outer whorls of the flower, with homeotic conversion of perianth organs to reproductive organs. In this example, loss of AP2 activity allows C-function expression to expand into the perianth. As a consequence, C-function activity would also result in the repression of *AP1* and possibly other negative regulators of the C-function, enhancing the effect of activators in this domain. In *Antirrhinum* and *Petunia*, loss of *FIS/BL* activity also results in expansion of C-function expression into the perianth. However, in this case, C-function expression levels at the center of the flower go unchecked, resulting in higher levels of gene product than in wild type and expansion into the outer floral whorls. Thus, the balance of activators and repressors at the center of the flower also determines the C-function expression domain. Loss of a C-function activator (*bottom panel*), such as *PAN* in *Arabidopsis*, shifts the balance in favor of C-function repressors at the center of the flower. This is exacerbated by expansion of the *AP1* expression domain, allowing formation of the AP1–LUG–SEU repression complex. In addition, SEP is no longer recruited into a complex with the C-function factor, resulting in increased levels of the SEP–LUG–SEU repressor. C-function expression is only partially diminished in the *Arabidopsis pan* mutant, resulting in defects in meristem determinacy. *Green arrows* represent activation. *Red barred lines* represent repression. *Dashed lines* represent partial activity. *Greyed* out elements represent loss of activity. In the central bar of each panel the gradient of *black* represents the levels of C-function gene expression, with *black* being the greatest whorl [32, 35]. *SUP* and *OCT* are activated by the B-function, suggesting that a regulatory loop exists in the control of the B-function domain. Exclusion of the B-function from the fourth whorl is critical for carpel development and for ensuring that the floral meristem terminates correctly [32, 37, 38].

The fine balance between activation and repression in the establishment of floral homeotic gene expression domains has been best characterized for the C-function (Fig. 2). In *Arabidopsis*, *LFY*, *SEP3*, the meristem maintenance gene *WUSCHEL* (*WUS*), and *PERIANTHIA* (*PAN*) are among the activators of *AG* expression [39–43]. While WUS acts specifically at the center of the flower to activate *AG*, factors such as LFY and SEP3 have the potential to activate *AG* in all floral whorls. Thus, several repressors act to prevent precocious *AG* expression in the perianth, including AP2, RABBIT EARS (RBE), BELLRINGER (BLR), CURLY LEAF (CLF), and LUG–SEU acting through AP1 complexed with AGL24, SVP, or SEP3 [44–50]. Similar factors have been demonstrated to regulate *PLE* expression in developing flowers of *Antirrhinum*, suggesting that regulation of the C-function is highly conserved. For example, FLO, like its *Arabidopsis* counterpart LFY, activates floral homeotic genes in *Antirrhinum* floral meristems [51].

Analysis of novel *Antirrhinum* mutants in which C-function activity is lost in the third whorl but not in the fourth demonstrated that timing of C-function activation was crucial for establishing the correct C-function expression domain [52]. Many of the regulators of *AG* expression appear to operate through *cis-elements* located within the second intron of the *AG* gene, although until recently it was not clear whether the second intron of C-genes also plays a regulatory role in other species [7, 39, 41, 44, 45, 52–55]. Among these regulatory elements is a LFY-binding site that is conserved in terms of sequence and position within the intron of C-function genes from all angiosperms examined. Deletion of a small region of the *PLE* intron containing the conserved LFY-binding site significantly delayed the onset of *PLE* expression resulting in the novel loss of stamen phenotype, suggesting that this small sequence motif was important in the evolution of the C-function [52].

One problem with models of C-function expression domain control is that many of the negative regulators of C-genes are also expressed in the center of the flower, overlapping with that of the C-function (Fig. 2). Thus, the balance between activation and repression is critical in establishing the C-function expression domain. *Antirrhinum FISTULATA* (*FIS*) and *Petunia BLIND* (*BL*) are conserved microRNA-encoding genes (related to miRNA169) that are important in regulating the C-function in these species [56]. *fis* and *bl* mutants resemble classic A-function mutants due to ectopic expression of the C-function in the perianth, converting these organs to stamens and carpels [57–59]. In *Antirrhinum*, outward expansion of the C-function in *fis* correlates with increased *PLE* at the center of the flower, suggesting that the level of C-function expression in the center of the flower can influence the C-function expression domain [56]. In agreement with this model, artificial enhancement of *AG* expression at the center of the *Arabidopsis* flower using the stem cell-specific *CLAVATA3* promoter also resulted in expansion of C-function gene expression into the perianth of *Arabidopsis* flowers [60]. Thus, the fine-tuning of C-function gene expression levels, which is also a product of the correct onset of expression, appears to be a conserved mechanism for controlling the C-function expression domain in eudicots [52, 60]. Components of the miRNA169-mediated circuit are conserved in *Arabidopsis* although it is not yet clear whether these regulate C-function expression [56].

AG expression in *Arabidopsis* is partly regulated by switching the activity of the SEP3 transcription factor between repressor and activator of *AG* transcription [61]. In the perianth, SEP3 recruits the LUG–SEU corepressor module to inhibit *AG* expression. However, in the developing third and fourth whorls of the flower, positive regulators of *AG* expression antagonize the repression by the SEP3–LUG–SEU complex. *AG* expression is then enhanced and reinforced by an autoregulatory mechanism that involves a complex between the SEP3 and AG proteins, which weakens the SEP3–LUG–SEU repression complex, resulting in derepression of *AG* [61] (Fig. 2). In *Antirrhinum*, STYLOSA (STY), the orthologue of LUG, also prevents expression of the C-function in the perianth [59, 62], although further analyses will be required to determine whether the mechanism of LUG/STY repression of the C-function is conserved between *Arabidopsis* and *Antirrhinum*.

In the classic ABC model, the role of the A-function was to specify sepal and petal organ identities and to repress C-function expression in the perianth [30]. The (A)BC model, developed from comparative analyses of several eudicot species, better reflects our new understanding of the complexity and evolutionary variability of these processes. The new (A)-function effectively comprises all the factors that enable the B- and C-functions to correctly specify floral organ identity [24].

3.2 The B-Function

Petal and stamen development requires the activity of B-function genes, which encode MADS-box transcription factors, expressed in floral whorls 2 and 3. B-function proteins form obligate heterodimers, which recruit E-function partners into tetrameric higher order complexes [63]. In *Arabidopsis* the APETALA3 (AP3) and PISTILLATA (PI) proteins constitute the heterodimer, which also operates in an autoregulatory loop that enhances and then maintains expression of *AP3* and *PI* in whorls 2 and 3 [64]. A similar interdependence has also been described for the *Antirrhinum* B-function genes *GLOBOSA* (*GLO*) (the orthologue of *PI*) and

DEFICIENS (*DEF*) (the *AP3* orthologue) [65]. The autoregulatory loop between *DEF* and *GLO* may only be critical during early stages of organ development. Later in petal development, the relationship between the *DEF* and *GLO* transcripts is not equal and is dynamic. In addition, downregulation of *GLO* in petals does not result in reduced *DEF* expression, and vice versa, suggesting that additional mechanisms of B-function regulation exist that allow for local variations in *DEF* and *GLO* levels which specify the complex petal shape [66]. In other species, including *Petunia* and *Gerbera*, the number of B-function genes can be higher than in *Arabidopsis* or *Antirrhinum*, with more complex interdependencies [32, 67].

In plants genome duplications have resulted in large families of MADS-box genes that have contributed to developmental diversity. Before the emergence of the angiosperms, the ancestral B-function gene duplicated giving rise to two B-function gene lineages known as paleo*AP3* and *PI* [68–70]. Later, at the base of the core eudicots, the paleo*AP3* gene duplicated to produce eu*AP3* and *TM6* B-gene lineages [69]. TM6 proteins share a C-terminal motif with paleoAP3 proteins, which has been replaced in euAP3 proteins by a new euAP3 motif due to a frameshift mutation [69, 71]. Proteins with the paleoAP3 motif are found throughout the angiosperms, while euAP3 proteins are unique to the core eudicots (reviewed in [19]). The evolution of the B-function has been traced using a variety of approaches (reviewed in [72]), which has revealed that the *TM6* lineage has been lost from both the *Arabidopsis* and *Antirrhinum* genomes, which can be clearly seen in synteny studies [72]. Consequently, the contribution of TM6 proteins to the classic B-function was only established when other eudicot species were examined.

The *Petunia* genome has two *PI/GLO* genes, *PhGLO1* and *PhGLO2*, the expression of which is mainly confined to the second and third floral whorls. Only *phglo1 phglo2* double mutants show homeotic conversions in whorls 2 and 3, suggesting that these genes act redundantly in specifying petal and stamen development in accordance with the (A)BC model [73]. Similarly, the *Petunia AP3/DEF* gene *PhDEF* is also expressed in whorls 2 and 3, but *phdef* mutants show a phenotype only in the second whorl where petals become sepalloid [74]. Thus, *PhDEF* is required for specifying petal identity but may act redundantly with other factors in the third whorl to promote stamen identity [73]. A good candidate for the redundant factor was the *Petunia* TM6 protein, PhTM6. Interestingly, unlike the classic B-function proteins, *PhTM6* is expressed predominantly in the third and fourth floral whorls where stamens and carpels form, a feature which it shares with *TM6* genes from other eudicots (reviewed in [19]). *PhTM6* shares a similar expression pattern to C-function genes and is regulated, at least in part by *BL*, which restricts expression of *Petunia* C-genes to the center of the flower [56, 73]. PhTM6 functions redundantly with PhDEF in the formation of stamens in the third whorl [75], where it forms a heterodimer with the PhGLO2 protein. In contrast, PhDEF interacts with both PhGLO1 and PhGLO2. A similar story has emerged for other angiosperms that carry both eu*AP3* and *TM6* genes. In tomato, TM6 acts predominantly in stamen development and not in petal development [76], as is the case for the *Gerbera TM6* gene *GDEF1* [77]. However, the function of TM6 is not absolutely required for normal flower development provided euAP3 is present, as shown by the loss of TM6 from *Antirrhinum* and *Arabidopsis* genomes. Similarly, euAP3 can be lost from the genome, as seen in papaya, without detriment to flower development, provided TM6 is present [72]. In these cases, loss of *AP3/DEF* genes from the genome must have occurred before subfunctionalization, since in species where both the *TM6* and eu*AP3* lineages are maintained, specification of stamens appears to be partitioned between the *TM6* and eu*AP3* genes (reviewed in [72]).

3.3 The C-Function

The C-function is critical for the development of male and female reproductive organs at the center of the flower, to which the C-function gene expression domain is restricted. It is also responsible for terminating further development of the floral meristem once the carpels have been specified. Loss-of-function mutations in C-function genes have a similar phenotypic effect in all dicots studied to date. For example, loss of *AGAMOUS* (*AG*) function in *Arabidopsis* or *PLENA* (*PLE*) function in *Antirrhinum* results in reproductive organs being replaced by nonreproductive perianth organs and a loss of floral meristem determinacy that produces a reiteration of the mutant flower [78, 79]. In both species, *ag* and *ple* are the only mutants that display these characteristic homeotic changes. Despite this, other genes that show strong homology to *AG* and *PLE* exist in both species: *SHATTERPROOF* genes (*SHP1* and *SHP2*) in *Arabidopsis* and *FARINELLI* (*FAR*) in *Antirrhinum*. These gene pairs are the product of a genome duplication that occurred in the ancestral dicots following divergence of the monocots and dicots, which gave rise to the *AG* and *PLE* gene lineages. In *Arabidopsis* and *Antirrhinum*, the main C-function gene homeotic activity was undertaken by different lineages: the *AG* lineage in *Arabidopsis* and the *PLE* lineage in *Antirrhinum*, illustrating the random nature of evolution [80, 81]. The expression of *PLE* lineage genes of *Arabidopsis*, *SHP1* and *SHP2*, became restricted to the fourth whorl where the genes play a role in fruit development and ovule identity [82–85]. Regulatory elements conserved in the second intron of C-function genes since before the monocot–dicot split are missing from the *SHP* genes, which may provide an explanation for the modified expression of the *SHPs* [52]. Interestingly, however, the overexpression phenotype of *SHPs* has shown that the encoded proteins have retained the ability to specify reproductive organs [84].

In contrast, while the expression of the *AG* lineage gene of *Antirrhinum*, *FAR*, shares a similar pattern to that of *PLE*, the proteins do not have the same biological function [86]. This is demonstrated by differences in the phenotypes of 35S::*PLE* and 35S::*FAR* in *Arabidopsis*, *Antirrhinum*, or tobacco [81, 86–88]. While overexpression of *PLE* converts sepals to carpels and petals to stamens, as is seen in 35S::*AG* flowers, 35S::*FAR* flowers show homeotic conversions only in the second whorl, with sepals remaining unaltered. The limited function of the FAR protein in specifying reproductive identity is the result of naturally occurring variation at a single amino acid residue that limits the repertoire of SEP proteins with which FAR interacts [86, 88, 89]. Thus, a single amino acid change in the FAR protein was sufficient to alter the contribution that the protein makes to reproductive organ development in *Antirrhinum*.

Duplicate C-function genes are present in the genomes of many eudicots studied to date, although generally only one of the duplicates functions to specify reproductive organ identity and floral meristem determinacy—the classic roles of the C-function. However, some exceptions have been reported. For example, downregulation of either of the *Gerbera GAGA1* and *GAGA2* genes, which are expressed in whorls 3 and 4 from early stages, results in *ag*-like flowers [90]. In contrast, downregulation of either *Petunia* C-function duplicate (*PMADS3* and *FBP6*) resulted in mild homeotic defects in floral whorls 3 and 4, and an *ag*-like phenotype was only generated when both *PMADS3* and *FBP6* were silenced [91]. Similarly, downregulation of both *Nicotiana benthamiana* C-genes is required to cause homeotic conversions of reproductive organs to perianth organs and to disrupt floral determinacy in this species [92], suggesting functional redundancy between duplicate C-function genes at least in *Petunia* and *Nicotiana*. The relative contribution to morphological evolution of changes to protein sequence and/or regulatory elements has been the subject of recent debate (reviewed in [88]). The duplication and independent evolution of C-function genes following speciation have provided an excellent tool to examine this. What is emerging is that, within the context of the regulatory network, the separation of biological activities and innovation of new functions are shaped by both coding and expression pattern changes.

3.4 The D-Function

Ovule identity in *Arabidopsis* and *Petunia* requires the so-called D-function class of genes closely related to the C-function genes, from which they diverged early in angiosperm evolution [80, 83, 84, 93, 94]. The D-function was originally based on *Petunia* lines in which the *FLORAL BINDING PROTEIN7* (*FBP7*) and *FBP11* genes were both downregulated, resulting in the conversion of ovules to carpelloid structures [93]. In *Arabidopsis*, the D-function gene *SEEDSTICK* (*STK*) functions redundantly with the C-class genes *AG*, *SHP1*, and *SHP2* in ovule identity [84], suggesting that the D-function may be viewed as specialized C-function [19].

3.5 The E-Function

The ability of the B-, C-, and D-function genes to specify floral organ identity is dependent upon an additional genetic function, known as the E-function. The role of the E-function in floral organ identity was first observed in transgenic *Petunia* and tomato plants in which the SEP-like MADS-box genes *FBP2* and *TM5*, respectively, were silenced. In these lines, which had a similar phenotype, floral organs acquired sepal-like identity [95–97], indicating a decreased influence of B- and C-function on floral organ identity. Both species, in common with all dicots examined to date, have multiple E-function or *SEP* genes [98]. *Petunia* has six such genes (reviewed in [16]), two of which have been functionally characterized. Interestingly, the *fbp2 fbp5* double mutant is less severe than transgenic *FBP2* cosuppression lines, suggesting that multiple E-function genes must be downregulated in these lines and that the *Petunia* E-function genes act redundantly. Conversion of floral organs into sepals is also a feature of *Arabidopsis* mutants in which three of the four E-function genes, *SEP1*, *SEP2*, and *SEP3*, are mutated [99]. A more severe phenotype is observed when the fourth E-function gene (*SEP4*) is also mutated, where all floral organs are converted to leaflike structures, suggesting that floral meristem identity is completely lost in the quadruple mutant [27]. A common feature of the E-function mutants of various species is floral reversion, where floral meristem identity is replaced by inflorescence identity [15, 27, 100].

The single *sep* mutants of *Arabidopsis* are reported to have a very subtle or no phenotype, indicating that they act almost completely redundantly [27, 99]. However, in other species, mutations in single E-function genes often result in floral defects. For example, the *Petunia fbp2* mutant shows partial petal-to-sepal conversions and secondary inflorescences in the third whorl [15]. In *Gerbera*, four E-function genes are known, two of which (*GRCD1* and *GRCD2*) appear to show non-redundant functions in either stamen or carpel development [101, 102]. Thus it seems more probable that E-function genes are partially redundant, but that they also have separate functions in flower development.

The E-function factors specify floral meristem identity and floral organ identity through the formation of protein complexes with other MADS-box factors. *Arabidopsis* SEP proteins interact with proteins involved in floral meristem identity, including AP1 and FUL [103]. Similarly, the *Antirrhinum* E-function proteins DEFH72 and DEFH200 interact with the floral meristem identity factor SQUA [104]. Specific E-function interactions are also conserved among eudicots for the specification of floral organs.

Specification of petals and reproductive organs relies on various B-, C-, and E-function complexes, which have been observed for *Arabidopsis*, *Petunia*, and tomato proteins [29, 97, 105]. Various interactions between C-, D-, and E-function proteins, which are required for ovule development, have also been described for *Arabidopsis* and *Petunia* [83, 97, 103, 106, 107].

In summary, floral organ development is reliant on the E-function proteins establishing the floral context. Initially the E-function forms part of the mechanism that specifies floral meristem identity, by interacting with meristem identity MADS-box factors. Later E-function proteins associate in tetrameric complexes with organ identity proteins (known as floral quartets [108]) that bind to regulatory regions of target genes, controlling their expression and driving appropriate spatial development of floral organs. Studies in a range of eudicots have helped to define the E-function and shown that their specific protein–protein interactions are conserved. It will be interesting to establish the degree to which these conserved transcription factor interactions lead to regulation of the same sets of target genes in the different floral organs of different species.

4 Floral Organ Development and Control of Floral Form

We have already seen that the underlying mechanisms that control floral meristem and floral organ identities are largely conserved among the angiosperms. However, in nature flowers come in a bewildering variety of shapes, colors, and sizes, and it is thought that such traits coevolved with insect pollinators [109]. Compared to many flowers, those of *Arabidopsis* are rather small and of no horticultural value. They are radially symmetrical, with small colorless petals. Not surprisingly then, *Arabidopsis* does not rely on insect pollinators but is mostly self-pollinated. However, *Arabidopsis* flowers do emit monoterpenes and sesquiterpenes that may attract small insect pollinators, such as hoverflies, which would correlate with the low frequency of outcrossing observed in natural populations of the plant [110]. In contrast, the flowers of other model eudicots, including *Antirrhinum* and *Petunia*, are large, highly colored, and in some cases asymmetric. A number of genes control the development of the dorsoventrally asymmetric (zygomorphic) flowers of *Antirrhinum*, as revealed by the analysis of floral mutants in which symmetry is disrupted. Single mutants of the *CYCLOIDEA* (*CYC*) and *DICHOTOMA* (*DICH*) genes result in the loss of dorsal identity, while the *cyc dich* double mutant forms radially symmetrical flowers where all petals have ventral identity [111, 112]. *CYC* and *DICH* encode TCP transcription factor proteins and are expressed in the dorsal regions of floral meristems. Interestingly, although *Arabidopsis* flowers are symmetrical, *TCP1*,

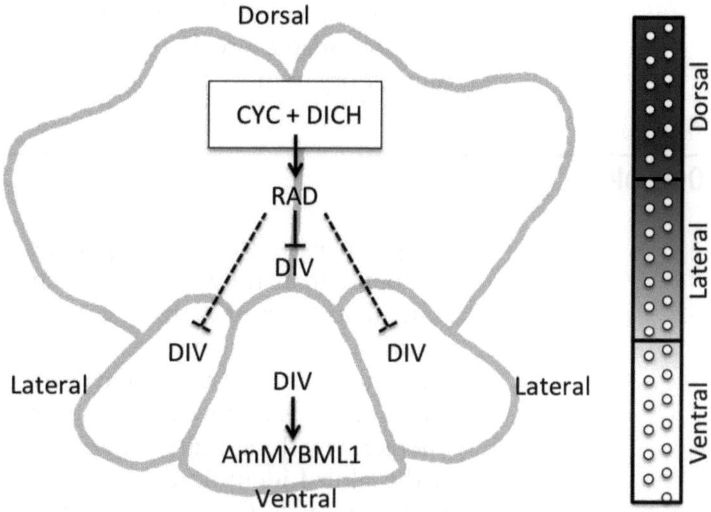

Fig. 3 The genetic control of floral symmetry in *Antirrhinum*. *Antirrhinum* flowers are bilaterally symmetrical and are a good model to study control of floral symmetry. Dorsal identity is promoted by expression of the TCP factors *CYC* and *DICH*. These activate the MYB-encoding gene *RAD*. RAD acts to antagonize the function of the related MYB factor DIV (represented by *barred lines*, where *dashed lines* represent partial antagonism). *DIV* is expressed in dorsal, lateral, and ventral organs (represented by *white circles* in the panel to the *left*). In contrast, *RAD* is expressed in dorsal organs but may move into lateral organs, thus establishing a gradient of RAD (represented by the *gradient pattern* in the *left-hand panel*, where *black* represents the highest RAD levels). Thus, the relative amounts of DIV and RAD contribute to the positioning of the dorsoventral axis of the *Antirrhinum* flower. Expression of *DIV* alone in the ventral petal activates *AmMYBML1*, which specifies ventral petal cell types

which is most similar to *CYC*, is also expressed at early stages in the dorsal part of floral meristems, suggesting that the common ancestor of *Antirrhinum* and *Arabidopsis* had the components to generate asymmetric flowers [113]. While *CYC* and *DICH* provide the dorsalizing function, ventral identity is supplied by *DIVARICATA* (*DIV*), which encodes a MYB family transcription factor [114] (Fig. 3). *DIV* is expressed throughout the flower, including regions where *CYC* and *DICH* are expressed. A fourth gene, *RADIALIS* (*RAD*), encoding a MYB factor similar to DIV, is induced in the dorsal domain by *CYC* and *DICH*. It has been proposed that RAD antagonizes the function of the DIV protein in the dorsal and lateral domains, thus modulating DIV activity to generate the lateral and ventral petal identities of wild-type flowers [115]. In the ventral petal DIV regulates expression of the AmMYBML1 transcription factor, which specifies ventral petal cell types [116] (Fig. 3). The control of *Antirrhinum* flower shape by these genes has led to the development of quantitative models, based on experimental and computational approaches, for the analysis of the genetic control of organ shape [117]. These studies are beginning to reveal how evolution adapts preexisting transcriptional modules to generate complex organ morphologies [118].

5 Outlook

Comparative studies of the genetic networks controlling flower development in numerous dicot species have shown how subtle refinements to an underlying conserved circuit have resulted in the evolution of the rich diversity of floral form seen in nature. Although many questions remain to be answered, we now have a toolkit of genes that we can use to modify flower development and generate new floral traits. In *Arabidopsis*, recent work has focused on the global identification of target genes of some of the floral regulators, including AP3 and PI [38, 119], AG [120], SEP3 [121], AP1 [122], SOC1 [123], LFY [5, 6], and FLC [124]. In other species, with the exception of the genome-wide identification of *Antirrhinum* B-function petal targets [125], very few target genes have been identified. It is clear that these studies should be expanded to more species, to identify both conserved and novel targets that will help us to understand how floral form evolved. Such comparative studies of target genes may also provide options for crop improvement.

Another challenge concerns the issue of gene redundancy, such as we see for the E-function genes. Why do plant genomes retain duplicate copies of certain genes? Can they really be acting in a completely redundant fashion? One possibility is that retention of duplicate genes provides the ability to respond and adapt to environmental changes in the wild. Mechanistically, it is possible that multiple proteins acting within a complex (such as MADS-domain transcription factors) buffer the fluctuation or loss of a single component [126, 127]. It will be important to determine what effect the loss of a component will have on complex composition and stoichiometry and what this means for target gene regulation. Current technologies will allow us to address such questions on a global scale and provide insights into the true degree of functional redundancy.

Comparisons across the eudicots have contributed greatly to our understanding and formulation of a model for flower development. However, despite this progress there remain many unanswered questions, including how the regulatory genes achieve such precision of spatial and temporal expression, how organ identity is linked to the development of organ primordia and their boundaries, and how the activation and repression of multiple target genes produce the complex organs of the flower. Previous discoveries show that the answers to these questions will require continued study using *Arabidopsis* and other model eudicots.

Acknowledgements

We are grateful to Chiara Airoldi for comments on the manuscript. Our research is funded by grants from the Biotechnology and Biological Sciences Research Council.

References

1. Wikström N, Savolainen V, Chase MW (2001) Evolution of the angiosperms: calibrating the family tree. Proc Biol Sci 268: 2211–2220
2. Siriwardana NS, Lamb RS (2012) The poetry of reproduction: the role of *LEAFY* in *Arabidopsis thaliana* flower formation. Int J Dev Biol 56:207–221
3. Wagner D, Sablowski RWM, Meyerowitz EM (1999) Transcriptional activation of *APETALA1* by LEAFY. Science 285: 582–584
4. Wagner D, Wellmer F, Dilks K et al (2004) Floral induction in tissue culture: a system for the analysis of LEAFY-dependent gene regulation. Plant J 39:273–282
5. William DA, Su Y, Smith MR et al (2004) Genomic identification of direct target genes of LEAFY. Proc Natl Acad Sci U S A 101:1775–1780
6. Winter CM, Austin RS, Blanvillain-Baufumé S et al (2011) LEAFY target genes reveal floral regulatory logic, cis motifs, and a link to biotic stimulus. Dev Cell 20:430–443
7. Moyroud E, Minguet EG, Ott F et al (2011) Prediction of regulatory interactions from genome sequences using a biophysical model for the *Arabidopsis* LEAFY transcription factor. Plant Cell 23:1293–1306
8. Benlloch R, Berbel A, Serrano-Mislata A, Madueño F (2007) Floral initiation and inflorescence architecture: a comparative view. Ann Bot 100:659–676
9. Weigel D, Alvarez J, Smyth DR et al (1992) LEAFY controls floral meristem identity in Arabidopsis. Cell 69:843–859
10. Coen ES, Romero JM, Doyle S et al (1990) Floricaula: a homeotic gene required for flower development in Antirrhinum majus. Cell 21:1311–1322
11. Souer E, van der Krol A, Kloos D et al (1998) Genetic control of branching pattern and floral identity during Petunia inflorescence development. Development 125:733–742
12. Bowman JL, Smyth DR, Meyerowitz EM (1991) Genetic interactions among floral homeotic genes of Arabidopsis. Development 112:1–20
13. Huijser P, Klein J, Lönnig WE et al (1992) Bracteomania, an inflorescence anomaly, is caused by the loss of function of the MADS-box gene squamosa in Antirrhinum majus. EMBO J 11:1239–1249
14. Immink R, Hannapel D, Ferrario S et al (1999) A petunia MADS box gene involved in the transition from vegetative to reproductive development. Development 126: 5117–5126
15. Vandenbussche M, Zethof J, Souer E et al (2003) Toward the analysis of the Petunia MADS box gene family by reverse and forward transposon insertion mutagenesis approaches: B, C and D floral organ identity functions require SEPALLATA-like MADS box genes in Petunia. Plant Cell 15: 2680–2693
16. Rijpkema A, Gerats T, Vandenbussche M (2006) Genetics of floral development in Petunia. Adv Bot Res 44:237–278
17. Levin JZ, Meyerowitz EM (1995) UFO: an Arabidopsis gene involved in both floral meristem and floral organ development. Plant Cell 7:529–548
18. Wilkinson MD, Haughn GW (1995) UNUSUAL FLORAL ORGANS Controls Meristem Identity and Organ Primordia Fate in Arabidopsis. Plant Cell 7:1485–1499
19. Rijpkema AS, Vandenbussche M, Koes R et al (2010) Variations on a theme: changes in the floral ABCs in angiosperms. Semin Cell Dev Biol 21:100–107
20. Ingram GC, Goodrich J, Wilkinson MD et al (1995) Parallels between UNUSUAL FLORAL ORGANS and FIMBRIATA, genes controlling flower development in Arabidopsis and Antirrhinum. Plant Cell 7:1501–1510
21. Allen KD, Sussex IM (1996) Falsiflora and anantha control early stages of floral meristem development in tomato (Lycopersicon esculentum Mill.). Planta 200:254–264
22. Lippman ZB, Cohen O, Alvarez JP et al (2008) The making of a compound inflorescence in tomato and related nightshades. PloS Biol 6:e288
23. Souer E, Rebocho AB, Bliek M et al (2008) Patterning of inflorescences and flowers by

the F-Box protein DOUBLE TOP and the LEAFY homolog ABERRANT LEAF AND FLOWER of petunia. Plant Cell 20: 2033–2048
24. Causier B, Schwarz-Sommer Z, Davies B (2010) Floral organ identity: 20 years of ABCs. Semin Cell Dev Biol 21:73–79
25. Keck E, McSteen P, Carpenter R, Coen E (2003) Separation of genetic functions controlling organ identity in flowers. EMBO J 22:1058–1066
26. Maes T, Van de Steene N, Zethof J et al (2001) Petunia Ap2-like genes and their role in flower and seed development. Plant Cell 13:229–244
27. Ditta G, Pinyopich A, Robles P et al (2004) The SEP4 gene of Arabidopsis thaliana functions in floral organ and meristem identity. Curr Biol 14:1935–1940
28. Pelaz S, Tapia-López R, Alvarez-Buylla ER, Yanofsky MF (2001) Conversion of leaves into petals in Arabidopsis. Curr Biol 11: 182–184
29. Honma T, Goto K (2001) Complexes of MADS-box proteins are sufficient to convert leaves into floral organs. Nature 409:525–529
30. Coen ES, Meyerowitz EM (1991) The war of the whorls: genetic interactions controlling flower development. Nature 353:31–37
31. Schwarz-Sommer Z, Huijser P, Nacken W, Saedler H, Sommer H (1990) Genetic control of flower development by homeotic genes in Antirrhinum majus. Science 250:931–936
32. Davies B, Cartolano M, Schwarz-Sommer Z (2006) Flower development: the Antirrhinum perspective. Adv Bot Res 44:279–321
33. Litt A (2007) An evaluation of the A-function: evidence from the APETALA1 and APETALA2 gene lineages. Int J Plant Sci 168:73–91
34. Gregis V, Sessa A, Dorca-Fornell C, Kater MM (2009) The Arabidopsis floral meristem identity genes AP1, AGL24 and SVP directly repress class B and C floral homeotic genes. Plant J 60:626–637
35. Bowman JL, Sakai H, Jack T et al (1992) SUPERMAN, a regulator of floral homeotic genes in Arabidopsis. Development 114: 599–615
36. Simon R, Carpenter R, Doyle S, Coen E (1994) Fimbriata controls flower development by mediating between meristem and organ identity genes. Cell 78:99–107
37. Yellina AL, Orashakova S, Lange S et al (2010) Floral homeotic C function genes repress specific B function genes in the carpel whorl of the basal eudicot California poppy (Eschscholzia californica). Evodevo 1:13
38. Wuest SE, O'Maoileidigh DS, Rae L et al (2012) Molecular basis for the specification of floral organs by APETALA3 and PISTILLATA. PNAS 109:13452–13457
39. Busch MA, Bomblies K, Weigel D (1999) Activation of a floral homeotic gene in Arabidopsis. Science 285:585–587
40. Lenhard M, Bohnert A, Jürgens G, Laux T (2001) Termination of stem cell maintenance in Arabidopsis floral meristems by interactions between WUSCHEL and AGAMOUS. Cell 105:805–814
41. Lohmann JU, Hong RL, Hobe M et al (2001) A molecular link between stem cell regulation and floral patterning in Arabidopsis. Cell 105:793–803
42. Das P, Ito T, Wellmer F et al (2009) Floral stem cell termination involves the direct regulation of AGAMOUS by PERIANTHIA. Development 136:1605–1611
43. Maier AT, Stehling-Sun S, Wollmann H et al (2009) Dual roles of the bZIP transcription factor PERIANTHIA in the control of floral architecture and homeotic gene expression. Development 136:1613–1620
44. Bao X, Franks RG, Levin JZ, Liu Z (2004) Repression of AGAMOUS by BELLRINGER in floral and inflorescence meristems. Plant Cell 16:1478–1489
45. Bomblies K, Dagenais N, Weigel D (1999) Redundant enhancers mediate transcriptional repression of AGAMOUS by APETALA2. Dev Biol 216:260–264
46. Drews GN, Bowman JL, Meyerowitz EM (1991) Negative regulation of the Arabidopsis homeotic gene AGAMOUS by the APETALA2 product. Cell 65:991–1002
47. Franks RG, Wang C, Levin JZ, Liu Z (2002) SEUSS, a member of a novel family of plant regulatory proteins, represses floral homeotic gene expression with LEUNIG. Development 129:253–263
48. Goodrich J, Puangsomlee P, Martin M et al (1997) A Polycomb-group gene regulates homeotic gene expression in Arabidopsis. Nature 386:44–51
49. Krizek BA, Lewis MW, Fletcher JC (2006) RABBIT EARS is a second-whorl repressor of AGAMOUS that maintains spatial boundaries in Arabidopsis flowers. Plant J 45:369–383
50. Liu Z, Meyerowitz EM (1995) LEUNIG regulates AGAMOUS expression in Arabidopsis flowers. Development 121:975–991
51. Hantke SS, Carpenter R, Coen ES (1995) Expression of Floricaula in single cell layers of periclinal chimeras activates downstream homeotic genes in all layers of floral meristems. Development 121:27–35
52. Causier B, Bradley D, Cook H, Davies B (2009) Conserved intragenic elements were critical for the evolution of the floral C-function. Plant J 58:41–52

53. Sieburth LE, Meyerowitz EM (1997) Molecular dissection of the AGAMOUS control region shows that cis elements for spatial regulation are located intragenically. Plant Cell 9:355–365
54. Deyholos MK, Sieburth LE (2000) Separable whorl-specific expression and negative regulation by enhancer elements within the AGAMOUS second intron. Plant Cell 12: 1799–1810
55. Dinh TT, Girke T, Liu X et al (2012) The floral homeotic protein APETALA2 recognizes and acts through an AT-rich sequence element. Development 139:1978–1986
56. Cartolano M, Castillo R, Efremova N et al (2007) A conserved microRNA module exerts homeotic control over Petunia hybrida and Antirrhinum majus floral organ identity. Nat Genet 39:901–905
57. Tsuchimoto S, van der Krol AR, Chua NH (1993) Ectopic expression of pMADS3 in transgenic petunia phenocopies the petunia blind mutant. Plant Cell 5:843–853
58. McSteen PC, Vincent CA, Doyle S et al (1998) Control of floral homeotic gene expression and organ morphogenesis in Antirrhinum. Development 125:2359–2369
59. Motte P, Saedler H, Schwarz-Sommer Z (1998) STYLOSA and FISTULATA: regulatory components of the homeotic control of Antirrhinum floral organogenesis. Development 125:71–84
60. Cartolano M, Efremova N, Kuckenberg M et al (2009) Enhanced AGAMOUS expression in the centre of the Arabidopsis flower causes ectopic expression over its outer expression boundaries. Planta 230:857–862
61. Sridhar VV, Surendrarao A, Liu Z (2006) APETALA1 and SEPALLATA3 interact with SEUSS to mediate transcription repression during flower development. Development 133:3159–3166
62. Navarro C, Efremova N, Golz JF et al (2004) Molecular and genetic interactions between STYLOSA and GRAMINIFOLIA in the control of Antirrhinum vegetative and reproductive development. Development 131: 3649–3659
63. Melzer R, Theissen G (2009) Reconstitution of 'floral quartets' in vitro involving class B and class E floral homeotic proteins. Nucleic Acids Res 37:2723–2736
64. Jack T, Fox GL, Meyerowitz EM (1994) Arabidopsis homeotic gene APETALA3 ectopic expression: transcriptional and posttranscriptional regulation determine floral organ identity. Cell 76:703–716
65. Schwarz-Sommer Z, Hue I, Huijser P et al (1992) Characterization of the Antirrhinum floral homeotic MADS-box gene deficiens: evidence for DNA binding and autoregulation of its persistent expression throughout flower development. EMBO J 11:251–263
66. Manchado-Rojo M, Delgado-Benarroch L, Roca MJ et al (2012) Quantitative levels of Deficiens and Globosa during late petal development show a complex transcriptional network topology of B function. Plant J 72: 294–307
67. Airoldi CA (2010) Determination of sexual organ development. Sex Plant Reprod 23: 53–62
68. Purugganan MD, Rounsley SD, Schmidt RJ, Yanofsky MF (1995) Molecular evolution of flower development: diversification of the plant MADS-box regulatory gene family. Genetics 140:345–356
69. Kramer EM, Dorit RL, Irish VF (1998) Molecular evolution of genes controlling petal and stamen development: duplication and divergence within the APETALA3 and PISTILLATA MADS-box gene lineages. Genetics 149:765–783
70. Kim S, Yoo MJ, Albert VA et al (2004) Phylogeny and diversification of B-function MADS-box genes in angiosperms: evolutionary and functional implications of a 260-million-year-old duplication. Am J Bot 91:2102–2118
71. Vandenbussche M, Theissen G, Van de Peer Y, Gerats T (2003) Structural diversification and neo-functionalization during floral MADS-box gene evolution by C-terminal frameshift mutations. Nucleic Acids Res 31:4401–4409
72. Causier B, Castillo R, Xue Y et al (2010) Tracing the evolution of the floral homeotic B- and C-function genes through genome synteny. Mol Biol Evol 27:2651–2664
73. Vandenbussche M, Zethof J, Royaert S et al (2004) The duplicated B-class heterodimer model: whorl-specific effects and complex genetic interactions in Petunia hybrida flower development. Plant Cell 16: 741–754
74. van der Krol AR, Brunelle A, Tsuchimoto S, Chua NH (1993) Functional analysis of petunia floral homeotic MADS box gene pMADS1. Genes Dev 7:1214–1228
75. Rijpkema AS, Royaert S, Zethof J, van der Weerden G, Gerats T, Vandenbussche M (2006) Analysis of the Petunia TM6 MADS box gene reveals functional divergence within the DEF/AP3 lineage. Plant Cell 18: 1819–1832
76. de Martino G, Pan I, Emmanuel E, Levy A, Irish VF (2006) Functional analyses of two tomato APETALA3 genes demonstrate

diversification in their roles in regulating floral development. Plant Cell 18:1833–1845
77. Broholm SK, Pöllänen E, Ruokolainen S et al (2010) Functional characterization of B class MADS-box transcription factors in Gerbera hybrida. J Exp Bot 61:75–85
78. Yanofsky MF, Ma H, Bowman JL et al (1990) The protein encoded by the Arabidopsis homeotic gene agamous resembles transcription factors. Nature 346:35–39
79. Bradley D, Carpenter R, Sommer H et al (1993) Complementary floral homeotic phenotypes result from opposite orientations of a transposon at the plena locus of Antirrhinum. Cell 72:85–95
80. Kramer EM, Jaramillo MA, Di Stilio VS (2004) Patterns of gene duplication and functional evolution during the diversification of the AGAMOUS subfamily of MADS box genes in angiosperms. Genetics 166: 1011–1023
81. Causier B, Castillo R, Zhou J et al (2005) Evolution in action: following function in duplicated floral homeotic genes. Curr Biol 15:1508–1512
82. Liljegren SJ, Ditta GS, Eshed Y et al (2000) SHATTERPROOF MADS-box genes control seed dispersal in Arabidopsis. Nature 404:766–770
83. Favaro R, Pinyopich A, Battaglia R et al (2003) MADS-box protein complexes control carpel and ovule development in Arabidopsis. Plant Cell 15:2603–2611
84. Pinyopich A, Ditta GS, Savidge B et al (2003) Assessing the redundancy of MADS-box genes during carpel and ovule development. Nature 424:85–88
85. Colombo M, Brambilla V, Marcheselli R et al (2010) A new role for the SHATTERPROOF genes during Arabidopsis gynoecium development. Dev Biol 337:294–302
86. Davies B, Motte P, Keck E et al (1999) PLENA and FARINELLI: redundancy and regulatory interactions between two Antirrhinum MADS-box factors controlling flower development. EMBO J 18:4023–4034
87. Davies B, Di Rosa A, Eneva T et al (1996) Alteration of tobacco floral organ identity by expression of combinations of Antirrhinum MADS-box genes. Plant J 10:663–677
88. Airoldi CA, Bergonzi S, Davies B (2010) Single amino acid change alters the ability to specify male or female organ identity. Proc Natl Acad Sci U S A 107:18898–18902
89. Davies B, Egea-Cortines M, de Andrade Silva E et al (1996) Multiple interactions amongst floral homeotic proteins. EMBO J 15: 4330–4343
90. Yu D, Kotilainen M, Pöllänen E et al (1999) Organ identity genes and modified patterns of flower development in Gerbera hybrida (Asteraceae). Plant J 17:51–62
91. Heijmans K, Ament K, Rijpkema AS et al (2012) Redefining C and D in the Petunia ABC. Plant Cell 24:2305–2317
92. Fourquin C, Ferrándiz C (2012) Functional analyses of AGAMOUS family members in Nicotiana benthamiana clarify the evolution of early and late roles of C-function genes in eudicots. Plant J 71:990–1001
93. Angenent GC, Franken J, Busscher M et al (1995) A novel class of MADS box genes is involved in ovule development in petunia. Plant Cell 7:1569–1582
94. Colombo L, Franken J, Koetje E et al (1995) The petunia MADS-box gene FBP11 determines ovule identity. Plant Cell 7:1859–1868
95. Pnueli L, Hareven D, Broday L et al (1994) The TM5 MADS box gene mediates organ differentiation in the three inner whorls of tomato flowers. Plant Cell 6:175–186
96. Angenent GC, Franken J, Busscher M et al (1994) Co-suppression of the petunia homeotic gene fbp2 affects the identity of the generative meristem. Plant J 5:33–44
97. Ferrario S, Immink RG, Shchennikova A (2003) et al The MADS box gene FBP2 is required for SEPALLATA function in petunia. Plant Cell 15:914–925
98. Zahn LM, Kong H, Leebens-Mack JH et al (2005) The evolution of the SEPALLATA subfamily of MADS-box genes: a preangiosperm origin with multiple duplications throughout angiosperm history. Genetics 169:2209–2223
99. Pelaz S, Ditta GS, Baumann E et al (2000) B and C floral organ identity functions require SEPALLATA MADS-box genes. Nature 405:200–203
100. Ampomah-Dwamena C, Morris BA, Sutherland P et al (2002) Down-regulation of TM29, a tomato SEPALLATA homolog, causes parthenocarpic fruit development and floral reversion. Plant Physiol 130:605–617
101. Kotilainen M, Elomaa P, Uimari A et al (2000) GRCD1, an AGL2-like MADS box gene, participates in the C function during stamen development in Gerbera hybrida. Plant Cell 12:1893–1902
102. Uimari A, Kotilainen M, Elomaa P et al (2004) Integration of reproductive meristem fates by a SEPALLATA-like MADS-box gene. Proc Natl Acad Sci U S A 101:15817–15822
103. de Folter S, Immink RG, Kieffer M et al (2005) Comprehensive interaction map of the Arabidopsis MADS Box transcription factors. Plant Cell 17:1424–1433
104. Causier B, Cook H, Davies B (2003) An Antirrhinum ternary complex factor specifically interacts with C-function and SEPALLATA-like

MADS-box factors. Plant Mol Biol 52: 1051–1062
105. Leseberg CH, Eissler CL, Wang X et al (2008) Interaction study of MADS-domain proteins in tomato. J Exp Bot 59:2253–2265
106. Immink RGH, Gadella TWJ Jr, Ferrario S et al (2002) Analysis of MADS box protein-protein interactions in living plant cells. Proc Natl Acad Sci U S A 99:2416–2421
107. Immink RG, Tonaco IA, de Folter S et al (2009) SEPALLATA3: the 'glue' for MADS box transcription factor complex formation. Genome Biol 10:R24
108. Theissen G, Saedler H (2001) Plant biology. Floral quartets. Nature 409:469–471
109. Galliot C, Stuurman J, Kuhlemeier C (2006) The genetic dissection of floral pollination syndromes. Curr Opin Plant Biol 9:78–82
110. Chen F, Tholl D, D'Auria JC et al (2003) Biosynthesis and emission of terpenoid volatiles from Arabidopsis flowers. Plant Cell 15:481–494
111. Luo D, Carpenter R, Vincent C et al (1996) Origin of floral asymmetry in Antirrhinum. Nature 383:794–799
112. Luo D, Carpenter R, Copsey L et al (1999) Control of organ asymmetry in flowers of Antirrhinum. Cell 99:367–376
113. Cubas P, Coen E, Zapater JM (2001) Ancient asymmetries in the evolution of flowers. Curr Biol 11:1050–1052
114. Galego L, Almeida J (2002) Role of DIVARICATA in the control of dorsoventral asymmetry in Antirrhinum flowers. Genes Dev 16:880–891
115. Corley SB, Carpenter R, Copsey L, Coen E (2005) Floral asymmetry involves an interplay between TCP and MYB transcription factors in Antirrhinum. Proc Natl Acad Sci U S A 102:5068–5073
116. Perez-Rodriguez M, Jaffe FW, Butelli E et al (2005) Development of three different cell types is associated with the activity of a specific MYB transcription factor in the ventral petal of Antirrhinum majus flowers. Development 132:359–370
117. Green AA, Kennaway JR, Hanna AI et al (2010) Genetic control of organ shape and tissue polarity. PloS Biol 8:e1000537
118. Cui ML, Copsey L, Green AA et al (2010) Quantitative control of organ shape by combinatorial gene activity. PloS Biol 8:e1000538
119. Zik M, Irish VF (2003) Global identification of target genes regulated by APETALA3 and PISTILLATA floral homeotic gene action. Plant Cell 15:207–222
120. Gómez-Mena C, de Folter S, Costa MM et al (2005) Transcriptional program controlled by the floral homeotic gene AGAMOUS during early organogenesis. Development 132: 429–438
121. Kaufmann K, Muiño JM, Jauregui R et al (2009) Target genes of the MADS transcription factor SEPALLATA3: integration of developmental and hormonal pathways in the Arabidopsis flower. PLoS Biol 7:e1000090
122. Kaufmann K, Wellmer F, Muiño JM et al (2010) Orchestration of floral initiation by APETALA1. Science 328:85–89
123. Immink RG, Posé D, Ferrario S et al (2012) Characterization of SOC1's central role in flowering by the identification of its upstream and downstream regulators. Plant Physiol 160:433–449
124. Deng W, Ying H, Helliwell CA et al (2011) FLOWERING LOCUS C (FLC) regulates development pathways throughout the life cycle of Arabidopsis. Proc Natl Acad Sci U S A 108:6680–6685
125. Bey M, Stüber K, Fellenberg K et al (2004) Characterization of antirrhinum petal development and identification of target genes of the class B MADS box gene DEFICIENS. Plant Cell 16:3197–3215
126. Cui R, Han J, Zhao S et al (2010) Functional conservation and diversification of class E floral homeotic genes in rice (Oryza sativa). Plant J 61:767–781
127. Airoldi CA, Davies B (2012) Gene duplication and the evolution of plant MADS-box transcription factors. J Genet Genomics 39: 157–165

Chapter 3

Grass Flower Development

Hiro-Yuki Hirano, Wakana Tanaka, and Taiyo Toriba

Abstract

Grasses bear unique flowers lacking obvious petals and sepals in special inflorescence units, the florets and the spikelet. Despite this, grass floral organs such as stamens and lodicules (petal homologs) are specified by ABC homeotic genes encoding MADS domain transcription factors, suggesting that the ABC model of eudicot flower development is largely applicable to grass flowers. However, some modifications need to be made for the model to fit grasses well: for example, a *YABBY* gene plays an important role in carpel specification. In addition, a number of genes are involved in the development of the lateral organs that constitute the spikelet. In this review, we discuss recent progress in elucidating the genes required for flower and spikelet development in grasses, together with those involved in fate determination of the spikelet and flower meristems.

Key words Floral organ specification, Flower development, Grass flower, Maize, Meristem, Rice, Sex determination, Spikelet, Stem cell maintenance

1 Introduction

Grasses constitute the large *Poaceae* family, which contains more than 10,000 species. The grass family diverged from its most closely related families around 55–70 million years ago and has expanded during evolution [1]. Several agronomically important crops such as rice, maize, and wheat are included in this family, and they support human life throughout the world by supplying seeds for food and materials for food processing. Grasses are also a familiar part of our landscape, in not only the countryside but also the suburbs, covering a large portion of the earth.

Angiosperm flowers are diverse, ranging from beautiful and entomophilous flowers to inconspicuous and anemophilous ones. The grasses bear the latter type of flower lacking obvious petals and sepals: floral organs, such as a pistil and a stamen, are enclosed by bract-like organs, a lemma and a palea, as described in detail below. The inflorescences of grasses are also distinct from those of eudicots and other monocots, and are composed of characteristic units, a spikelet and a floret.

José Luis Riechmann and Frank Wellmer (eds.), *Flower Development: Methods and Protocols*, Methods in Molecular Biology, vol. 1110, DOI 10.1007/978-1-4614-9408-9_3, © Springer Science+Business Media New York 2014

The molecular mechanisms of flower development have been intensively studied in *Arabidopsis* thaliana and other eudicots in the last two decades [2, 3], beginning with the proposal of the ABC model based on genetic analyses [4], as described in previous chapters. In this chapter, we will focus on the genes that regulate the development of grass flowers, the morphology of which is different from that of eudicot flowers. In particular, we concentrate on two model grasses, rice (*Oryza sativa*) and maize (*Zea mays*). Both plants have advantages for study by molecular genetic approaches, allowing the identification of important genes that regulate flower development and the elucidation of their functions, via analysis of mutants that lack their activities. Rice and maize diverged at an initial period of grass evolution [1], and their spikelet structures and inflorescence architectures are distinct even from each other. Therefore, the outcomes from developmental studies in both rice and maize may contribute to understanding the functional evolution of developmental genes not only between distantly related species (eudicots and grasses) that bear divergent flowers, but also between relatively close species (rice and maize) that have only slightly different flowers.

In this chapter, we first describe genes that regulate the fate of reproductive meristems and the specification of floral organs, which are common to eudicots, in both rice and maize. We then focus on genes involved in the development of organs specific to grass spikelets mainly in rice, and finally describe the sex determination of maize flowers. There are several excellent reviews describing flower development in grasses [5–8]. However, progress in this field is very rapid and much information has accumulated recently. We have tried to incorporate recent studies and cite original papers for rice and maize research. With respect to *Arabidopsis* research, however, we mainly cite review articles due to space restrictions.

2 Structure of the Flower and the Spikelet

Grass flowers are formed in a special inflorescence unit called the spikelet. The spikelet is formed directly on the main axis of the inflorescence or on the branches, which are derived from the inflorescence meristem (IM). The pattern of spikelet formation is variable depending on species and on sexual inflorescence even within a species. The spikelet is composed of a pair of glumes and several florets, consisting of floral organs (pistil, stamen and lodicule) and outer organs (palea and lemma) enclosing the floral organs (Fig. 1). In this review, we define the pistil, stamen, and lodicule (the petal homolog) as floral organs for several reasons, as described below. The number of florets formed in a spikelet is variable depending on species. Thus, grass inflorescences are unique and complex.

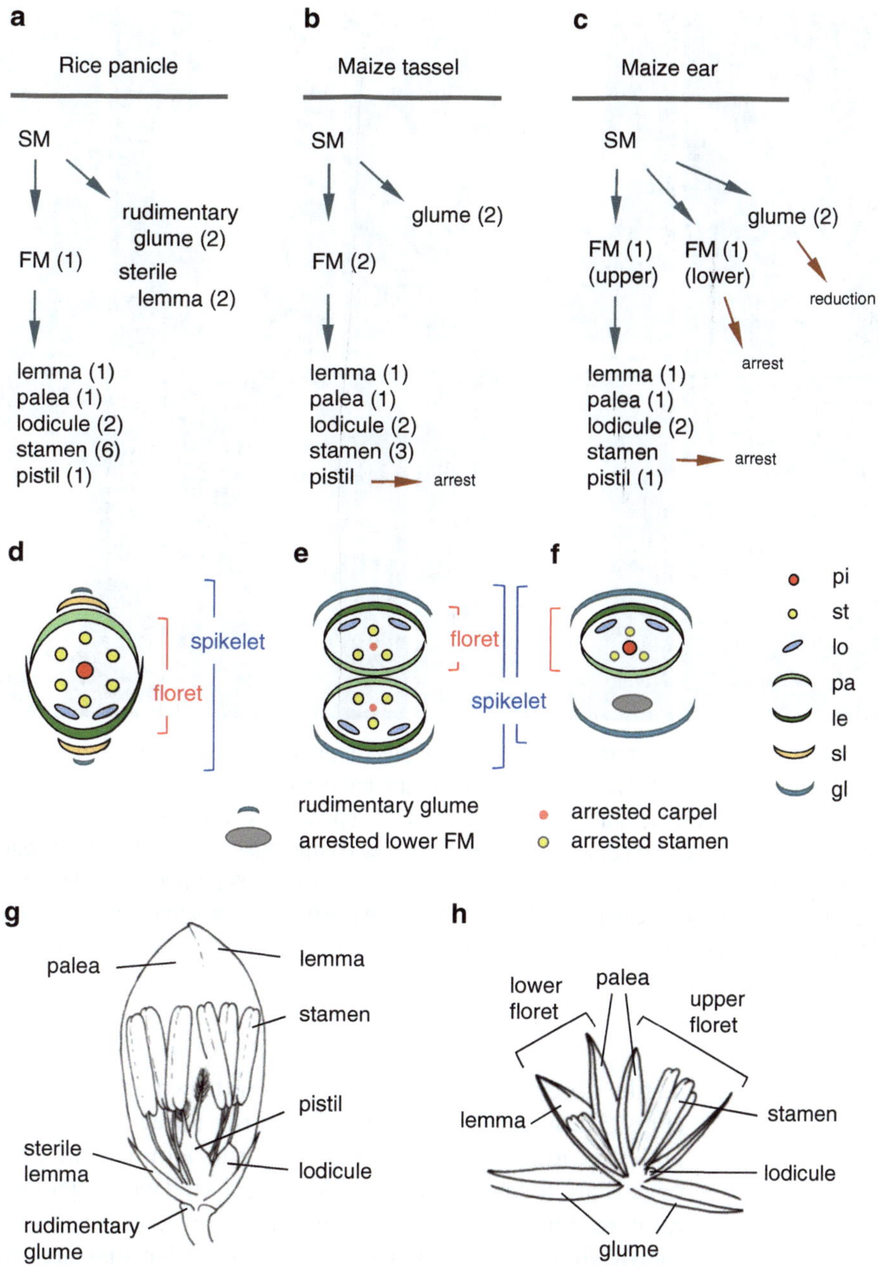

Fig. 1 Spikelet development in rice and maize. (**a**–**c**) Meristem transition and lateral organ differentiation. (**d**–**f**) Schematic representation of the bisexual spikelet in rice (**d**) and the male (**e**) and female (**f**) spikelets in maize. (**g** and **h**) Illustration of rice (**g**) and maize male (**h**) spikelets. *FM* flower meristem, *gl* glume, *le* lemma, *lo* lodicule, *pa* palea, *pi* pistil, *SM* spikelet meristem, *sl* sterile lemma, *st* stamen

In rice, the IM initiates the primary branch meristems (pBMs), which then initiate the secondary branch meristems (sBM) and the spikelet meristems (SMs) [5, 9, 10]. The SM initiates the floret (flower) meristem (FM) (Fig. 1a). The rice spikelet consists of one

Fig. 2 Mutants of the rice spikelet. (**a**) Wild type. (**b**) *dl-sup1* mutant. A pistil (carpel) is homeotically transformed into stamens. (**c**) *spw1-11* mutant. Stamens are homeotically transformed into pistils (carpels). (**d**) *lhs1-11* mutant. Florets with lemma/palea-like organs are repeatedly formed. (**e**) *mof1-11* mutant. Ectopic lemma/palea-like organs are formed, and FM determinacy is compromised. (**f**) Wild type showing a normal lemma and palea. (**g**) *tob1-1* mutant. A seamless lemma/palea-like organ is formed instead of the lemma and palea. (**h**) *dp1* mutant with a reduced palea (rpa). *g/lf* glume/leaf-like organ, *le* lemma, *le/pa* lemma/palea-like organ, *lo* lodicule, *pa* palea, *pi* pistil, *rg* rudimentary glume, *rpa* reduced palea, *sl* sterile lemma, *so* seamless organ, *st* stamen. Bars = 1 mm

fertile floret, two sterile lemmas, and two rudimentary glumes (Figs. 1d, g and 2a, f) [5, 8, 11]. In the floret, a pistil, six stamens, and two lodicules are enclosed by the lemma and palea. These organs are formed from the FM (Fig. 1a). Two lodicules, homologous to petals in eudicots and other monocots, are located asymmetrically on the lemma side. The sterile lemmas are small white leaf-like organs that are not observed in other grasses such as maize and wheat (Figs. 1d, g and 2a, f). A classical comparative morphological study [12] suggested that the sterile lemmas are derived from a reduced lemma of the lateral two florets, and a recent molecular genetic study [13] confirms this idea (*see* Subheading 6.5). The rudimentary glumes are highly degenerated and are observed as very tiny projections (Figs. 1d, g and 2a, f). The rudimentary glumes and the sterile lemmas are formed from the SM (Fig. 1a). After initiation of the sterile lemmas, the fate of the SM probably converts to that of the FM, which then initiates lemma, palea and floral organs. Morphological changes as result of this conversion are not clearly observed, because one SM forms only one FM in rice.

Whereas rice bears bisexual flowers, maize is a monoecious plant, and bears female flowers on the ear and male flowers on the tassel (Fig. 1b, c, e, f, h). In the ear, the IM initiates the spikelet pair meristems (SPMs), which subsequently divide into two SMs [7, 10]. In the tassel, the IM first initiates the BMs, and both the IM and BM initiate the SPMs. The SM forms two FMs in both the ear and tassel, and two FMs normally develop into two florets in the tassel (Fig. 1b, e, h). In the ear, however, the upper FM normally develops into the floret, whereas the lower FM is arrested during development (Fig. 1c, f). During floret development, all floral organs initiate, but the stamen primordium is arrested in the female flower, whereas the carpel primordia are arrested in the male flower [14, 15].

3 Regulation of Fate and Maintenance of Reproductive Meristems

3.1 Transition of Meristem Fate

To elaborate the complex inflorescence architecture, a number of genes contribute to determining and maintaining the fates of the various types of meristems. In this section, we will focus on the genes that regulate fate determination and determinacy of the SM and FM. Hormonal control of inflorescence development and genes involved in regulation of BM fate and branching patterns are described in other excellent reviews [5, 6, 10, 16, 17].

Specification of SM identity is regulated by the *branched silkless1* (*bd1*) and *FRIZZY PANICLE* (*FZP*) genes in maize and rice, respectively [18, 19]. In the maize *bd1* mutant, indeterminate branches are formed in place of spikelets in the ear, whereas spikelets are produced repeatedly in the tassel [18]. In the rice *fzp* mutant, spikelets are also replaced by indeterminate branches [19]. Thus, both maize *bd1* and rice *FZP* are required for SM determinacy by repressing indeterminate branching. The primordia of the glume/rudimentary glume are formed in both maize *bd1* and rice *fzp* mutants, suggesting that SM fate is acquired but fails to be maintained. *bd1* and *FZP* encode transcription factors of the AP2/ERF family [18, 19] (Table 1), and are orthologous to *PUCHI* in *Arabidopsis* [20]. Both *bd1* and *FZP* are expressed in the narrow region between the SM and the initiation site of the glume, but not expressed in the meristem per se [18, 19]. This expression pattern implies that the junction region of the SM and the glume initiation site might have important functions in establishing the identity and determinacy of the SM. Unlike maize *bd1* and rice *fzy*, the *Arabidopsis puchi* mutant does not show such a severe inflorescence phenotype [20]; however, close examination indicates that *PUCHI* is required for meristem identity also in *Arabidopsis*, suggesting that the fundamental function of these genes is conserved in grasses and eudicots.

Table 1
Genes involved in the regulation of meristem identity and flower development

Rice	Maize	*Arabidopsis*	Protein domain
Meristem identity			
FZP	*bd1*	*PUCHI*	AP2/ERF
SNB	*sid1*	*AP2/TOE*	AP2/ERF
OsIDS1	*ids1* (*Ts6*)	*AP2/TOE*	AP2/ERF
miR172a	*ts4*	*miR172*	
Flower development			
SPW1/OsMADS16	*si*	*AP3*	MADS
OsMADS2	*zmm16*	*PI*	MADS
OsMADS4	*zmm18, zmm29*	*PI*	MADS
OsMADS3	*zmm2, zmm23*	*AG*	MADS
OsMADS58	*zag1*	*AG*	MADS
OsMADS13	*zag2*	*STK*	MADS
LHS1/OsMADS1	*zmm8, zmm14*	*SEP1, SEP2, SEP4*	MADS
MFO1/OsMADS6	*bde*	*AGL6, AGL13*	MADS
DL	*ZmDL*	*CRC*	YABBY
TOB1	*zyb14*	*FIL, YAB3*	YABBY
RFL/APO2	*zfl1, zfl2*	*LFY*	Plant-specific
APO1		*UFO*	F-box
OPB/SL1		*JAG*	Zinc finger

The transition from SM to FM is regulated by genes encoding another class of AP2/ERF transcription factors, both in maize and rice [21–24] (Table 1). In maize *indeterminate spikelet1* (*ids1*) mutants, a few extra florets are formed within the spikelets both in the tassel and the ear [21]. When the *ids1* mutation is combined with a mutation of its close paralog *sister of indeterminate spikelet1* (*sid1*), glumes (bracts) are repetitively formed [22]. These phenotypes suggest that the transition from SM to FM is delayed or inhibited, and that SM determinacy is compromised. In rice, the *supernumerary bract* (*snb*) mutant exhibits repetitious production of rudimentary glumes (bracts) or extra lemma/palea and, in rare cases, two florets in a spikelet [23]. Although *osids1* mutant shows a weak defect in the spikelet, the *snb osids1* double mutant displays an enhanced *snb* phenotype [24]. *SNB* and *OsIDS1* encode members of a class of AP2/ERF transcription factors closely related to *IDS1* and *SID1* [23, 24]. Thus, *ids1*-like genes redundantly act to control the transition from SM to FM and the determinacy of the SM in both maize and rice.

These *ids1*-like genes are regulated by microRNA *miR172*. In maize, *tasselseed4* (*ts4*) and *Tasselseed6* (*Ts6*), which have been known as mutants defective in sex determination [14, 15], exhibit increased branching in the inflorescence in addition to failure of carpel abortion in the male flower [25]. *ts4* encodes *miR172*, which targets *ids1* mRNA, whereas the dominant *Ts6* mutant has a mutation in the binding site of *miR172* in the *ids1* gene [25]. *ids1* is broadly expressed in the *ts4* mutant, compared with wild type, and the branching and sex determination phenotypes of *ts4* are almost completely suppressed by mutation of *ids1*. Thus, the defects in *ts4* and *Ts6* are the consequence of ectopic expression of *ids1*. In rice, overexpression of the *miR172* genes largely mimics the phenotype of the *snb osids1* double mutant [24, 26]. Thus, proper regulation of *ids1*-like genes by *miR172* is important for specifying SM fate in both maize and rice.

In rice, *LEAFY HULL STERILE1* (*LHS1*)/*OsMADS1*, a class E MADS-box gene (*see* Chapter 1), has a crucial role in SM and FM determinacy [27, 28]. A mutation in *LHS1* causes pleiotropic phenotypes that range from weak to severe. In a weak phenotype, the lemma and palea show leaf-like characteristics, suggesting that *LHS1* regulates the identity of these two organs (*see* Subheading 6.2). In a severe phenotype, a secondary floret/spikelet develops within a spikelet, and leaf-like organs are repetitively formed [27, 28] (Fig. 2d). This suggests that *LHS1* regulates the identity and determinacy of the SM/FM by repressing reversion of meristem fate. *LHS1* is expressed in the meristem at an early developmental stage of the spikelet [29, 30]. Subsequently, the expression is restricted to the lemma and palea primordia and is excluded from the inner floral organs except for weak expression in the carpel primordia. A few reports claim that the repetitive formation of leaf-like organs in the *lhs1* mutant (or in a knockdown line of multiple class E genes) is an outcome of homeotic transformation of inner floral organs [28, 31] and that this *lhs1* phenotype is analogous to that of the *Arabidopsis sep1/2/3* mutant, in which inner organs are transformed into sepals [2, 3]. However, it is unlikely that rice *LHS1* has a function similar to that of *Arabidopsis SEP* genes. Firstly, the *SEP* genes are expressed in all floral organs in *Arabidopsis*, although the expression patterns differ among the members. By contrast, no *LHS1* expression is detected in lodicule and stamen in rice. Secondly, the phenotype of the *sep1/2/3* mutant is a consequence of the loss of function of these genes. But loss of lodicule and stamen in the *lhs1* mutant is not explained in a similar manner, because this gene is not expressed in these organs. As alternative hypotheses, the repetition of the leaf-like organ in *lhs1* could be regarded as the repeated formation of spikelets/florets with leafy lemma/palea because of loss-of-determinacy of the meristem, or as a delay (or failure) in the timing of initiation of inner floral organs because of defects in FM identity. In the latter interpretation, the repeated formation of a leafy palea/lemma-like organ resembles the repetitious formation of rudimentary glumes in the *snb* mutant, in which

SM identity is defective. Further molecular genetic analysis and careful interpretation of the results are required to elucidate the precise function of *LHS1*.

3.2 Stem Cell Maintenance

In the meristem, stem cells are continuously self-maintained at the apex and supply cells to the peripheral region to differentiate lateral organs. A proper balance between self-maintenance and differentiation is required for homeostasis of the stem cell population. In *Arabidopsis*, this homeostasis is regulated by a negative feedback loop consisting of *WUSCHEL* (*WUS*) and *CLAVATA* (*CLV*) [32, 33]. *WUS* promotes stem cell identity. The CLV-pathway, consisting of the CLV3 signaling molecule and its receptor complexes including CLV1, negatively regulates stem cell proliferation by repressing *WUS* expression, whereas *WUS* positively regulates *CLV3* expression.

This mechanism underlying negative regulation of stem cell maintenance is largely conserved in grasses. Mutations in the *thick tassel dwarf* (*td1*) and *fasciated ear2* (*fea2*) genes cause an enlargement of the IM and the FM in maize, resulting in fasciation of the inflorescences and increased floral organ number [34, 35]. The *td1 fea2* double mutant shows an enhancement of the phenotype of each single mutant, suggesting that these two genes act in different genetic pathways [34]. In two rice *floral organ number* (*fon*) mutants, *fon1* and *fon2* (*fon4*), the FM is enlarged, resulting in more floral organs [36–39]. *FON1* and *FON2* act in the same genetic pathway, as evidenced by the fact that the double mutant of *fon1* and *fon2* shows the same phenotype as that of each single mutant [38]. Maize *td1* and rice *FON1* encode CLV1-like LRR receptor kinases [34, 37], whereas maize *fea2* encodes a CLV2-like LRR protein that lacks an intercellular kinase domain [35]. Rice *FON2* encodes a small secreted protein with a CLE domain, and the CLE domain of FON2 is highly similar to that of CLV3 [38, 39]. *FON1* and *td1* are expressed throughout the meristem [34, 37], whereas *FON2*, like *CLV3*, is expressed in the apical region of the meristem [38, 39]. The *FON2* expression domain is highly expanded in the enlarged FM of the *fon1* mutant [38]. These observations suggest that a CLV-like pathway that negatively regulates stem cell maintenance is conserved in both maize and rice.

In addition to this conservation, genes specific to rice have also been identified. The maintenance of the FM is also regulated by *FON2 SPARE1* (*FOS1*), which is functionally redundant to *FON2* [40]. *FOS1* was identified as a suppressor of the *fon2* mutation through analysis of the F2 progeny of a *fon2* mutant (*japonica*) and Kasalath (*indica*) cross, and it was determined that *FOS1* encodes a CLE protein. Genetic analysis revealed that the *japonica FOS1* allele carries a mutation at a putative signal peptide processing site, and that the *fon2* and *fos1* single mutants do not show the *fon* phenotype [40]. Because the *fos1* mutation is present only in *japonica* rice, the maintenance of FM is redundantly regulated by *FON2* and *FOS1* in other rice, such as *indica* rice and wild *Oryza* species [40]. Unlike in *Arabidopsis clv* mutants, no abnormalities are observed in the vegetative meristem of grass mutants defective in maintenance of the reproductive meristems [34, 35, 37–40]. This implies that other genes are required for the maintenance of the vegetative SAM in grasses. One candidate is *FON2-LIKE CLE PROTEIN1* (*FCP1*), which has a CLE domain similar to that of FON2 [41]. Constitutive expression of *FCP1* leads to failure to maintain the vegetative SAM in regenerating shoots. This effect is also observed if *FCP1* is over-expressed in regenerating shoots of *fon1*, suggesting that *FCP1* acts through a receptor other than *FON1*. Similar effects on the vegetative SAM are also observed in *FOS1*-expressing shoots [40], but not in *FON2*-expressing shoots [38, 41]. These observations demonstrate that stem cell maintenance in rice is regulated by at least three related negative pathways, and each pathway is likely to be differently involved in this regulation depending on the type of the meristem.

4 Floral Organ Development

4.1 Carpel Specification

Carpel identity is specified by a class C MADS box gene, *AGAMOUS* (*AG*), in *Arabidopsis* [2, 3]. By contrast, carpel specification is controlled in rice by a *YABBY* gene, *DROOPING LEAF* (*DL*) [42] (Fig. 3). In a loss-of-function mutant of *DL*, carpels are homeotically converted into stamens in the central whorl of the rice flower [42, 43] (Fig. 2b). *DL* encodes a *YABBY* transcription factor related to *CRABS CLAW* (*CRC*) in *Arabidopsis* [44] (Table 1), and is expressed in the presumptive region of carpel initiation in the FM and the carpel primordia [42]. *DL* transcripts appear to accumulate in three roughly separate domains in a cross-section of the developing carpel [42], implying that the rice pistil is formed from three congenitally fused carpel primordia. This inference is consistent with the concept that monocot flowers generally consist of trimerous floral organs [1, 45].

Mutation in the *DL* gene causes another defect, loss of the midrib in leaves, resulting in a drooping leaf phenotype (the origin of the gene name) [42, 43]. *DL* is specifically expressed in the central region of leaf primordia (P1–P4) [42, 46]. In situ hybridization analyses show that the expression patterns of *DL* orthologs in other grasses, such as maize, wheat and sorghum, are very similar to that of *DL* in rice in both flower and leaf development [47]. In addition, mutants similar to rice *dl* are reported in other grass species [48, 49]. Therefore, the function of *DL*-related genes, which is required for carpel specification and midrib formation, is conserved in grasses. *Arabidopsis CRC* is partially involved in the

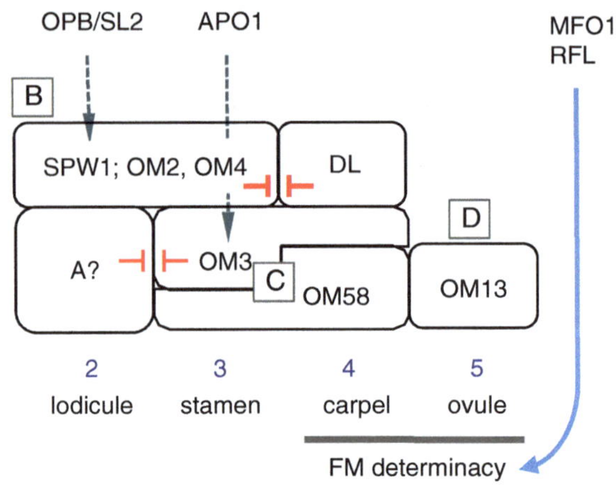

Fig. 3 Model of floral organ specification and genetic interaction in rice. Genes in *boxes* are involved in specification of the floral organ in each whorl or regulation of FM determinacy. *OPB/SL1* and *APO1* positively regulate the expression of *SPW1* and *OsMADS3*, respectively. *MFO1* and *RFL* are responsible for the FM determinacy. *OM* OsMADS

regulation of carpel identity, but does not have homeotic gene function [44]. Therefore, *DL* and its orthologs may have gained a crucial function to specify carpel identity during grass evolution.

Carpels do develop in the *zag1* mutant, a class C mutant in maize [50]. In rice, knockdown of *OsMADS58* in the *osmads3* mutant does not affect carpel specification [51]. In contrast, a literature reported that carpel identity is lost in the *osmads3* and *osmads58* double mutant [52]. However, green carpel-like organs are observed in this double mutant. The authors regarded these green organs as palea-like organs, from the observation that they have trichome-like structures. An alternative plausible explanation, however, is that these trichome-like structures are formed by the function of genes that are ectopically expressed in this green organ due to loss of function of two class C genes. In addition, *DL* is expressed in this green organ in the double mutant [52]. Therefore, it is likely that this green organ is a morphologically abnormal carpel, which would be specified by *DL*, but not a palea-like organ. Based on this idea, the conclusions in this literature [52], i.e., (i) class C genes are required to specify carpel identity in rice and that (ii) *DL* is not involved in carpel specification, should be reconsidered. Further detailed studies are required to clarify this issue.

4.2 Ovule Development

In *Arabidopsis*, ovules differentiate from the placenta after establishment of the carpel, and the *SEEDSTICK* (*STK*), *SHATTERPROOF1* (*SHP1*), *SHP2*, and *AG* genes are responsible for this process [3]. By contrast, a single ovule develops directly from the FM in grasses, resulting in a single ovule within a single ovary [9, 53, 54]. In rice, *OsMADS13*, a D-lineage MADS-box gene, plays a key role in ovule development (Table 1; Fig. 3). In the *OsMADS13* knockout mutant, the ovule is replaced by carpeloid organs, and determinacy of the FM is compromised [53, 54]. Because the ovule is the final lateral organ developing from the FM, failure in ovule specification is probably associated with meristem determinacy. Yamaki et al. [54] proposed that there is an "ovule founder region," a part of the FM remaining after initiation of the carpel primordia, and pointed out the similarity in function between *OsMADS13* in rice and *STK*, *SHP1*, and *SHP2* in *Arabidopsis*.

A weak allele (*log-3*) of *log lonely guy* (*log*) [55], which is responsible for active cytokinin synthesis, fails to form the ovule and shows reduced expression of *OSH1* in the FM after carpel initiation [54]. The ovule-less phenotype of *log-3* is suppressed by the *fon1* mutation, which causes excess stem cells in the FM. These observations confirm the importance of meristem activity in forming the ovule in rice.

4.3 Stamen Specification

In contrast to carpel specification, genes required for stamen specification are conserved between grasses and eudicots. Combinatorial activity of class B genes, *APETALA3* (*AP3*) and *PISTILATA* (*PI*), and the class C gene *AG* specifies stamen identity in *Arabidopsis* [2–4]. Loss of function of any of these genes results in homeotic conversion of the stamens, either into petals (*ag* mutant) or into carpels (*ap3* and *pi* mutants).

Mutation of the maize *silky1* (*si1*) gene and the rice *SUPERWOMAN1* (*SPW1*) gene results in homeotic conversion of stamens and lodicules into carpels and glume-like structures, respectively [43, 56] (Fig. 2c). These homeotic changes resemble those observed in the *ap3* and *pi* mutants in *Arabidopsis*. Both *si1* and *SPW1* encode a MADS domain protein similar to *Arabidopsis AP3* [43, 56] (Table 1). Rice has two *PI*-like genes, *OsMADS2* and *OsMADS4*. Double knockdown of both genes also causes a homeotic transformation similar to that observed in the *spw1* mutant [57]. By contrast, each single knockdown mutant does not affect stamen development, suggesting that *OsMADS2* and *OsMADS4* have a redundant function similar to that of *PI*. These grass class B genes are expressed in whorl 2 (the lodicule whorl) and whorl 3 (the stamen whorl) in the FM and in the primordia of the lodicule and stamen [43, 56, 58–60]. Thus, the function of class B genes is conserved in both maize and rice, and is required for both stamen and lodicule specification (Fig. 3).

In *Arabidopsis*, *AP3* and *PI* form a heterodimer and promote the expression of target genes by binding to their promoter regions [2, 3]. It has been reported that gene products of maize *si* and *zmm16* (*PI* ortholog) form heterodimers and bind in vitro to the

CArG-box sequence present in the *Arabidopsis AGL5* promoter [61]. Heterodimerization between rice *SPW1* and *OsMADS2/OsMADS4* is also observed in the yeast two-hybrid system [57]. Furthermore, maize *si* and *zmm16* can rescue the phenotypes of the corresponding class B mutant in *Arabidopsis* [61]. These observations indicate that the biochemical function of class B MADS domain proteins is conserved between grasses and eudicots.

The stamen is homeotically transformed into a lodicule in the rice *osmads3* mutant [51]. Conversely, ectopic expression of *OsMADS3* in wild type results in homeotic transformation of the lodicule into a stamen [62]. By contrast, stamens develop normally in the *osmads58* knockdown line, although the number of stamens is reduced [51]. When the expression of *OsMADS58* is silenced in the weak *osmads3-2* mutant, which produces stamens, homeotic transformation of the stamen is also observed. These observations indicate that class C genes redundantly regulate stamen specification in rice, and the contribution of *OsMADS3* is much higher than that of *OsMADS58* (Fig. 3). Together, the data indicate that stamen identity in grasses is regulated by B and C class genes, as in *Arabidopsis* and other eudicots. (In other reports [52], an insertion knockout line of *osmads58* was described as showing no obvious phenotype, probably because of a different genetic background. Using this mutant, it has recently been shown that *OsMADS58* was involved in flower patterning together with *SPW1*, as *OsMADS3* was [117].)

4.4 Elaboration of the Anther and the Filament in Stamen Development

The stamen is composed of two parts: the anther in the distal region and the filament in the proximal region. The anther consists of two thecae and the connective between them. The theca has two pollen sacs, in which pollen differentiates. The pollen sac and the filament are cylindrical and rod-like structures, respectively. Because the floral organs are thought to be modified leaves, it is an interesting question how cylindrical or rod-like structures are formed during stamen development from lateral organ primordia, which should have characteristics common to leaves and other floral organs.

Recent analyses in rice have revealed that morphological changes in stamen development are closely associated with the establishment of adaxial–abaxial polarity [63]. Gene markers for adaxial and abaxial identity are initially expressed in stamen primordia in a pattern similar to that in leaf primordia. Subsequently, however, the expression patterns are rearranged, and the markers are expressed in specific regions, where adaxial and abaxial identity seems to be respecified in a new developmental unit, the theca primordium. The regions between the new adaxial and abaxial domains protrude, probably due to accelerated cell division, and differentiate into four pollen sacs. In the proximal part, expression patterns of marker genes suggest that this part is completely abaxialized, and is formed as a rod-like structure, the filament.

Thus, anther patterning and filament formation seem to be closely related to the establishment of adaxial–abaxial polarity [63]. This inference obtained from analysis of wild type is confirmed by observations of stamen development and similar expression analyses in the *rod-like lemma* (*rol*) mutant, which is partially defective in establishment of adaxial–abaxial polarity in flower development [63]. Thus, it is likely that the two parts that constitute the stamen, the anther and the filament, are formed by independently adopting the regulatory mechanism underlying the establishment of adaxial–abaxial polarity.

In *Arabidopsis*, there is no report describing stamen development from this viewpoint. However, several lines of evidence suggest that a similar conclusion would be drawn. Firstly, expression patterns of genes involved in adaxial and abaxial identity rearrange during anther development of *Arabidopsis*, as in rice (see references cited in Toriba et al. [63]). Secondly, *Arabidopsis* mutants defective in adaxial–abaxial establishment show stamen phenotypes similar to those observed in the rice *rol* mutant. Therefore, the idea that anther patterning and filament formation are associated with the mechanism of adaxial–abaxial establishment could be widely applicable to stamen development in angiosperms.

4.5 Lodicule Development

The lodicule is an organ characteristic of grass flowers and plays a role in opening the flower by swelling and pushing the lemma and palea. From comparative morphological studies, the lodicule is a homolog of petals in grasses.

As described in Subheading 4.3, mutations in *AP3*-like genes result in transformation of lodicules into glume/leaf-like organs in maize and rice [43, 56] (Fig. 2c). Double knockdown of two rice *PI*-like genes, *OsMADS2* and *OsMADS4*, has a similar effect on lodicule identity [57] (Table 1). Thus, two types of class B genes are required for specification of lodicules in grass flowers (Fig. 3), as in the case of *Arabidopsis*, in which class B genes are responsible for petal identity. Although *OsMADS2* and *OsMADS4* redundantly act in the specification of both the stamen and the lodicule in rice, the contribution of the two genes differs in the two processes. *OsMADS2* plays a larger role in lodicule development than *OsMADS4* does, as single knockdown of *OsMADS2* produces abnormal lodicules, but that of *OsMADS4* does not [57].

Lodicules develop asymmetrically: lodicules (two) are formed only on the lemma side. *OsMADS3* is likely to be involved in suppression of lodicule initiation: ectopic lodicules are formed on the palea side in whorl 2 of the *osmads3* mutant, in addition to a homeotically transformed lodicule in whorl 3, suggesting that *OsMADS3* inhibits lodicule initiation [51]. In wild type flowers, *OsMADS3* is asymmetrically expressed, such that the expression domain of *OsMADS3* is expanded to the region close to the palea primordia. Therefore it is possible that loss of the lodicule at the palea side results from the expansion of *OsMADS3* expression near the palea primordium [51]. Many genes are involved in lodicule development, and these are summarized in [64].

A partial defect in lodicule development is associated with cleistogamy, which promotes self-pollination and prevents outcrossing. Cleistogamy is important for agronomic uses, because it allows prevention of gene flow from genetically modified crops. In a barley mutant, cleistogamy is related to an enhancement of the expression of an *AP2*-like gene [65]. Mutants showing cleistogamy have mutations in the target site of miR172 in this gene. In rice, a weak missense mutation in *SPW1* causes cleistogamy without defects in flower development other than in lodicule formation [60].

5 Genetic Interaction in Flower Development and Meristem Determinacy

5.1 Genes That Regulate Floral Homeotic Genes

LEAFY (*LFY*) has a central role in flower development in *Arabidopsis*: it specifies identity of the FM and controls the expression of ABC floral homeotic genes [2, 3]. The *RFL/ABERRANT PANICLE ORGANIZATION2* (*APO2*) gene in rice and the *zfl1* and *zfl2* genes in maize encode plant-specific transcription factors orthologous to *Arabidopsis LFY* [66–68] (Table 1). The rice *apo2* mutant generates flowers with aberrant floral organ identity and small panicles with reduced branching [67, 68]. In the maize *zfl1 zfl2* double mutant, severe defects in the fate and determinacy of the meristem are observed, in addition to abnormal floral organ identity and reduced branching [66]. Thus, with respect to the control of floral homeotic genes, the function of *LFY* and its orthologs are likely to be partially conserved between *Arabidopsis* and grasses. By contrast, reduced branching in the grass mutants suggests that grass *LFY* orthologs have roles different from that of *Arabidopsis LFY* in inflorescence development, because shoot branching increases in the *lfy* mutant.

A mutation in the *ABERRANT PANICLE ORGANIZATION1* (*APO1*) gene causes pleiotropic defects in flower and inflorescence development in rice [69, 70]. In *apo1* flowers, extra lodicules are formed at the expense of stamens, and the carpels are formed indeterminately. Whereas *SPW1* is normally expressed, *OsMADS3* expression is reduced, suggesting that *APO1* is a positive regulator of *OsMADS3* (Fig. 3). Thus reduction in the expression level of a class C gene is probably associated with partial transformation of the stamen. *APO1* encodes an F-box protein similar to the *Arabidopsis* UNUSUAL FLORAL ORGANS (UFO) protein (Table 1), and is expressed throughout the meristem and in primordia of lateral organs [69]. IMs are enlarged in gain-of-function mutants of *APO1*, resulting in increased branch and spikelet numbers and enhanced culm strength and thickness [71, 72]. *Arabidopsis UFO* is expressed in a restricted region of the

FM, whorls 3 and 4, and positively regulates *AP3* expression [2, 3]. Therefore, the functions of *APO1* and *UFO* seem to be divergent between grasses and eudicots.

Double mutant analysis indicated that *RFL/APO2* and *APO1* act in the same genetic pathway with respect to vegetative and inflorescence development [68]. Overexpression of *APO1* in the *rfl/apo2* mutant does not affect panicle phenotypes, suggesting that *RFL/APO2* is required for *APO 1* function. Consistent with these observations, RFL/APO2 and APO1 physically interact with each other. By contrast, a synergetic effect is observed on floral organ identity and FM determinacy in the *apo1 rfl/apo2* double mutant, and *APO1* expression is upregulated in the *rfl/apo2* mutant [68]. Thus, the genetic interaction of *RFL/APO2* and *APO1* is complex, and more detailed examination is needed.

SPW1 is positively regulated by *OPEN BEAK* (*OPB*)/ *STAMENLESS1* (*SL1*) in rice [73, 74]. In *opb/sl1* flowers, stamen and lodicule identities are misspecified, resulting in the formation of mosaic floral organs, whereas lemma and palea are malformed (*see* Subheading 6.4). Downregulation of class B genes is likely to be related to formation of glume-like lodicules and carpeloid stamens. Class 1 *knox* genes are ectopically expressed in *opb*, causing abnormal cell proliferation in both spikelets and leaves [73]. *OPB/SL1* encodes a protein with a C2H2 zinc finger domain (Table 1), and is closely related to *JAGGED* and *NUBBIN* [73, 74], which are responsible for morphogenesis of the stamen and carpel in *Arabidopsis* [75–77].

5.2 Antagonistic Interaction Between DL and Class B Genes

In *Arabidopsis*, class A and class C genes repress each other's expression [2, 3]. Therefore, removal of class A genes causes ectopic expression of class C genes in whorls 1 and 2, resulting in homeotic transformation, and vice versa.

A similar antagonistic interaction is observed between *DL* and class B genes. In rice, carpels are ectopically formed in whorl 3 in the *spw1* mutant, whereas stamens are ectopically formed in whorl 4 in the *dl* mutant [42, 43]. Consistent with the ectopic organ formation, *DL* and *SPW1* are ectopically expressed in whorls 3 and 4 in the *spw1* and *dl* mutant, respectively. With respect to *PI* homologs, double knockdown of *OsMADS2* and *OsMADS4* results in the production of ectopic carpels in whorl 3 [57].

Constitutive expression of *SPW1/OsMADS16* results in partial homeotic transformation of carpels into stamens [78]. This suggests that *SPW1*, which is ectopically expressed in whorl 4, represses *DL* expression and promotes stamen identity. In the *sl1* mutant, *DL* is misexpressed in abnormal organs formed in whorl 3 [74]. This misexpression seems to result from reduced expression of class B genes. These observations also support the idea that *DL* and class B genes interact antagonistically. This antagonistic interaction between a *DL* ortholog and a class B gene is also observed in the pistillody mutant of wheat [47, 79].

5.3 Flower Meristem Determinacy

Unlike indeterminate meristems such as the SAM and IM, the FM is determinate, and the stem cells are consumed by the final floral organ initiated from the FM. In *Arabidopsis*, *AG*, which encodes a class C MADS-domain transcription factor, is a key gene responsible for meristem determinacy [2, 3]. Loss of function of *AG* results in repetitive formation of a set of floral organs consisting of sepals, petals, and ectopic petals.

In grasses, class C genes also play an important role in meristem determinacy (Fig. 3). A severe defect is observed in knockdown lines of *OsMADS58*: a set of floral organs (lodicules–stamens–partial carpels) is repeatedly formed in the flower, and a meristem-like structure is observed in the mature flower [51]. Loss of function of other class C genes, such as maize *zag1* and rice *OsMADS3*, show repetitive formation of the carpel, a partially indeterminate phenotype [50, 51]. Although the *osmads58* mutation has little effect on the floral phenotype in a different genetic background, it dramatically enhances the phenotype of the *osmads3* single mutant [52]. Thus, class C MADS-box genes play crucial roles in regulating FM determinacy in both maize and rice. In addition, two class C genes have diversified functionally: *OsMADS58* is predominantly involved in meristem determinacy, whereas *OsMADS3* is involved in stamen specification [51].

As described in Subheading 4.2, the ovule is the final organ differentiated from the FM in grasses. Loss of function of *OsMADS13*, which specifies the ovule, results in partial loss of determinacy in the FM: repeated formation of carpels and prolonged expression of *OSH1* are observed in the *osmads13* mutant [53, 54]. Mutation of *osmads13* enhances the indeterminate phenotype observed in the *osmads3* or *osmads58* mutant [52].

In the loss-of-function mutant of *DL*, meristem determinacy is also partially compromised [42]. In addition, *dl* mutation enhances the indeterminate phenotypes of *apo1* [69], *osmads3* [80], and *mosaic floral organ1* (*mfo1*) (see below) [81, 82].

MADS-box genes of the *AGL6* subfamily, including maize *bearded-ear* (*bde*)/*zag3* [83] and rice *MFO1/OsMADS6*, are also responsible for FM determinacy [81, 84] (Fig. 3). In the maize *bde* mutant, the upper FM forms extra floral organs, whereas the lower FM initiates additional FMs [83]. BDE protein physically interacts with ZAG1 protein, suggesting that *bde* function in the regulation of FM determinacy is associated with class C gene activity. Extra carpels and spikelets are formed in the center of the flower in the rice *mfo1/osmads6* mutant [81, 84] (Fig. 2e). This loss-of-determinacy of the FM in *mfo1/osmads6* is enhanced in combination with the *lhs1* mutation [81, 84]. Expression patterns of *AGL6*-like genes in other grasses suggest that these genes are commonly involved in the regulation of FM determinacy in grasses [85].

As described in Subheading 5.1, *zfl1* and *zfl2* of maize and *RFL/APO2* in rice are also involved in the regulation of FM determinacy [66, 68] (Fig. 3). Thus, a number of genes that regulate floral organ development are involved in this regulation. In *Arabidopsis*, FM determinacy is achieved through the *AG* function that represses the expression of *WUS*, which promotes stem cell identity [2, 3, 86]. In grasses, however, a key gene that positively regulates stem cell identity has not yet been identified. (Recently, it was reported that the *OsWOX4* in the *WOX* gene family has a role to promote the undifferentiated states of the meristem in rice [118]). Therefore, identification of a gene like *Arabidopsis WUS* remains an important step for understanding the genetic mechanism underlying regulation of meristem determinacy in grasses.

6 Development of Organs Specific to the Spikelet

6.1 Origin of the Lemma and the Palea

The lemma and palea are unique to grasses, enclosing the floral organs (lodicules, stamens, and a pistil), and several hypotheses have been proposed for their. The lemma and palea are regarded as a bract and a prophyll, respectively: the palea is the first leaf born on the axil branch of the lemma [1, 87]. By contrast, a molecular genetic study on maize floral mutants supports that both organs are modified sepals, based on the observation that removal of class B activity results in homeotic transformation of lodicules into palea/lemma-like structures [56]. This transformation is similar to that of petals into sepals in *Arabidopsis* class B mutants [2, 3]. In this view, the lemma and palea, like *Arabidopsis* sepals, should be specified by class A gene activities. However, a recent rice study showed that triple knockdown of *OsMADS14, 15,* and *18* genes, which belong to A-class subfamily of MADS-box genes, causes no defects in the development of the lemma and palea [88]. Other genetic studies also demonstrate that development of the lemma and the palea is regulated differently (*see* Subheadings 6.2 and 6.3), supporting a different origin of the two organs.

In rice, the glumes are highly reduced, forming rudimentary glumes, and cannot enclose the florets, unlike in other grasses. Therefore, the lemma and palea are the most conspicuous organs, enabling identification of a number of mutants with abnormal lemma and palea morphology. The rice lemma and palea are hard, and their abaxial surface is very rough, with convex structures called tubercles arranged in parallel. Although their shapes are similar, the lemma and palea are distinguished by several characteristics, such as the morphology of the peripheral region and the number of vascular bundles [13]. A thin and long structure, called an awn, is formed at the top of the lemma, although the growth of the awn is suppressed in *japonica* rice. In maize, the lemma and palea are thin and fragile and are enclosed by glumes. With respect to lemma/palea identity, we will focus mainly on rice genes because there are few reports describing these organs in maize.

6.2 Genes That Regulate Both Lemma/Palea Identity and the Floral Meristem in Rice

LHS1/OsMADS1 is well studied in rice. *LHS1* is predominantly expressed in the primordia of the lemma and palea at the organ differentiation stage, in addition to the meristem at an early stage of spikelet development, as described in Subheading 3.1. The lemma and palea are partially transformed into leaf-like structures with smooth abaxial surfaces in both the *lhs1* mutant and in RNA silencing lines of *LHS1* [27, 28, 30]. Conversely, constitutive expression of *LHS1* results in the formation of convex structures on the surface of the sterile lemma [30]. The effect of loss (or gain) of function of *LHS1* is more severe in the lemma than in the palea. Thus, *LHS1* is likely to be involved in specification of the lemma and palea in rice, in addition to meristem determinacy as described in Subheading 3.1.

Although no mutant has yet been isolated in maize, two *LHS1-like* genes, *zmm8* and *zmm14* (Table 1), are expressed throughout the SM at early developmental stages [89]. Later, the expression of these genes is restricted to the floral organ primordia in the upper floret, whereas no expression is detected in the lower floret. In other grasses, *LHS1*-like *SEP* genes are commonly expressed in the SM and the lemma/palea primordia, but at later stages show various expression patterns among species [90]. Therefore, *LHS1-like* genes have general roles in specifying SM determinacy and lemma/palea identity in grasses, whereas they also have diverse roles in spikelet development depending on the species.

Loss of function of the rice *TONGARI-BOUSHI1* (*TOB1*) gene causes pleiotropic defects in both lemma and palea, such as formation of a seamless lemma/palea-like organ (Fig. 2g), failure or reduction in lemma/palea growth, and elongation of the awn [91]. In addition, maintenance and organization of the meristem are compromised in the *tob1* mutant. For example, the FM is prematurely terminated, whereas two florets are formed within a spikelet, probably because of a disorganized FM [91, 92]. *TOB1* encodes a YABBY transcription factor, similar to *Arabidopsis FILAMENTOUS FLOWER* (*FIL*) and *YABBY3* [91, 93] (Table 1). Unexpectedly, *TOB1* is strongly expressed in the primordia of lateral organs, such as the lemma and palea, but not expressed in the meristem per se. Therefore, *TOB1* activity in the lemma and palea acts non-cell autonomously on the FM and is required for communication between the lateral organs and the meristem. This non-cell autonomous function of *TOB1* is similar to that of *FIL* in vegetative development of *Arabidopsis* [94].

6.3 Genes Affecting Mainly Palea Identity in Rice

The development of the palea is regulated by the *RETARDED PALEA1* (*REP1*) and *DEPRESSED PALEA1* (*DP1*) genes in rice [95, 96]. In the *rep1* mutant, palea growth is delayed, resulting in a small abnormal palea [95]. In *dp1* mutant, the development of the palea is compromised, such that it lacks the central region of the palea [96] (Fig. 2h). Because no obvious defect is observed in the lemma in either mutant, these genes are specifically required for palea development. *REP1* and *DP1* encode a TCP-family transcription factor and an AT-hook DNA binding protein, respectively [95, 96]. Both genes are specifically expressed in palea primordia at the palea/lemma initiation stage. *REP1* expression is significantly reduced in the *dp1* mutant, suggesting that *DP1* acts upstream of *REP1* [96]. In contrast to the *dp1* mutant, the marginal regions of the palea are lost in the *mfo1/osmads6* mutant and appear to be replaced by those of the lemma [81, 84]. In addition, the *DL* gene, a marker of the lemma, is ectopically expressed in the palea of *mfo1* [81]. Thus, *MFO1* is also responsible for palea identity.

6.4 Other Genes Involved in Lemma/Palea Morphology in Rice

The lemma is changed into a rod-like structure in loss-of-function mutants of *shoot organization1* (*sho1*), *sho2*, and *wavy leaf1* (*waf1*) [97–99] and in a weak allele of *shootless2* [63]. The genes responsible for these mutations encode proteins involved in the synthesis of trans-acting small interfering RNA (ta-siRNA), which is required for the establishment of adaxial–abaxial polarity. Morphological and molecular analyses have revealed that the rod-like structure is a consequence of abaxialization of the lemma [63]. In contrast to the severe defects in the lemma, palea morphology is less affected in these mutants, suggesting that lemma development is probably more sensitive to reduction in ta-siRNA production.

Differentiation of the lateral region is suppressed in both the lemma and palea in the *opb/sl1* mutant, resulting in a failure to enclose the inner floral organs [73, 74]. The lemma and palea are replaced by semitransparent thin leaf-like or filamentous organs in the *degenerated hull1* (*dh1*) mutant [100]. *DH1* encodes a protein with a LOB domain. The morphology and epidermal characteristics of the lemma and palea are affected by mutation in the *TRIANGULAR HULL1* (*TH1*) gene [101] or by RNA silencing of the gene similar to maize *CRINKLY4* [102, 103]. Despite the name of the mutant, extra lemma-like structures are formed in the *extra glume1* (*eg1*) mutant [104]. *EG1* encodes a phospholipase and is expressed in the IM, BM, FM, and lateral organ primordia in the spikelet.

6.5 The Sterile Lemma and Rudimentary Glume in Rice

The sterile lemma is formed between the floret and the rudimentary glume in the rice spikelet. This organ is unique to species in the *Oryza* genus and is not seen in grasses such as maize and wheat. Therefore, the sterile lemma is an organ that deviates from the general floral formula of grass species. More than 70 years ago, Arber [12] proposed that this unique organ could be considered a modified lemma that is evolutionary derived from the lemma of two degenerated lateral florets (sterile florets).

Mutation in *LONG STERILE LEMMA* (*G1*) results in a homeotic transformation of the sterile lemma into a lemma in rice

[13]. The transformed lemma in the *g1* mutant has the same morphological characteristics as the wild-type lemma and differs from the palea. Therefore, *G1* functions to specify sterile lemma identity by repressing lemma identity. *G1* encodes a nuclear protein with an ALOG domain, which is specific to land plants [13].

As described in Subheading 6.1, constitutive expression of *LHS1* promotes lemma-like characteristics in the sterile lemma [30]. Thus, both the *g1* mutant and *LHS1* constitutive expression demonstrate that the sterile lemma has the potential to transform into a lemma and supports the idea that the sterile lemma is a modified lemma [12]. Thus, the name "sterile lemma" should be used for this organ, although different names such as glume and empty glume are still used in some papers.

The sterile lemmas and rudimentary glumes are affected in *aberrant panicle and spikelet1* (*asp1*) [105] and *panicle phytomer2* (*pap2*)/*osmads34* mutants [106, 107]. In the *asp1* mutant, both organs are elongated, and their epidermal surface is also affected [105]. In addition, the developmental program of the spikelet is compromised, probably due to a defect in maintenance of the meristem. Close examinations have revealed that the *asp1* mutation affects the fate of various types of meristems, such as the vegetative SAM, the IM, BM, and SM, and causes pleiotropic phenotypes throughout the rice life cycle. *ASP1* encodes a transcriptional co-repressor related to *Arabidopsis* TOPLESS [105, 108]. Expression of a number of genes is probably deregulated temporally and spatially, leading to disruption of developmental programs in *asp1*. In *pap2*, the sterile lemmas are often highly elongated, whereas its identity is unchanged [106]. In addition to the elongation of the rudimentary glume, an ectopic filamentous organ is formed in the axil of the elongated rudimentary glume. *PAP2* also has a role in specifying SM identity, in addition to controlling the transition from vegetative meristem to IM [88].

6.6 Glume Development in Maize and Teosinte

The *teosinte glume architecture1* (*tga1*) gene is responsible for the difference in the glume and cupule between maize and its ancestor teosinte [109]. In teosinte, the kernel is enclosed by a hard fruitcase formed from the glume and cupule, and the epidermal cells of the fruitcase are lignified and filled with silica. By contrast, in maize, the glume and cupule are reduced in size, and accumulation of lignin and silica is reduced. Genetic introduction of the teosinte allele of *tga1* into maize results in an enlargement of the glume and cupule and silica accumulation. Conversely, the maize *tga1* allele promotes reduction of the sizes of the glume and cupule in teosinte. Thus, *tga1* plays a role in glume and cupule development. The *tga1* gene encodes an SBP-domain transcription factor [109]. Reduced hardness of the fruitcase is an important trait in agronomy. Several nucleotide differences, including one resulting in an amino acid substitution, are found between the *tga1* alleles of maize and teosinte, suggesting that these mutations are associated with maize domestication.

7 Sex Determination in Maize

Maize is a monoecious plant and bears pistillate flowers in the ear and staminate flowers in the tassel. As described in Subheading 2, two male florets are formed in a spikelet in the tassel. By contrast, only the upper floret develops to maturity in the ear, whereas the development of the lower floret arrests at an early stage. In both female and male flowers of maize, the carpel and stamen primordia initiate as in bisexual flowers of other species, but subsequent development of the stamen and the carpel is arrested in the female and male flowers, respectively.

There are a number of mutants that are defective in this process [14, 15]. In *tasselseed* (*ts*) mutants, the female floret develops instead of the male floret or the carpel fails to abort in the tassel. As described in Subheading 3.1, in *ts4* and *Ts6* mutants, the *ids1* gene is misexpressed in developing spikelets, suggesting that proper expression of *ids1* is required for carpel abortion in the male flower [25]. The *ts2* gene was first identified at the molecular level about 20 years ago [110]. *ts2* encodes a short-chain dehydrogenase and is expressed in the carpel primordia, suggesting that this gene is involved in carpel abortion. *ts1* encodes a lipoxygenase responsible for jasmonic acid (JA) biosynthesis [111]. JA concentration in the tassel is much reduced in the *ts1* mutant compared to wild type, and exogenous application of JA rescues the feminized floret of the *ts1* mutant. These findings suggest that JA is required for carpel abortion in the tassel. JA application also rescues the tasselseed phenotype of the *ts2* mutant [111]. This observation may shed light on the role of the enzyme encoded by *ts2*, which has not yet been uncovered.

A dwarf mutant, *nana plant1* (*na1*), also shows feminization of the floret in the tassel. *na1* encodes protein similar to *Arabidopsis* DET2, which is involved in brassinosteroid (BR) biosynthesis [112]. Treatment of wild type with BR biosynthetic inhibitors mimics both the tasselseed and dwarf phenotypes of *na1*. The *na1* gene is expressed in the stamen throughout its development, suggesting that BR is required for stamen maturation.

Mutations in *anther ear1* (*an1*) and *Dwarf8* (*D8*) are associated with masculinization of the flower in the ear [14, 15, 113]. In the ear of the *an1* mutant, a perfect flower with stamens and a pistil is formed in the upper floret, whereas stamens develop also in the lower floret, which forms no mature floral organs in wild type [113]. The *an1* gene encodes ent-kaurene synthase, which catalyzes an early step of gibberellic acid (GA) biosynthesis.

The dominant *D8* mutant is insensitive to GA [114]. *D8* encodes a DELLA protein related to the *Arabidopsis* GAI, which is a negative regulator of GA responses, and the mutation in *D8* is associated with a dominant negative form of this repressor. These observations indicate that GA signaling is involved in sex determination in maize by suppressing stamen development in the ear.

Thus, several phytohormones play important roles in sex determination in maize, by directing abortion of reproductive organ development. *terminal ear* and *Tunicate1*, which are known as mutants defective in meristem fate, are also compromised in sex determination [115, 116]. In the *zfl1* and *zfl2* double mutant, a few branches are replaced by the ear, the whole female inflorescence, suggesting that BM fate in the tassel is transformed into female IM fate [66]. It is of great interest to know how sexual fate is associated with the regulation of meristem fate in maize.

8 Concluding Remarks

The ABC model of floral organ specification, which was established from the study of model eudicots [4], is likely to be largely applicable to grass flowers despite their morphological differences from those of eudicots (Fig. 3). For example, the stamen is specified by the combinatorial activities of class B and class C genes in the grasses, as in eudicots [43, 56, 57]. Class B genes are also required for the specification of a whorl 2 organ, the lodicule, in grasses, as for the petal in eudicots. Although the lodicule and the petal are evolutionary homologs, they are morphologically and functionally different from each other. Therefore, the function of class B genes that specify whorl 2 organs may have been established before the divergence of monocots from related basal angiosperms. The differences in the morphology of whorl 2 organs are probably associated with differences in the target genes that are regulated by class B genes. The rescue of a class B mutant of *Arabidopsis* by introduction of a maize class B gene suggests that protein functions encoded by class B genes are conserved in *Arabidopsis* and maize [61]. The putative regulation of different target genes in each species may result from differences in the sequences of regulatory regions of the genes.

In contrast to these instances of conservation, carpel specification is regulated by a YABBY transcription factor encoded by *DL* in rice [42], unlike the class C MADS transcription factor in *Arabidopsis*. This difference may be associated with the different structures of the pistils/gynoecia in rice and *Arabidopsis*, as pointed out by Yamaguchi et al. [42]. The *Arabidopsis* gynoecium is a complex organ composed of independent tissues such as the ovary wall, replum, septum, and placenta, whereas the rice pistil is formed simply from congenitally fused carpels, and the ovule develops directly from the FM.

In addition, extra genes are required for the development of organs specific to spikelets in grasses. *Arabidopsis* has genes similar to the *LHS1/OsMADS1* gene required for lemma/palea identity in rice and the *tga1*-like SPB-box gene responsible for glume development in teosinte [27, 109]. Therefore, these genes may have been recruited to regulate the development of spikelet-specific organs during grass evolution.

Further studies of model plants in both eudicots and grasses will provide information to elucidate the developmental mechanism of flowers in the plants themselves and the evolution and diversification of the genetic mechanism of flower development in angiosperms.

Acknowledgement

We thank Dr. H. Yoshida for critical reading of the manuscript. W. T. is supported by a Research Fellowship for Young Scientists from the Japan Society for the Promotion of Science.

References

1. Kellogg EA (2001) Evolutionary history of the grasses. Plant Physiol 125:1198–1205
2. Lohmann JU, Weigel D (2002) Building beauty: the genetic control of floral patterning. Dev Cell 2:135–142
3. Jack T (2004) Molecular and genetic mechanisms of floral control. Plant Cell 16:S1–S17
4. Coen ES, Meyerowitz EM (1991) The war of the whorls: genetic interactions controlling flower development. Nature 353:31–37
5. Bommert P, Satoh-Nagasawa N, Jackson D, Hirano H-Y (2005) Genetics and evolution of inflorescence and flower development in grasses. Plant Cell Physiol 46:69–78
6. Bortiri E, Hake S (2007) Flowering and determinacy in maize. J Exp Bot 58:909–916
7. Thompson BE, Hake S (2008) Translational biology: from Arabidopsis flowers to grass inflorescence architecture. Plant Physiol 149:38–45
8. Yoshida H, Nagato Y (2011) Flower development in rice. J Exp Bot 62:4719–4730
9. Itoh J-I, Nonomura K-I, Ikeda K, Yamaki S, Inukai Y, Yamagishi H, Kitano H, Nagato Y (2005) Rice plant development: from zygote to spikelet. Plant Cell Physiol 46:23–47
10. Tanaka W, Pautler M, Jackson D, Hirano H-Y (2013) Grass meristems II—inflorescence architecture, flower development and meristem fate. Plant Cell Physiol 54:313–324
11. Hirano H-Y (2008) Genetic regulation of meristem maintenance and organ specification in rice flower development. In: Hirano H-Y, Hirai A, Sano Y, Sasaki T (eds) Rice biology in the genomics era, vol 62. Springer, Heidelberg, pp 69–79
12. Arber A (1934) The gramineae: a study of cereal, bamboo, and grasses. University Press, Cambridge
13. Yoshida A, Suzaki T, Tanaka W, Hirano H-Y (2009) The homeotic gene *LONG STERILE LEMMA* (*G1*) specifies sterile lemma identity in the rice spikelet. Proc Natl Acad Sci USA 106:20103–20108
14. Dellaporta SL, Calderon-Urrea A (1994) The sex determination process in maize. Science 266:1501–1505
15. Irish EE (1996) Regulation of sex determination in maize. Bioessays 18:363–369
16. McSteen P (2009) Hormonal regulation of branching in grasses. Plant Physiol 149:46–55
17. Pautler M, Tanaka W, Hirano H-Y, Jackson D (2013) Grass meristem I—shoot apical meristem maintenance, axillary meristem determinacy, and the floral transition. Plant Cell Physiol 54:302–312

18. Chuck G, Muszynski M, Kellogg E, Hake S, Schmidt RJ (2002) The control of spikelet meristem identity by the *branched silkless1* gene in maize. Science 298:1238–1241
19. Komatsu M, Chujo A, Nagato Y, Shimamoto K, Kyozuka J (2003) *FRIZZY PANICLE* is required to prevent the formation of axillary meristems and to establish floral meristem identity in rice spikelets. Development 130: 3841–3850
20. Karim MR, Hirota A, Kwiatkowska D, Tasaka M, Aida M (2009) A role for *Arabidopsis PUCHI* in floral meristem identity and bract suppression. Plant Cell 21:1360–1372
21. Chuck G, Meeley RB, Hake S (1998) The control of maize spikelet meristem fate by the APETALA2-like gene *indeterminate spikelet1*. Genes Dev 12:1145–1154
22. Chuck G, Meeley R, Hake S (2008) Floral meristem initiation and meristem cell fate are regulated by the maize *AP2* genes *ids1* and *sid1*. Development 135:3013–3019
23. Lee D-Y, Lee J, Moon S, Park SY, An G (2007) The rice heterochronic gene *SUPERNUMERARY BRACT* regulates the transition from spikelet meristem to floral meristem. Plant J 49:64–78
24. Lee DY, An G (2012) Two AP2 family genes, *SUPERNUMERARY BRACT* (*SNB*) and *OsINDETERMINATE SPIKELET 1* (*OsIDS1*), synergistically control inflorescence architecture and floral meristem establishment in rice. Plant J 69:445–461
25. Chuck G, Meeley R, Irish E, Sakai H, Hake S (2007) The maize *tasselseed4* microRNA controls sex determination and meristem cell fate by targeting *Tasselseed6/indeterminate spikelet1*. Nat Genet 39:1517–1521
26. Zhu QH, Upadhyaya NM, Gubler F, Helliwell CA (2009) Over-expression of miR172 causes loss of spikelet determinacy and floral organ abnormalities in rice (*Oryza sativa*). BMC Plant Biol 9:149
27. Jeon JS, Jang S, Lee S, Nam J, Kim C, Lee SH, Chung YY, Kim SR, Lee YH, Cho YG, An G (2000) *leafy hull sterile1* is a homeotic mutation in a rice MADS box gene affecting rice flower development. Plant Cell 12:871–884
28. Agrawal GK, Abe K, Yamazaki M, Miyao A, Hirochika H (2005) Conservation of the E-function for floral organ identity in rice revealed by the analysis of tissue culture-induced loss-of-function mutants of the *OsMADS1* gene. Plant Mol Biol 59:125–135
29. Prasad K, Sriram P, Kumar CS, Kushalappa K, Vijayraghavan U (2001) Ectopic expression of rice *OsMADS1* reveals a role in specifying the lemma and palea, grass floral organs analogous to sepals. Dev Genes Evol 211: 281–290
30. Prasad K, Parameswaran S, Vijayraghavan U (2005) *OsMADS1*, a rice MADS-box factor, controls differentiation of specific cell types in the lemma and palea and is an early-acting regulator of inner floral organs. Plant J 43: 915–928
31. Cui R, Han J, Zhao S, Su K, Wu F, Du X, Xu Q, Chong K, Theißen G, Meng Z (2010) Functional conservation and diversification of class E floral homeotic genes in rice (*Oryza sativa*). Plant J 61:767–781
32. Ha CM, Jun JH, Fletcher JC (2010) Shoot apical meristem form and function. Curr Top Dev Biol 91:103–140
33. Aichinger E, Kornet N, Friedrich T, Laux T (2012) Plant stem cell niches. Annu Rev Plant Biol 63:615–636
34. Bommert P, Lunde C, Nardmann J, Vollbrecht E, Running M, Jackson D, Hake S, Werr W (2005) *thick tassel dwarf1* encodes a putative maize ortholog of the *ArabidopsisCLAVATA1* leucine-rich repeat receptor-like kinase. Development 132: 1235–1245
35. Taguchi-Shiobara F, Yuan Z, Hake S, Jackson D (2001) The *fasciated ear2* gene encodes a leucine-rich repeat receptor-like protein that regulates shoot meristem proliferation in maize. Genes Dev 15:2755–2766
36. Nagasawa N, Miyoshi M, Kitano H, Satoh H, Nagato Y (1996) Mutations associated with floral organ number in rice. Planta 198: 627–633
37. Suzaki T, Sato M, Ashikari M, Miyoshi M, Nagato Y, Hirano H-Y (2004) The gene *FLORAL ORGAN NUMBER1* regulates floral meristem size in rice and encodes a leucine-rich repeat receptor kinase orthologous to *Arabidopsis* CLAVATA1. Development 131: 5649–5657
38. Suzaki T, Toriba T, Fujimoto M, Tsutsumi N, Kitano H, Hirano H-Y (2006) Conservation and diversification of meristem maintenance mechanism in *Oryza sativa*: function of the *FLORAL ORGAN NUMBER2* gene. Plant Cell Physiol 47:1591–1602
39. Chu H, Qian Q, Liang W, Yin C, Tan H, Yao X, Yuan Z, Yang J, Huang H, Luo D, Ma H, Zhang D (2006) The *FLORAL ORGAN NUMBER4* gene encoding a putative ortholog of *Arabidopsis* CLAVATA3 regulates apical meristem size in rice. Plant Physiol 142: 1039–1052
40. Suzaki T, Ohneda M, Toriba T, Yoshida A, Hirano H-Y (2009) *FON2 SPARE1* redundantly regulates floral meristem maintenance

with *FLORAL ORGAN NUMBER2* in rice. PLoS Genet 5:e1000693

41. Suzaki T, Yoshida A, Hirano H-Y (2008) Functional diversification of CLAVATA3-related CLE proteins in meristem maintenance in rice. Plant Cell 20:2049–2058
42. Yamaguchi T, Nagasawa N, Kawasaki S, Matsuoka M, Nagato Y, Hirano H-Y (2004) The *YABBY* gene *DROOPING LEAF* regulates carpel specification and midrib development in *Oryza sativa*. Plant Cell 16:500–509
43. Nagasawa N, Miyoshi M, Sano Y, Satoh H, Hirano H-Y, Sakai H, Nagato Y (2003) *SUPERWOMAN 1* and *DROOPING LEAF* genes control floral organ identity in rice. Development 130:705–718
44. Bowman JL, Smyth DR (1999) *CRABS CLAW*, a gene that regulates carpel and nectary development in *Arabidopsis*, encodes a novel protein with zinc finger and helix-loop-helix domains. Development 126:2387–2396
45. Kiesselbach TA (1999) The structure and reproduction of corn—50th anniversary edition. Cold Spring Harbor Laboratory Press, New York
46. Ohmori Y, Toriba T, Nakamura H, Ichikawa H, Hirano H-Y (2011) Temporal and spatial regulation of *DROOPING LEAF* gene expression that promotes midrib formation in rice. Plant J 65:77–86
47. Ishikawa M, Ohmori Y, Tanaka W, Hirabayashi C, Murai K, Ogihara Y, Yamaguchi T, Hirano H-Y (2009) The spatial expression patterns of *DROOPING LEAF* orthologs suggest a conserved function in grasses. Genes Genet Syst 84:137–146
48. Rao SA, Mengesha MH, Reddy CR (1988) Characteristics, inheritance, and allelic relationships of *midribless* mutants in pearl millet. J Hered 79:18–20
49. Fladung M, Bossinger G, Roeb GW, Salamini F (1991) Correlated alterations in leaf and flower morphology and rate of leaf photosynthesis in a *midribless* (*mbl*) mutant of *Panicum maximum* Jacq. Planta 184:356–361
50. Mena M, Ambrose BA, Meeley RB, Briggs SP, Yanofsky MF, Schmidt RJ (1996) Diversification of C-function activity in maize flower development. Science 274:1537–1540
51. Yamaguchi T, Lee YD, Miyao A, Hirochika H, An G, Hirano H-Y (2006) Functional diversification of the two C-class genes, *OSMADS3* and *OSMADS58*, in *Oryza sativa*. Plant Cell 18:15–28
52. Dreni L, Pilatone A, Yun D, Erreni S, Pajoro A, Caporali E, Zhang D, Kater MM (2011) Functional analysis of all AGAMOUS subfamily members in rice reveals their roles in reproductive organ identity determination and meristem determinacy. Plant Cell 23: 2850–2863
53. Dreni L, Jacchia S, Fornara F, Fornari M, Ouwerkerk PBF, An G, Colombo L, Kater MM (2007) The D-lineage MADS-box gene *OsMADS13* controls ovule identity in rice. Plant J 52:690–699
54. Yamaki S, Nagato Y, Kurata N, Nonomura K-I (2011) Ovule is a lateral organ finally differentiated from the terminating floral meristem in rice. Dev Biol 351:208–216
55. Kurakawa T, Ueda N, Maekawa M, Kobayashi K, Kojima M, Nagato Y, Sakakibara H, Kyozuka J (2007) Direct control of shoot meristem activity by a cytokinin-activating enzyme. Nature 445:652–655
56. Ambrose BA, Lerner DR, Ciceri P, Padilla CM, Yanofsky MF, Schmidt RJ (2000) Molecular and genetic analyses of the *silky1* gene reveal conservation in floral organ specification between eudicots and monocots. Mol Cell 5:569–579
57. Yao S-G, Ohmori S, Kimizu M, Yoshida H (2008) Unequal genetic redundancy of rice *PISTILLATA* orthologs, *OsMADS2* and *OsMADS4*, in lodicule and stamen development. Plant Cell Physiol 49:853–857
58. Kyozuka J, Kobayashi T, Morita M, Shimamoto K (2000) Spatially and temporally regulated expression of rice MADS box genes with similarity to *Arabidopsis* Class A, B and C Genes. Plant Cell Physiol 41: 710–718
59. Yadav SR, Prasad K, Vijayraghavan U (2007) Divergent regulatory *OsMADS2* functions control size, shape and differentiation of the highly derived rice floret second-whorl organ. Genetics 176:283–294
60. Yoshida H, Itoh J-I, Ohmori S, Miyoshi K, Horigome A, Uchida E, Kimizu M, Matsumura Y, Kusaba M, Satoh H, Nagato Y (2007) *superwoman1-cleistogamy*, a hopeful allele for gene containment in GM rice. Plant Biotechnol J 5:835–846
61. Whipple CJ, Ciceri P, Padilla CM, Ambrose BA, Bandong SL, Schmidt RJ (2004) Conservation of B-class floral homeotic gene function between maize and *Arabidopsis*. Development 131:6083–6091
62. Kyozuka J, Shimamoto K (2002) Ectopic expression of *OsMADS3*, a rice ortholog of *AGAMOUS*, caused a homeotic transformation of lodicules to stamens in transgenic rice plants. Plant Cell Physiol 43:130–135
63. Toriba T, Suzaki T, Yamaguchi T, Ohmori Y, Tsukaya H, Hirano H-Y (2010) Distinct

regulation of adaxial-abaxial polarity in anther patterning in rice. Plant Cell 22:1452–1462

64. Yoshida H (2012) Is the lodicule a petal: molecular evidence? Plant Sci 184:121–128
65. Nair SK, Wang N, Turuspekov Y, Pourkheirandish M, Sinsuwongwat S, Chen G, Sameri M, Tagiri A, Honda I, Watanabe Y, Kanamori H, Wicker T, Stein N, Nagamura Y, Matsumoto T, Komatsuda T (2010) Cleistogamous flowering in barley arises from the suppression of microRNA-guided *HvAP2* mRNA cleavage. Proc Natl Acad Sci USA 107:490–495
66. Bomblies K, Wang R-L, Ambrose BA, Schmidt RJ, Meeley RB, Doebley J (2003) Duplicate *FLORICAULA/LEAFY* homologs *zfl1* and *zfl2* control inflorescence architecture and flower patterning in maize. Development 130:2385–2395
67. Rao NN, Prasad K, Kumar PR, Vijayraghavan U (2008) Distinct regulatory role for *RFL*, the rice *LFY* homolog, in determining flowering time and plant architecture. Proc Natl Acad Sci USA 105:3646–3651
68. Ikeda-Kawakatsu K, Maekawa M, Izawa T, Itoh J-I, Nagato Y (2012) *ABERRANT PANICLE ORGANIZATION 2/RFL*, the rice ortholog of Arabidopsis *LEAFY*, suppresses the transition from inflorescence meristem to floral meristem through interaction with *APO1*. Plant J 69:168–180
69. Ikeda K, Ito M, Nagasawa N, Kyozuka J, Nagato Y (2007) Rice *ABERRANT PANICLE ORGANIZATION 1*, encoding an F-box protein, regulates meristem fate. Plant J 51:1030–1040
70. Ikeda K, Nagasawa N, Nagato Y (2005) *Aberrant panicle organization 1* temporally regulates meristem identity in rice. Dev Biol 282:349–360
71. Ikeda-Kawakatsu K, Yasuno N, Oikawa T, Iida S, Nagato Y, Maekawa M, Kyozuka J (2009) Expression level of *ABERRANT PANICLE ORGANIZATION1* determines rice inflorescence form through control of cell proliferation in the meristem. Plant Physiol 150:736–747
72. Ookawa T, Hobo T, Yano M, Murata K, Ando T, Miura H, Asano K, Ochiai Y, Ikeda M, Nishitani R, Ebitani T, Ozaki H, Angeles ER, Hirasawa T, Matsuoka M (2010) New approach for rice improvement using a pleiotropic QTL gene for lodging resistance and yield. Nat Commun 1:132
73. Horigome A, Nagasawa N, Ikeda K, Ito M, Itoh J-I, Nagato Y (2009) Rice *OPEN BEAK* is a negative regulator of class 1 *knox* genes and a positive regulator of class B floral homeotic gene. Plant J 58:724–736
74. Xiao H, Tang J, Li Y, Wang W, Li X, Jin L, Xie R, Luo H, Zhao X, Meng Z, He G, Zhu L (2009) *STAMENLESS 1*, encoding a single C2H2 zinc finger protein, regulates floral organ identity in rice. Plant J 59:789–801
75. Dinneny JR, Yadegari R, Fischer RL, Yanofsky MF, Weigel D (2004) The role of *JAGGED* in shaping lateral organs. Development 131: 1101–1110
76. Ohno CK, Reddy GV, Heisler MGB, Meyerowitz EM (2004) The *Arabidopsis JAGGED* gene encodes a zinc finger protein that promotes leaf tissue development. Development 131:1111–1122
77. Dinneny JR, Weigel D, Yanofsky MF (2006) *NUBBIN* and *JAGGED* define stamen and carpel shape in *Arabidopsis*. Development 133:1645–1655
78. Lee S, Jeon J-S, An K, Moon Y-H, Lee S, Chung Y-Y, An G (2003) Alteration of floral organ identity in rice through ectopic expression of *OsMADS16*. Planta 217:904–911
79. Hama E, Takumi S, Ogihara Y, Murai K (2004) Pistillody is caused by alterations to the class-B MADS-box gene expression pattern in alloplasmic wheats. Planta 218: 712–720
80. Li H, Liang W, Yin C, Zhu L, Zhang D (2011) Genetic interaction of *OsMADS3*, *DROOPING LEAF*, and *OsMADS13* in specifying rice floral organ identities and meristem determinacy. Plant Physiol 156:263–274
81. Ohmori S, Kimizu M, Sugita M, Miyao A, Hirochika H, Uchida E, Nagato Y, Yoshida H (2009) *MOSAIC FLORAL ORGANS1*, an *AGL6*-like MADS box gene, regulates floral organ identity and meristem fate in rice. Plant Cell 21:3008–3025
82. Li H, Liang W, Hu Y, Zhu L, Yin C, Xu J, Dreni L, Kater MM, Zhang D (2011) Rice *MADS6* interacts with the floral homeotic genes *SUPERWOMAN1*, *MADS3*, *MADS58*, *MADS13*, and *DROOPING LEAF* in specifying floral organ identities and meristem fate. Plant Cell 23:2536–2552
83. Thompson BE, Bartling L, Whipple C, Hall DH, Sakai H, Schmidt R, Hake S (2009) *bearded-ear* encodes a MADS box transcription factor critical for maize floral development. Plant Cell 21:2578–2590
84. Li H, Liang W, Jia R, Yin C, Zong J, Kong H, Zhang D (2010) The *AGL6*-like gene *OsMADS6* regulates floral organ and meristem identities in rice. Cell Res 20:299–313

85. Reinheimer R, Kellogg EA (2009) Evolution of *AGL6*-like MADS box genes in grasses (Poaceae): ovule expression is ancient and palea expression is new. Plant Cell 21: 2591–2605
86. Sun B, Xu Y, Ng K-H, Ito T (2009) A timing mechanism for stem cell maintenance and differentiation in the Arabidopsis floral meristem. Genes Dev 23:1791–1804
87. Clifford HT (1987) Spikelet and floral morphology. In: Soderstrom TR, Hilu KW, Campbell CS, Barkworth ME (eds) Grass systematics and evolution. Smithsonian Institution Press, Washington, DC, pp 21–30
88. Kobayashi K, Yasuno N, Sato Y, Yoda M, Yamazaki R, Kimizu M, Yoshida H, Nagamura Y, Kyozuka J (2012) Inflorescence meristem identity in rice is specified by overlapping functions of three *AP1/FUL*-Like MADS box genes and *PAP2*, a *SEPALLATA* MADS box gene. Plant Cell 24:1848–1859
89. Cacharron J, Saedler H, Theissen G (1999) Expression of MADS box genes *ZMM8* and *ZMM14* during inflorescence development of *Zea mays* discriminates between the upper and the lower floret of each spikelet. Dev Genes Evol 209:411–420
90. Malcomber ST, Kellogg EA (2004) Heterogeneous expression patterns and separate roles of the *SEPALLATA* gene *LEAFY HULL STERILE1* in grasses. Plant Cell 16: 1692–1706
91. Tanaka W, Toriba T, Ohmori Y, Yoshida A, Kawai A, Mayama-Tsuchida T, Ichikawa H, Mitsuda N, Ohme-Takagi M, Hirano H-Y (2012) The *YABBY* gene *TONGARI-BOUSHI1* is involved in lateral organ development and maintenance of meristem organization in the rice spikelet. Plant Cell 24:80–95
92. Tanaka W, Toriba T, Ohomori Y, Hirano H-Y (2012) Formation of two florets within a single spikelet in the rice *tongari-boushi1* mutant. Plant Signal Behav 7:793–795
93. Siegfried KR, Eshed Y, Baum SF, Otsuga D, Drews GN, Bowman JL (1999) Members of the *YABBY* gene family specify abaxial cell fate in *Arabidopsis*. Development 126:4117–4128
94. Goldshmidt A, Alvarez JP, Bowman JL, Eshed Y (2008) Signals derived from *YABBY* gene activities in organ primordia regulate growth and partitioning of *Arabidopsis* shoot apical meristems. Plant Cell 20:1217–1230
95. Yuan Z, Gao S, Xue D-W, Luo D, Li L-T, Ding S-Y, Yao X, Wilson ZA, Qian Q, Zhang D-B (2009) *RETARDED PALEA1* controls palea development and floral zygomorphy in rice. Plant Physiol 149:235–244
96. Jin Y, Luo Q, Tong H, Wang A, Cheng Z, Tang J, Li D, Zhao X, Li X, Wan J, Jiao Y, Chu C, Zhu L (2011) An AT-hook gene is required for palea formation and floral organ number control in rice. Dev Biol 359: 277–288
97. Itoh J-I, Kitano H, Matsuoka M, Nagato Y (2000) *SHOOT ORGANIZATION* genes regulate shoot apical meristem organization and the pattern of leaf primordium initiation in rice. Plant Cell 12:2161–2174
98. Liu B, Chen Z, Song X, Liu C, Cui X, Zhao X, Fang J, Xu W, Zhang H, Wang X, Chu C, Deng X, Xue Y, Cao X (2007) *Oryza sativa dicer-like4* reveals a key role for small interfering RNA silencing in plant development. Plant Cell 19:2705–2718
99. Abe M, Yoshikawa T, Nosaka M, Sakakibara H, Sato Y, Nagato Y, Itoh J-I (2010) *WAVY LEAF1*, an ortholog of Arabidopsis *HEN1*, regulates shoot development by maintaining microRNA and trans-acting small interfering RNA accumulation in rice. Plant Physiol 154: 1335–1346
100. Li A, Zhang Y, Wu X, Tang W, Wu R, Dai Z, Liu G, Zhang H, Wu C, Chen G, Pan X (2008) DH1, a LOB domain-like protein required for glume formation in rice. Plant Mol Biol 66:491–502
101. Li X, Sun L, Tan L, Liu F, Zhu Z, Fu Y, Sun X, Sun X, Xie D, Sun C (2012) *TH1*, a DUF640 domain-like gene controls lemma and palea development in rice. Plant Mol Biol 78:351–359
102. Becraft PW, Stinard PS, McCarty DR (1996) CRINKLY4: A TNFR-like receptor kinase involved in maize epidermal differentiation. Science 273:1406–1409
103. Pu C-X, Ma Y, Wang J, Zhang Y-C, Jiao X-W, Hu Y-H, Wang L-L, Zhu Z-G, Sun D, Sun Y (2012) Crinkly4 receptor-like kinase is required to maintain the interlocking of the palea and lemma, and fertility in rice, by promoting epidermal cell differentiation. Plant J 70:940–953
104. Li H, Xue D, Gao Z, Yan M, Xu W, Xing Z, Huang D, Qian Q, Xue Y (2009) A putative lipase gene *EXTRA GLUME1* regulates both empty-glume fate and spikelet development in rice. Plant J 57:593–605
105. Yoshida A, Ohmori Y, Kitano H, Taguchi-Shiobara F, Hirano H-Y (2012) *ABERRANT SPIKELET AND PANICLE1*, encoding a TOPLESS-related transcriptional co-repressor, is involved in the regulation of meristem fate in rice. Plant J 70:327–339
106. Kobayashi K, Maekawa M, Miyao A, Hirochika H, Kyozuka J (2010) *PANICLE*

PHYTOMER2 (*PAP2*), encoding a SEPALLATA subfamily MADS-box protein, positively controls spikelet meristem identity in rice. Plant Cell Physiol 51:47–57

107. Gao X, Liang W, Yin C, Ji S, Wang H, Su X, Guo C, Kong H, Xue H, Zhang D (2010) The *SEPALLATA*-like gene *OsMADS34* is required for rice inflorescence and spikelet development. Plant Physiol 153:728–740
108. Long JA, Ohno C, Smith ZR, Meyerowitz EM (2006) TOPLESS regulates apical embryonic fate in *Arabidopsis*. Science 312: 1520–1523
109. Wang H, Nussbaum-Wagler T, Li B, Zhao Q, Vigouroux Y, Faller M, Bomblies K, Lukens L, Doebley JF (2005) The origin of the naked grains of maize. Nature 436:714–719
110. DeLong A, Calderon-Urrea A, Dellaporta SL (1993) Sex determination gene *TASSELSEED2* of maize encodes a short-chain alcohol dehydrogenase required for stage-specific floral organ abortion. Cell 74:757–768
111. Acosta IF, Laparra H, Romero SP, Schmelz E, Hamberg M, Mottinger JP, Moreno MA, Dellaporta SL (2009) *tasselseed1* is a lipoxygenase affecting jasmonic acid signaling in sex determination of maize. Science 323: 262–265
112. Hartwig T, Chuck GS, Fujioka S, Klempien A, Weizbauer R, Potluri DP, Choe S, Johal GS, Schulz B (2011) Brassinosteroid control of sex determination in maize. Proc Natl Acad Sci USA 108:19814–19819
113. Bensen RJ, Johal GS, Crane VC, Tossberg JT, Schnable PS, Meeley RB, Briggs SP (1995) Cloning and characterization of the maize *An1* gene. Plant Cell 7:75–84
114. Peng J, Richards DE, Hartley NM, Murphy GP, Devos KM, Flintham JE, Beales J, Fish LJ, Worland AJ, Pelica F, Sudhakar D, Christou P, Snape JW, Gale MD, Harberd NP (1999) 'Green revolution' genes encode mutant gibberellin response modulators. Nature 400:256–261
115. Veit B, Briggs SP, Schmidt RJ, Yanofsky MF, Hake S (1998) Regulation of leaf initiation by the *terminal ear 1* gene of maize. Nature 393:166–168
116. Han JJ, Jackson D, Martienssen R (2012) Pod corn is caused by rearrangement at the *Tunicate1* locus. Plant Cell 24:2733–2744
117. Yun D, Liang W, Dreni L, Yin C, Zhou Z, Kater M, Zhang D (2013) *OsMADS16* interacts with *OsMADS3* and *OsMADS58* in specifying floral patterning in rice. Mol Plant 6(3): 743–756
118. Ohmori Y, Tanaka W, Kojima M, Sakakibara H, Hirano, H-Y (2013) *WUSCHEL-RELATED HOMEOBOX4* is involved in meristem maintenance and is negatively regulated by the CLE gene *FCP1* in rice. Plant Cell 25:229–241

Chapter 4

Flower Diversity and Angiosperm Diversification

Pamela S. Soltis and Douglas E. Soltis

Abstract

The flower itself, which comprises most of the evolutionary innovations of flowering plants, bears special significance for understanding the origin and diversification of angiosperms. The sudden origin of angiosperms in the fossil record poses unanswered questions on both the origins of flowering plants and their rapid spread and diversification. Central to these questions is the role that the flower, and floral diversity, played. Recent clarifications of angiosperm phylogeny provide the foundation for investigating evolutionary transitions in floral features and the underlying genetic mechanisms of stasis and change. The general features of floral diversity can best be addressed by considering key patterns of variation: an undifferentiated versus a differentiated perianth; elaboration of perianth organs in size and color; merosity of the flower; and phyllotaxy of floral organs. Various models of gene expression now explain the regulation of floral organization and floral organ identity; the best understood are the ABC(E) model and its modifications, but other gene systems are important in specific clades and require further study. Furthermore, the propensity for gene and genome duplications in angiosperms provides abundant raw material for novel floral features—emphasizing the importance of understanding the conservation and diversification of gene lineages and functions in studies of macroevolution.

Key words Flower evolution, Phylogeny, Angiosperms

1 Introduction

The flower is the single most distinguishing feature of angiosperms (flowering plants), despite a suite of other synapomorphies (shared derived features) that characterize this clade of roughly 300,000 species. The actual innovations that mark this group are double fertilization, enclosure of seeds in a carpel, ovules with two integuments, features of the stamen, reduced male and female gametophytes, and phloem sieve tube elements. As all of these features except the latter are associated with reproduction and are housed together in the flower, the flower itself has taken on special significance for understanding the origin and diversification of angiosperms. Other typical features of angiosperms, such as the presence of vessel elements in wood and

José Luis Riechmann and Frank Wellmer (eds.), *Flower Development: Methods and Protocols*, Methods in Molecular Biology, vol. 1110, DOI 10.1007/978-1-4614-9408-9_4, © Springer Science+Business Media New York 2014

congenital fusion of the carpel, do not characterize all of the early lineages of angiosperms and originated early in angiosperm evolution rather than in their common ancestor.

Evolutionary biologists have long been intrigued by the sudden origin of angiosperms in the fossil record and have sought both their origins and explanations for their rapid spread and diversification. Darwin himself referred to the "origin and early diversification of angiosperms" as "an abominable mystery," and the origin of the flower—and therefore flowering plants—is still a question (*see* ref. 1, and accompanying series of papers in the *American Journal of Botany*), although progress is being made in understanding both the homologies of floral structures to structures in nonflowering plants and the genes that control the development and expression of these structures.

The angiosperms, although a young clade by comparison with other major groups of land plants, with a fossil record that dates back to only 132 million years ago [2, 3], comprise the vast majority of plant species, with estimates ranging from 257,000 [4] to over 400,000 [5] of the approximately half a million species of extant green plants. Most of this diversity is diagnosed by variation in floral features, and in fact, speciation often involves shifts in floral characters, such as size, shape, color, and phenology. Often associated with variation in floral features is a shift in pollinators, such that animal pollinators often drive patterns of floral diversity [6, 7]. Throughout angiosperm phylogeny we see shifts in pollinators associated with floral traits: for example, red tubular flowers, regardless of their phylogenetic position, are typically pollinated by hummingbirds, whereas open, blue flowers are generally pollinated by bees. Patterns of floral variation in specific clades, such as Polemoniaceae (the phlox family), are certainly related to pollination biology, and their investigation on morphological, anatomical, developmental, and genetic levels may be key to understanding the role of floral diversity in species diversification. However, what features and forces account for the origin of the flower and the early diversification of flowering plants? To address this question, we must consider the placement of angiosperms within the green plants (Viridiplantae), the overall structure of angiosperm phylogeny, and the patterns of floral diversity across the angiosperms.

In this chapter, we provide an overview of plant phylogeny, with an emphasis on the angiosperms, describe general patterns of floral diversity, and review what is known about the genes underlying these general patterns. Our emphasis is on basal angiosperms—those lineages that arose before the divergence of monocots and eudicots approximately 125 million years ago.

2 Overview of Plant Phylogeny

2.1 Viridiplantae

To place the angiosperms in the proper historical perspective, and to understand the history of the genes that contribute to floral structure and diversity, we must consider the ancestry of the angiosperm clade. The Viridiplantae (green plants) are a clade of perhaps 500,000 species or more, dating back nearly one billion years. All green plants share a common cyanobacterial endosymbiotic origin with red algae and glaucophytes and can be diagnosed by chlorophyll *b*, starch as the storage product for photosynthate, and a stellate flagellar structure (e.g., [4]). Viridiplantae comprise two clades, the chlorophytes (mostly marine "green algae") and streptophytes (which include freshwater "green algae" and embryophytes). The sister to all other streptophytes is *Mesostigma*, a freshwater "alga." Early clades of streptophytes include a number of aquatic groups: Klebsormidiales, Zygnematales, Coleochaetales, and Charales, the limits of which are uncertain. Although many studies have identified *Chara* (and relatives) and Coleochaetales as possible sister group(s) of the embryophytes, or land plants, recent evidence suggests that perhaps Zygnematales may be the sister group, raising questions about phylogenetic relationships among the early branches of streptophyte evolution. Embryophytes have a fossil history that extends at least back to the Ordovician, and they began to diversify extensively in the Silurian and Devonian. Morphological and anatomical synapomorphies of the embryophytes are a multicellular sporangium, thick-walled spores, multicellular gametangia, an embryo, and a cuticle.

Within the embryophytes, the phylogeny of the major clades is not completely clear (*see* Fig. 1). Although traditionally considered a single taxonomic group, the bryophytes (consisting of mosses, liverworts, and hornworts) are not monophyletic, and their branching order relative to the tracheophytes (*Tracheophyta* sensu Cantino et al. [8]; vascular plants) is not clear (*see* ref. 4). The moss *Physcomitrella*, with its sequenced genome, provides a reference point for comparisons of genes controlling the development and morphology of flowers with those involved in reproduction in an early lineage of land plants.

Tracheophyta comprises two major clades, lycophytes (*Lycopodiophyta*) and euphyllophytes (*Euphyllophyta*). *Lycopodiophyta* is composed of *Isoetes*, *Selaginella*, and Lycopodiaceae and has an extensive fossil record. The extant *Euphyllophyta* contains two major clades, *Monilophyta* (Psilotales, Ophioglossales, Equisetales, Marattiales, and the leptosporangiate ferns (*Leptosporangiatae*)) and *Lignophyta* (several fossil lineages and the *Spermatophyta*, the seed plants). The monilophytes—essentially a clade of traditionally recognized ferns and their relatives—were recognized on the basis of stem anatomy [9] and later supported by molecular data [10].

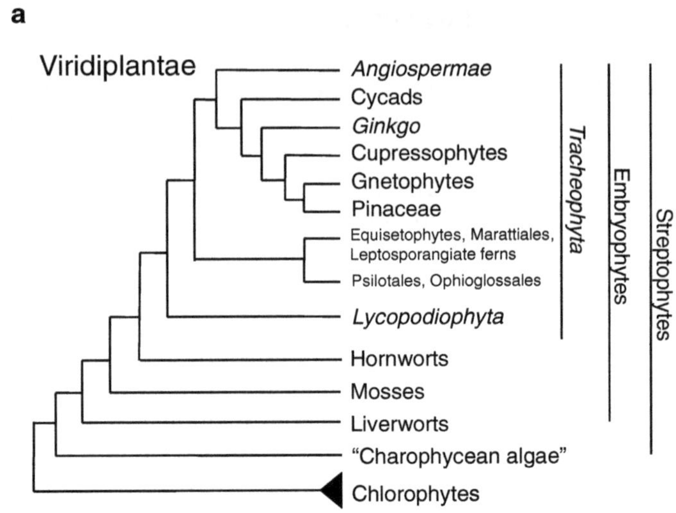

Fig. 1 Overview of the phylogeny of Viridiplantae. (**a**) All major clades of Viridiplantae, summarized from many studies. (**b**) *Angiospermae*, based on ref. 12

Relationships among seed plants are perhaps the most challenging in plant phylogeny, with massive extinction having contributed to the difficulty of phylogeny reconstruction. The "gymnosperms" as typically recognized are not monophyletic and include four clades with extant members (cycads, *Ginkgo*, conifers, and gnetophytes) plus several extinct groups (*Medullosa*, seed ferns, glossopterids, *Caytonia*, Bennetittales). Molecular-based trees typically (but not always) recover reciprocally monophyletic gymnosperms and angiosperms. However, extant gymnosperms are not monophyletic when fossils are included in reconstructions, and the sister group to the angiosperms then becomes unclear. When fossils are included in phylogenetic analyses of seed plants, disagreements exist with regard to both the placement of many of the nonflowering seed plants and in the sister group of the angiosperms, and there is little consensus on the overall phylogeny of all seed plants. Glossopterids, *Caytonia*, and Bennetittales consistently appear more closely related to angiosperms than to other gymnosperms; the *Acrogymnospermae*, composed of conifers, gnetophytes, *Ginkgo*, and cycads, is sister to the clade of angiosperms, glossopterids, *Caytonia*, and Bennetittales (*see* ref. 11). However, these relationships remain controversial, and this lack of resolution is unfortunate for studies that attempt to address the evolutionary history of angiosperm characters and the origins of angiosperm features themselves. Thus, much attention has focused on the early lineages of angiosperms and their traits as reference points for the evolution of more recently derived groups of angiosperms and their features.

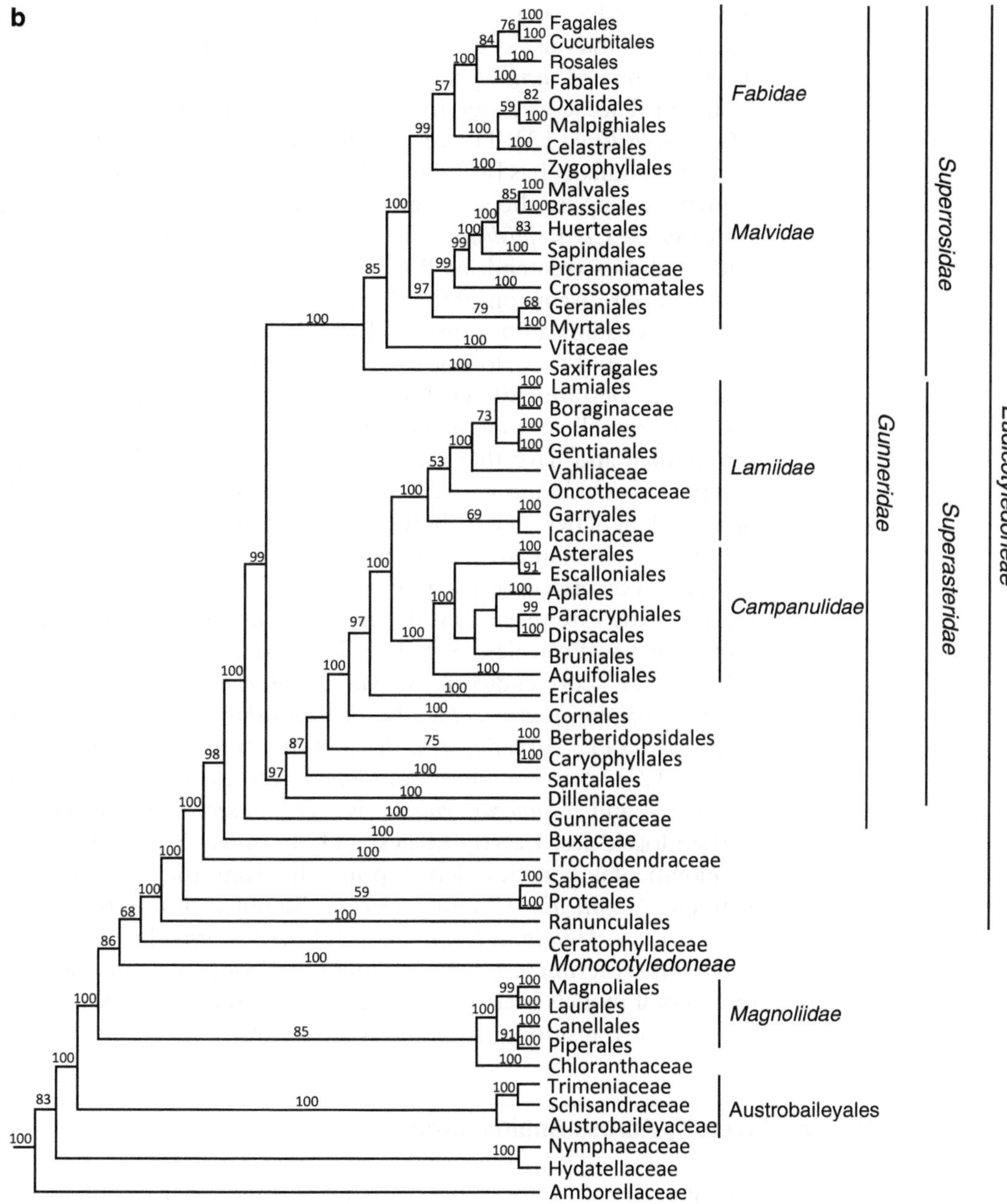

Fig. 1 (continued)

2.2 Angiospermae

Large collaborative efforts at angiosperm phylogeny reconstruction have been underway for two decades. Here we briefly summarize the results of those studies, with an emphasis on the most recent such analysis [12]. Nearly all large-scale, molecular-based analyses of the past decade have identified *Amborella* as the sister to all other extant angiosperms, most often alone or occasionally with Nymphaeales (*see* Fig. 1). All analyses then place

Austrobaileyales (comprising *Austrobaileya*, *Trimenia*, *Illicium*, and Schisandraceae) as the sister group to all other extant angiosperms, the *Mesangiospermae*, which comprises *Magnoliidae* + Chloranthaceae as sister to *Monocotyledoneae* + *Eudicotyledoneae* + *Ceratophyllum* [13]. The relationship among magnoliids, monocots, and eudicots has been difficult to resolve, but plastid genome sequences support the sister-group relationship of monocots and eudicots (+ *Ceratophyllum*). The *Magnoliidae*, although a small clade, exhibits tremendous floral diversity and will figure prominently in our discussion of floral variation below.

Relationships among major clades of monocots are now clear (*see* Fig. 1), and they do not follow traditional taxonomic circumscriptions. Despite striking flowers within many groups of monocots, nearly all monocot flowers follow the same trimerous groundplan and have therefore received relatively little attention, except for orchids and grasses, from the perspective of floral developmental genetics. We will therefore not describe relationships within monocots in any detail here.

The eudicots comprise approximately 75 % of all angiosperm species, with extensive floral diversity. A basal grade consisting of Ranunculales, Proteales, Sabiales, Trochodendrales, and Buxales subtends the "core eudicots," or *Gunneridae*. Gunnerales are sister to the remaining *Gunneridae*, the *Pentapetalae*, which fall into two major clades [12, 14]: *Superrosidae* and *Superasteridae* (sensu Soltis et al. [12]). *Superrosidae* comprises Saxifragales, Vitaceae, and *Rosidae*, and *Superasteridae* is composed of Santalales, Berberidopsidales, Caryophyllales, and *Asteridae*. Models for floral developmental genetics derive primarily from the eudicots: for example, *Aquilegia* (Ranunculaceae, Ranunculales), *Arabidopsis* (Brassicaceae, *Malvidae*, *Rosidae*), and *Antirrhinum* (Plantaginaceae, *Lamiidae*, *Asteridae*). Understanding the phylogenetic positions of these floral models is crucial for proper interpretation of both floral and genetic transitions across the angiosperms.

3 Patterns of Floral Diversity Across Angiosperms

Flowers vary in many ways, with many innovations confined to single clades, such as orchids (Orchidaceae), grasses (Poaceae), and milkweeds (Apocynaceae). However, the general features of floral diversity can perhaps best be addressed by considering the following: an undifferentiated versus a differentiated perianth; elaboration of perianth organs in terms of size and color; merosity of the flower; and phyllotaxy of floral organs. And of course, any of the main floral organs—sepals, petals, stamens, carpels—may be missing in any given group of flowering plants. Here we provide a primer of general patterns of floral diversity in these features.

3.1 Undifferentiated Versus Differentiated Perianth

The perianth is the collective term for the sepals and petals of the flowers, which function primarily in protection of the reproductive organs and attraction of pollinators. The distinction between sepals and petals is typically based on morphology, with sepals typically green and leaf-like and petals colorful and more elaborate in form. However, they also differ in position, with sepals located in the outer whorl of the flower and petals located to the inside; furthermore, the position of sepals often alternates with that of the petals, so that identical structures may occupy different positions in the flower, for example, a tulip, with three outer perianth parts (sometimes called sepals) and three inner perianth parts (sometimes called petals). In this case, the distinction between the perianth organs is based strictly on position rather than on morphology. In general, however, discussions of differentiated versus undifferentiated perianths refer to the morphology of the organs and not simply their positions.

In general, extant basal angiosperms have an undifferentiated perianth, in which all perianth organs are identical in morphology and are therefore often referred to as tepals. Examples include *Magnolia* (Magnoliaceae, *Magnoliidae*) and *Persea* (which includes the avocado; Lauraceae, *Magnoliidae*). Most other *Magnoliidae*, as well as the earliest branches of angiosperms also have an undifferentiated perianth [15, 16]. However, in some cases, such as *Amborella* and *Nymphaea* (a water lily), there is a gradual transition from outer tepals to inner tepals, during which the outer tepals are greenish and more "sepal-like" and the inner tepals are more pigmented and "petal-like" (e.g., [17–19]). Character reconstructions of perianth form across the angiosperms suggest that an undifferentiated perianth was the ancestral condition [16].

Other perianth features certainly vary prominently across angiosperms. For example, petals are typically distinguished from sepals by differences in cell types and surfaces. Enlargement of petals, dramatic shifts in coloration, or modification of petals or sepals into spurs or other structures, such as coronas, have occurred repeatedly in angiosperm evolution. These sorts of shifts tend to occur on more localized phylogenetic scales than across all angiosperms, and we will therefore not consider them further in this review.

3.2 Merosity

Merosity refers to the number of floral parts and whether or not the parts are in regular multiples. For example, we think of monocots as having flowers based on a groundplan with parts in 3 s or multiples thereof—a case referred to as trimery. Most *Pentapetalae* have parts in 5 s or multiples thereof, or occasionally 4 s. However, many groups of angiosperms do not have flowers with regular numbers of parts or based on a predictable groundplan. For example, *Amborella* and *Illicium* (Austrobaileyales) have irregular numbers of flower parts, in both cases with perianth parts ranging from perhaps 6–10 and an irregular number of stamens and carpels. Even within the eudicots are examples of irregular numbers of stamens and carpels, such as in some Ranunculaceae. In some cases, the merosity of the perianth may differ from that of the reproductive structures, such that there is a regular and predictable number of perianth organs but numerous stamens and/or carpels. However, despite the typical presence of many stamens and carpels in many basal angiosperm groups, many basal angiosperms, such as Magnoliaceae, have a trimerous groundplan, and trimery has even been reconstructed as a possible ancestral state for angiosperms (*see* ref. 16).

3.3 Phyllotaxy

Phyllotaxy (or phyllotaxis) is the arrangement of floral organs around the central axis of the flower. Floral organs may be arranged spirally around the floral axis or in regular whorls. In flowers with whorled phyllotaxy, there may be a single whorl of each organ, or there may be multiple whorls of petals or stamens, for example. Spiral phyllotaxy is found in many basal angiosperms, such as *Amborella*, *Illicium*, and *Magnolia*, for example, but because Nymphaeales (water lilies) have whorled phyllotaxy, reconstructions of the ancestral angiosperm condition are equivocal [16].

4 Flowers in the Fossil Record

Floral diversity among the hundreds of thousands of extant angiosperms is remarkable, but even the fossil record shows extensive variation in flower size and morphology, even among the very earliest fossils [3]. Although longstanding views suggested that a flower such as a *Magnolia* flower, with its large size, undifferentiated perianth, large and irregular number of floral organs, and spiral phyllotaxy, represented the prototypic flower, paleobotanical research during the past two decades has dramatically altered this view. The earliest flowers, such as a tiny water lily [2, 3, 20] and others, are all very small, although they tend to exhibit some of the other features of *Magnolia* flowers, such as irregular merosity, but both spiral and whorled phyllotaxy (*see* review by Soltis et al. [21]). Interestingly, these small flowers have been found in early Cretaceous deposits in both North America and Europe and seem to represent the general features of flowers at this early point in their evolution. By the mid-Cretaceous, floral diversity was extensive, with many modern groups recognizable. It was likely the fossil record at this point in time that Darwin found so perplexing—angiosperms seem to have burst on the scene with little warning with a spectacular array of diversity. The recent discoveries of the early Cretaceous fossils relied on using novel methods for both searching for and analyzing the fossil flowers.

5 Genes Involved in Floral Morphology and Variation

The flower has received extensive attention from morphological, developmental, and genetic perspectives, and syntheses of developmental genetics have emerged, based on prominent models and crops (*see*, e.g., reviews in refs. 22, and 21, 23–27). Developmental genetics is providing a host of candidate genes with which to test hypothesized effects and infer causation of floral diversity. For example, the genetic control of floral organ identity—via the ABC model [28, 29]—has been thoroughly investigated in *Arabidopsis* and *Antirrhinum* and extended, in one form or another, to most other angiosperms (*see* refs. 21, 25–27, for reviews). Furthermore, information on other gene systems, such as the transcription factors that regulate floral symmetry, allows for testable hypotheses on the shifts between radially and bilaterally symmetrical flowers. Sequenced genomes from a florally diverse array of angiosperms are providing opportunities for addressing candidate gene activity in phylogenetically disparate species. Below we review examples of floral diversity with specific hypotheses that may explain divergent evolutionary forms, with an emphasis on overall floral form, and especially the perianth. Such studies are in their infancy, in an evolutionary context, particularly for basal angiosperms, and the examples presented here are simply illustrative of the types of studies that have been and can be conducted.

MADS-box genes code for transcription factors, and they play extensive roles in plant development, particularly in the flower. These genes therefore represent excellent candidates for many possible roles in both flower development and other developmental processes in plants. Some of them are master regulators of floral organization, and evolutionary shifts in their expression have yielded fundamental changes in floral morphology. The following sections will review various models of gene expression and function that regulate floral organization and floral organ identity.

5.1 The ABC(E) Model of Floral Organ Identity

The ABC(E) model derives from studies of genetic mutants in *Arabidopsis* and *Antirrhinum* and describes the activities of transcription factors that regulate floral organ identity. In this combinatorial model of gene activity, A function specifies sepals, the combined activity of A and B functions specifies petals, B and C functions together specify stamens, and C function alone specifies carpels. In *Arabidopsis*, *APETALA1* (*AP1*) and *APETALA2* (*AP2*) control the A function, *APETALA3* (*AP3*) and *PISTILLATA* (*PI*) are B-function genes, and *AGAMOUS* (*AG*) is the C-function gene. The *SEPALLATA* (*SEP*) genes are now known to play important roles in the specification of floral organ identities (E function; [30–32]), leading to the revision of the "ABC" model as the "ABCE" or "ABC(E)" model (*see* Fig. 2a). Note, however, that the A function of *AP1* has not been found outside of

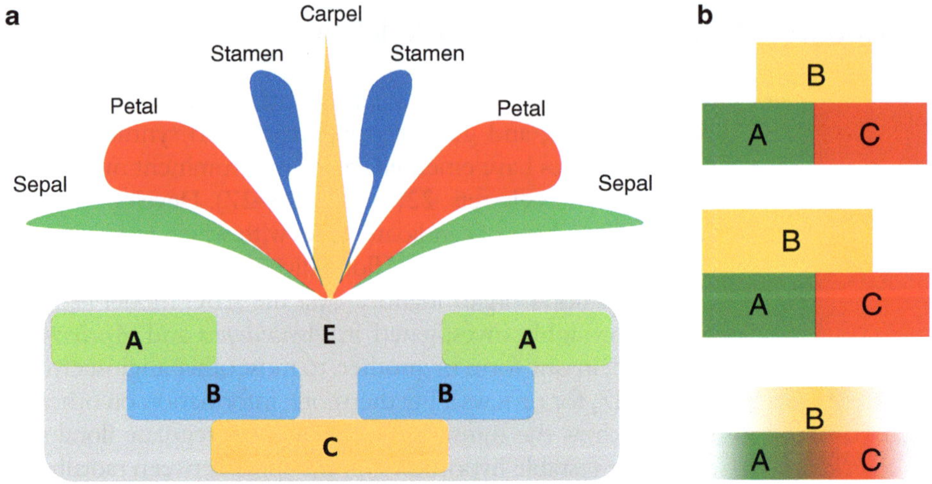

Fig. 2 The ABC(E) model of floral organ identity and its variants. (**a**) The ABC(E) model, which yields flowers with four discrete organs: sepals, petals, stamens, and carpels. (**b**) Variants of the ABC(E) model (with E function removed for simplicity). *Top*: ABC model, as in (**a**). *Middle*: Sliding boundary model. Note that expression of B-function has shifted to the *left*, extending petaloidy into the first whorl of the flower; this model has been proposed to account for two whorls of petaloid tepals, as in tulips, lilies, and some members of Ranunculaceae. *Bottom*: Fading borders model. Here, gradients of gene expression occur across the floral meristem, resulting in flowers with organs of intergrading morphologies

Arabidopsis. Ovule identity is controlled by the D-function, which in *Arabidopsis* is provided by *SEEDSTICK* (*STK*) [33, 34]. All of these genes, except *AP2*, are MADS-box genes.

The ABC(E) genes represent excellent candidates for the genetic control of floral differences when those divergent floral morphologies are due to alterations in organ specification. Examples of floral morphologies that might be controlled by shifts in ABC(E) gene expression include flowers with perianth organs that are morphologically similar but positionally distinct, as in lilies and tulips, and those with morphologically intergrading and spirally arranged perianth organs, as in *Amborella* and *Illicium*.

5.2 Modifications of the ABC(E) Model

The sliding/shifting boundary model [35] posits that a spatial outward shift in the expression of B-function genes in the floral meristem will yield a flower that lacks sepals but instead has two whorls of petals. This is exactly the case in lilies and tulips, and such shifts in B-class gene expression have been documented (*see* review by Soltis et al. [21]) (Fig. 2b).

This simple shift in B-class genes into a new region of the meristem cannot explain the floral morphology of many basal angiosperm groups that are characterized by morphologically intergrading floral organs. In *Amborella*, for example, there is an intergradation from extra-floral bracts to outer tepals with tinges of green, thus somewhat resembling the bracts, to inner tepals that lack green pigmentation, to either laminar stamens in male flowers or laminar staminodes in female flowers, to broad carpels that developmentally resemble stamens in the female flowers [17]. This pattern of morphological intergradation can be explained by a gradient in expression of floral organ identity genes, with overlapping zones of activity (*see* Fig. 2b) and is described by the fading borders model. In fact, analysis of B- and C-class homologs in *Amborella* and *Illicium*, as well as other basal angiosperms, has demonstrated broad and overlapping regions of expression across the floral meristem, resulting in floral organs of intergrading morphology [36]. The fading borders model applies to a number of basal angiosperms that exhibit this characteristic morphology of intergrading floral organs.

Although the ABC(E) model (with or without E function) in its essence can describe the organ identities in many eudicots and monocots, with their regular, whorled arrangements of floral organs, it does not adequately explain the flowers of most basal angiosperms (including *Magnoliidae*), with their undifferentiated perianths, or many basal eudicots. Given the typically broad expression patterns of B and C homologs in most basal angiosperms studied to date, the ABC(E) model of *Arabidopsis* and *Antirrhinum* appears to have been derived from an ancestral genetic program that expressed floral regulators broadly across the floral meristem (*see* ref. 36). Evolutionary diversification of this ancestral program appears to have occurred through localized expression of different regulators to different regions of the meristem. In the core eudicots, restricted gene expression results in flowers with distinct and differentiated sepals, petals, stamens, and carpels, as described for *Arabidopsis*. Similar restrictions in *Asimina* (Annonaceae, *Magnoliidae*) have likewise resulted in flowers with these same features—obviously an independent derivation of the whorled floral structure in *Asimina* from that of core eudicots [36]. Spatial shifts in B function in *Ranunculus* and *Aquilegia* (Ranunculaceae), governed by duplicate B-function genes, are correlated with changes in perianth morphology [37, 38]. Different alterations to the ancestral program exhibited by basal angiosperms may be responsible for having generated much of the floral diversity that exists, or has existed.

5.3 Extending the Fading Borders Model

The fading borders model explicitly applies to the ABC(E)-function genes, particularly the B- and C-function genes. However, what genes control the action of these master regulators, and what are their downstream targets? Perhaps the differences in floral morphologies between basal angiosperms and core eudicots are determined not by only a few key transcription factors but by an entire cascade of differently expressed genes. Although the components of such cascades are poorly known even in model species, this question can be addressed through analysis of genes expressed in the relevant tissues. Analysis of microarrays for a phylogenetically selected set of species with different floral morphologies (*Nuphar*

Fig. 3 Transcriptional profiles in floral organs of, from *left*, a water lily (*Nuphar advena*, Nymphaeaceae), avocado (*Persea americana*, Lauraceae), California poppy (*Eschscholzia californica*, Papaveraceae), and *Arabidopsis thaliana* (Brassicaceae) based on microarray analysis of floral unigenes. Broadly overlapping expression domains of the ABC genes in *Nuphar* and *Persea* set up correspondingly broad downstream transcriptional programs, as revealed by the blurred borders between adjacent organs. In contrast, the eudicots *Eschscholzia* and *Arabidopsis* exhibit discrete patterns of gene expression, with little overlap between adjacent organs. *S* sepals, *P* petals, *St* stamens, *C* carpels. From ref. 39

advena, a water lily, Nymphaeales; *Persea americana*, the avocado, Lauraceae, *Magnoliidae*; *Eschscholzia californica*, California poppy, Ranunculaceae, Ranunculales, a basal eudicot; *Arabidopsis thaliana*, Brassicaceae, *Malvidae*, *Rosidae*) addressed this question with approximately 5,000 genes each for the first three species and nearly 13,000 for *Arabidopsis*. For *Nuphar* and *Persea*, entire transcriptional cascades had broadly overlapping spatial domains across the morphologically intergrading organs, resembling the patterns observed for the B- and C-function genes ([39]; *see* Fig. 3). In contrast, the transcriptional programs for *Eschscholzia* and *Arabidopsis*, with their distinct floral organs, were spatially discrete in their expression. These gene sets now provide opportunities to investigate the causes of floral diversity at a much finer scale than previously possible. Other aspects of floral diversity would benefit from this approach.

6 Gene and Genome Duplication and Flower Evolution

Genome duplication is a fundamental process in plant evolution, with all angiosperms now known to be descended from an ancestral genome duplication [40] and most lineages characterized by additional duplications. Additionally, gene duplications also arise via other mechanisms, potentially making plant genomes highly variable in gene copy number. These duplications therefore provide both evolutionary "challenges" and "opportunities" for plant genomes. For example, how do polyploid plants "deal" with multiple copies of floral regulators, and what effects might these multiple copies have on floral diversification?

Many genes involved in floral development have undergone duplication during angiosperm evolution. For example, the MADS-box genes have been duplicated multiple times in the history of flowering plants (reviewed in ref. 21). Phylogenetic reconstructions suggest that the MADS-box gene family has diversified extensively in angiosperms, relative to other clades of life, with duplications of many gene lineages in angiosperm ancestors or within specific angiosperm clades [41]. Based on functional studies in *Arabidopsis*, it is clear that ancient duplications of MADS-box genes have had multiple outcomes, such as retention of ancestral roles by the duplicate genes, role swapping, and the acquisition of novel roles in floral development.

Here we present a single example—that of the B-function genes, given that they play such an important role in overall floral morphology through their effect(s) on the perianth—but we refer the reader to the larger review of gene duplication and floral genes by Soltis et al. [42]. B-function in the ABC(E) model is conferred in *Arabidopsis* by two genes, *AP3* and *PI*, whose proteins interact, but these homologs are part of a dynamic gene lineage that is characterized by a history of gene duplication. *AP3* and *PI* belong to two paralogous gene lineages that resulted from a duplication that occurred prior to the origin of the angiosperms [43]. The *AP3* lineage subsequently underwent another duplication in the common ancestor of the core eudicots, giving rise to two *AP3* sublineages: the eu*AP3* and the *TOMATO MADS BOX GENE6* (*TM6*) gene lineages [44]. Those angiosperms that diverged prior to the origin of the core eudicots have so-called *paleoAP3* genes, and these genes share greater sequence similarity with *TM6* than with eu*AP3*. Most core eudicots have both paralogs, but *TM6* genes have been lost independently in *Arabidopsis* and *Antirrhinum*, leaving a single copy of this duplicate pair in each. Functional studies suggest functional diversification in these paralogous lineages, as well as sequence divergence. In tomato (Solanaceae, *Lamiidae*, *Asteridae*), the ortholog of *AP3* regulates both stamen and petal development whereas *TM6* is involved in the development of stamens but not petals [45]. In contrast, in petunia (also in Solanaceae), petals are transformed into sepals with loss of the *AP3* ortholog, with little effect on stamens, and stamens are only affected when the orthologs of both *AP3* and *TM6* are nonfunctional [46].

Duplications have also occurred in the *paleoAP3* gene lineage, with duplicate copies having diverged in expression, if not function, in poppy and *Aquilegia* [37]. *PaleoAP3* genes in rice and maize are not duplicated, itself interesting given whole-genome duplication(s) in the common ancestor of grasses [47, 48], and function in both stamen and petal (= lodicule in grasses) development [49, 50]. *PaleoAP3* genes in basal angiosperms appear to have played a role in stamen and perianth identity [18, 19, 26, 36, 51, 52]; however, duplications may have resulted in neo-functionalization or subfunctionalization. For example, in *Nuphar advena*, both *paleoAP3* paralogs are expressed throughout the flower in mature developmental stages, but during early development, one of the two paralogs is restricted to stamens while the other is expressed in both stamens and inner tepals (petals) [36]. Expression shifts of three *paleoAP3* paralogs in *Illicium floridanum* also suggest possible functional divergence. One paralog is expressed in the outer tepals, inner tepals, and stamens (the typical *paleoAP3* expression), the second is restricted to the inner tepals and stamens, and the third is limited to the inner tepals [36].

The sister to the *AP3* lineage, the *PI* lineage, in contrast, was not duplicated in the common ancestor of the core eudicots, despite at least one whole-genome duplication in the immediate common ancestor of this clade [53]. However, ancient duplications of *PI* have occurred in specific clades of angiosperms, such as the *Magnoliidae* [54], in which duplicate copies have retained similar expression patterns [51, 55]; likewise, duplications in Ranunculaceae and monocots [56] do not seem to have led to functionally divergent paralogs. Thus, gene duplication, whether resulting from polyploidization or a local process, does not necessarily yield functional diversification in gene families. Similar patterns of gene duplication and occasional divergence in expression and/or function of duplicates occur in the other major lineages of floral MADS-box genes: in the *AP1* lineage (A-function), yielding the paralogs *AP1*, *CAULIFLOWER* (*CAL*), and *FRUITFUL* (*FUL*) in *Arabidopsis* (e.g., [57, 58]); in the *AG* lineage (C-function), resulting in four paralogs in *Arabidopsis*, *AG*, *SEEDSTICK* (*STK*), and *SHATTERPROOF1* (*SHP1*) and *SHATTERPROOF2* (*SHP2*); and in the *SEP* lineage (E-function), which itself arose through duplication in the *AG* lineage, generating *SEP1-4* [59]. Incidentally, many of these duplications, like the event that produced the split between *AP3* and *PI*, occurred in the common ancestor of angiosperms, after the split from extant gymnosperms.

7 Conclusions and Future Directions

Floral diversity is best understood when placed in its phylogenetic perspective. Although the genetic control of most floral variants remains uncertain, models of gene action have been proposed to explain key aspects of floral variation. Most attention has been focused on variants of the ABC(E) model, but ongoing research on floral symmetry is also leading to new discoveries. Floral symmetry in *Antirrhinum* is controlled by *CYCLOIDEA* (*CYC*) of the TCP family, which specifies dorsal floral identity ([60–62]; reviewed in ref. 63), and *DIVARICATA* and *RADIALIS* of the MYB family. Homologs of *CYC* have been recruited independently in legumes for the same function, setting the stage for further comparisons of these genetic systems. Study of the genetic differences between floral variants angiosperm-wide, between distant plant groups such as *Antirrhinum* and legumes, and between close relatives will continue to provide evidence of the evolution of the floral genetic program.

At deeper levels, analyses of gene phylogenies, coupled with comparisons of gene expression and function, inform on the origin of the flower itself. The transcriptional cascades identified for *Nuphar* and *Persea*, with their broadly overlapping spatial domains, resemble the hypothetical ancestral gymnosperm genetic program, based on comparisons with the cycad *Zamia*. Further, the genetic programs of putatively homologous floral organs can be traced to those of gymnosperm cones, demonstrating deep evolutionary conservation of these genetic systems. For example, shared genetic features of the carpels of angiosperms and the female cones of gymnosperms may represent a conserved genetic program underlying the development of the seed. Such analyses of thousands of genes across millions of years of evolution may yield new insight into the homologies of reproductive programs across land plants.

Gene and genome duplications are ubiquitous features of plant genomes, providing both new raw material for evolution and also a possible catalyst for morphological innovation. Shifts in the location and/or timing of expression of duplicate genes, as well as neofunctionalization and subfunctionalization, could lead to new morphologies. Such detailed genetic analyses have been conducted in the context of understanding genes duplicated through polyploidy (reviewed for cotton by Wendel et al. [64]), but extending these studies to genes involved in specifying floral features would provide more explicit assessment of the role that gene duplications play in morphological innovation. For example, petaloidy in the highly modified *Aquilegia* flower, with its novel petal spurs, is controlled by divergent expression of three *AP3* paralogs and *PI* [37, 38], contributing to the novel features of columbine flowers. Similar studies of innovations in other clades could be equally enlightening.

Acknowledgments

We thank the many students and postdocs in our lab who have contributed to our understanding of floral diversity and genetics: S. Brockington, M. Buzgo, A. Chanderbali, S. Kim, J. Koh, and M.-J. Yoo. This chapter reports on research supported by NSF grants PGR-0115684, PGR-0638595, and EF-0431266.

References

1. Friedman WE (2009) The meaning of Darwin's 'abominable mystery'. Am J Bot 96(1):5–21
2. Friis EM, Crane PR, Pedersen KR (2011) Early flowers and angiosperm evolution. Cambridge University Press, Cambridge
3. Friis EM, Pedersen KR, Crane PR (2010) Diversity in obscurity: fossil flowers and the early history of angiosperms. Philos Trans R Soc Lond B Biol Sci 365(1539):369–382
4. Judd WS, Campbell CS, Kellogg EA, Stevens PF, Donoghue MJ (2008) Plant systematics—a phylogenetic approach, 3rd edn. Sinauer, Sunderland, MA
5. Bramwell D (2002) How many plant species are there? Plant Talk 28:32–34
6. Fenster CB, Armbruster WS, Wilson P, Dudash MR, Thomson JD (2004) Pollination syndromes and floral specialization. Annu Rev Ecol Evol Syst 35:375–403
7. Waser NM (2006) Specialization and generalization in plant-pollinator interactions: a historical perspective. In: Waser NM, Ollerton J (eds) Plant-pollinator interactions from a specialization to generalization. University of Chicago Press, Chicago, IL, pp 3–17
8. Cantino P, Doyle J, Graham S, Judd W, Olmstead R, Soltis DE, Soltis PS, Donoghue M (2007) Towards a phylogenetic nomenclature of Tracheophyta. Taxon 56:822–846
9. Kenrick P, Crane PR (1997) The origin and early diversification of land plants. Smithsonian Institution Press, Washington, DC
10. Pryer KM, Schneider H, Smith AR, Cranfill R, Wolf PG, Hunt JS, Sipes SD (2001) Horsetails and ferns are a monophyletic group and the closest living relatives to seed plants. Nature 409(6820):618–622
11. Doyle JA (2012) Molecular and fossil evidence on the origin of angiosperms. Rev Earth Planet Sci 40:301–326
12. Soltis DE, Smith SA, Cellinese N, Wurdack KJ, Tank DC, Brockington SF, Refulio-Rodriguez NF, Walker JB, Moore MJ, Carlsward BS, Bell CD, Latvis M, Crawley S, Black C, Diouf D, Xi Z, Rushworth CA, Gitzendanner MA, Sytsma KJ, Qiu YL, Hilu KW, Davis CC, Sanderson MJ, Beaman RS, Olmstead RG, Judd WS, Donoghue MJ, Soltis PS (2011) Angiosperm phylogeny: 17 genes, 640 taxa. Am J Bot 98(4):704–730
13. Moore MJ, Bell CD, Soltis PS, Soltis DE (2007) Using plastid genome-scale data to resolve enigmatic relationships among basal angiosperms. Proc Natl Acad Sci USA 104(49):19363–19368
14. Moore MJ, Soltis PS, Bell CD, Burleigh JG, Soltis DE (2010) Phylogenetic analysis of 83 plastid genes further resolves the early diversification of eudicots. Proc Natl Acad Sci USA 107(10):4623–4628
15. Ronse deCraene LP, Soltis PS, Soltis DE (2003) Evolution of floral structures in basal angiosperms. Int J Plant Sci 164:S329–S363
16. Soltis DE, Soltis PS, Endress PK, Chase MW (2005) Angiosperm phylogeny and evolution. Sinauer Associates, Sunderland, MA
17. Buzgo M, Soltis PS, Soltis DE (2004) Floral developmental morphology of *Amborella trichopoda* (Amborellaceae). Int J Plant Sci 165:925–947
18. Yoo M-J, Soltis PS, Soltis DE (2010) Expression of floral MADS-box genes in two divergent water lilies: Nymphaeales and Nelumbo. Int J Plant Sci 171:121–146
19. Yoo MJ, Chanderbali AS, Altman NS, Soltis PS, Soltis DE (2010) Evolutionary trends in the floral transcriptome: insights from one of the basalmost angiosperms, the water lily *Nuphar advena* (Nymphaeaceae). Plant J 64(4):687–698
20. Friis EM, Pedersen KR, Crane PR (2001) Fossil evidence of water lilies (Nymphaeales) in the early cretaceous. Nature 410(6826):357–360
21. Soltis PS, Brockington SF, Yoo MJ, Piedrahita A, Latvis M, Moore MJ, Chanderbali AS, Soltis DE (2009) Floral variation and floral genetics in basal angiosperms. Am J Bot 96(1):110–128

22. Soltis DE, Leebens-Mack J, Soltis PS (eds) (2006) Developmental genetics of the flower, vol 44, Advances in botanical research. Academic, San Diego
23. Endress PK (2006) Angiosperm floral evolution: morphological developmental framework. In: Soltis DE, Leebens-Mack J, Soltis PS (eds) Developmental genetics of the flower, vol 44, Advances in botanical research. Elsevier, New York, pp 1–61
24. Frohlich MW (2006) Recent developments regarding the evolutionary origins of flowers. In: Soltis DE, Leebens-Mack J, Soltis PS (eds) Developmental genetics of the flower, vol 44, Advances in botanical research. Elsevier, New York, pp 63–127
25. Irish V (2006) Duplication, diversification, and comparative genetics of angiosperm MADS-box genes. In: Soltis DE, Leebens-Mack J, Soltis PS (eds) Developmental genetics of the flower, vol 44, Advances in botanical research. Elsevier, New York, pp 127–159
26. Soltis PS, Soltis DE, Kim S, Chanderbali A, Buzgo M (2006) Expression of floral regulators in basal angiosperms and the origin and evolution of the ABC model. In: Soltis DE, Leebens-Mack J, Soltis PS (eds) Developmental genetics of the flower, vol 44, Advances in botanical research. Elsevier, New York, pp 483–506
27. Theissen G, Melzer R (2007) Molecular mechanisms underlying origin and diversification of the angiosperm flower. Ann Bot (Lond) 100(3):603–619
28. Bowman JL, Smyth DR, Meyerowitz EM (1991) Genetic interactions among floral homeotic genes of Arabidopsis. Development 112(1):1–20
29. Coen ES, Meyerowitz EM (1991) The war of the whorls: genetic interactions controlling flower development. Nature 353(6339):31–37
30. Ditta G, Pinyopich A, Robles P, Pelaz S, Yanofsky MF (2004) The *SEP4* gene of *Arabidopsis thaliana* functions in floral organ and meristem identity. Curr Biol 14(21):1935–1940
31. Pelaz S, Ditta GS, Baumann E, Wisman E, Yanofsky MF (2000) B and C floral organ identity functions require *SEPALLATA* MADS-box genes. Nature 405(6783):200–203
32. Theiben G (2001) Development of floral organ identity: stories from the MADS house. Curr Opin Plant Biol 4(1):75–85
33. Colombo L, Franken J, Koetje E, van Went J, Dons HJ, Angenent GC, van Tunen AJ (1995) The petunia MADS box gene FBP11 determines ovule identity. Plant Cell 7(11):1859–1868
34. Pinyopich A, Ditta GS, Savidge B, Liljegren SJ, Baumann E, Wisman E, Yanofsky MF (2003) Assessing the redundancy of MADS-box genes during carpel and ovule development. Nature 424(6944):85–88
35. Bowman JL (1997) Evolutionary conservation of angiosperm flower development at the molecular and genetic levels. J Biosci 22:515–527
36. Kim S, Koh J, Yoo MJ, Kong H, Hu Y, Ma H, Soltis PS, Soltis DE (2005) Expression of floral MADS-box genes in basal angiosperms: implications for the evolution of floral regulators. Plant J 43(5):724–744
37. Kramer EM, Holappa L, Gould B, Jaramillo MA, Setnikov D, Santiago PM (2007) Elaboration of B gene function to include the identity of novel floral organs in the lower eudicot Aquilegia. Plant Cell 19(3):750–766
38. Rasmussen DA, Kramer EM, Zimmer EA (2009) One size fits all? Molecular evidence for a commonly inherited petal identity program in Ranunculales. Am J Bot 96(1):96–109
39. Chanderbali AS, Yoo MJ, Zahn LM, Brockington SF, Wall PK, Gitzendanner MA, Albert VA, Leebens-Mack J, Altman NS, Ma H, dePamphilis CW, Soltis DE, Soltis PS (2010) Conservation and canalization of gene expression during angiosperm diversification accompany the origin and evolution of the flower. Proc Natl Acad Sci USA 107(52):22570–22575
40. Jiao Y, Wickett NJ, Ayyampalayam S, Chanderbali AS, Landherr L, Ralph PE, Tomsho LP, Hu Y, Liang H, Soltis PS, Soltis DE, Clifton SW, Schlarbaum SE, Schuster SC, Ma H, Leebens-Mack J, dePamphilis CW (2011) Ancestral polyploidy in seed plants and angiosperms. Nature 473(7345):97–100
41. Becker A, Theissen G (2003) The major clades of MADS-box genes and their role in the development and evolution of flowering plants. Mol Phylogenet Evol 29(3):464–489
42. Soltis PS, Burleigh JG, Chanderbali AS, Yoo M-J, Soltis DE (2010) Gene and genome duplication in plants. In: Dittmar K, Liberles D (eds) Evolution after genome duplication. Wiley, Hoboken, pp 369–398
43. Kim S, Yoo MJ, Albert VA, Farris JS, Soltis PS, Soltis DE (2004) Phylogeny and diversification of B-function MADS-box genes in angiosperms: evolutionary and functional implications of a 260-million-year-old duplication. Am J Bot 91(12):2102–2118
44. Kramer EM, Dorit RL, Irish VF (1998) Molecular evolution of genes controlling petal and stamen development: duplication and divergence within the APETALA3 and PISTILLATA MADS-box gene lineages. Genetics 149(2):765–783

45. de Martino G, Pan I, Emmanuel E, Levy A, Irish VF (2006) Functional analyses of two tomato APETALA3 genes demonstrate diversification in their roles in regulating floral development. Plant Cell 18(8):1833–1845
46. Rijpkema A, Gerats T, Vandenbussche M (2006) Genetics of floral development in Petunia. In: Soltis DE, Leebens-Mack J, Soltis PS (eds) Developmental genetics of the flower, vol 44, Advances in botanical research. Elsevier, New York, pp 237–278
47. Paterson AH, Bowers JE, Chapman BA (2004) Ancient polyploidization predating divergence of the cereals, and its consequences for comparative genomics. Proc Natl Acad Sci USA 101(26):9903–9908
48. Paterson AH, Wang X, Li J, Tang H (2012) Ancient and recent polyploidy in monocots. In: Soltis PS, Soltis DE (eds) Polyploidy and genome evolution. Springer, Heidelberg, pp 93–108
49. Whipple CJ, Ciceri P, Padilla CM, Ambrose BA, Bandong SL, Schmidt RJ (2004) Conservation of B-class floral homeotic gene function between maize and Arabidopsis. Development 131(24):6083–6091
50. Whipple CJ, Zanis MJ, Kellogg EA, Schmidt RJ (2007) Conservation of B class gene expression in the second whorl of a basal grass and outgroups links the origin of lodicules and petals. Proc Natl Acad Sci USA 104(3):1081–1086
51. Chanderbali AS, Kim S, Buzgo M, Zheng P, Oppenheimer DG, Soltis DE, Soltis PS (2006) Genetic footprints of stamen ancestors guide perianth evolution in Persea (Lauraceae). Int J Plant Sci 167:1075–1089
52. Soltis DE, Chanderbali AS, Kim S, Buzgo M, Soltis PS (2007) The ABC model and its applicability to basal angiosperms. Ann Bot 100(2):155–163
53. Jiao Y, Leebens-Mack J, Ayyampalayam S, Bowers JE, McKain MR, McNeal J, Rolf M, Ruzicka DR, Wafula E, Wickett NJ, Wu X, Zhang Y, Wang J, Carpenter EJ, Deyholos MK, Kutchan TM, Chanderbali AS, Soltis PS, Stevenson DW, McCombie R, Pires JC, Wong GK, Soltis DE, Depamphilis CW (2012) A genome triplication associated with early diversification of the core eudicots. Genome Biol 13(1):R3
54. Stellari GM, Jaramillo MA, Kramer EM (2004) Evolution of the APETALA3 and PISTILLATA lineages of MADS-box-containing genes in the basal angiosperms. Mol Biol Evol 21(3):506–519
55. Chanderbali AS, Albert VA, Leebens-Mack J, Altman NS, Soltis DE, Soltis PS (2009) Transcriptional signatures of ancient floral developmental genetics in avocado (*Persea americana*; Lauraceae). Proc Natl Acad Sci USA 106(22):8929–8934
56. Winter KU, Weiser C, Kaufmann K, Bohne A, Kirchner C, Kanno A, Saedler H, Theissen G (2002) Evolution of class B floral homeotic proteins: obligate heterodimerization originated from homodimerization. Mol Biol Evol 19(5):587–596
57. Litt A (2007) An evaluation of A-function: Evidence from the APETALA1 and APETALA2 gene lineages. Int J Plant Sci 168(1):73–91
58. Litt A, Irish VF (2003) Duplication and diversification in the APETALA1/FRUITFULL floral homeotic gene lineage: implications for the evolution of floral development. Genetics 165(2):821–833
59. Zahn LM, Kong H, Leebens-Mack JH, Kim S, Soltis PS, Landherr LL, Soltis DE, dePamphilis CW, Ma H (2005) The evolution of the *SEPALLATA* subfamily of MADS-box genes: a preangiosperm origin with multiple duplications throughout angiosperm history. Genetics 169(4):2209–2223
60. Almeida J, Rocheta M, Galego L (1997) Genetic control of flower shape in *Antirrhinum majus*. Development 124(7):1387–1392
61. Luo D, Carpenter R, Copsey L, Vincent C, Clark J, Coen E (1999) Control of organ asymmetry in flowers of Antirrhinum. Cell 99(4):367–376
62. Luo D, Carpenter R, Vincent C, Copsey L, Coen E (1996) Origin of floral asymmetry in Antirrhinum. Nature 383(6603):794–799
63. Howarth DG, Donoghue MJ (2006) Phylogenetic analysis of the "ECE" (CYC/TB1) clade reveals duplications predating the core eudicots. Proc Natl Acad Sci USA 103(24):9101–9106
64. Wendel JF, Flagel LE, Adams KL (2012) Jeans, genes, and genomes: cotton as a model for studying polyploidy. In: Soltis PS, Soltis DE (eds) Polyploidy and genome evolution. Springer, Heidelberg, pp 181–208

Chapter 5

Flower Development: Open Questions and Future Directions

Frank Wellmer, John L. Bowman, Brendan Davies, Cristina Ferrándiz, Jennifer C. Fletcher, Robert G. Franks, Emmanuelle Graciet, Veronica Gregis, Toshiro Ito, Thomas P. Jack, Yuling Jiao, Martin M. Kater, Hong Ma, Elliot M. Meyerowitz, Nathanaël Prunet, and José Luis Riechmann

Abstract

Almost three decades of genetic and molecular analyses have resulted in detailed insights into many of the processes that take place during flower development and in the identification of a large number of key regulatory genes that control these processes. Despite this impressive progress, many questions about how flower development is controlled in different angiosperm species remain unanswered. In this chapter, we discuss some of these open questions and the experimental strategies with which they could be addressed. Specifically, we focus on the areas of floral meristem development and patterning, floral organ specification and differentiation, as well as on the molecular mechanisms underlying the evolutionary changes that have led to the astounding variations in flower size and architecture among extant and extinct angiosperms.

Key words Flower development, Floral meristems, ABC model, Floral evolution

1 Introduction

Over two decades ago, the ABC model was proposed as a genetic framework for a key aspect of flower development, floral organ identity specification. Since then, extensive genetic and molecular analyses of this and other processes have resulted in tremendous progress in our understanding of flower development (as reviewed in Chapters 1–4). In particular, all of the floral organ identity genes, whose combinatorial activities are described in the ABC model, have been identified and detailed insights into their functions have been obtained. Similarly, key gene products and molecular mechanisms that underlie how stem cells in the flower are established, maintained and ultimately terminated have been characterized. While flower development has been studied mainly in

José Luis Riechmann and Frank Wellmer (eds.), *Flower Development: Methods and Protocols*, Methods in Molecular Biology, vol. 1110, DOI 10.1007/978-1-4614-9408-9_5, © Springer Science+Business Media New York 2014

the model plant *Arabidopsis thaliana*, as well as in a few other species (e.g., snapdragon, petunia, and rice), expanding molecular studies to additional species and new model organisms is providing a better understanding of flower evolution and diversity. Despite these advances, many unanswered questions or poorly understood topics remain. In this chapter, we discuss some of those open questions and, whenever pertinent, outline possible experimental strategies with which they could be addressed. While this list of questions is far from complete, we believe that the ones described in this chapter are among the most important problems that have to be solved in order to obtain a comprehensive view of the molecular processes underlying flower development and evolution.

2 Floral Meristem Development

The formation of flowers commences shortly after the so-called floral transition, which, in *Arabidopsis*, is controlled by a complex regulatory network of flowering time genes and pathways and depends on both internal and external cues (reviewed in: refs. 1–3). Floral transition involves the transformation of the shoot apical meristem (SAM) into an inflorescence meristem (IM). While the SAM initiates leaves that subtend branches formed by axillary meristems, during flowering, initially several cauline leaves and associated axillary meristems are formed before the IM generates floral meristems (FMs) on its flanks. FMs are typically subtended by bracts, although the outgrowth of these modified leaves is suppressed in several plant species, including members of the Brassicaceae (e.g., *Arabidopsis*) and the Poaceae [4]. FMs are therefore developmentally similar to axillary meristems as both types of meristems form a primordium consisting of a leaf-competent domain and a meristem domain, possibly as a consequence of a common developmental program [4, 5]. The fate of these two domains depends on the developmental stage of the plant, leading to rosette leaves with an axillary meristem during the vegetative phase and bracts with flowers during reproductive development. Notably, cauline leaves and the associated axillary secondary IMs display properties of both phases [6], suggesting that their formation is a direct result of the transition process, which leads from vegetative to reproductive development. A careful experiment using photoinduction of flowering further showed that newly formed primordia, which would otherwise develop into leaves with axillary meristems, can be induced into flowers or flower/leaf chimeric shoots, indicating that the fate of primordia can be modified after their initiation [7]. The genetic basis of the transitions between these different meristematic activities and the molecular mechanisms underlying primordia specification are largely unknown, and future work will have to be directed towards unraveling these important questions.

The development of FMs can be divided into distinct phases [8]. The first phase is characterized by the formation of a region (the so-called floral *Anlage*) within an IM from which the FM ultimately arises. Subsequently, outgrowth of the floral meristem results in the formation of a floral primordium, which rapidly increases in size. At early floral stages, the formation of organ primordia commences and is followed by their differentiation and maturation. A key question that remains largely unanswered is how these 3 distinct phases, i.e., *Anlagen* formation, FM initiation and floral organ differentiation are interconnected. For instance, floral *Anlagen* are specified in a very precise position of the IM. It is known that a complex interplay between the plant hormones auxin and cytokinin is responsible for FM initiation and outgrowth. Auxin maxima are necessary for FM initiation, which is evidenced by the auxin efflux carrier *pin1* mutant, which develops a naked inflorescence without FMs [9, 10]. However, how auxin controls the definition of the floral *Anlagen* is still largely unknown although recent evidence suggests that the control of *LEAFY* (*LFY*) expression by the Auxin Response Factor MONOPTEROS (MP) plays an important role in this process [11]. *LFY* is one of the key genes controlling floral meristem identity, which is expressed in floral *Anlagen* [12]. *LFY* encodes a transcription factor specific to the plant kingdom, and in *lfy* mutants, inflorescence-like structures develop instead of flowers [13].

In addition to auxin, floral meristem outgrowth is also controlled by a complex network of transcription factors that specify meristem identity and prevent the precocious differentiation of floral organs. The genes encoding these factors are the so-called floral meristem identity genes, of which *LFY* is the central regulator. LFY is able to directly activate other floral meristem identity genes such as *LATE MERISTEM IDENTITY1* (*LMI1*), *CAULIFLOWER* (*CAL*), and *APETALA1* (*AP1*) [14–18]. Subsequently, both AP1 and CAL are responsible for the maintenance of *LFY* expression in a positive regulatory loop [19, 20]. AP1 and CAL are MADS-domain transcription factors that are highly similar in sequence and partially redundant. In fact, the *ap1 cal* double mutant develops mainly IMs instead of flowers [21, 22]. Interestingly, in this background another MADS-box gene, closely related to both *AP1* and *CAL*, *FRUITFULL* (*FUL*), which is normally not expressed in wild-type FMs, appears to act as a floral meristem identity gene, since *ap1 cal ful* triple mutants show an enhanced phenotype when compared to an *ap1 cal* double mutant [19]. Thus, *FUL* is not a floral meristem identity gene but can act as such when it is ectopically expressed in the FM. These types of regulatory interactions can shroud the genetic interactions and render the genetic analyses and the interpretations of the phenotypes of higher order mutants arduous.

Recently, two other MADS-box genes, *SHORT VEGETATIVE PHASE* (*SVP*) and *AGAMOUS-LIKE24* (*AGL24*), have been shown to be involved in conferring floral meristem identity, as *ap1 agl24 svp* triple mutants are phenocopies of the *ap1 cal* double mutant [23]. The role of *AGL24* and *SVP* in promoting FM identity was recently confirmed by demonstrating genetic interactions between these genes and *LFY* [24]. Moreover, they are part of the complex gene regulatory network that ensures floral meristem commitment and prevents floral reversion, and they are able to directly activate both *LFY* and *AP1* [24]. At stage 3 of flower development, when the formation of floral organs commences, *SVP* expression is lost and the expression of the *SEPALLATA* (*SEP*) genes is activated. Based on these and other observations, it has been suggested that the SEP proteins (which, like AP1, CAL, FUL, SVP, and AGL24, are MADS-domain transcription factors) compete with AGL24 and SVP for binding to AP1, and that AP1-SEP dimers eventually silence *AGL24* and *SVP* and promote the formation of floral organs [23, 25]. Although this working model is likely correct, at least in part, it is currently unclear what ultimately induces the differentiation of FMs. Is this switch triggered when FMs reach a certain size, or is it rather dependent on hormone levels or transcription factor concentrations? Answering these questions will be a major step forward in our understanding of the early processes of flower development.

One of the difficulties in studying the three phases of FM development and how they are interconnected is the extremely small size of these structures in *Arabidopsis*. However, the development of new technologies has already facilitated research in this area and will likely contribute to major breakthroughs in our understanding of FM development. For instance, it is now feasible to study the expression of key regulators of FM development and their protein–protein interactions with high spatial and temporal resolution using confocal microscopy (*see* Chapter 25). Furthermore, analysis of the transcriptomes of buds of different developmental stages, and of the regulatory processes that take place in these flowers, will be aided by techniques such as laser capture microdissection (*see* Chapter 19) in combination with RNA-Seq (*see* Chapter 23), or through the use of a floral induction system (*see* Chapter 16), which allows the collection of large quantities of early-stage floral buds for genomic and proteomic studies.

Another largely unanswered question regarding FMs is how the cells that give rise to individual floral organs (i.e., founder cells) are specified. Multiple lines of evidence demonstrate that the specification of organ founder cells precedes floral organ identity determination. However, the number and position of floral organs that arise in each whorl, as well as the number of founder cells is variable in different parts of the flower. Although auxin and the gene *DORNRÖSCHEN-LIKE* (*DRNL*) are involved in specifying where floral organs will arise [10, 26–28], the precise interplay of how these factors establish a complex pattern of floral organ primordia from an undifferentiated FM is unclear. Unlike the SAM, which leads to a single organ type on its flanks, either leaves (during vegetative growth) or flowers (during reproductive growth), the FM generates a diversity of organ types that arise in very close proximity to one another. Founder cell specification is intrinsically linked to the establishment of boundaries between floral organs. While several genes involved in boundary formation have been identified [29–34], their precise functions at the molecular level remain to be elucidated.

To prevent the overgrowth of FMs, plants have evolved multiple mechanisms to terminate meristematic activity once floral organ development has been correctly initiated. A key regulator of one of these mechanisms in *Arabidopsis* is the floral organ identity MADS-domain factor AGAMOUS (AG), which integrates stem cell regulation with floral patterning events [35, 36]. *AG* expression is induced by the stem cell determinant WUSCHEL (WUS) in early floral primordia [37, 38]. AG then specifies reproductive organ identity and eventually terminates *WUS* expression. Precise temporal induction of the C2H2-type zinc finger repressor KNUCKLES (KNU) is essential to balance proliferation and differentiation in the AG-WUS feedback loop [39]. *KNU* is a direct target of AG and functions as an upstream repressor of *WUS* [39]. Although *knu* mutant flowers exhibit extra floral organs inside carpels [40], this indeterminacy is weaker than that typically observed in *ag* mutant flowers. Therefore, AG likely controls additional genes that control FM determinacy together or in parallel with *KNU*. One candidate for such a gene is *CRABS CLAW* (*CRC*), which encodes a YABBY-type transcription factor, and is involved in the regulation of carpel and nectary development [41]. *CRC* is a direct target of AG [42] and *knu crc* double-mutant flowers show stronger indeterminacy [43]. Furthermore, when *ag* mutants are combined with *superman (sup)* loss-of-function alleles, the *ag sup* double-mutant flowers continue to grow, reaching a size several times bigger than that of wild-type or *ag* single mutant flowers, often with a fasciated meristem in the center [44]. *SUP* starts to be expressed earlier than *KNU* and *CRC* in flower development [39–41, 45]. These results indicate that floral meristem activity is terminated by multiple genetic pathways functioning at different developmental stages. However, how many regulators act in this process, and when, where and through which mechanisms they function, are largely unanswered questions.

3 Understanding the Molecular Nature of the (A)BCE Model

The work to decipher the molecular mechanisms underlying flower development has been driven, at each stage, by advances in technology. The application of molecular biology to classical mutants, transgenic and other approaches to the analysis of gene function, methods to study protein–protein and protein–DNA interactions, and improvements in DNA sequencing and microscopy have all contributed to our current state of understanding. However, even if we just put to one side the small matter that we do not really understand exactly how the multiple actors in the (A)BCE model of floral organ identity specification (*see* Chapter 1) come to be present at the right levels and the right time and place, how much do we really know? There is accumulating evidence that the space between the model and reality is highly complex and that we are going to need several more technological advances to make the connections between the high-level regulators identified using the homeotic mutants and the proteins that ultimately make one cell different from another.

Focusing on the MADS-box organ identity genes highlights this challenge. We have known for several years that the MADS-domain transcription factors dimerize and heterodimerize [46, 47], form higher order complexes, especially when bound to DNA [48], and that they bind DNA as interacting heterodimer pairs to form tetramers, or quartets [49]. There have also been several reports in the literature, largely overlooked, of interactions between MADS-domain factors and other transcription factors or regulators (e.g., [25, 50–58]). These interactions clearly suggest that the (A)BCE MADS-domain factors do not work alone. Recently, a proteomics approach to study the composition of A, B, and C function complexes present in the developing meristem has shown that they are likely to be even more complicated than was imagined [59]. Whilst this novel approach validated previous MADS–MADS interactions, it also showed that the complexes are far larger than predicted; 670 kDa in the case of SEP3. The floral MADS-domain factor complexes include other transcription factors, co-repressors and, notably, several chromatin-associated factors such as nucleosome remodeling factors and histone demethylase. The composition of these complexes is likely to dictate the effect on the expression of their target genes (i.e., to repress or to activate), effects that will also be dynamic in development. For example, AP1, as one of the first of the (A)BCE genes to be expressed, acts to repress inflorescence identity genes such as *SUPPRESSOR OF OVEREXPRESSION OF CO 1* (*SOC1*) and to promote floral meristem identity genes like *LFY* [20, 60]. Understanding the effects of the multiple possible (A)BCE-factor complexes, in terms of transcriptional regulation of their target genes, all in a dynamic developmental context, just became even more difficult.

Not knowing how a specific (A)BCE complex variant affects the expression of a specific target gene at a particular point in development leads on to a bigger and more intractable question: how do we understand and quantify the contribution of downstream targets to building a flower? Recently there has been tremendous progress in identifying targets of the floral organ identity factors, made possible by advances in the techniques for the genome-wide localization of transcription factor binding sites (*see* Chapter 24). Using this approach, targets have been identified for the MADS-domain proteins SEP3, AP1, APETALA3 (AP3), PISTILLATA (PI), FLOWERING LOCUS C (FLC) and SOC1 [20, 61–64]. These results have highlighted three serious obstacles to filling the space between the homeotic regulators and the ultimate effector proteins. Firstly, it is clear that targets of the (A) BCE-function transcription factors are more numerous than had been thought [60]. Secondly, the organ identity factors are not strictly hierarchical and act at all levels in the network, making network models highly complex. Thirdly, the genes that are directly, indirectly and in combination downstream of them are often redundant and do not lend themselves to genetic analysis by displaying the kind of informative mutant phenotypes observed for the homeotic regulators. Overcoming these challenges will require advances in target gene detection, verification and network modeling. More significantly, it will also necessitate new approaches to the analysis of gene function and phenotypic analyses to detect the diverse influences of the target genes on the development of the flower. These new approaches will require more sophisticated morphometric analyses to quantify subtle variation [65].

4 Floral Organ Formation

Flowers are highly complex organs that consist of many specialized tissues. Their development must therefore be tightly controlled in order to generate a functional structure that attracts pollinators, orchestrates both male and female gametogenesis, and generates and disperses the seeds of the subsequent generation. Yet the intercellular signaling landscape that mediates the coordination of floral organogenesis and the integration of flower-specific cellular activities is a largely uncharted territory that awaits discovery. Similarly, we know very little about the genes that control the size and shape of the floral organs. Genes involved in hormone biosynthesis and response are enriched in the transcriptome of early flower buds [66] and as targets of floral organ identity genes [20, 42, 62, 64, 67]. Some progress has been made in identifying roles for gibberellins, auxins, jasmonates, and brassinosteroids in regulating aspects of stamen and carpel development [68, 69], including anther differentiation [70], stamen maturation [71], and gynoecium patterning [72]. Similarly, a variety of small polypeptide signaling genes are expressed in developing flowers [73, 74], and several components of ligand-mediated signal transduction pathways are known to control facets of anther and pollen formation [75–78]. However, these investigations represent only an initial foray into the functional analysis of these intercellular signals and the developmental events they direct and shape. Much work remains to characterize the intracellular constituents of each of the signaling pathways, to link the upstream signals with the downstream effectors that mediate specific cellular events, and to distinguish between direct and indirect regulatory interactions. It will be an even more ambitious undertaking to elucidate how gradients of signaling molecules are interpreted at the cellular level, and to integrate the inputs and outputs of the various signaling pathways into a comprehensive system that accurately reflects the intricacy of these beautiful but complex structures.

Despite the recent identification of target genes of the floral organ identity factors (*see* above), little is known about how these genes shape each floral organ and how they contribute to their modular structure by defining their specific functional compartments and tissues [79]. Many mutants related to floral organ patterning and tissue differentiation have been identified, but the connection of the genes affected by these mutations and the floral organ identity factors is poorly understood, and the genetic hierarchies of the network in the floral context are unclear as well. To further complicate things, it is frequently observed that these mutations cause defects in more than one whorl and are usually also affected in leaf morphology, supporting the long standing idea of the common basic developmental plan of the lateral organs, on top of which floral organ identity is laid [50, 80–82]. Thus, an open question remains on how these "general" factors acquire, within the floral context, new functions to direct the differentiation of highly specialized tissues. We can propose again a scenario based on the combinatorial nature of transcriptional regulators. The "leaf" factors could physically interact with ABC factors, or with the few floral-specific factors directly downstream of them, such as, for example, CRC [41], and thus generate new functional domains. In addition, specificity could be also given by the unique overlapping pattern of some of the general factors in the floral organs. In this sense, we could envision the existence of another floral combinatorial code, directing not organ specification but organ patterning. To explore this hypothesis, much work would be needed, beginning with the comprehensive identification of truly floral-specific factors, as well as a better definition of the protein interaction network among transcription factors in the flower.

A key aspect of floral organ formation is the development of the male and female reproductive structures of the plant. Male gametogenesis takes place in anthers, which usually have four similarly structured portions called lobes. In most flowering plants, each lobe contains at least four layers of somatic cells: the epidermis, the endothecium, the middle layer, and the tapetum, which surrounds the male meiotic cells, or pollen mother cells (PMCs). Anther development starting from a primordium with cells from the three meristematic layers has been described for several plants, providing a morphological basis for understanding gene functions from mutant phenotypes [83–86]. Genetic studies of male sterile mutants and molecular analyses of the corresponding genes have identified regulators of early anther division and differentiation encoding likely cell-cell signaling components, such as putative receptor-like protein kinases and their ligands; additional genes essential for proper tapetum formation and function often encode putative transcription factors that are required for normal expression of thousands of genes, some of which are enzymes for synthesis of pollen wall components [83, 84].

Although it is certain that many genes important for anther development will continue to be uncovered by a combination of genetic, genomic, and biochemical approaches, greater challenges lie ahead in the comprehensive understanding of how these genes function at the molecular level to orchestrate the formation of the functional anther and to ensure proper pollen development. For example, the importance of cell-cell signaling has been strongly supported by the requirement for several genes encoding receptor-like protein kinases and intracellular kinases, but how they perceive extracellular signals and transmit them to regulate gene expression and other cellular processes are not known. At the same time, key transcription factors have been demonstrated to be important for tapetum function, with some evidence for genetic and physical interactions at multiple levels, yet how the tapetum transcriptome is precisely controlled spatially and temporally is far from understood. Furthermore, little is known about the earliest aspects of anther differentiation, even at the morphological or cellular levels, because of the lack of molecular and cellular markers that can distinguish the earliest cell types. Moreover, compared with the tapetum, the control of differentiation of other cell types is much less known.

As mentioned above, the male meiotic cells are formed interior of the tapetum in the anther locules, and they have been the source of most of the information regarding plant meiosis, including classic cytological studies that used plants with large genomes, such as maize and lily [87, 88]. *Arabidopsis* has been a recent member of this small club of plants [84, 88–91], yet its advantage in molecular genetics has allowed a relatively large number of genes to be studied using forward and reverse genetics, uncovering both conserved mechanisms and features unique to plants, from chromosome cohesion to meiotic recombination, from spindle function to meiotic cytokinesis. Functional homologs of well-known meiotic genes, such as *SPO11* and *RAD51*, first discovered in yeast and other organisms, have been analyzed in *Arabidopsis* [90, 92–94]. Reverse genetic studies have also been very powerful in demonstrating the importance of gene functions that are also required for mitotic growth using hypomorphic alleles or meiosis-specific gene knock-down approaches. Perhaps more significantly, novel genes have been identified by forward genetics that have homologs in other organisms but were not recognized as having a meiotic function [84, 89, 95].

A central question of how germ line specification is controlled has recently seen a very exciting new development: the lack of oxygen, or hypoxia, has been implicated in promoting the male germ line fate [96]. How a low oxygen level is translated into molecular programs specifying the male meiocytes is not yet known, but there have been reports of genes involved in redox status of the cell being important for male fertility. Clearly, the molecular basis of homolog pairing is still largely a mystery, which might be solved with advances in both molecular cell biology and live imaging technology. Also not understood is how meiotic recombination is regulated at the chromosome level, where plants might offer new insights not possible from yeasts because of their very small genomes.

5 Evolution of Flower Development

Angiosperms exhibit astounding differences in floral architecture, size and color. While the molecular basis to these morphological variations is largely unknown, the recent advances in the technologies used for genome sequencing and transcriptome profiling will likely result in significant and rapid progress in the understanding of floral evolution. To date, evolutionary studies on flower development have focused mainly on the floral organ identity genes and their functions during organ specification and development. In order to obtain a more comprehensive understanding of the differences in the gene networks controlling flower development in different species, these studies will have to be expanded to genes involved processes such as floral patterning, growth and differentiation.

Analyses of phylogenetically diverse angiosperms showed that B and C functions specifying reproductive floral organ identity are highly conserved. For example, C class orthologs act to promote reproductive organ development in the monocot maize and eudicot *Antirrhinum*, although some subfunctionalization is evident in maize due to gene duplication [97, 98]. B function activity to specify petal identity also appears largely conserved across angiosperms from core eudicots to grasses, providing evidence that lodicules are homologous to petals [99–102]. Again, gene duplication events leading to subfunctionalization of B class activities have occurred in several taxa, for example in the Ranunculaceae and orchids [103, 104]. The roles of B and C class genes in specifying reproductive organ identity likely predate angiosperms. For example, in gymnosperms (e.g., *Picea*, *Pinus*, *Gnetum*) C class genes are expressed in both male and female cones while B class genes are expressed in male cones [105–108]. Thus, a BC model of reproductive organ specification is likely ancestral for seed plants.

In contrast to B and C function, there has not been a consensus for a broadly conserved A class function, with models for non-Brassicaceae often consisting of a BC model [99, 100], and some authors arguing that A class is phylogenetically restricted to the Brassicaceae [109]. However, since outer perianth organs (sepals in many angiosperms) differ from leaves in many respects, there should exist an activity specifying them as such. In *Arabidopsis*, A class was originally defined as an activity that (1) promoted sepals and petal identity and (2) prevented C activity in the perianth whorls. The latter function is fulfilled by APETALA2 (AP2), which acts to repress *AG* transcription in first and second whorls [110]. Based on the floral phenotypes of weak *ap2* alleles and those in which *AP2* and all potential C class genes are eliminated, *AP2* also contributes to sepal identity [111, 112]. In contrast, the A class gene *AP1* does not act to restrict C class activity, but contributes to sepal and petal development [21, 113–116]. *AP1* and its close paralog, *CAL*, also act earlier in flower development in the specification of flower meristem identity [21, 22] (*see* above). It could be argued that the recent genome duplication within the Brassicaceae producing the *AP1* and *CAL* paralogs facilitated the identification of *AP1* as an A class gene. *AP1* promotes sepal identity, but how prominent a role *AP1* plays in petal specification is obscured by the loss of second whorl organs in the *ap1 cal* double mutant. Targeted loss-of-function of both paralogs in petals after their initiation would be useful to address that question.

In the ancestor of the core eudicots, a gene duplication resulted in the *AP1* and *FUL* paralogs, whereas the basal eudicots and more basal angiosperms possess only a single paralog. The roles of *AP1* orthologs in floral meristem and perianth identity specification seem to be conserved in the core eudicots [117–119]. In poppies, a basal eudicot, the single paralog performs the combined functions that *AP1* and *FUL* perform in *Arabidopsis*, suggesting subfunctionalization within the core eudicots [120]. *AP1/FUL* orthologs have not been fully functionally examined in basal angiosperms. In contrast, *AP2* orthologs in other eudicots, such as *Antirrhinum*, have been shown to play a role in perianth development, but there is no evidence for a role in repression of C class activity [121]. However, in both *Antirrhinum* and *Petunia* the microRNA miR169 acts to restrict, at least in part, C function from the outer whorls via repression of a C function gene activator [122]. Thus, in the Brassicaceae (rosids) and in two asterids, different mechanisms have evolved to restrict C function from the perianth whorls.

MIKC^c gene family members are present in the genomes of all land plants [123, 124]. Several aspects of the evolutionary history of this gene family are revealed in a phylogenetic analysis of land plant MIKC^c genes (*see* Fig. 1). First, the common ancestor of land plants possessed a single MIKC^c gene. The presence of multiple genes in moss species is due to gene duplications with the moss lineage. Second, the common ancestor of vascular plants also likely possessed a single MIKC^c gene. As in the mosses, the presence to multiple gene family members in lycophyte species is due to gene duplications within the lycophyte lineage. Third, the common ancestor of the euphyllophytes possessed multiple (2–3?) MIKC^c genes. The tree topology suggests that the ancestors of the B and C genes might have diverged prior to the divergence of ferns from seed plants. Fourth, the common ancestor of seed plants possessed several (8–12?) MIKC^c genes. Several of these ancestral lineages expanded within the angiosperms, gymnosperms, or both. At least 10 MIKC^c genes are present in the *Amborella trichopoda* genome and most derived flowering plant genomes harbor 20 or more, suggesting they have been preferentially retained following genome/gene duplications. Fifth, orthologs for each of the A (*AP1/FUL*), B (*AP3/PI*), C (*AG*), and E (*SEP*) classes can be found in extant gymnosperms. The tree topology prompts a few questions posed below.

Knowledge of the functions of MIKC^c genes in other lineages of land plants might help elucidate the ancestral functions of the gene family and provide clues to the evolution of their more derived functions. For example, fern genes of two different clades, one more closely related to the B genes and the other to the A/C/E genes of seed plants, are broadly expressed in both vegetative and reproductive tissues, with some genes expressed in both sporophytic and gametophytic generations [125, 126]. What are the functions of these genes, and do they play any role in specification of tissues during reproductive development? Similar questions may be asked about the lycophyte genes, which are also broadly expressed during the sporophytic generation [127], and genes, such as the single MIKC^c gene in the liverwort *Marchantia* that is

Fig. 1 (continued) another and may be more closely related to different fern homologs. (3) There exist gymnosperm orthologs for each of the A, B, C, and E classes of angiosperm genes. Bayesian phylogenetic analysis was performed using Mr. Bayes 3.1 [140]. This version of the software conducts two independent analyses simultaneously. The mixed model option (aamodelpr = mixed) was used to estimate the appropriate amino acid fixed rate model. The analysis was run for 3,000,000 generations, which was sufficient for the standard deviation of the split frequencies to drop below 0.02. To allow for the burn in phase, the first 500 trees (10 % of the total number of saved trees) were discarded. The tree was rooted with the single *Marchantia* MIKC^c sequence resulting in a tree that broadly reflects land plant phylogeny. However, the positions of the fern genes relative to the seed plant genes are ambiguous. Directed analyses of well-supported clades or additional taxa and/or sequences may help resolve some aspects; however, this is not guaranteed [141, 142]

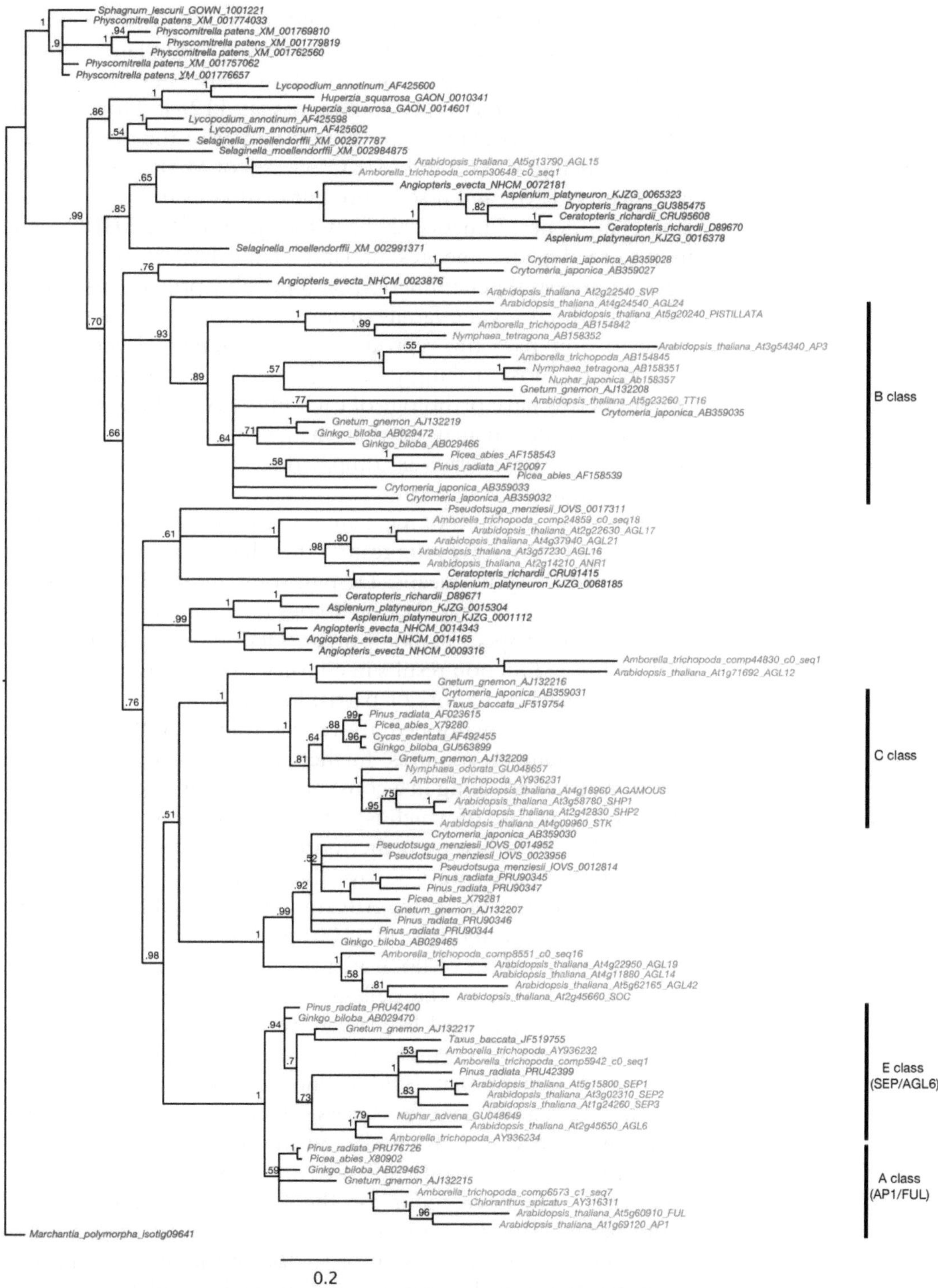

Fig. 1 Bayesian phylogram of land plant MIKC^c genes based on alignment of 163 characters (amino acids). Numbers above the branches indicate posterior probability values. While there is a lack of resolution in many branches of the tree, three aspects are relevant to the evolution of the ABC genes. (1) There are no one-to one orthologs of the ABC genes in non-seed plant taxa. (2) The B and C class genes are only distantly related to one

expressed in the gametophyte. In contrast to the broad expression patterns of MIKC^c genes in basal lineages of land plants, where it has been investigated, expression of MIKC genes in Charophytes is restricted to haploid reproductive tissues [128]. However, the Charophycean genes may have their origin prior the gene duplication producing MIKC^c and MIKC* classes of MADS-box genes in land plants [128], and if so, represent orthologs to both classes of land plant genes. Functional characterization of the MIKC^c genes in non-seed plant lineages of land plants has the potential to uncover ancestral gene functions and possibly provide insight into how this obscure gene family, from a non-seed plant perspective, came to dominate the reproductive developmental programs in seed plants, most strikingly in angiosperms. Furthermore, since gymnosperms have orthologs of each of the A, B, C, and E classes of MIKC^c genes, it may be that changes in how these genes are regulated in angiosperms as compared to gymnosperms facilitated the evolution of the flower.

The outer perianth of most angiosperm flowers is morphologically distinct from leaves suggesting that their identity as such must be specified by some means. Despite some initial skepticism, it is becoming apparent that *AP1* orthologs have a relatively conserved function with respect to floral organ identity in eudicots: loss of AP1 ortholog function results in conversion of sepals into bract- or leaf-like organs and a loss or partial loss of flower meristem identity. However, as mentioned above, the role of *AP1* orthologs in petal development requires further examination, with loss-of-function alleles resulting in a conversion of petals into a more foliar identity, the nature of which is not clear, but is not conspicuously carpelloid in most cases. Furthermore, the function of *AP1* orthologs as A class genes must be investigated in basal angiosperms.

One idea to reconcile the ABC model proposed for *Arabidopsis* [129] with the BC models [99, 100] proposed for *Antirrhinum* is that the specification of floral identity and that of sepals are intimately linked, that is, the default state of organs produced by a floral meristem is a sepal-like organ, whose identity can be modified by the additional expression of B and C class genes. For example, one BC scenario that could account for floral organ specification without the requirement for A class if E class genes have some role in organ identity, with B + E class promoting petal identity and E class alone outer perianth identity. Such possibilities should be considered when searching for the "missing" A class in other angiosperms. The status of the other aspect of *Arabidopsis* A class, that responsible for repression of C class in the perianth, in other angiosperms is not clear. Given the disparity in regulatory networks excluding C function from the perianth in *Arabidopsis* (*AP2*) [110] as compared to *Antirrhinum* and Petunia (*miR169*) [122], and the potential for regulatory mechanisms to evolve quickly, functional analyses are required to determine which, if either, is found more broadly throughout the angiosperms.

6 Further Considerations and Conclusions

Degeneracy, a partial functional overlap of structurally distinct elements, is a common feature of biological systems [130–132]. This is quite in contrast to systems designed by human engineers that seek to minimize degenerate functions. Degeneracy supports robustness within biological processes and thus is a key feature of evolvable systems. Theoretical treatments of evolvability suggest that degeneracy is a key source of robustness and is required for a system to be both robust and yet adaptive [131, 132]. Future explorations into the role of degenerate functional modules during development will provide insight into the evolution of developmental structures, as well as the ability of plants to develop properly under fluctuating environmental conditions. Genome-wide analyses of degenerate functional elements offer an opportunity to study the functional integration and independence within the complex hierarchy that supports the development of critical structures such as flowers. Also, modeling approaches that consider "imperfectly engineered" biological solutions will likely provide additional insight into the emergent properties of complex adaptive systems.

The application of current methods—RNA-seq, ChIP-seq, and TRAP-seq (*see* Chapters 18, 23 and 24)—will allow, after considerable effort, an outline of the gene regulatory networks in operation in each cell type of the developing flower. This in turn will enable a much more detailed understanding of the evolution of the regulatory networks that control organ specification and organ growth. It seems likely, as well, that continued application of methods such as live imaging (*see* Chapter 25) and computational modeling (*see* Chapter 26) will result in satisfactory models for the developmental origin of organ numbers and positions, much as such models have been developed for phyllotaxis (e.g., [133, 134]). What current methods and experiments will not supply is the link between gene expression networks and the size and shape of floral organs and of flowers, and in this area—predicting morphological phenotype from genotype—lies a great challenge for the future.

To make this connection we need a better understanding of the consequences of gene network activation on cellular activities, including cell expansion, anisotropy of cell expansion, and cell division. We also need an understanding of the communication between cells that allows the coordination of their expansion and division across tissues, and that limits the direction and amount of these activities to create final size and shape of organs. Recent evidence ties together the chemical signaling that controls domains of gene activity in meristems (e.g., [135, 136]) with the physical interactions of cells as they push on their neighbors due to anisotropic cellular expansion and division (e.g., [133, 137, 138]).

Chemical signals, such as auxin, control cell expansion, and cell expansion controls levels and directions of auxin transport. In the center of this set of feedbacks is the cell wall, as it creates the anisotropy that controls direction of expansion, and therefore, future signaling. This in turn indicates that another critical control is the cytoskeleton, as the microtubule cytoskeleton controls the pattern of cell wall deposition (e.g., [139]).

The path to the future, and to a full understanding of flower development, thus lies in achieving a new level of understanding of cytoskeletal dynamics and cell wall synthesis, and in relating cellular function and cell division to the feedback control of cellular processes through chemical and physical signaling in groups of cells. This new synthesis will require a far more dynamic view of cellular behavior than we have now, and development of computational modeling environments in which the detailed activities of genes, cells, and tissues can interact at multiple levels. If this challenge is met, we will not only understand flower development, but also have a toolbox with which we can understand all of development, plant and animal.

References

1. Amasino R (2010) Seasonal and developmental timing of flowering. Plant J 61(6):1001–1013
2. Srikanth A, Schmid M (2011) Regulation of flowering time: all roads lead to Rome. Cell Mol Life Sci 68(12):2013–2037
3. Andres F, Coupland G (2012) The genetic basis of flowering responses to seasonal cues. Nat Rev Genet 13(9):627–639
4. Long J, Barton MK (2000) Initiation of axillary and floral meristems in Arabidopsis. Dev Biol 218(2):341–353
5. Grbic V, Bleecker AB (2000) Axillary meristem development in Arabidopsis thaliana. Plant J 21(2):215–223
6. Hempel FD, Feldman LJ (1994) Bi-directional inflorescence development in Arabidopsis thaliana: acropetal initiation of flowers and basipetal initiation of paraclades. Planta 192:276–286
7. Hempel FD, Zambryski PC, Feldman LJ (1998) Photoinduction of flower identity in vegetatively biased primordia. Plant Cell 10(10):1663–1676
8. Smyth DR, Bowman JL, Meyerowitz EM (1990) Early flower development in Arabidopsis. Plant Cell 2(8):755–767
9. Galweiler L, Guan C, Muller A, Wisman E, Mendgen K, Yephremov A, Palme K (1998) Regulation of polar auxin transport by AtPIN1 in Arabidopsis vascular tissue. Science 282(5397):2226–2230
10. Reinhardt D, Pesce ER, Stieger P, Mandel T, Baltensperger K, Bennett M, Traas J, Friml J, Kuhlemeier C (2003) Regulation of phyllotaxis by polar auxin transport. Nature 426(6964): 255–260
11. Yamaguchi N, Wu MF, Winter CM, Berns MC, Nole-Wilson S, Yamaguchi A, Coupland G, Krizek BA, Wagner D (2013) A molecular framework for auxin-mediated initiation of flower primordia. Dev Cell 24(3):271–282
12. Weigel D, Alvarez J, Smyth DR, Yanofsky MF, Meyerowitz EM (1992) *LEAFY* controls floral meristem identity in Arabidopsis. Cell 69(5):843–859
13. Schultz EA, Haughn GW (1991) LEAFY, a homeotic gene that regulates inflorescence development in Arabidopsis. Plant Cell 3(8): 771–781
14. Parcy F, Nilsson O, Busch MA, Lee I, Weigel D (1998) A genetic framework for floral patterning. Nature 395(6702):561–566
15. Saddic LA, Huvermann B, Bezhani S, Su Y, Winter CM, Kwon CS, Collum RP, Wagner D (2006) The LEAFY target LMI1 is a meristem identity regulator and acts together with LEAFY to regulate expression of CAULIFLOWER. Development 133(9): 1673–1682
16. Wagner D, Sablowski RW, Meyerowitz EM (1999) Transcriptional activation of *APETALA1* by LEAFY. Science 285(5427): 582–584
17. William DA, Su Y, Smith MR, Lu M, Baldwin DA, Wagner D (2004) Genomic identification

of direct target genes of LEAFY. Proc Natl Acad Sci U S A 101(6):1775–1780
18. Winter CM, Austin RS, Blanvillain-Baufume S, Reback MA, Monniaux M, Wu MF, Sang Y, Yamaguchi A, Yamaguchi N, Parker JE, Parcy F, Jensen ST, Li H, Wagner D (2011) LEAFY target genes reveal floral regulatory logic, cis motifs, and a link to biotic stimulus response. Dev Cell 20(4):430–443
19. Ferrándiz C, Gu Q, Martienssen R, Yanofsky MF (2000) Redundant regulation of meristem identity and plant architecture by FRUITFULL, APETALA1 and CAULIFLOWER. Development 127(4): 725–734
20. Kaufmann K, Wellmer F, Muino JM, Ferrier T, Wuest SE, Kumar V, Serrano-Mislata A, Madueno F, Krajewski P, Meyerowitz EM, Angenent GC, Riechmann JL (2010) Orchestration of floral initiation by APETALA1. Science 328(5974):85–89
21. Bowman JL, Alvarez J, Weigel D, Meyerowitz EM, Smyth D (1993) Control of flower development in *Arabidopsis thaliana* by *APETALA1* and interacting genes. Development 119:721–743
22. Kempin SA, Savidge B, Yanofsky MF (1995) Molecular basis of the cauliflower phenotype in Arabidopsis. Science 267(5197):522–525
23. Gregis V, Sessa A, Colombo L, Kater MM (2008) AGAMOUS-LIKE24 and SHORT VEGETATIVE PHASE determine floral meristem identity in Arabidopsis. Plant J 56(6): 891–902
24. Grandi V, Gregis V, Kater MM (2012) Uncovering genetic and molecular interactions among floral meristem identity genes in Arabidopsis thaliana. Plant J 69(5):881–893
25. Liu C, Xi W, Shen L, Tan C, Yu H (2009) Regulation of floral patterning by flowering time genes. Dev Cell 16(5):711–722
26. Chandler JW, Jacobs B, Cole M, Comelli P, Werr W (2011) DORNROSCHEN-LIKE expression marks Arabidopsis floral organ founder cells and precedes auxin response maxima. Plant Mol Biol 76(1–2):171–185
27. Nag A, Yang Y, Jack T (2007) DORNROSCHEN-LIKE, an AP2 gene, is necessary for stamen emergence in Arabidopsis. Plant Mol Biol 65(3):219–232
28. Reinhardt D, Mandel T, Kuhlemeier C (2000) Auxin regulates the initiation and radial position of plant lateral organs. Plant Cell 12(4):507–518
29. Aida M, Ishida T, Fukaki H, Fujisawa H, Tasaka M (1997) Genes involved in organ separation in Arabidopsis: an analysis of the cup-shaped cotyledon mutant. Plant Cell 9(6):841–857
30. Bell EM, Lin WC, Husbands AY, Yu L, Jaganatha V, Jablonska B, Mangeon A, Neff MM, Girke T, Springer PS (2012) Arabidopsis lateral organ boundaries negatively regulates brassinosteroid accumulation to limit growth in organ boundaries. Proc Natl Acad Sci U S A 109(51):21146–21151
31. Lampugnani ER, Kilinc A, Smyth DR (2012) PETAL LOSS is a boundary gene that inhibits growth between developing sepals in Arabidopsis thaliana. Plant J 71(5):724–735
32. Laufs P, Peaucelle A, Morin H, Traas J (2004) MicroRNA regulation of the CUC genes is required for boundary size control in Arabidopsis meristems. Development 131(17):4311–4322
33. Sieber P, Wellmer F, Gheyselinck J, Riechmann JL, Meyerowitz EM (2007) Redundancy and specialization among plant microRNAs: role of the MIR164 family in developmental robustness. Development 134(6):1051–1060
34. Weir I, Lu J, Cook H, Causier B, Schwarz-Sommer Z, Davies B (2004) CUPULIFORMIS establishes lateral organ boundaries in Antirrhinum. Development 131(4):915–922
35. Mizukami Y, Ma H (1992) Ectopic expression of the floral homeotic gene AGAMOUS in transgenic Arabidopsis plants alters floral organ identity. Cell 71(1):119–131
36. Yanofsky MF, Ma H, Bowman JL, Drews GN, Feldmann KA, Meyerowitz EM (1990) The protein encoded by the Arabidopsis homeotic gene agamous resembles transcription factors. Nature 346(6279):35–39
37. Lenhard M, Bohnert A, Jurgens G, Laux T (2001) Termination of stem cell maintenance in Arabidopsis floral meristems by interactions between *WUSCHEL* and *AGAMOUS*. Cell 105(6):805–814
38. Lohmann JU, Hong RL, Hobe M, Busch MA, Parcy F, Simon R, Weigel D (2001) A molecular link between stem cell regulation and floral patterning in Arabidopsis. Cell 105:793–803
39. Sun B, Xu Y, Ng KH, Ito T (2009) A timing mechanism for stem cell maintenance and differentiation in the Arabidopsis floral meristem. Genes Dev 23(15):1791–1804
40. Payne T, Johnson SD, Koltunow AM (2004) *KNUCKLES* (*KNU*) encodes a C2H2 zinc-finger protein that regulates development of basal pattern elements of the Arabidopsis gynoecium. Development 131(15): 3737–3749
41. Bowman JL, Smyth DR (1999) *CRABS CLAW*, a gene that regulates carpel and nectary development in Arabidopsis, encodes a novel protein with zinc finger and helix-loop-helix domains. Development 126(11):2387–2396

42. Gomez-Mena C, de Folter S, Costa MM, Angenent GC, Sablowski R (2005) Transcriptional program controlled by the floral homeotic gene *AGAMOUS* during early organogenesis. Development 132(3): 429–438
43. Sun B, Ito T (2010) Floral stem cells: from dynamic balance towards termination. Biochem Soc Trans 38(2):613–616
44. Bowman JL, Sakai H, Jack T, Weigel D, Mayer U, Meyerowitz EM (1992) SUPERMAN, a regulator of floral homeotic genes in Arabidopsis. Development 114(3): 599–615
45. Sakai H, Medrano LJ, Meyerowitz EM (1995) Role of SUPERMAN in maintaining Arabidopsis floral whorl boundaries. Nature 378(6553):199–203
46. Davies B, Egea-Cortines M, de Andrade SE, Saedler H, Sommer H (1996) Multiple interactions amongst floral homeotic MADS box proteins. EMBO J 15(16):4330–4343
47. Riechmann JL, Krizek BA, Meyerowitz EM (1996) Dimerization specificity of Arabidopsis MADS domain homeotic proteins APETALA1, APETALA3, PISTILLATA, and AGAMOUS. Proc Natl Acad Sci U S A 93(10):4793–4798
48. Egea-Cortines M, Saedler H, Sommer H (1999) Ternary complex formation between the MADS-box proteins SQUAMOSA, DEFICIENS and GLOBOSA is involved in the control of floral architecture in *Antirrhinum majus*. EMBO J 18(19):5370–5379
49. Melzer R, Theissen G (2009) Reconstitution of 'floral quartets' in vitro involving class B and class E floral homeotic proteins. Nucleic Acids Res 37(8):2723–2736
50. Honma T, Goto K (2001) Complexes of MADS-box proteins are sufficient to convert leaves into floral organs. Nature 409(6819): 525–529
51. Gamboa A, Paez-Valencia J, Acevedo GF, Vazquez-Moreno L, Alvarez-Buylla RE (2001) Floral transcription factor AGAMOUS interacts in vitro with a leucine-rich repeat and an acid phosphatase protein complex. Biochem Biophys Res Commun 288(4): 1018–1026
52. Pelaz S, Gustafson-Brown C, Kohalmi SE, Crosby WL, Yanofsky MF (2001) *APETALA1* and *SEPALLATA3* interact to promote flower development. Plant J 26(4):385–394
53. Causier B, Ashworth M, Guo W, Davies B (2012) The TOPLESS Interactome: a framework for gene repression in Arabidopsis. Plant Physiol 158(1):423–438
54. Causier B, Cook H, Davies B (2003) An antirrhinum ternary complex factor specifically interacts with C-function and SEPALLATA-like MADS-box factors. Plant Mol Biol 52(5):1051–1062
55. Gregis V, Sessa A, Colombo L, Kater MM (2006) AGL24, SHORT VEGETATIVE PHASE, and APETALA1 redundantly control AGAMOUS during early stages of flower development in Arabidopsis. Plant Cell 18(6):1373–1382
56. Hill K, Wang H, Perry SE (2008) A transcriptional repression motif in the MADS factor AGL15 is involved in recruitment of histone deacetylase complex components. Plant J 53(1):172–185
57. Masiero S, Imbriano C, Ravasio F, Favaro R, Pelucchi N, Gorla MS, Mantovani R, Colombo L, Kater MM (2002) Ternary complex formation between MADS-box transcription factors and the histone fold protein NF-YB. J Biol Chem 277(29):26429–26435
58. Sridhar VV, Surendrarao A, Liu Z (2006) APETALA1 and SEPALLATA3 interact with SEUSS to mediate transcription repression during flower development. Development 133(16):3159–3166
59. Smaczniak C, Immink RG, Muino JM, Blanvillain R, Busscher M, Busscher-Lange J, Dinh QD, Liu S, Westphal AH, Boeren S, Parcy F, Xu L, Carles CC, Angenent GC, Kaufmann K (2012) Characterization of MADS-domain transcription factor complexes in Arabidopsis flower development. Proc Natl Acad Sci U S A 109(5):1560–1565
60. Kaufmann K, Pajoro A, Angenent GC (2010) Regulation of transcription in plants: mechanisms controlling developmental switches. Nat Rev Genet 11(12):830–842
61. Deng W, Ying H, Helliwell CA, Taylor JM, Peacock WJ, Dennis ES (2011) FLOWERING LOCUS C (FLC) regulates development pathways throughout the life cycle of Arabidopsis. Proc Natl Acad Sci U S A 108(16):6680–6685
62. Kaufmann K, Muino JM, Jauregui R, Airoldi CA, Smaczniak C, Krajewski P, Angenent GC (2009) Target genes of the MADS transcription factor SEPALLATA3: integration of developmental and hormonal pathways in the Arabidopsis flower. PLoS Biol 7(4):e1000090
63. Immink RG, Pose D, Ferrario S, Ott F, Kaufmann K, Valentim FL, de Folter S, van der Wal F, van Dijk AD, Schmid M, Angenent GC (2012) Characterization of SOC1's central role in flowering by the identification of its upstream and downstream regulators. Plant Physiol 160(1):433–449
64. Wuest SE, O'Maoileidigh DS, Rae L, Kwasniewska K, Raganelli A, Hanczaryk K, Lohan AJ, Loftus B, Graciet E, Wellmer F (2012) Molecular basis for the specification of floral organs by APETALA3 and

PISTILLATA. Proc Natl Acad Sci U S A 109(33):13452–13457
65. Kieffer M, Master V, Waites R, Davies B (2011) TCP14 and TCP15 affect internode length and leaf shape in Arabidopsis. Plant J 68(1):147–158
66. Wellmer F, Alves-Ferreira M, Dubois A, Riechmann JL, Meyerowitz EM (2006) Genome-wide analysis of gene expression during early Arabidopsis flower development. PLoS Genet 2(7):e117
67. Yant L, Mathieu J, Dinh TT, Ott F, Lanz C, Wollmann H, Chen X, Schmid M (2010) Orchestration of the floral transition and floral development in Arabidopsis by the bifunctional transcription factor APETALA2. Plant Cell 22(7):2156–2170
68. Ge X, Chang F, Ma H (2010) Signaling and transcriptional control of reproductive development in Arabidopsis. Curr Biol 20(22): R988–R997
69. Yu H, Ito T, Zhao Y, Peng J, Kumar P, Meyerowitz EM (2004) Floral homeotic genes are targets of gibberellin signaling in flower development. Proc Natl Acad Sci U S A 101(20):7827–7832
70. Cheng H, Qin L, Lee S, Fu X, Richards DE, Cao D, Luo D, Harberd NP, Peng J (2004) Gibberellin regulates Arabidopsis floral development via suppression of DELLA protein function. Development 131(5):1055–1064
71. Ito T, Ng KH, Lim TS, Yu H, Meyerowitz EM (2007) The homeotic protein AGAMOUS controls late stamen development by regulating a jasmonate biosynthetic gene in Arabidopsis. Plant Cell 19(11): 3516–3529
72. Nemhauser JL, Feldman LJ, Zambryski PC (2000) Auxin and ETTIN in Arabidopsis gynoecium morphogenesis. Development 127(18):3877–3888
73. Fletcher JC, Brand U, Running MP, Simon R, Meyerowitz EM (1999) Signaling of cell fate decisions by CLAVATA3 in Arabidopsis shoot meristems. Science 283(5409):1911–1914
74. Jun J, Fiume E, Roeder AH, Meng L, Sharma VK, Osmont KS, Baker C, Ha CM, Meyerowitz EM, Feldman LJ, Fletcher JC (2010) Comprehensive analysis of CLE polypeptide signaling gene expression and overexpression activity in Arabidopsis. Plant Physiol 154(4):1721–1736
75. Canales C, Bhatt AM, Scott R, Dickinson H (2002) EXS, a putative LRR receptor kinase, regulates male germline cell number and tapetal identity and promotes seed development in Arabidopsis. Curr Biol 12(20): 1718–1727
76. Hord CL, Chen C, Deyoung BJ, Clark SE, Ma H (2006) The BAM1/BAM2 receptor-like kinases are important regulators of Arabidopsis early anther development. Plant Cell 18(7): 1667–1680
77. Yang SL, Xie LF, Mao HZ, Puah CS, Yang WC, Jiang L, Sundaresan V, Ye D (2003) Tapetum determinant1 is required for cell specialization in the Arabidopsis anther. Plant Cell 15(12):2792–2804
78. Zhao DZ, Wang GF, Speal B, Ma H (2002) The excess microsporocytes1 gene encodes a putative leucine-rich repeat receptor protein kinase that controls somatic and reproductive cell fates in the Arabidopsis anther. Genes Dev 16(15):2021–2031
79. Sablowski R (2010) Genes and functions controlled by floral organ identity genes. Semin Cell Dev Biol 21(1):94–99
80. Goto K, Kyozuka J, Bowman JL (2001) Turning floral organs into leaves, leaves into floral organs. Curr Opin Genet Dev 11(4): 449–456
81. Pelaz S, Tapia-Lopez R, Alvarez-Buylla ER, Yanofsky MF (2001) Conversion of leaves into petals in Arabidopsis. Curr Biol 11(3): 182–184
82. Sundberg E, Ferrándiz C, Østergaard, L., ed (2009) Gynoecium patterning in Arabidopsis: a basic plan behind a complex structure. In: Annual plant reviews volume 38: fruit development and seed dispersal. Oxford, UK: Blackwell Publishing Ltd pp 35–69
83. Chang F, Wang Y, Wang S, Ma H (2011) Molecular control of microsporogenesis in Arabidopsis. Curr Opin Plant Biol 14(1):66–73
84. Ma H (2005) Molecular genetic analyses of microsporogenesis and microgametogenesis in flowering plants. Annu Rev Plant Biol 56:393–434
85. Sanders PM, Bui AQ, Weterings K, McIntire KN, Hsu YC, Lee PY, Truong MT, Beals TP, Goldberg RB (1999) Anther developmental defects in Arabidopsis thaliana male-sterile mutants. Sex Plant Reprod 11:297–322
86. Singh MB, Bhalla PL (2007) Control of male germ-cell development in flowering plants. Bioessays 29(11):1124–1132
87. Golubovskaya IN (1979) Genetic control of meiosis. Int Rev Cytol 58:247–290
88. Hamant O, Ma H, Cande WZ (2006) Genetics of meiotic prophase I in plants. Annu Rev Plant Biol 57:267–302
89. Ma H (2006) A molecular portrait of Arabidopsis meiosis. Arabidopsis Book 4:e0095
90. Mercier R, Grelon M (2008) Meiosis in plants: ten years of gene discovery. Cytogenet Genome Res 120(3–4):281–290
91. Osman K, Higgins JD, Sanchez-Moran E, Armstrong SJ, Franklin FC (2011) Pathways to meiotic recombination in Arabidopsis thaliana. New Phytol 190(3):523–544

92. Li W, Chen C, Markmann-Mulisch U, Timofejeva L, Schmelzer E, Ma H, Reiss B (2004) The Arabidopsis AtRAD51 gene is dispensable for vegetative development but required for meiosis. Proc Natl Acad Sci U S A 101(29):10596–10601
93. Li W, Yang X, Lin Z, Timofejeva L, Xiao R, Makaroff CA, Ma H (2005) The AtRAD51C gene is required for normal meiotic chromosome synapsis and double-stranded break repair in Arabidopsis. Plant Physiol 138(2):965–976
94. Zickler D, Kleckner N (1999) Meiotic chromosomes: integrating structure and function. Annu Rev Genet 33:603–754
95. Wijeratne AJ, Chen C, Zhang W, Timofejeva L, Ma H (2006) The Arabidopsis thaliana PARTING DANCERS gene encoding a novel protein is required for normal meiotic homologous recombination. Mol Biol Cell 17(3):1331–1343
96. Kelliher T, Walbot V (2012) Hypoxia triggers meiotic fate acquisition in maize. Science 337(6092):345–348. doi:10.1126/science.1220080
97. Bradley D, Carpenter R, Sommer H, Hartley N, Coen E (1993) Complementary floral homeotic phenotypes result from opposite orientations of a transposon at the plena locus of Antirrhinum. Cell 72(1):85–95
98. Mena M, Ambrose BA, Meeley RB, Briggs SP, Yanofsky MF, Schmidt RJ (1996) Diversification of C-function activity in maize flower development. Science 274(5292):1537–1540
99. Carpenter R, Coen ES (1990) Floral homeotic mutations produced by transposon-mutagenesis in Antirrhinum majus. Genes Dev 4(9):1483–1493
100. Schwarz-Sommer Z, Huijser P, Nacken W, Saedler H, Sommer H (1990) Genetic control of flower development by homeotic genes in Antirrhinum majus. Science 250(4983):931–936
101. Ambrose BA, Lerner DR, Ciceri P, Padilla CM, Yanofsky MF, Schmidt RJ (2000) Molecular and genetic analyses of the silky1 gene reveal conservation in floral organ specification between eudicots and monocots. Mol Cell 5(3):569–579
102. Kyozuka J, Kobayashi T, Morita M, Shimamoto K (2000) Spatially and temporally regulated expression of rice MADS box genes with similarity to Arabidopsis class A, B and C genes. Plant Cell Physiol 41(6):710–718
103. Kramer EM, Holappa L, Gould B, Jaramillo MA, Setnikov D, Santiago PM (2007) Elaboration of B gene function to include the identity of novel floral organs in the lower eudicot Aquilegia. Plant Cell 19(3):750–766
104. Mondragon-Palomino M, Theissen G (2011) Conserved differential expression of paralogous DEFICIENS- and GLOBOSA-like MADS-box genes in the flowers of Orchidaceae: refining the 'orchid code'. Plant J 66(6):1008–1019
105. Mouradov A, Hamdorf B, Teasdale RD, Kim JT, Winter KU, Theissen G (1999) A DEF/GLO-like MADS-box gene from a gymnosperm: Pinus radiata contains an ortholog of angiosperm B class floral homeotic genes. Dev Genet 25(3):245–252
106. Shindo S, Ito M, Ueda K, Kato M, Hasebe M (1999) Characterization of MADS genes in the gymnosperm Gnetum parvifolium and its implication on the evolution of reproductive organs in seed plants. Evol Dev 1(3):180–190
107. Sundstrom J, Carlsbecker A, Svensson ME, Svenson M, Johanson U, Theißen G, Engstrom P (1999) MADS-box genes active in developing pollen cones of Norway spruce (Picea abies) are homologous to the B-class floral homeotic genes in angiosperms. Dev Genet 25(3):253–266
108. Tandre K, Svenson M, Svensson ME, Engstrom P (1998) Conservation of gene structure and activity in the regulation of reproductive organ development of conifers and angiosperms. Plant J 15(5):615–623
109. Litt A (2007) An evaluation of A-function: evidence from the APETALA1 and APETALA2 gene lineages. Int J Plant Sci 168(1):73–91
110. Drews GN, Bowman JL, Meyerowitz EM (1991) Negative regulation of the Arabidopsis homeotic gene AGAMOUS by the APETALA2 product. Cell 65(6):991–1002
111. Bowman JL, Smyth DR, Meyerowitz EM (1989) Genes directing flower development in Arabidopsis. Plant Cell 1(1):37–52
112. Pinyopich A, Ditta GS, Savidge B, Liljegren SJ, Baumann E, Wisman E, Yanofsky MF (2003) Assessing the redundancy of MADS-box genes during carpel and ovule development. Nature 424(6944):85–88
113. Gustafson-Brown C, Savidge B, Yanofsky MF (1994) Regulation of the arabidopsis floral homeotic gene APETALA1. Cell 76(1):131–143
114. Irish VF, Sussex IM (1990) Function of the apetala-1 gene during Arabidopsis floral development. Plant Cell 2(8):741–753
115. Mandel MA, Gustafson-Brown C, Savidge B, Yanofsky MF (1992) Molecular characterization

of the Arabidopsis floral homeotic gene *APETALA1*. Nature 360(6401):273–277
116. Mandel MA, Yanofsky MF (1995) A gene triggering flower formation in Arabidopsis. Nature 377(6549):522–524
117. Benlloch R, d'Erfurth I, Ferrandiz C, Cosson V, Beltran JP, Canas LA, Kondorosi A, Madueno F, Ratet P (2006) Isolation of mtpim proves Tnt1 a useful reverse genetics tool in Medicago truncatula and uncovers new aspects of AP1-like functions in legumes. Plant Physiol 142(3):972–983
118. Shalit A, Rozman A, Goldshmidt A, Alvarez JP, Bowman JL, Eshed Y, Lifschitz E (2009) The flowering hormone florigen functions as a general systemic regulator of growth and termination. Proc Natl Acad Sci U S A 106(20):8392–8397
119. Vrebalov J, Ruezinsky D, Padmanabhan V, White R, Medrano D, Drake R, Schuch W, Giovannoni J (2002) A MADS-box gene necessary for fruit ripening at the tomato ripening-inhibitor (rin) locus. Science 296(5566):343–346
120. Pabon-Mora N, Ambrose BA, Litt A (2012) Poppy APETALA1/FRUITFULL orthologs control flowering time, branching, perianth identity, and fruit development. Plant Physiol 158(4):1685–1704
121. Keck E, McSteen P, Carpenter R, Coen E (2003) Separation of genetic functions controlling organ identity in flowers. EMBO J 22(5):1058–1066
122. Cartolano M, Castillo R, Efremova N, Kuckenberg M, Zethof J, Gerats T, Schwarz-Sommer Z, Vandenbussche M (2007) A conserved microRNA module exerts homeotic control over Petunia hybrida and Antirrhinum majus floral organ identity. Nat Genet 39(7):901–905
123. Alvarez-Buylla ER, Pelaz S, Liljegren SJ, Gold SE, Burgeff C, Ditta GS, Ribas de Pouplana L, Martinez-Castilla L, Yanofsky MF (2000) An ancestral MADS-box gene duplication occurred before the divergence of plants and animals. Proc Natl Acad Sci U S A 97(10):5328–5333
124. Theißen G, Kim JT, Saedler H (1996) Classification and phylogeny of the MADS-box multigene family suggest defined roles of MADS-box gene subfamilies in the morphological evolution of eukaryotes. J Mol Evol 43(5):484–516
125. Hasebe M, Wen CK, Kato M, Banks JA (1998) Characterization of MADS homeotic genes in the fern *Ceratopteris richardii*. Proc Natl Acad Sci U S A 95(11):6222–6227
126. Münster T, Faigl W, Saedler H, Theissen G (2002) Evolutionary aspects of MADS-box genes in the eusporangiate fern Ophioglossum. Plant Biol 4(4):474–483
127. Svensson ME, Engström P (2002) Closely related MADS-box genes in club moss (Lycopodium) show broad expression patterns and are structurally similar to, but phylogenetically distinct from, typical seed plant MADS-box genes. New Phytol 154(2): 439–450
128. Tanabe Y, Hasebe M, Sekimoto H, Nishiyama T, Kitani M, Henschel K, Munster T, Theissen G, Nozaki H, Ito M (2005) Characterization of MADS-box genes in charophycean green algae and its implication for the evolution of MADS-box genes. Proc Natl Acad Sci U S A 102(7):2436–2441
129. Bowman JL, Smyth DR, Meyerowitz EM (1991) Genetic interactions among floral homeotic genes of Arabidopsis. Development 112(1):1–20
130. Edelman GM, Gally JA (2001) Degeneracy and complexity in biological systems. Proc Natl Acad Sci U S A 98(24):13763–13768
131. Whitacre JM (2010) Degeneracy: a link between evolvability, robustness and complexity in biological systems. Theor Biol Med Model 7:6
132. Whitacre J, Bender A (2010) Degeneracy: a design principle for achieving robustness and evolvability. J Theor Biol 263(1):143–153
133. Heisler MG, Hamant O, Krupinski P, Uyttewaal M, Ohno C, Jonsson H, Traas J, Meyerowitz EM (2010) Alignment between PIN1 polarity and microtubule orientation in the shoot apical meristem reveals a tight coupling between morphogenesis and auxin transport. PLoS Biol 8(10):e1000516
134. Jönsson H, Heisler MG, Shapiro BE, Meyerowitz EM, Mjolsness E (2006) An auxin-driven polarized transport model for phyllotaxis. Proc Natl Acad Sci U S A 103(5): 1633–1638
135. Chickarmane VS, Gordon SP, Tarr PT, Heisler MG, Meyerowitz EM (2012) Cytokinin signaling as a positional cue for patterning the apical-basal axis of the growing Arabidopsis shoot meristem. Proc Natl Acad Sci U S A 109(10):4002–4007
136. Heisler MG, Ohno C, Das P, Sieber P, Reddy GV, Long JA, Meyerowitz EM (2005) Patterns of auxin transport and gene expression during primordium development revealed by live imaging of the Arabidopsis inflorescence meristem. Curr Biol 15(21): 1899–1911
137. Hamant O, Heisler MG, Jonsson H, Krupinski P, Uyttewaal M, Bokov P, Corson F, Sahlin P, Boudaoud A, Meyerowitz EM, Couder Y, Traas J (2008) Developmental

patterning by mechanical signals in Arabidopsis. Science 322(5908):1650–1655
138. Nakayama N, Smith RS, Mandel T, Robinson S, Kimura S, Boudaoud A, Kuhlemeier C (2012) Mechanical regulation of auxin-mediated growth. Curr Biol 22(16):1468–1476
139. Paredez AR, Somerville CR, Ehrhardt DW (2006) Visualization of cellulose synthase demonstrates functional association with microtubules. Science 312(5779):1491–1495
140. Huelsenbeck JP, Ronquist F (2001) MRBAYES: Bayesian inference of phylogenetic trees. Bioinformatics 17(8):754–755
141. Philippe H, Brinkmann H, Lavrov DV, Littlewood DT, Manuel M, Worheide G, Baurain D (2011) Resolving difficult phylogenetic questions: why more sequences are not enough. PLoS Biol 9(3):e1000602
142. Yang Z, Rannala B (2012) Molecular phylogenetics: principles and practice. Nat Rev Genet 13(5):303–314

Part II

Genetic and Phenotypic Analyses

Chapter 6

Genetic Screens for Floral Mutants in *Arabidopsis thaliana*: Enhancers and Suppressors

Thanh Theresa Dinh, Elizabeth Luscher, Shaofang Li, Xigang Liu, So Youn Won, and Xuemei Chen

Abstract

The flower is a hallmark feature that has contributed to the evolutionary success of land plants. Diverse mutagenic agents have been employed as a tool to genetically perturb flower development and identify genes involved in floral patterning and morphogenesis. Since the initial studies to identify genes governing processes such as floral organ specification, mutagenesis in sensitized backgrounds has been used to isolate enhancers and suppressors to further probe the molecular basis of floral development. Here, we first describe two commonly employed methods for mutagenesis (using ethyl methanesulfonate (EMS) or T-DNAs as mutagens), and then describe three methods for identifying a mutation that leads to phenotypic alterations—traditional map-based cloning, TAIL-PCR, and deep sequencing in the plant model *Arabidopsis thaliana*.

Key words EMS, T-DNA, Floral development, Mutagenesis, Genetic screen, *Arabidopsis*, Map-based cloning, TAIL-PCR

1 Introduction

Plants, being sessile organisms, have to cope with many different stresses, both biotic and abiotic. The flower is a major contributing factor to the success and robustness of angiosperms. Until the 1980s, studies on plants had been largely limited to agriculturally important species; however, it is often difficult to work with these plants due to their large, in some cases, polyploid genomes. The small weed *Arabidopsis thaliana* was first purported as a genetic model system for plants by Friedrich Laibach [1]. *Arabidopsis* is amenable to genetic analyses due to its small size, small genome (125 Mb), rapid generation time (5–6 weeks from seed to seed), high fecundity (up to 10,000 seeds per plant), low repetitive sequences, and the ability to self-fertilize [1–5]. The power of *Arabidopsis* as a model system was evidenced, in part, by the identification of a series of genes involved in floral development in the

José Luis Riechmann and Frank Wellmer (eds.), *Flower Development: Methods and Protocols*, Methods in Molecular Biology, vol. 1110, DOI 10.1007/978-1-4614-9408-9_6,

late 1980s and early 1990s [6–14]. Since then, *Arabidopsis* has been the genetic model organism for plants and has been used to study diverse biological pathways.

Chemical mutagenesis has been used extensively to genetically perturb biological pathways in *Arabidopsis* in order to identify components of these pathways. Initial genetic screens are often conducted in a wild-type background to identify major, nonredundant factors. However, such screens often fail to identify regulatory genes that have overlapping functions with other (often closely related) genes. Genetic modifier screens can be conducted to circumvent this genetic redundancy. A genetic modifier screen is performed in a mutant background, usually a weak allele in a major player in a biological process, in order to isolate other genes in that biological pathway. The second-site mutations resulting from this screen can either lead to an enhancement or to a suppression of the phenotype of the mutant plants that are being mutagenized.

There have been several notable examples in which modifier screens have allowed researchers to effectively dissect complex genetic pathways in floral development. For instance, *AGAMOUS* (*AG*) controls carpel and stamen development in *Arabidopsis*, and loss of *AG* results in floral patterning defects and a reiterative flower-in-flower phenotype indicative of loss of floral determinacy [7, 14, 15]. Many alleles of *AG* have been identified, ranging from plants with a weak (*ag-10*) to strong (*ag-1*) phenotype. One of the alleles, *ag-4*, generates a partially functional protein that is able to confer stamen identity, but floral determinacy or carpel identity is compromised [16]. Dr. Xuemei Chen in Dr. Elliot Meyerowitz's laboratory performed an EMS mutagenesis in the *ag-4* background and identified a mutant with an enhanced phenotype, reminiscent of *ag-1* (or flowers in which the stamens have been converted to petals). Map-based cloning revealed that the enhanced phenotype was due to mutations in two genes, which she named *HUA1* and *HUA2* (*HUA* means flower in Chinese) [17]. *HUA1* is an RNA-binding protein and *HUA2* is a novel protein involved in RNA processing [17–19]. *hua1 hua2* double mutants have a weak phenotype, while the single mutants are phenotypically normal [17]. Using the *hua1 hua2* double mutant, Dr. Chen and her group performed another EMS mutagenesis screen to identify mutants with enhanced floral determinacy and organ identity defects. From this screen, many mutants were isolated, including another weak allele of *AG*, *ag-10*; mutant alleles of *HEN4*, a gene encoding a KH domain RNA-binding protein; mutants in *HEN1*, a gene involved small RNA biogenesis; and a mutant in *ARGONAUTE10* (*AGO10*), an effector in microRNA function [20]. Studies with the *hua1 hua2 hen4* mutants revealed the redundant functions of the three genes in promoting the transcription elongation or splicing of *AG* pre-mRNA [18]. Subsequent work on HEN1 eventually led to the discovery that miR172 promotes floral determinacy by repressing its target gene *APETALA2* (*AP2*) [21].

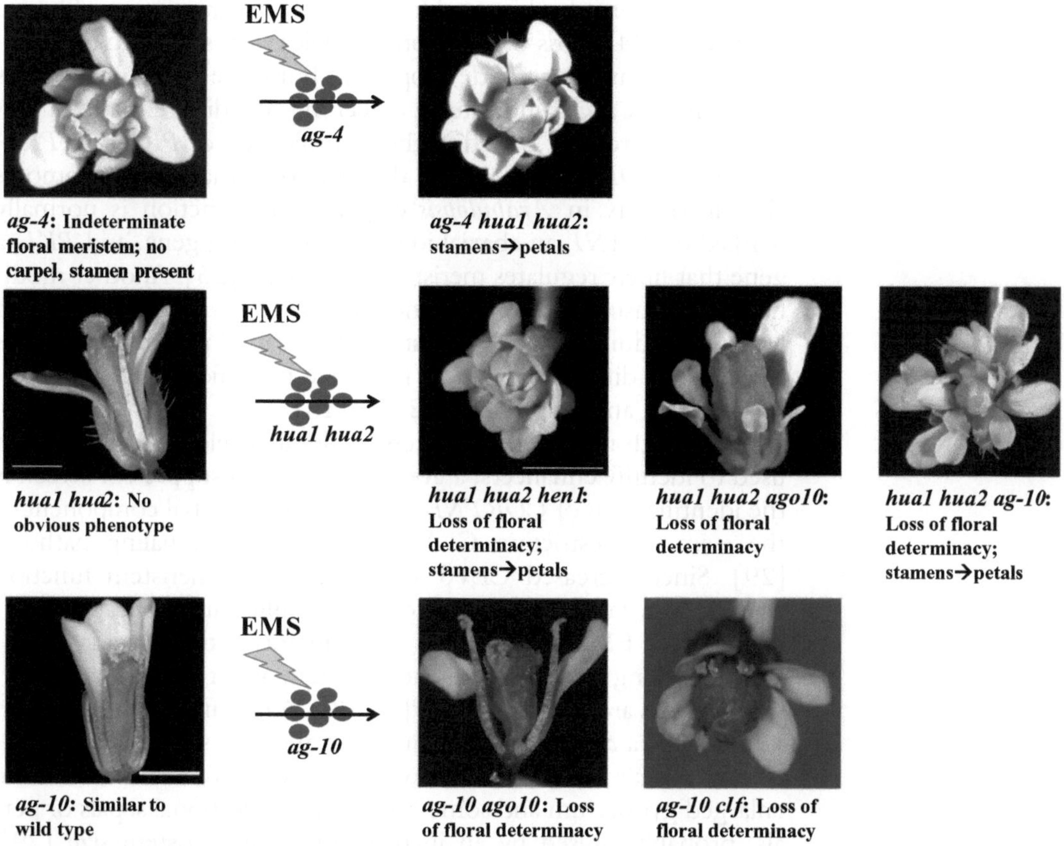

Fig. 1 EMS Mutagenesis is a Powerful Tool to Dissect Pathways Governing Floral Development. An outline of several EMS-based enhancer screens used to dissect the floral determinacy pathway governed by *AG*. A relatively weak allele of *AG*, *ag-4*, is defective in carpel identity and floral determinacy; however, stamens are still present. EMS mutagenesis was performed on *ag-4* seeds, resulting in the identification of the genes *HUA1* and *HUA2*, mutations in which convert the stamens in ag-4 to petals. The *hua1 hua2* mutant has weak organ identity defects but is normal in floral determinacy. Another EMS mutagenesis experiment was performed on *hua1 hua2* and another *ag* allele was isolated (*ag-10*), as well as mutations in *HEN1* and *AGO10*. *ag-10* has a very weak phenotype, only a few siliques in a plant are slightly bulged, indicative of loss of determinacy. Another EMS mutagenesis experiment was performed on *ag-10*, and mutants in genes such as *CLF* and *AGO10* were isolated. Thus, *CLF* and *AGO10* are involved in floral determinacy

Furthermore, Dr. Xigang Liu, a postdoctoral scholar in Dr. Chen's laboratory, performed an enhancer screen using the weak *ag-10* allele. From that screen, *CURLY LEAF* and *AGO10* were among the few genes identified and shown to be involved in floral determinacy [20, 22]. Figure 1 depicts the discovery of players that specify the identities of the reproductive organs or confer floral determinacy through genetic screens.

Another well-known example of the usefulness of modifier screens involves the gene *CRABS CLAW* (*CRC*) and the pathway concerning polarity in *Arabidopsis* carpels. *CRC* is the founding member of the YABBY family of transcription factors, in which members are involved in establishing polar differentiation of lateral organs [23, 24]. Loss-of-function *crc* alleles were shown to be defective in carpel and nectary development, but none of the phenotypes in the mutant alleles indicated a defect in polar differentiation [15]. A modifier screen of *crc* led to the identification of *PICKLE* (*PKL*) and *KANADI1* (*KAN1*), and the discovery that *CRC* promotes abaxial identity in *Arabidopsis* carpels. This function is normally masked by *KAN1*, an abaxial identity promoting gene, and *PKL*, a gene that finely regulates meristematic activity [25]. Another modifier screen (using EMS) was done on *kan1-2 pkl-12* plants to search for other redundant genes in establishing carpel polarity. From this screen, four different enhancer loci were identified: *crc*, *hasty* (*hst*), *splayed* (*spl*), and *kanadi2* (*kan2*) [26–28].

Although most genetic screens in floral development have been used to identify enhancers, a good example of a suppressor screen is the identification of *CORYNE* (*CRN*) as an essential component of the stem-cell restricting CLAVATA3 (CLV3) signaling pathway [29]. Since increased CLV3 signaling arrests meristem function leading to a facile phenotypic output, Muller and colleagues performed an EMS mutagenesis experiment with a mild *CLV3*-overexpressing line such that the meristem arrest is not that severe, so the plants are able to set seed [29, 30]. A mutant was isolated and found to be a *crn-1* allele, which has an aberrant silique shape reminiscent of *clv* mutants. Interestingly, the flowers in *crn-1* have enlarged gynoecium and some flowers have additional sepals or petals, probably caused by an increase in floral meristem size [29]. Initially, CRN was thought to be a receptor kinase that aids in the transmission of the CLV3 signal [29]; however, it was later found to be a pseudokinase and is hypothesized to play a scaffolding role, perhaps to aid the export of CLV2 to the plasma membrane and/or to assemble higher-order CLV1 or CLV1-substrate complexes [31].

In addition to the three pathways mentioned above, modifier screens have been used for the isolation of mutants in *DORNRÖSCHEN-LIKE* from the *pi-5* background [32]; mutants in three *FUSED FLORAL ORGANS* genes from a *ufo* (*unusual floral organs*) allele [33]; and mutants in *SPLAYED* (*SYD*) from a weak *leafy* mutant background [27]. Thus, the aforementioned experiments highlight the power of modifier screens in teasing apart the molecular mechanisms governing different aspects of floral development. There have been several diverse reviews [34, 35] as well as protocols for EMS mutagenesis [36]; the present chapter therefore offers a comprehensive view of both EMS and insertional mutagenesis and outlines methods used to identify the mutation(s) responsible for the aberrant floral phenotypes. The chapter is divided into two parts. First, we provide facile protocols for two methods of mutagenesis using EMS and T-DNA as mutagens. Second, we describe three protocols used to identify the molecular lesion responsible for the mutant phenotypes.

1.1 Mutagenesis of Arabidopsis

Due to its small genome size, fecundity, and ease of transformation, *Arabidopsis* is the ideal organism to perform genetic mutagenesis screens. First, we discuss two major, divergent agents of mutagenesis: EMS and T-DNA.

1.1.1 EMS Mutagenesis of Arabidopsis

EMS is an alkylating agent that, ~90 % of the time, induces C/G to T/A substitutions [37, 38]. At a low frequency, EMS can also generate G/C to C/G or G/C to T/A transversions by 7-ethylguanine hydrolysis or A/T to G/C transition by 3-ethyladenine pairing errors [37]. In *Arabidopsis*, the frequencies of EMS-induced stop codon and missense mutations are about ~5 % and ~65 %, respectively [39]. EMS mutagenesis allows the identification of loss- and gain-of-function mutants and can help researchers to understand the function of specific amino acids within a protein.

EMS mutagenesis tends to lead to plants containing more than one mutation (this issue will be addressed later). However, to ensure adequate coverage of the genome, meaning that one will get a mutation in every single gene, a large screening population must be obtained. A study showed that saturation could theoretically be achieved if 135,000M1 lines (five-fold coverage) are obtained [40]. In addition, the desired phenotypic output plays a role in saturation levels as well as the amount and time the seeds were exposed to EMS.

1.1.2 T-DNA Insertional Mutagenesis

Another common mode of mutagenesis is the use of *Agrobacterium-mediated* transformation via the floral dip method [41]. Transfer-DNA (T-DNA) transformation is a phenomenon by which *Agrobacterium tumefaciens* inserts a portion of its Ti plasmid into the host (plant) genome, usually causing infection. Scientists have used this natural machinery and manipulated it in order to transfer their gene of interest or in cases of random mutagenesis, a selectable marker placed in lieu of the normal T-DNA, into *Arabidopsis*. The T-DNA region is flanked by two 25-bp repeat regions called the left border (LB) and right border (RB) (*see* Fig. 2). T-DNA integration occurs through illegal recombination by utilizing the plant DNA double-stranded break repair system. Several models of integration have been proposed; however, there is a consensus that integration is random and is mediated by the LB and RB [42]. T-DNA integration is higher at gene-rich regions than centromeric regions. Moreover, in actively transcribed genes, integration is higher around the transcription initiation and termination sites than in coding regions [43–45].

Both EMS and T-DNA insertional mutagenesis have been widely used as tools to probe diverse biological pathways in *Arabidopsis*. However, there are advantages as well as caveats for each technique. EMS mutagenesis is performed on seeds, making the procedure easy to perform; however, there is a low statistical probability that any one gene will be mutated in any one plant. Therefore, a copious amount of seeds

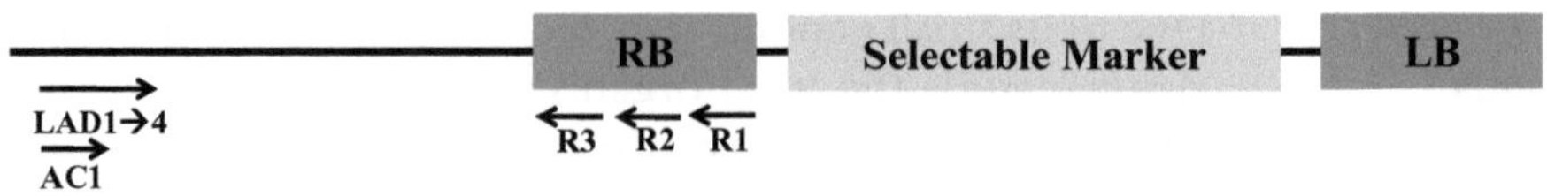

Fig. 2 Schematic diagram depicting the identification of a T-DNA insertion site. A basic schematic of the T-DNA insertion region and location of all primers used to identify the site of integration is shown. The *dark grey boxes* represent the right and left border (RB and LB, respectively). The *light grey portion* represents the region where the selectable marker can be placed. The T-DNA region (the region that will be integrated into the host genome) encompasses everything from the LB to RB (with the selectable marker). The LAD1→4 primers are each used with R1 in the four initial Tail-PCR reactions. AC1 in conjunction with R2 and R3 primers are used in subsequent nested PCR reactions

need to be processed in each mutagenesis. Furthermore, the kill rate increases exponentially with dose, while the mutation rate for any single gene rises linearly. Hence, achieving saturation in a genetic screen is quite difficult. In addition, multiple mutations may incur within a single plant, so that backcrossing is necessary to ensure the elimination of background mutations unrelated to the phenotypes of interest. One advantage of EMS mutagenesis is that it can produce viable, weak alleles in genes whose function is essential to the plant. In contrast, T-DNA mutagenesis often produces loss-of-function mutants as integration of the T-DNA usually perturbs the gene's function. T-DNA mutagenesis can also be used to overexpress genes that are located close to the T-DNA insertion if the T-DNA harbors transcriptional enhancers [46]. An advantage of T-DNA mutagenesis compared to the use of EMS is that T-DNA insertion sites can often be readily identified using PCR-based approaches. A possible disadvantage is that it can result in chromosomal rearrangements, such as inversions or deletions. Moreover, multiple insertions and complex T-DNA loci may incur [47]. In addition, not all mutations generated through T-DNA transformation are actually caused by T-DNA insertion events [47]. Linkage studies or rescue experiments are necessary to determine whether a mutant phenotype is caused by a T-DNA insertion. After mutagenesis and isolation of a mutant with the desired phenotype, the next step is to identify the mutation(s) responsible for the phenotype.

1.2 Identification of the Mutation Responsible for the Observed Mutant Phenotype

There are several experimental strategies that can be employed for the identification of a mutation, which causes a mutant phenotype. These include: map-based cloning (mainly used for EMS-induced mutations), TAIL-PCR (for T-DNA insertion mutants), as well as genome re-sequencing (for both EMS and T-DNA induced mutations).

1.2.1 Map-Based Positional Cloning

Map-based positional cloning is the most common method used to identify mutations after mutagenesis. It is a PCR-based method relying on markers, such as simple sequence length polymorphisms (SSLPs); cleaved amplified polymorphic sequences (CAPS); and derived CAPS (dCAPS). Many such markers have been described

[48], but several new markers have been designed by our laboratory, which allow a better distinction between sequences derived from *Ler* and Col-0 accessions (*see* Table 1). The underlying premise of this technique is to cross the mutant of interest into a different accession. In the F2 generation, in which the mutant phenotype is observed, one can determine where the mutation lies due to innate sequence differences between the accessions as markers in genetic mapping. This technique is often laborious and time-consuming. Theoretically, 300–400 plants are sufficient to narrow down the region where a mutated gene is located. However, in practice, thousands of plants may need to be processed.

1.2.2 Thermal Asymmetric Interlaced Polymerase Chain Reaction (TAIL-PCR)

One of the advantages of T-DNA insertional mutagenesis is that since the RB and LB sequences are known, TAIL-PCR can be performed to discover the genomic position of the T-DNA insertion. TAIL-PCR was first described in 1995 by Liu and Whittier [49]. In summary, TAIL-PCR utilizes nested primers in three consecutive reactions in order to find the insertion site [50]. Four initial PCR reactions are conducted, each with a "longer," arbitrary degenerate (LAD) primer (one each from LAD1→4) in combination with primer R1. Products from these reactions are then utilized as a template in a second round of PCR reactions with a primer termed AC1 (AC1 partially overlaps with the LAD primers, except it does not contain the degenerate sequences) and the primer R2 from the T-DNA region (*see* Fig. 2 and Table 2). A third round of PCR reactions are conducted with products from the second round as templates and the AC1/R3 primer pair (*see* Fig. 2 and Table 2). The products from the second and third rounds of PCR are resolved in an agarose gel and products containing a single band *and* with the correct band patterning (the product from the third round of PCR should run faster than that from the second round of PCR) are sequenced. The advantages of TAIL-PCR are that it is a simple, sensitive, efficient, and fast method; that it has high specificity; and that the PCR product can be directly sequenced [49]. However, the major disadvantage is that this procedure will not work if multiple copies of the T-DNA are inserted in different positions in the genome, are tandemly inverted at the same position, or if the T-DNA is only partially inserted. Thus, it is important that genetic analysis be performed to ensure single locus insertion prior to performing TAIL-PCR. However, this does not help if multiple copies are inserted within a single locus and/or the copies are inverted. Further, the vector backbone may also get transferred or the insertion may cause chromosomal rearrangements. In addition, there may not be linkage between the T-DNA and the phenotype, or the T-DNA may be partially inserted [47]. Thus, in these cases, map-based positional cloning or deep sequencing is utilized to identify the gene that, when mutated, leads to the observed phenotype.

Table 1
Oligonucleotide sequences for map-based positional cloning

	Primer name	Sequence (5′→3′)	Primer name	Sequence (5′→3′)	Position
CHR1	F21M12-F	GGCTTTCTCGAAATCTGTCC	F21M12-R	TTACTTTTTGCCTCTTGTCATTG	AT1G09880
	1CER458676F	CGTTTGAAACACCTACAGGATTAC	1CER458676R	AGCTCAAACTTAAGCATAGAAACC	AT1G27520
	F28H19-F	TGCGGGAGTGTGATAGAATA	F28H19-R	TCCTCGAAAGATTCATTGAT	AT1G43763
	F24O1-F	TCACAAAAATGCAACATTTA	F24O1-R	TAATGGCTCCAATCAATA	AT1G62370
	1CER470018F	GATCATATTCTATTGCACCCATCAG	1CER470018R	CTTACGCGTTTACTGTCTCATGTTT	AT1G71930
CHR2	T23K3-F	CGTGTTTACCGGGTCGGA	T23K3-R	AAAACCCTTGAAGAATACG	AT2G01860
	2CER449854F	GACGACTTCGAGAAAGTTACAAAAC	2CER449854R	AACGGAGTAATAACTCCAATCTCATC	AT2G13851
	T20P8-F	TCCGATTCGATTAAACTC	T20P8-R	TTATTTCCTATTTCAAGACT	AT2G27130
	T3D7-F	GGTATCGATTGAGCAAATAA	T3D7-R	ACATGCGTCTGCTTGGAG	AT2G47160
CHR3	F24P17	ACTGCACATTGCACGAACA	F24P17	GGATGGCAACTTAGGCTGAA	AT3G06400
	K1G2-F	ATGAGCTTTAGGAGTGTGTA	K1G2-R	AATTTTGTCCCAAAAGAATA	AT3G27460
	CIW4-F	GTTCATTAAACTTGCGTGTGT	CIW4-R	TACGGTCAGATTGAGTGATTC	AT3G50820
	T20O10-F	AAATGCCAGGGGAATAGA	T20O10-R	CAAACCATGCAATGATGC	AT3G62988
CHR4	CIW5-F	GGTTAAAAATTAGGGTTACGA	CIW5-R	AGATTTACGTGGAAGCAAT	AT4G01710
	CIW6-F	CTCGTAGTGCACTTTCATCA	CIW6-R	CACATGGTTAGGGAAACAATA	AT4G13575
	4CER450255F	CACAAGACAACACCAAAAAC	4CER450255R	AGAAGGAATGGCTTCATCTA	AT4G20095
	AP22-F	ATTATGTTAGGAAAATGAGAT	AP22-R	GCGTTGTAAGAATTAAGAA	AT4G36780
CHR5	T31P16-F	TCGAAGTAACTTACTTTCTA	T31P16-R	AATGTCGCAAAGACTTCC	AT5G10040
	CIW8-F	TAGTGAAACCTTTCTCAGAT	CIW8-R	TTATGTTTTCTTCAATCAGTT	AT5G22550
	ATPHYC-F	CTCAGAGAATTCCCAGAAAAATCT	ATPHYC-R	AAACTCGAGAGTTTTGTCTAGATC	AT5G35840
	MJB21-38F	CGATGCTCAGGTTCTACATT	MJB21-38R	ACTAAAATATCATCTCGTTGTA	AT5G42720
	MRB17-60F	CGAGCAAATGAATCTGAAGG	MRB17-60R	GTGATAATTCGTAAATATGGACT	AT5G54630

Table 2
Oligonucleotide sequences for TAIL-PCR

Name	Sequence (5′→3′)
LAD1	ACGATGGACTCCAGAGCGGCCGC(G/C/A)N(G/C/A)NNNGGAA
LAD2	ACGATGGACTCCAGAGCGGCCGC(G/C/T)N(G/C/T)NNNGGTT
LAD3	ACGATGGACTCCAGAGCGGCCGC(G/C/A)(G/C/A)N(G/C/A)NNNCCAA
LAD4	ACGATGGACTCCAGAGCGGCCGC(G/C/T)(G/A/T)N(G/C/T)NNNCGGT
AC1	ACGATGGACTCCAGAG
R1	cct cta gag tcg acc tgc agg cat g
R2	acg atg gac tcc agt ccg gcc aac tta
R3	tga atg gcg aat gct aga gca gct t

1.2.3 Mapping Using Deep Sequencing Technology

As mentioned above, the map-based cloning method is commonly used to identify mutations causing a phenotype. However, there are some problems that prevent this method from being used in all cases. One, if the phenotype studied is very sensitive to the genetic background, then map-based cloning would not be applicable. For example, the phenotype may become suppressed when crossed to another accession, so that F2 mutant plants cannot be identified. Also, if the mutation occurs in a region where the recombination rate is low, thousands of plants will be needed to locate the mutation, which makes the process very tedious and time-consuming. With technological advances, another method is available to identify genes from mutagenesis experiments—high-throughput DNA sequencing. Though it is presently costly, with technological advances, this procedure should become more affordable.

The identification of EMS-induced mutations responsible for the observed phenotype can be accomplished in three manners using high-throughput sequencing. Firstly, bulk analyses of an F2 mapping population can be utilized by following the same rules as map-based cloning. DNA from pooled F2 plants of the mutant phenotype is used to produce a library for deep sequencing, which provides information on the segregation of various SNPs relative to the mutation. Linkage to certain SNPs helps to narrow down the region containing the mutation. Searching for genes with point mutations in this region then helps identify candidate genes. Two pipelines, SHOREMAP (http://1001genomes.org/software/shoremap.html) and NGM (http://bar.utoronto.ca/NGM/), have been developed to locate the mutation caused by EMS using bulk analysis of an F2 mapping population [51, 52]. It is suggested to use around ~500 plants when using SHOREMAP and around 80 plants when using NGM.

Secondly, bulk analysis of an F2 backcross population can also be used to identify the mutation in question. Locating the position of the mutation responsible for the mutant phenotype relies on the fact that other EMS-generated mutations in the genome segregate according to their linkage (or not) with the mutation in question. SHOREMAP provides an algorithm to locate a mutation using an F2 backcross population. To rule out the non-causal mutations, it is suggested to sequence pooled DNA from wild-type-looking plants as well [53].

Finally, the phenotype-causing mutation can also be found through deep sequencing of the mutant itself, which has already been used in our and several other laboratories [54]. Since the rough location of the mutation is not known, there may be too many candidates. Several practices may help to filter the candidates. Backcrosses prior to sequencing can reduce the number of non-causal mutations. Preliminary rough mapping information helps to limit the causal mutation to a region. Sequencing the parental line from which the mutant was derived also helps to eliminate non-phenotype-causing mutations. Sequencing multiple mutant alleles in the same gene also helps pinpointing the gene of interest.

To identify the mutations caused by T-DNA transformation, there are two possible scenarios. For mutations caused by the T-DNA insertion, the affected gene could be easily located by the reads containing both genomic DNA and T-DNA sequences. To confidently identify such chimeric reads, it is better to sequence at 100 cycles (rather than at 50). For mutations caused by T-DNA transformation but that are not linked to a T-DNA insertion, the identification of candidate mutations is similar but more complicated than for EMS-generated mutations since three types of mutations: big INDELs (insertion and deletion), small INDELs, or SNPs (single nucleotide polymorphisms) need to be considered. To locate the region containing the mutation, the mapping can be done in the same manner as that for EMS-generated mutations. For the identification of the mutation from the region of interest, the three types of potential changes all need to be considered.

Using deep sequencing technology to find a mutation is not limited to the accessions with sequenced genomes. As long as the species has a reference genome, deep sequencing can be applied to mutants derived from un-sequenced accessions. Which methods to use and how to analyze the data depend on the situation and the specific experiment at hand. Despite these variables, the process of building the DNA sequencing libraries can be done in the same way, which will be described in this chapter.

2 Materials

2.1 Mutagenesis, Mutant Screening, and Initial Mapping of Mutations

2.1.1 EMS Mutagenesis of Arabidopsis

1. 1 mL of Arabidopsis seeds (*see* **Note 1**).
2. Twenty trays of soil with 12 pots per tray.
3. Disposable 50 mL conical tubes.
4. Parafilm.
5. 5 M NaOH. Prepare 0.5 M NaOH by mixing 146.1 g of NaOH with 400 mL of water in a beaker containing a mag-

netic stir bar. Stir until dissolved, and bring volume up to 500 mL with water. Autoclave the solution for 25 min.

6. Tween-20 solution: 0.1 % (v/v) Tween-20. Prepare 50 mL using autoclaved H2O.
7. Ethyl methanesulfonate (EMS) solution: 0.2 % EMS (e.g., from Sigma-Aldrich, M0880) in autoclaved H2O (*see* **Note 2**).
8. 0.1 % agar. Add 0.9 g of agar to 90 mL of autoclaved H2O. Microwave to dissolve the agar and leave to cool at room temperature prior to use (*see* **Note 3**).

2.1.2 Planting EMS Mutagenized M1, M2, and Mapping Population Seeds

1. Soil.

2.1.3 Preparing DNA for Map-Based Positional Cloning

1. Toothpicks and tape for labeling.
2. Mortar and pestle.
3. Liquid nitrogen.

2.1.4 CTAB DNA Extraction

1. CTAB extraction buffer: 2 % (w/v) cetyltrimethylammonium bromide, 1.4 M NaCl, 20 mM EDTA, 0.2 % (v/v) β-mercaptoethanol, 200 mM Tris–HCl pH 8.0. For 500 mL buffer: 10 g CTAB, 100 mL Tris–HCl pH 8.0 (to make a 500 mL 1 M Tris–HCl stock solution, add 60.55 g Tris base to 400 mL of H_2O, add ~21.1 mL concentrated HCl (pH 8.0), add water to a final volume of 500 mL and autoclave for 25 min), 40.95 g of NaCl, 20 mL of 0.5 M EDTA (to make 500 mL 0.5 M EDTA stock solution, add 84.05 g EDTA to 250 mL H_2O, add 5 M NaOH (~71 mL) slowly while stirring until the EDTA dissolves, pH should be 8, bring up to volume with H_2O and sterilize by autoclaving for 25 min). Prior to use, calculate how much CTAB buffer you will need, transfer that volume to another tube and add 1/500 volume of β-mercaptoethanol (*see* **Note 4**).
2. Chloroform.
3. Isopropanol.
4. Nanodrop spectrophotometer or any means to quantify the DNA.

2.1.5 Quick and Dirty PCR

1. Quick and Dirty PCR Extraction Buffer: 200 mM Tris–HCl pH 7.6 (to make a 500 mL 1 M Tris–HCl stock solution, add 60.55 g Tris base to 400 mL of H2O, add ~28.5 mL concentrated HCl (pH 7.6), add water to final volume of 500 mL and autoclave for 25 min), 250 mM NaCl, 25 mM EDTA, 0.5 % (w/v) SDS (to make a 10 % (w/v) SDS stock solution, add 50 g in 500 mL of distilled water, stir overnight) (*see* **Note 5**).
2. Small pestles for grinding tissue in 1.5 mL centrifuge tubes.

2.1.6 T-DNA Insertional Mutagenesis: Agrobacterium-Mediated Transformation

1. Flowering *Arabidopsis thaliana* (age is dependent on growth conditions) (*see* **Note 6**).
2. Electrically competent *Agrobacterium tumefaciens* cells.
3. Plant transformation vector containing the T-DNA construct of interest.
4. Plates for plant growth, and antibiotics/herbicides needed for selection of transgenic lines.
5. 5 % (w/v) sucrose solution.
6. Silwet L-77.
7. Trays for floral dipping.
8. Spectrophotometer or other method of measuring cell densities.
9. MS media, LB media, and agar or other growth media.
10. Dark area and/or cover to keep plants away from bright light.
11. Gene Pulser™ (Bio-Rad) or another suitable instrument for *Agrobacterium* transformation, and corresponding electroporation cuvettes.

2.2 Pinpointing the Mutation that Causes the Phenotype

2.2.1 Map-Based Positional Cloning: PCR

1. 3–4 % agarose gels (*see* **Note 7**).
2. 10× PCR buffer: 15 mM MgCl2, 500 mM KCl, 100 mM Tris–HCl (pH 8.3).
3. 2.5 mM dNTPs.
4. Taq DNA polymerase (*see* **Note 8**).
5. 10× DNA loading buffer: 3.9 mL glycerol, 500 μL 10 % (w/v) SDS, 200 μL 0.5 M EDTA, 0.025 g bromophenol blue, 0.025 g xylene cyanol, bring to 10 mL final volume with distilled H_2O.

2.2.2 TAIL-PCR

1. Ex-Taq Polymerase kit (Takara).
2. Gel DNA Recovery Kit (Zymo cat#D4001).

2.2.3 Map-Based Cloning via Deep Sequencing

1. 5 μg RNase-treated, high-quality genomic DNA (*see* **Note 9**).
2. Biorupter (Diagenode, #UCD-200 TO).
3. E-Gel® iBase™ Power System (Invitrogen, G6400).
4. E-Gel® 1.2 % General Purpose Agarose 18-Pak (Invitrogen, G501801).
5. E-Gel® 1 kb Plus DNA Ladder (Invitrogen, 10488-090).
6. Agencourt AMPure XP 60 mL kit (Beckman Coulter Genomics, # A63881).
7. TruSeq DNA Sample Preparation kit (Illumina, FC-121-2001).

8. Magnetic stand-96 (Invitrogen, # AM10027).
9. 96-well thermal cycler with heated lid.
10. 96-well 0.3 mL DNase-free PCR plate.
11. Ethanol (200 proof).

3 Methods

3.1

3.1.1 EMS Mutagenesis of Arabidopsis

1. Place seeds (100 μL; corresponding to ~1,000 seeds) in a 50 mL conical tube and wash in 30 mL of 0.1 % (v/v) Tween-20 for 15 min at room temperature (RT). Keep the tube shaking constantly (*see* **Note 10**).
2. Spin down at 400 × *g* for 1 min at RT.
3. Remove as much solution as possible first using a transfer pipet and then a PIPETMAN.
4. Add up to 15 mL of autoclaved water.
5. Add 30 μL of EMS to the tube and wrap the cap with Parafilm (*see* **Note 11**).
6. Mix by inverting the tube several times, and incubate at RT (while shaking) for 8–12 h.
7. Spin the tube at 400 × *g* for 1 min at RT.
8. Remove the EMS solution, and put it in a separate 50 mL conical tube.
9. Add 5 M NaOH to the tube so that the final concentration is 0.5 M NaOH. Put Parafilm on the tube to seal it and incubate, shaking at RT, overnight. Discard the Parafilm and solution into a biohazardous waste container.
10. Rinse the seeds ten times in 20 mL autoclaved water each (*see* **Note 12**).
11. After the last rinse, add 10 mL of autoclaved water and incubate the seeds at RT, while shaking, for 2–4.5 h.
12. Pour off the water, add an additional 10 mL of water to the tube, and using a transfer pipet, transfer water and seeds to a flask containing 90 mL of 0.1 % agar (*see* **Note 13**).
13. Mix and sow the seeds at 500 μL per pot with a PIPETMAN using a 1,000 μL tip.
14. Cover the trays of soil with a lid and leave at 4 °C for 1 week.
15. Transfer the trays to the growth chamber and grow at 23 °C under continuous light.
16. After the plants have developed two leaves, remove the lid (*see* **Note 14**).

3.1.2 Planting EMS Mutagenized M1, M2, and Mapping Population Seeds

1. As the M1 plants are growing, check the plants and look for dominant mutations. If a dominant mutant is identified, proceed to **step 10**.
2. Harvest M2 seeds from the individual M1 plants. Place seeds from each plant into a separate tube (*see* **Note 15**).
3. Poke a small hole in the lid of the tube and leave the seeds to dry for 1 week at RT.
4. Plant the seeds one by one (24–30 seeds per pot), one M2 family for one pot.
5. Place trays at 4 °C for 1 week so the seeds can stratify.
6. Transfer the trays to the growth chamber at 23 °C under continuous light (*see* **Note 16**).
7. When the plants grow to the two-leaf stage, remove the lid.
8. As the plants flower, assay the plants for the desired floral phenotype (either enhancer or suppressor).
9. Once a mutant phenotype is observed, note the segregation ratio of that phenotype in the pot (*see* **Note 17**).
10. Backcross the mutant plants to the parental line at least twice before any molecular analysis is performed (*see* **Note 18**).
11. Cross the mutant plant to another plant (from a different ecotype) to generate the mapping population (*see* **Note 19**).
12. Generate a "Het" (heterozygous) sample as a control. For example, if the screen was performed in the L*er* background, cross the mutant plant with another accession (that you will use for mapping, such as Col-0). With the F1 plants, collect the tissue (several leaves) and extract DNA via the CTAB method or the "quick and dirty method" (for Quick and Dirty PCR prep) (*see* **Note 20**). This sample will be included in subsequent mapping reactions as a control.
13. Using the F2 seeds [from the cross with another accession (same accession that you chose for **step 12**)], plant (24 seeds per pot) the mapping population. Keep in mind that you will, on average, get only 1 out of 4 plants that exhibit the phenotype. For a population of 50 mutant plants, which is needed for a rough mapping of the mutation, at least 200 seeds need to be planted.
14. Place trays at 4 °C for 1 week so the seeds can stratify.
15. Transfer trays to the growth chamber at 23 °C under continuous light.
16. When the plants grow to the two-leaf stage, remove the lid.

3.1.3 Preparing DNA for Map-Based Positional Cloning

1. Within the F2 mapping population, look for plants with your desired phenotype. If it is a dominant mutation, look for and perform the map-based positional cloning on plants without the desired phenotype (*see* **Note 21**).

2. Label the plants with toothpicks, and number them. For the rough mapping, 50 plants are needed, which can be pooled (put small samples from many plants in one tube and isolate DNA at the same time). For the pooled method, CTAB DNA extraction is performed for rough mapping. Otherwise, use the Quick and Dirty method on individual plants. For fine mapping, collect tissue from each plant separately, using the Quick and Dirty method.
3. Collect one leaf and place it either in a 15 mL conical tube for the CTAB DNA extraction (in liquid nitrogen) or individually in 1.5 mL centrifuge tubes.
4. Remember to collect tissue from both ecotypes for use as controls (*see* **Note 22**).

3.1.4 CTAB DNA Extraction

1. Cool the mortar and pestle in liquid nitrogen.
2. Add frozen tissue to the cold mortar and pestle.
3. Grind the tissue until the powder is light green.
4. Add the tissue to 10 mL of CTAB extraction buffer (with β-mercaptoethanol added) and invert tube several times (*see* **Note 23**).
5. Place the tubes at 65 °C for 40–60 min. Label two sets of tubes while waiting.
6. Centrifuge for 7 min at 11,648 × *g* at RT.
7. Transfer the supernatant into a new 15 mL tube.
8. Add an equal volume of chloroform and vortex for 10 s (*see* **Note 24**).
9. Centrifuge for 15 min at RT at 11,648 × *g*.
10. Transfer the upper phase to a new tube (*see* **Note 25**).
11. Add an equal volume of cold isopropanol (*see* **Note 26**).
12. Place the tubes in −80 °C for 30 min or −20 °C overnight (*see* **Note 27**).
13. Centrifuge tubes for 30 min at 4 °C at 11,648 × *g*.
14. Decant the supernatant into a liquid waste container.
15. Wash pellet twice with 4 mL of 70 % EtOH (*see* **Note 28**).
16. Invert the tube and leave the tube in the hood to air-dry (*see* **Note 29**).
17. Once the pellet is dry, add 100 μL of autoclaved water for resuspension (*see* **Note 30**).
18. If you wish to remove RNA, add RNase A to a final concentration of 20 μg/mL and incubate at 37 °C for 1 h. If not, continue with **step 29**.
19. Add water to a final volume of 300 μL.

20. Add 150 μL phenol (high pH) and 150 μL chloroform.
21. Briefly vortex.
22. Spin in a tabletop microfuge at 16,100 × *g* for 15 min.
23. Transfer supernatant to a new, clean tube and add an equal volume of isopropanol.
24. Leave at −80 °C for 30 min (*see* **Note 31**).
25. Spin in a tabletop microfuge at 16,100 × *g* for 25 min at 4 °C.
26. Wash twice with 70 % EtOH, as in **step 15**.
27. Invert and leave the tube to air-dry in the hood.
28. Resuspend in 100 μL autoclaved water, as in **step 17**.
29. Quantify the DNA.
30. Store DNA at −20 °C (*see* **Note 32**).

3.1.5 Quick and Dirty Extraction

1. Grind tissue in an 1.5 mL Eppendorf tube (*see* **Note 33**).
2. Add 400 μL Quick and Dirty extraction buffer to the tube.
3. Grind a little more until you do not see chunks of tissue.
4. Spin in a tabletop microfuge at 16,100 × *g* for 5 min at RT.
5. Transfer the supernatant to a new tube (*see* **Note 34**).
6. Add an equal volume of isopropanol.
7. Leave the tubes at RT for 15 min (*see* **Note 35**).
8. Spin down the samples in a tabletop microfuge for 10 min at RT at 16,100 × *g*.
9. Decant the supernatant.
10. Wash pellet twice with 150 μL 70 % EtOH.
11. For the last wash, spin the tubes in a tabletop microfuge for 5 min at RT at 16,100 × *g*.
12. Decant the supernatant and leave the inverted tubes/plates to dry in the hood.
13. Add 50 μL of autoclaved water to the pellet.
14. Leave DNA at RT for 1 h if performing the PCR on the same day. Use 1 μL for a PCR reaction with a 10 μL final volume.
15. Store the DNA at −20 °C.

3.1.6 T-DNA Insertional Mutagenesis: Transformation and Selection of Agrobacterium

1. Add 0.5 mL (2–10 ng) plasmid to 50 mL of electrically competent *Agrobacterium* cells on ice (*see* **Note 36**).
2. Cool a 0.2 cm electroporation cuvette (Bio-Rad) on ice.
3. Add the entire plasmid and competent cell mixture to the cuvette.
4. Place the electroporation cuvette in the Gene Pulser™ (Bio-Rad). Make sure to select the setting for bacteria, and press the pulse button. Remove the cuvette from the machine and place it on ice.

5. Pipette the mixture carefully out of the cuvette and into a 1.5 mL tube containing media (LB or other). This should be done in the flow hood to help reduce contamination.
6. Shake for 1–3 h in a 28 °C shaker.
7. Pipette the mixture carefully out of the tube and onto a selection plate containing the appropriate media (LB or other) and antibiotic. Make sure the mixture is spread evenly on the surface of the plate. Let it air-dry in the flow hood.
8. Incubate this plate overnight at 28 °C or until colonies form (*see* **Note 37**).
9. Prepare a test tube of 5 mL media of choice and antibiotic. Inoculate a colony into the liquid media using a pipet tip by touching the colony. Then, dip the tip end into the media, pipetting if desired. The tip can also be placed into the media if necessary.
10. Incubate the 5 mL tube on a 28 °C shaker overnight.
11. The presence of the desired DNA construct can be verified by colony PCR (*see* **Note 38**).
12. Store *Agrobacterium* if needed (*see* **Note 39**).

3.1.7 T-DNA Insertional Mutagenesis: Preparation of Agrobacterium

1. The next morning, pour the 5 mL *Agrobacterium* culture into 500 mL (1:100) LB, or other growth media. Make sure to add the antibiotics for selection. Grow at 28 °C until OD600 = 0.8 (*see* **Note 40**).
2. Spin down the *Agrobacterium* (5,500 × g, 15 min at RT) and resuspend in a 5 % sucrose, 0.5× MS solution, pH 5.7. For example, if the OD_{600} = 0.8 in 500 mL LB, the OD_{600} will also be 0.8 in 500 mL of 5 % sucrose solution. However, only about 200–300 mL of solution is needed for dipping of 1–3 trays of Arabidopsis (*see* **Note 41**).
3. Add 0.05 % (v/v) of Silwet L-77 to the solution and mix well (*see* **Note 42**).
4. Pour the solution into a low, flat tray. This solution does not need to autoclaved if used quickly.

3.1.8 T-DNA Insertional Mutagenesis: Floral Dip

1. Dip the shoots of plants into the solution, and let sit for 5 min. Do not let the soil touch the solution (*see* **Note 43**).
2. If transforming L*er* plants, you must put the solution and pot (from **step 1**) into a vacuum for 5 min. Release the vacuum slowly after 5 min. This step can be skipped for Col-0 plants, which are easier to transform.
3. Spray the shoots of the plants with water, using a spray bottle, for several seconds to remove the Sucrose solution.
4. Cover the plants with a plastic dome or other covering for about 24 h. Plants are usually stored on their sides during this time.

5. Return plants to the upright position and grow them until they are ready for seed harvesting.
6. Collect seeds from the plants and seeds from many plants can be pooled.

3.1.9 T-DNA Insertional Mutagenesis: Selection of Transformants

1. Grow the seeds on soil and spray with herbicide such as ammonium-glufosinate ("BASTA") to select transformants. Remember to stratify the seeds for 2–5 days after planting.
2. To select transformants with an antibiotic, prepare plates of 0.5× MS and 0.6 % agar with the antibiotic added in the correct amount. Grow under continuous light for 7–10 days.
3. Positive transformants will appear larger than seedlings that were not transformed.
4. Transfer tranformants to soil, and collect seeds from either a single plant or a few plants (5–50).

3.2

3.2.1 Map-Based Positional Cloning: PCR

1. The PCR reaction mix is as follows for a 10 μL reaction: 1 μL of Quick and Dirty DNA (or 50 ng of CTAB DNA), 0.1 μL each of 10 mM forward and reverse primer, 0.4 μL of 2.5 mM dNTPs, 1 μL 10× PCR buffer (contains 18 mM MgCl2), 0.1 μL of Taq DNA polymerase, and 7.3 μL of autoclaved water (*see* **Note 44**).
2. The PCR reaction conditions are as follows: step 1: 94 °C for 3 min; step 2: 94 °C for 30 s; step 3: 54 °C for 30 s; step 4: 72 °C for 30 s; repeat steps 2–4 for an additional 34 times; step 5: 72 °C for 10 min.
3. Run the PCR product on a 3 % or 4 % agarose gel. Make sure to add EtBr before pouring the gel since EtBr is not easily absorbed into a higher percentage gel after the gel is made. Be careful when handling EtBr, it is a mutagen. The lanes should be set up as follows: pooled DNA from mutant plants; L*er*; Col-0; DNA from F1 plants or L*er* and Col DNA mixed in a 1:1 ratio.

3.2.2 Map-Based Positional Cloning: Analysis

1. For the rough mapping via the pooled method, if the mutant is in the L*er* background, one should look for the set of primers that show a predominant L*er* pattern (*see* **Note 45**).
2. Once the region that is linked to the mutation is identified, fine mapping can be started.
3. Isolate the DNA individually for each mutant (as described above).
4. Go to http://www.arabidopsis.org. Click "sequence viewer" and identify BACs flanking the region in which the mutation is linked.
5. Find or design SSLP, CAPS, or dCAPS markers in the BAC. One useful resource is http://amp.genomics.org.cn/, which

Table 3
Example of analysis of F2 mapping population

	Marker 1	Marker 2	Marker 3	Ciw8	Marker 4	Marker 5
Plant 1	Het	L*er*	L*er*	L*er*	L*er*	L*er*
Plant 2	Het	Het	L*er*	L*er*	L*er*	Het
Plant 3	Het	L*er*	L*er*	L*er*	Het	Het
Plant 4	Het	Het	Het	L*er*	Het	L*er*

lists such markers. Alternatively, use the SNP information from http://signal.salk.edu/atg1001/3.0/gebrowser.php. Design primers that flank the SNPs (*see* **Note 46**).

6. Run a PCR reaction and resolve the product on an agarose gel.
7. For the analysis, find the region flanked by recombinant markers. An example is shown in Table 3. Assume ciw8 was the marker to which the mutant was linked. Markers 1–5 are in the vicinity of the linked region. Results from plant 1 show that the mutation is below Marker 1. Results from plant 2 tell us that the mutation is between Marker 2 and Marker 5. Results from plant 3 tell us that the mutation is between Marker 1 and 4. Results from plant 4 tell us that the mutation is between Marker 3 and 4. In summary, the mutation is between Marker 3 and Marker 4. The mutation is localized to the region flanked by two markers, one on each side, with the least number of recombinants in the population.
8. Once a region of 200 kb or less is identified, go to http://www.arabidopsis.org, find the region in Seqviewer, and scan the region to see if there are any genes of interest. If so, sequence the gene from the mutant to find any mutations. To narrow down the region further, more plants will be needed for fine mapping.
9. If a mutation is identified, complementation analysis should be performed.

3.2.3 TAIL-PCR

The TAIL-PCR is comprised of three separate PCR reactions as outlined below.

1. The first reaction is the pre-amplification reaction. *see* Table 4 for the set-up (*see* **Note 47**).
2. The PCR conditions for the first reaction are shown in Table 5.
3. The second reaction is the Primary TAIL-PCR reaction. *see* Table 6 for the set-up (*see* **Note 48**).
4. The PCR condition for the second reaction is shown in Table 7.

Table 4
TAIL-PCR set-up for the pre-amplification reaction

Reagent	Final concentration	Amount to add (μL)
H_2O		12.7
10× PCR buffer		2
dNTP	200 μM	1.6
LAD 1–4 primer	1 μM (10 μM stock)	2
R1 primer	0.3 μM (10 μM stock)	0.6
ExTaq	0.5 units	0.1
DNA template	50 ng	1
Total		20

Table 5
TAIL-PCR conditions for the pre-amplification reaction

Step	Temperature (°C)	Time (min:sec)
1	93	2:00
2	95	1:00
3	94	0:30
4	60	1:00
5	72	3:00
6	Go to step 3	10 times
7	94	0:30
8	25	2:00
9	Ramping to 72	0.5 °C/s
10	72	3:00
11	94	0:20
12	58	1:00
13	72	3:00
14	Go to step 11	25 times
15	72	5:00
16	End	

5. The third reaction is the Secondary TAIL-PCR reaction. *see* Table 8 for the set-up (*see* **Note 49**).
6. The PCR condition for the third reaction is shown in Table 9.

Table 6
TAIL-PCR set-up for the primary reaction

Reagent	Final concentration	Amount to add (μL)
H_2O		17.88
10× PCR buffer		2.5
dNTP	200 μM	2
AC1 primer	0.3 μM (10 μM stock)	0.75
R2 primer	0.3 μM (10 μM stock)	0.75
ExTaq	0.6 units	0.12
DNA template	1/40 dilution of the pre-amplification product	1
Total		25

Table 7
TAIL-PCR conditions for the primary reaction

Step	Temperature (°C)	Time (min:sec)
1	94	0:20
2	65	1:00
3	72	3:00
4	Go to step 1	1 time
5	94	0:20
6	68	1:00
7	72	3:00
8	94	0:20
9	68	1:00
10	72	3:00
11	94	0:20
12	50	1:00
13	72	3:00
14	Go to step 5	13 times
15	72	5:00
16	End	

Table 8
TAIL-PCR set-up for the secondary reaction

Reagent	Final concentration	Amount to add (μL)
H_2O		35.8
10× PCR buffer		5
dNTP	200 μM	4
AC1 primer	0.3 μM (10 μM stock)	1.5
R3 primer	0.3 μM (10 μM stock)	1.5
ExTaq	1 unit	0.2
DNA template	1/10 dilution of the primary reaction PCR product	1
Total		50

Table 9
TAIL-PCR conditions for the secondary reaction

Step	Temperature (°C)	Time (min:sec)
1	94	0:20
2	68	1:00
3	72	3:00
4	94	0:20
5	68	1:00
6	72	3:00
7	94	0:20
8	50	1:00
9	72	3:00
10	Go to step 1	7 times
11	72	5:00
12	End	

7. Run the Primary and Secondary TAIL-PCR reactions on a 1.5 % agarose gel.
8. Choose the PCR product with clean bands and a proper ladder effect (*see* **Note 50**).
9. Weigh an empty tube, and record the weight (*see* **Note 51**).

10. Take an image of the gel (*see* **Note 52**).
11. Put the gel on top of plastic wrap, and place it on top of a UV light source (*see* **Note 53**).
12. Cut out the band (*see* **Note 54**).
13. Place the band into the pre-weighed 1.5 mL centrifuge tube.
14. Weigh the tube, and subtract the weight of the tube without the gel (*see* **Note 55**).
15. Multiply the weight of the agarose by three. Add that volume (in μL) of ADB buffer (*see* **Note 56**).
16. Leave the tube at 37–55 °C for 5–10 min (*see* **Note 57**).
17. Transfer the solution containing the melted gel to a Zymo-Spin™ Column in the collection tube.
18. Leave the solution in the column for 3 min.
19. Spin the column in a tabletop microfuge for 30 s at RT at 16,100 × *g*.
20. Discard the flow-through.
21. Add 200 μL of wash buffer to the column (*see* **Note 58**).
22. Let the column sit at RT for 3 min.
23. Centrifuge for 30 s at 16,100 × *g* in a tabletop microfuge. Discard the flow-through.
24. Repeat the wash step.
25. Centrifuge the empty column for 1 min at 16,100 × *g* in a tabletop microfuge.
26. Transfer the column to a new, labeled centrifuge tube.
27. Add 25 μL water to the column (*see* **Note 59**).
28. Let the column sit at RT for 3 min.
29. Centrifuge the column for 2 min at 16,100 × *g* in a tabletop microfuge.
30. Quantify the DNA, and run ~2–5 μL of DNA on the agarose gel to make sure there is a clean, single band.
31. Send the required amount to the sequencing facility per their specifications.

3.2.4 Map-Based Cloning via Deep Sequencing: Library Preparation

1. Obtain 5 μg RNase-treated genomic DNA via the CTAB method above with the exception of resuspending in 50 μL autoclaved water at the final step.
2. Use the bioruptor to sonicate the DNA. Perform two replicates of 15-min sonication with 30 s on at maximum speed and 30 s off (*see* **Note 60**).
3. Recycle the DNA between 200 and 300 bp through the E-gel system (*see* **Note 61**).

4. Build a DNA sequencing library using the TruSeq DNA Sample Preparation kit.
5. Sequence the library using the Illumina Genome Analyzer (*see* **Note 62**).

4 Notes

1. Since this protocol is written for modifier screens, use a mutant allele with a weak phenotype to identify enhancers and a mutant allele with a stronger phenotype to identify suppressors. A seed volume of 100 μL corresponds to ~1,000 seeds.
2. EMS is a highly hazardous and volatile compound. Wear double gloves, a lab coat, goggles, and closed-toe shoes when working with this compound. In addition, make sure to prepare all the solutions inside a fume hood. Decontaminate everything that has come in contact with EMS or is touched while working with EMS by washing it with 1 M NaOH and discard waste into the appropriate biohazard container.
3. Suspending the seeds in 0.1 % agar will allow for better dispersal of the seeds.
4. All reagents must be added in the order listed. EDTA will not dissolve quickly, so adding the NAOH slowly will allow the EDTA to dissolve when the pH reaches 8.0. Before using the solution, make sure it has cooled down after autoclaving. Make sure to add the β-mercaptoethanol in a hood.
5. SDS is potentially harmful to the respiratory system. Thus, wear a mask and weigh SDS in the hood.
6. It is optional to clip off the first bolts of flowering *Arabidopsis* in order to promote more secondary bolts. Four to six days after clipping, plants can be transformed. Since immature flowers are the organs transformed, plants can be dipped multiple times (2–3 times).
7. Depending on the size difference between the two bands, 3 % or 4 % gels can be used. The gels can be used up to four times by melting and re-pouring.
8. Any *Taq* DNA polymerase should be suitable.
9. Although the DNA will be fragmented afterwards, DNA integrity is important in order to prepare high quality DNA sequencing libraries.
10. The purpose of using Tween-20 is to allow the EMS to penetrate the seeds more easily. Therefore, this protocol requires less EMS than other protocols.
11. Wrapping with Parafilm helps prevent contamination.

12. Rinsing a copious amount of times helps to get rid of the EMS. Treat the water from the rinse by adding NaOH (to get a 0.5 M final concentration) using the 5 M stock solution.
13. The agar helps to suspend the seeds in the solution more evenly, making planting much easier. Remember to allow the 0.1 % agar to cool prior to adding the seeds.
14. The clear, dome-shaped lid maintains moisture to promote germination.
15. Dominant mutations are rare, so most plants will not exhibit an abnormal phenotype. Seeds are collected from individual plants so that mutations in the M2 can be easily maintained in the heterozygous situation if the homozygous plant is sterile.
16. Keeping plants under continuous light allows them to grow faster. Flowering can be accelerated by growing plants under continuous light, which will expedite the genetic screen aimed at the isolation of floral mutants. In addition, the plants should be grown in their own section so as to avoid contamination by other plants in the growth chamber.
17. In the M1 plant, usually only one of two cells that give rise to the germ line gets mutated. The segregation ratio for a recessive mutation in the M2 should be close to 1 out of 8.
18. This is very important as molecular analyses of the mutant should only be performed on a clean line.
19. For example, if our mutant were in the L*er* background, we would cross it to Col-0 to generate the mapping population. Oftentimes, mutants have a stronger or weaker phenotype in the Col-0 background. Thus, there could be an enhancer or suppressor in Col-0, so one could possibly map the modifier. Also, if the mutagenesis is performed in a given mutant background (such as *ag-10*), it is a good idea to introgress the mutation into a different ecotype several times as the mapping parental line (in our case, we crossed *ag-10* to Col-0 six times and used the *ag-10*$^{Col\text{-}0}$ as the other parent to generate the mapping population). This is important, as we want to keep two copies of *ag-10* in the subsequent population so that we are able to see the "enhancer" phenotype in a reasonable proportion in the F2 population.
20. This DNA is used as a control for future mapping experiments (called Het hereafter). Collect at least four samples for each method because a lot is used for positional cloning.
21. Positional cloning can also be used to map dominant mutations. In the F2 population, the plants with no mutant phenotypes should be wild type at both alleles, and thus are used for mapping.

22. For simplicity purposes, the controls will be referred to as *Ler* and Col-0. For pooled tissue, make sure the leaves are about the same size for each plant. The tissue can be stored at −80 °C (for the pooled tissue) or 4 °C (for the Quick and Dirty PCR) for several days prior to using.
23. The mixture should be dark green and viscous but not too thick. If it is too thick, the solution is hard to move when inverted. In this case, split the samples into different tubes and add more extraction buffer. If CTAB extraction is being performed on one plant (several leaves), no liquid nitrogen is needed. Use a blue pestle (for microcentrifuge tubes) and grind the tissue in a 1.5 mL centrifuge tube. All spins can be done in a tabletop microfuge at 16,100 × *g*.
24. The solution should be cloudy.
25. Be careful not to jostle the tube to disrupt the separation. If the tube is jostled, re-spin. This phase should be clear.
26. Be careful not to transfer any chloroform at this step.
27. Theoretically, the DNA can be saved at −20 °C indefinitely as it is stable; however, it is best to use the DNA as soon as possible.
28. For the first wash, add 4 mL of 70 % EtOH very slowly and invert carefully so as to avoid dislodging the pellet. If the pellet is dislodged, spin for 5 min and wash again. If you do not dislodge the pellet, decant 70 % EtOH, add more 70 % EtOH, and invert the tube vigorously. Spin down the tube at 11,648 × *g*, 4 °C for 5 min.
29. Since the pellet may disappear after drying, make a mark on the outside of the tube to show the pellet's location. That way, when the pellet is resuspended, autoclaved water can be added up to that point.
30. Use a tip to help resuspend the pellet. Warming up the water (37 °C) also helps with the resuspension. If there was a small amount of tissue used, resuspend in 30 μL of autoclaved water.
31. The solution can also be left at −20 °C for a few hours to overnight. If left overnight, add a 1/10 volume 3 M sodium acetate and 1 μL glycogen, as a carrier, to the solution.
32. If the DNA is used the same day, leave the tubes at room temperature for 1 h.
33. Keep elbow above the wrist when grinding. This will allow better grinding without using that much energy and/or causing wrist injury. When performing extraction for hundreds of plants, this helps.
34. If hundreds of preps are being performed, it might be easier to use the plate method. In this case, add 140 μL of the supernatant to each well of a 96-well PCR plate, which contain 140 μL

of isopropanol. Because this leads to the well being almost completely full, be careful to avoid cross-contaminations between individual wells. As a precaution and to allow you to repeat the experiment in case of a contamination, keep the tubes with the leftover supernatant.

35. Do not leave the tubes in the freezer or at 4 °C as this will result in DNA preparations with a high level of impurities.
36. There is an optional step of keeping the mixture of competent cells and plasmid on ice 30 min prior to electroporation.
37. Colonies should form within a couple days. If they do not form, the transformation was not successful and should be repeated. These plates can be stored at 4 °C for a few months at a time.
38. To make sure a positive colony is obtained, multiple colonies can be inoculated into test tubes (5 mL media with antibiotics). Multiple colonies can also be verified using colony PCR to save time.
39. To store bacteria, inoculate another 5 mL media and antibiotics in a test tube with a positive colony. Grow overnight in a 28 °C shaker. Pipette 150 μL sterile glycerol into a 1.5 mL plastic tube. Pipette 850 μL bacterial culture into the same tube and mix well. It will take some effort to mix the two completely. Label the tube and flash freeze it in liquid nitrogen. Store in −80 °C until needed for future transformations.
40. This OD_{600} requirement is not strict. Generally, 0.6–0.8 is used. However, as long as one can suspend the *Agrobacterium* to about an $OD_{600} = 0.8$ in a solution of 5 % sucrose, then efficiency of transformation will not be adversely effected. The time will be about 18–24 h of incubation. Alternatively, the *Agrobacterium* can be grown to 1.0–1.2 and diluted to 0.8.
41. Each tray is sized 21″ × 11″ × 2″ with 12 pots, in which multiple plants of the same genotype are grown.
42. Standard is 0.05 % (v/v) Silwet L-77, but as low as 0.005 % (v/v) can be used successfully. Less should only be used when toxicity is a risk to plants.
43. If the solution touches the soil, the high sugar concentration will change soil pH, leading to unhealthy plants.
44. Make sure to run both ecotypes and 1:1 mix of the two accessions as controls for all reactions.
45. If the mutant is in the Col-0 background, identify the marker that is linked to Col-0.
46. Go to the markers tab and insert BAC number. A set of primers will be displayed. The most commonly used polymorphism information between Col-0 and L*er* is http://www.arabidopsis.org/browse/Cereon/index.jsp; however, the one mentioned in the actual protocol is more comprehensive.

47. There will be a total of four different PCR tubes for this reaction (one for each one of the LAD primers).
48. Once again, there will be a total of four different PCR tubes for this reaction (one for each one of the LAD primers).
49. Once again, there will be a total of four different PCR tubes for this reaction (one for each one of the LAD primers).
50. The band from the Secondary PCR reaction should run faster than the band from the Primary PCR reaction. As the product will be used for sequencing, make sure everything is clean. For instance, clean the gel apparatus, use new running buffer, etc.
51. Use a balance that measures the weight to the precision of at least 1 mg.
52. Do this as fast as possible as long exposure to the UV light will induce thymine dimers that affect subsequent PCR reactions.
53. Make sure to wear a UV mask to protect eyes and skin from burns.
54. Make sure to use a new razor blade and get rid of as much agarose as possible.
55. This is to find the weight of the gel containing the band.
56. For example, if the weight of the agarose band is 100 mg, add 300 μL of ADB buffer.
57. Every few minutes, briefly vortex the tube to help the agarose dissolve. Running the PCR product on a higher percentage gel will cause the gel to be harder to dissolve. It may take longer than 10 min for the gel to fully dissolve.
58. Make sure to add the required amount of ethanol to the wash buffer prior to using.
59. Be careful not to let the tip touch the column.
60. The genomic DNA can be fragmented by any kind of sonication method as long as the DNA is sheared to 200–300 bp. Run a 1 % agarose gel to check. A smear is expected, but the majority of the band intensity should be between 200 and 300 bp.
61. This step can be skipped if the gel method is used to purify the ligation product before enrichment of the DNA fragment when preparing the DNA sequencing libraries. Also, this can be done through regular gel purification methods (as mentioned above with the Zymo kit). If purifying the DNA at this step, there is no need to purify the ligated DNA during the next part of the procedure.
62. Multiple samples can be barcoded in a single lane to lower costs of sequencing. However, the coverage for the genome should be higher than 10.

References

1. Laibach F (1943) Arabidopsis thaliana (L.) Heynh als objekt fur genetische und entwicklungsphysiologische Untersuchungen. Bot Archiv 44:439–455
2. Leutwiler LS, Hough-Evans BR, Meyerowitz EM (1984) The DNA of Arabidopsis thaliana. Mol Gen Genet 194:15–23
3. Meyerowitz EM (1989) Arabidopsis, a useful weed. Cell 56(2):263–269
4. Meyerowitz EM, Pruitt RE (1985) Arabidopsis thaliana and plant molecular genetics. Science 229(4719):1214–1218
5. Pruitt RE, Meyerowitz EM (1986) Characterization of the genome of Arabidopsis thaliana. J Mol Biol 187(2):169–183
6. Bowman JL, Meyerowitz EM (1991) Genetic control of pattern formation during flower development in Arabidopsis. Symp Soc Exp Biol 45:89–115
7. Bowman JL, Smyth DR, Meyerowitz EM (1989) Genes directing flower development in Arabidopsis. Plant Cell 1(1):37–52
8. Coen ES, Meyerowitz EM (1991) The war of the whorls: genetic interactions controlling flower development. Nature 353(6339):31–37
9. Meyerowitz EM et al (1991) A genetic and molecular model for flower development in Arabidopsis thaliana. Dev Suppl 1:157–167
10. Drews GN, Weigel D, Meyerowitz EM (1991) Floral patterning. Curr Opin Genet Dev 1(2):174–178
11. Jack T, Brockman LL, Meyerowitz EM (1992) The homeotic gene APETALA3 of Arabidopsis thaliana encodes a MADS box and is expressed in petals and stamens. Cell 68(4):683–697
12. Smyth DR, Bowman JL, Meyerowitz EM (1990) Early flower development in Arabidopsis. Plant Cell 2(8):755–767
13. Weigel D et al (1992) LEAFY controls floral meristem identity in Arabidopsis. Cell 69(5):843–859
14. Yanofsky MF et al (1990) The protein encoded by the Arabidopsis homeotic gene agamous resembles transcription factors. Nature 346(6279):35–39
15. Bowman JL, Smyth DR (1999) CRABS CLAW, a gene that regulates carpel and nectary development in Arabidopsis, encodes a novel protein with zinc finger and helix-loop-helix domains. Development 126(11):2387–2396
16. Sieburth LE, Running MP, Meyerowitz EM (1995) Genetic separation of third and fourth whorl functions of AGAMOUS. Plant Cell 7(8):1249–1258
17. Chen X, Meyerowitz EM (1999) HUA1 and HUA2 are two members of the floral homeotic AGAMOUS pathway. Mol Cell 3(3):349–360
18. Cheng Y et al (2003) Two RNA binding proteins, HEN4 and HUA1, act in the processing of AGAMOUS pre-mRNA in Arabidopsis thaliana. Dev Cell 4(1):53–66
19. Li J, Jia D, Chen X (2001) HUA1, a regulator of stamen and carpel identities in Arabidopsis, codes for a nuclear RNA binding protein. Plant Cell 13(10):2269–2281
20. Ji L et al (2011) ARGONAUTE10 and ARGONAUTE1 regulate the termination of floral stem cells through two microRNAs in Arabidopsis. PLoS Genet 7(3):e1001358
21. Chen X (2004) A microRNA as a translational repressor of APETALA2 in Arabidopsis flower development. Science 303(5666):2022–2025
22. Liu X et al (2011) AGAMOUS terminates floral stem cell maintenance in Arabidopsis by directly repressing WUSCHEL through recruitment of Polycomb Group proteins. Plant Cell 23(10):3654–3670
23. Siegfried KR et al (1999) Members of the YABBY gene family specify abaxial cell fate in Arabidopsis. Development 126(18):4117–4128
24. Sawa S et al (1999) FILAMENTOUS FLOWER, a meristem and organ identity gene of Arabidopsis, encodes a protein with a zinc finger and HMG-related domains. Genes Dev 13(9):1079–1088
25. Eshed Y, Baum SF, Bowman JL (1999) Distinct mechanisms promote polarity establishment in carpels of Arabidopsis. Cell 99(2):199–209
26. Telfer A, Poethig RS (1998) HASTY: a gene that regulates the timing of shoot maturation in Arabidopsis thaliana. Development 125(10):1889–1898
27. Wagner D, Meyerowitz EM (2002) SPLAYED, a novel SWI/SNF ATPase homolog, controls reproductive development in Arabidopsis. Curr Biol 12(2):85–94
28. Eshed Y et al (2001) Establishment of polarity in lateral organs of plants. Curr Biol 11(16):1251–1260
29. Muller R, Bleckmann A, Simon R (2008) The receptor kinase CORYNE of Arabidopsis transmits the stem cell-limiting signal CLAVATA3 independently of CLAVATA1. Plant Cell 20(4):934–946
30. Brand U et al (2000) Dependence of stem cell fate in Arabidopsis on a feedback loop regulated by CLV3 activity. Science 289(5479):617–619

31. Nimchuk ZL, Tarr PT, Meyerowitz EM (2011) An evolutionarily conserved pseudokinase mediates stem cell production in plants. Plant Cell 23(3):851–854
32. Nag A, Yang Y, Jack T (2007) DORNRO-SCHEN-LIKE, an AP2 gene, is necessary for stamen emergence in Arabidopsis. Plant Mol Biol 65(3):219–232
33. Levin JZ et al (1998) A genetic screen for modifiers of UFO meristem activity identifies three novel FUSED FLORAL ORGANS genes required for early flower development in Arabidopsis. Genetics 149(2):579–595
34. Page DR, Grossniklaus U (2002) The art and design of genetic screens: Arabidopsis thaliana. Nat Rev Genet 3(2):124–136
35. Papdi C et al (2010) Genetic screens to identify plant stress genes. Methods Mol Biol 639:121–139
36. Kim Y, Schumaker KS, Zhu JK (2006) EMS mutagenesis of Arabidopsis. Methods Mol Biol 323:101–103
37. Krieg DR (1963) Ethyl methanesulfonate-induced reversion of bacteriophage T4rII mutants. Genetics 48:561–580
38. Greene EA et al (2003) Spectrum of chemically induced mutations from a large-scale reverse genetic screen in Arabidopsis. Genetics 164(2):731–740
39. McCallum CM et al (2000) Targeted screening for induced mutations. Nat Biotechnol 18(4):455–457
40. Jander G et al (2003) Ethylmethanesulfonate saturation mutagenesis in Arabidopsis to determine frequency of herbicide resistance. Plant Physiol 131(1):139–146
41. Clough SJ, Bent AF (1998) Floral dip: a simplified method for Agrobacterium-mediated transformation of Arabidopsis thaliana. Plant J 16(6):735–743
42. Tzfira T et al (2004) Agrobacterium T-DNA integration: molecules and models. Trends Genet 20(8):375–383
43. Szabados L et al (2002) Distribution of 1000 sequenced T-DNA tags in the Arabidopsis genome. Plant J 32(2):233–242
44. Wang Y (2008) How effective is T-DNA insertional mutagenesis in Arabidopsis? J Biochem Tech 1(1):11–20
45. Li Y et al (2006) Analysis of T-DNA insertion site distribution patterns in Arabidopsis thaliana reveals special features of genes without insertions. Genomics 87(5):645–652
46. Weigel D et al (2000) Activation tagging in Arabidopsis. Plant Physiol 122(4):1003–1013
47. Radhamony RN, Prasad AM, Srinivasan R (2005) T-DNA insertional mutagenesis in Arabidopsis: a tool for functional genomics. Electr J Biotech 8(1). doi: 10.2225/vol8-issue1-fulltext-4
48. Lukowitz W, Gillmor CS, Scheible WR (2000) Positional cloning in Arabidopsis. Why it feels good to have a genome initiative working for you. Plant Physiol 123(3):795–805
49. Liu YG, Whittier RF (1995) Thermal asymmetric interlaced PCR: automatable amplification and sequencing of insert end fragments from P1 and YAC clones for chromosome walking. Genomics 25(3):674–681
50. Liu YG, Chen Y (2007) High-efficiency thermal asymmetric interlaced PCR for amplification of unknown flanking sequences. Biotechniques 43(5):649–654
51. Austin RS et al (2011) Next-generation mapping of Arabidopsis genes. Plant J 67(4):715–725
52. Schneeberger K et al (2009) SHOREmap: simultaneous mapping and mutation identification by deep sequencing. Nat Methods 6(8):550–551
53. Zhu Y et al (2012) Gene discovery using mutagen-induced polymorphisms and deep sequencing: application to plant disease resistance. Genetics 192(1):139–146
54. Sarin S et al (2008) Caenorhabditis elegans mutant allele identification by whole-genome sequencing. Nat Methods 5(10):865–867

Chapter 7

Genetic and Phenotypic Analysis of Shoot Apical and Floral Meristem Development

Mona M. Monfared and Jennifer C. Fletcher

Abstract

The shoot apical and floral meristems (SAM and FM, respectively) of *Arabidopsis thaliana* contain reservoirs of self-renewing stem cells that function as sources of progenitor cells for organ formation during development. The primary SAM produces all of the aerial structures of the adult plant, whereas the FMs generate the four types of floral organs. Consequently, aberrant SAM and FM activity can profoundly affect vegetative and reproductive plant morphology. The embedded location and small size of Arabidopsis meristems make accessing these structures difficult, so specialized techniques have been developed to facilitate their analysis. Microscopic, histological, and molecular techniques provide both qualitative and quantitative data on meristem organization and function, which are crucial for the normal growth and development of the entire plant.

Key words Shoot apical meristem, Floral meristem, Inflorescence meristem, Stem cells, Confocal laser scanning microscopy, Histology, In situ hybridization, Scanning electron microscopy

1 Introduction

One of the main distinctions between plant and animal development is that, unlike animals, the embryonic form of plants does not closely resemble the adult body plan. A plant embryo is a simple structure containing a pool of pluripotent stem cells at its shoot apex that post-embryonically generates all aerial organs, such as leaves and flowers [1]. The shoot apical meristem (SAM) maintains this central stem cell reservoir throughout the plant life cycle while simultaneously providing progeny cells for continuous organ formation on its flanks [2]. During early development leaves are produced from the vegetative SAM, which later transitions into an inflorescence meristem (IFM) when endogenous and environmental cues trigger flowering. Floral meristems (FM) form on the flanks of the IFM and contain transient stem cells as well as their progenitor cells that produce the four types of flower organs: sepals, petals, stamens, and carpels.

José Luis Riechmann and Frank Wellmer (eds.), *Flower Development: Methods and Protocols*, Methods in Molecular Biology, vol. 1110, DOI 10.1007/978-1-4614-9408-9_7, © Springer Science+Business Media New York 2014

Analyzing the morphology and histology of the SAM and FM and characterizing their gene expression patterns can shed light on many developmental processes [2]. Because the majority of development occurs post-embryonically, overall plant growth and architecture depends on the maintenance of stable apical meristems. Thus defects in growth rate, organ initiation, organ number, and/or stem size can often be traced back to perturbations in meristem structure or function that are undetectable by the naked eye [3, 4]. As the stable source of cells for organogenesis, plant meristems are an excellent model system for studying stem cell maintenance and termination, cell fate specification, organ morphogenesis, and pattern formation.

In our chapter we will explain methods and approaches for the analysis of SAM and FM development: (1 and 5) Confocal laser scanning microscopy (CLSM) permits the imaging of Arabidopsis SAM and FM meristem tissue from whole mount samples [5] (*see* Subheadings 3.1 and 3.5, respectively). CLSM is a powerful method for the examining internal cell patterns of embryonic and inflorescence SAMs and FMs without the need for physically sectioning the specimens, and the optical sections can be combined into a three-dimensional digital image of the sample. (2) Histological sectioning of resin-embedded samples allows the analysis of vegetative SAM tissues, which are inaccessible by other methods. Cell-specific stains permit high resolution imaging of the internal cellular morphology (*see* Subheading 3.2). (3) Meristem size is a key indicator of stem cell activity in Arabidopsis [6, 7]. The use of publicly available image analysis software provides a fast, convenient, and highly accurate method of measuring the height and width of vegetative meristem sections (*see* Subheading 3.3). (4) In situ hybridization uses a labeled probe to localize a specific RNA transcription pattern within a whole-mount or tissue section [8]. First introduced in the late 1960s for RNA detection in Xenopus oocytes [9], this technique has been successfully adapted to study gene transcription patterns in plants at all developmental stages [10, 11]. However, Arabidopsis vegetative SAMs are surrounded by leaf primordia, making it very difficult to achieve full tissue penetration and obtain a satisfactory signal-to-noise ratio. A modified RNA in situ hybridization protocol has been developed to effectively detect and localize mRNA transcripts to specific regions within vegetative shoot apices [12–14] (*see* Subheading 3.4). (6) Scanning electron microscopy (SEM) allows high resolution imaging by measuring the angle and energies of electrons scattered by atoms on the surface of a sample. The extreme level of surface detail that can be acquired using SEM is applicable to plants for analyzing the number and arrangement of floral meristems, the position and structure of floral organ primordia, and the cell surface morphology of individual flower organs [15] (*see* Subheading 3.6). (7) Floral organ number counting is a simple method to quantify and statistically analyze the phenotypes of mutants that display altered floral organ number, and/or produce mosaic floral organs. Such phenotypes are often caused by an underlying defect in the regulation of floral meristem size [6, 16] (*see* Subheading 3.7).

2 Materials

Arabidopsis thaliana seedlings are grown at 21 °C on plates containing Murashige and Skoog (MS) medium [17] under cool-white fluorescent lights (100–140 μmol/m^2 s) for 4–10 days.

MS medium: 4.33 g MS salts, 3 g sucrose, 900 mL distilled water. Adjust pH to 6.0 with 1 M KOH. Add 9 g bactoagar and adjust volume to 1 L with distilled water. For analysis of Arabidopsis plants during the reproductive phase, seeds are sown in a 1:1:1 mixture of perlite–vermiculite–topsoil and grown under cool-white fluorescent lights (100–140 μmol/m^2 s) at 21–22 °C. A minimum of ten samples from each genotype should be analyzed per experiment to obtain statistical robustness. For RNA in situ hybridization experiments, at least 50 samples per genotype or experimental batch should be prepared in order to achieve a good representation of the expression pattern of each gene of interest.

2.1 Confocal Laser Scanning Microscopy of the Embryonic Meristem

1. Glass scintillation vials or similar containers.
2. Pasteur pipets or micropipets.
3. Microscope depression slides.
4. Coverslips (24 mm × 60 mm).
5. Fine forceps (number 5, Ted Pella Inc.).
6. Fine needles.
7. Formaldehyde Acetic Acid (FAA) fixation solution: 3.7 % formaldehyde, 50 % ethanol, 5 % acetic acid (*see* **Note 1**). Prepare a fresh mixture of ~10 mL per vial.
8. Graded ethanol series: Prepare 500 mL solutions of 15, 30, 50, 70, 85, 90, and 95 % ethanol in distilled water.
9. Graded Histo-Clear series: Prepare 500 mL solutions of 25:75 Histo-Clear (National Diagnostics)–Ethanol, 50:50 Histo-Clear–Ethanol, and 75:25 Histo-Clear–Ethanol.
10. Propidium iodide stock solution: Prepare a 20 mL stock solution of 100 μg/mL propidium iodide in 0.1 M L-arginine and adjust the pH to 12.4 with 5 M NaOH (*see* **Note 2**). This solution is deep red and can be stored for months at 4 °C, wrapped in aluminum foil. When it turns dark orange it cannot be used anymore and should be disposed of as hazardous waste.
11. Propidium iodide staining solution: Prepare propidium iodide staining solution by diluting the 100 μg/mL propidium iodide

stock solution to 5 μg/mL in 0.1 M L-arginine and adjust the pH to 12.4 with 5 M NaOH.

12. Rinsing solution: 0.1 M L-arginine buffer in distilled water. Adjust the pH to 8 with HCl.
13. Immersion oil (Cargille Laboratories).
14. Nail polish.

2.2 Histological Sectioning of the Vegetative Meristem

1. Single-edge razor blades.
2. Scintillation vials.
3. Vacuum bell.
4. Microtome.
5. 42–55 °C slide-warmer.
6. Histoform S (Teflon embedding mold; Heraeus Kulzer, Wehrheim/Ts., Germany).
7. Histobloc (Heraeus Kulzer, Wehrheim/Ts., Germany).
8. Tungsten Carbide Microtome Knife.
9. FAA fixation solution: 3.7 % formaldehyde, 50 % ethanol, 5 % acetic acid. Prepare a fresh mixture of ~10 mL per vial.
10. Graded ethanol series: Prepare 500 mL solutions of 60, 70, 80, 90, and 95 % ethanol in distilled water.
11. Technovit Glycol Methacrylate (GMA) Kit 7100: Each kit contains 500 mL of GMA monomer, five 1 g packs of Hardener I, and 40 mL of Hardener II (*see* **Note 3**).
12. Graded Technovit series: Prepare 50 % (v/v), 70 % (v/v), 90 % (v/v) Technovit 7100 resin in 100 % ethanol.
13. Toluidine blue solution: Prepare 0.1 % (w/v) Toluidine blue O in 0.1 % aqueous sodium tetraborate (*see* **Note 4**).
14. Neutral Red solution: Prepare 0.01 % (w/v) solution by melting neutral red powder in Technovit 7100 Hardener I solution.

2.3 Meristem Size Measurement

1. Computer equipped with NCBI Image J software for image processing and analysis (*see* **Note 5**).
2. Digital images of shoot apical meristem sections (*see* **Note 6**).

2.4 Vegetative Meristem RNA In Situ Hybridization

1. Single-edge razor blades.
2. Scintillation vials.
3. Vacuum bell.
4. 56 °C oven for flasks.
5. Benchtop microscope slide warming table.
6. Weighing boats (8 cm × 8 cm).
7. Microtome.

8. Paraffin block holder.
9. Positively charged adhesion slides (Probe-on plus slides, FisherBrand).
10. 42 °C water bath for slides.
11. Wooden applicator sticks.
12. 42–55 °C bench for slides (slide-warmer).
13. Racks for slides or Copeland jars: Glassware (racks or Copeland jars) that will hold the slides must be baked (*see* **Note 7**).
14. 37–60 °C incubator with a shaker for the slide racks/Copeland jars.
15. UV cross-linker.
16. Plastic microscope slide boxes with latch, approximately 21 cm × 16.5 × 3 cm.
17. Stereomicroscope.
18. Eosin Y.
19. RNA molecular weight marker.
20. T7/T3/SP6 RNA Polymerases.
21. RNaseOUT/RNasin.
22. DNaseI.
23. Boehringer Blocking reagent (*see* **Note 8**): A 5× Blocking reagent (2.5 %) stock can be prepared in advance and stored at −20 °C.
24. Nitroblue Tetrazolium (NBT) (*see* **Note 8**).
25. 5-bromo-4-chloro-3-indolyl phosphate (BCIP) (*see* **Note 8**).
26. 10× Phosphate Buffer Saline (PBS): 1.3 M NaCl, 70 mM Na_2HPO_4, 30 mM NaH_2PO_4 pH 7 (NaOH). Prepare 1 L and autoclave it.
27. 4 % Paraformaldehyde fixation solution (*see* **Note 9**): Take 100 mL of 1× PBS and adjust pH to 11 with 1 M NaOH. Heat to 60 °C, add 4 g of fresh paraformaldehyde and stir to dissolve. Cool on ice and adjust pH to 7 with 1 M H_2SO_4. Add 1 mL of Triton X-100 and 1 mL of dimethyl sulfoxide (DMSO) (*see* **Note 10**). Make this solution fresh each time and keep on ice.
28. Prepare a graded series of 30, 50, 75, and 95 % ethanol in distilled water.
29. 2× Probe hydrolysis solution: Make fresh each time by mixing 30 μL of 21 mg/mL Na_2CO_3 and 20 μL of 16.8 mg/mL $NaHCO_3$ per probe.
30. Proteinase K buffer: Prepare 50 mL of 100 mM Tris–HCl, pH 8.0, 50 mM EDTA.

31. Proteinase K stock: 5 mg/mL proteinase K in 1 mL of proteinase K buffer. The stock can be aliquoted and stored at −20 °C for subsequent use.
32. 20× Sodium Salt Citrate (SSC) stock solution: 3 M NaCl, 0.3 M $Na_2C_6H_5O_7$ $2H_2O$. Prepare 1 L and autoclave solution.
33. Hybridization buffer: To prepare 4 mL of hybridization buffer, mix 2 mL of pure deionized formamide, 1.2 mL of 20× SSC, 600 μL of 20 % SDS, 40 μL of 10 mg/mL yeast tRNA, and 158 μL of RNAse-free Milli-Q water. Leftover buffer can be frozen at −20 °C for further use (*see* **Note 11**). The hybridization buffer precipitates at room temperature. Keep it at hybridization temperature (55 °C) before aliquoting it onto the slides.
34. 1× Tris Buffer Saline (TBS) solution: 0.4 M NaCl, 0.1 M Tris–HCl, pH 7.5. Prepare 2 L and autoclave. A 10× TBS stock can be prepared, although the powders will not dissolve easily.
35. Detection buffer: For 250 mL, mix 25 mL of 1 M Tris–HCl, pH 9.6, 6.25 mL of 4 M NaCl, 12.5 mL of 1 M $MgCl_2$, 207 mL of RNase-free Milli-Q water. Prepare the detection buffer with the substrates by mixing 1.6 μL of 50 mg/mL BCIP and 2.2 μL of 100 mg/mL NBT per mL of detection buffer.

2.5 Confocal Laser Scanning Microscopy of the Inflorescence Meristem

1. Glass scintillation vials or similar containers.
2. Microscope depression slides.
3. Coverslips (24 mm × 60 mm).
4. Fine forceps (number 5, Ted Pella Inc.).
5. Fine needles.
6. Formaldehyde Acetic Acid (FAA) fixation solution: 3.7 % formaldehyde, 50 % ethanol, 5 % acetic acid (*see* **Note 1**). Prepare a fresh mixture of ~10 mL per vial.
7. Graded ethanol series: Prepare 500 mL solutions of 15, 30, 50, 70, 85, 90, and 95 % ethanol in distilled water.
8. Graded Histo-Clear series: Prepare 500 mL solutions of 25:75 Histo-Clear (National Diagnostics)–Ethanol, 50:50 Histo-Clear–Ethanol, and 75:25 Histo-Clear–Ethanol.
9. Propidium iodide stock solution: Prepare a 20 mL stock solution of 100 μg/mL propidium iodide in 0.1 M L-arginine and adjust the pH to 12.4 with 5 M NaOH (*see* **Note 2**). This solution is deep red and can be stored for months at 4 °C, wrapped in aluminum foil. When it turns dark orange it cannot be used anymore and should be disposed of as hazardous waste.
10. Prepare propidium iodide staining solution by diluting the 100 μg/mL propidium iodide stock solution to 5 μg/mL in 0.1 M L-arginine and adjust the pH to 12.4 with 5 M NaOH.

11. Rinsing solution: Prepare a solution of 0.1 M L-arginine buffer in distilled water and adjust the pH to 8 with HCl.
12. Immersion oil (Cargille Laboratories).
13. Nail polish.

2.6 Scanning Electron Microscopy of the Inflorescence Meristem

1. Glass scintillation vials.
2. Plastic conical tubes.
3. Fine forceps (number 5, Ted Pella Inc.).
4. Dissecting microscope.
5. Cylinder mount gripper (Ted Pella, Inc.).
6. Mounting bases (*see* **Note 12**).
7. Mounting stub.
8. Conductive stickers.
9. White index cards.
10. Critical point dryer.
11. Sputter coater apparatus.
12. 0.1 M Sodium Phosphate Buffer (PB) buffer: combine 200 mL of 0.1 M Sodium Phosphate Monobasic NaH_2PO_4 (12 g/L) and 800 mL of 0.1 M Sodium Phosphate Dibasic Na_2HPO_4 (14.2 g/L). The pH should be between 7.2 and 7.4.
13. 25 mM Sodium Phosphate Buffer (PB) wash solution: dilute 100 mL of 0.1 M Sodium Phosphate Buffer (PB) to 25 mM in distilled water.
14. Glutaraldehyde fixation solution: freshly prepare 6 mL 0.1 M PB, 3 mL 25 % glutaraldehyde, 16 mL distilled water in a plastic conical tube in the hood (*see* **Note 13**).
15. Graded ethanol series: prepare 30, 50, 65, 75, 89, and 95 % ethanol in distilled water.

2.7 Floral Organ Number Counting

1. Fine forceps (number 5, Ted Pella Inc.).
2. Scissors with a sharp blade.
3. Dissecting microscope with 10× objective.
4. Statistical analysis software.

3 Methods

3.1 Confocal Laser Scanning Microscopy of the Embryonic Meristem

3.1.1 Embryo Dissection

1. Place two discs of Whatman or 3MM paper in a petri dish and let them absorb as much water as they can hold.
2. Distribute mature and dry Arabidopsis seeds on the surface of the paper, close the petri dish and let the seeds imbibe overnight at 4 °C in the dark (*see* **Note 14**).

3. The next day place a small piece of wet Whatman or 3MM paper on a slide. Transfer a few seeds onto the slide and dissect the embryos out of the seed coat under a dissecting microscope using fine forceps and needles (*see* **Note 15**). Hold the seeds with fine forceps by the micropylar end and make an incision on the other end of the seed coat with a needle or another pair of fine forceps. Apply gentle pressure on the micropylar side of the seed using the forceps slanted to one side. The embryo should pop out of the scar in the seed coat. Add more water if the paper on the slide dries.
4. With a fine needle immediately transfer the isolated embryo into a glass scintillation vial containing a few mL of cold water and keep the vial on ice (*see* **Note 16**).

3.1.2 Tissue Staining and Rinsing

1. Replace the water with 1 mL of the 5 μg/mL propidium iodide staining solution (*see* **Note 17**) for every 25 embryos in the vial. Stain for 6 h at room temperature. After the staining incubation the tissue appears pale orange.
2. Replace the staining solution with 2–3 mL of the L-arginine buffer rinsing solution. Leave in L-arginine buffer at 4 °C for 4 days, changing the rinsing solution once per day.

3.1.3 Tissue Clearing

1. Dehydrate the tissues through a graded ethanol series (15, 30, 50, 70, 85, 95, and 100 %) at room temperature, leaving the samples in each solution for 30 min.
2. Wash twice more with 100 % ethanol for 15 min each.
3. Incubate the tissues in a Histo-Clear series (75:25, 50:50, 25:75 Ethanol–Histo-Clear) to 100 % Histo-Clear. Leave the samples in each solution for at least 2 h to completely clear the tissue.
4. Wash with 100 % Histo-Clear three times for 2 h each and leave the samples overnight in the last change.

3.1.4 Mounting and Imaging

1. Pipet three or four separate drops of immersion oil on a microscope slide.
2. Pipet some embryos with a P100 micropipet (*see* **Note 18**) into the first drop of immersion oil.
3. Transfer the embryos serially from one oil drop to the next using fine forceps, in order to remove the Histo-Clear (*see* **Note 19**).
4. Place the embryos in the center of a coverslip. Each embryo should be lying on its side with the root and the two cotyledons touching the coverslip, and be fully covered with immersion oil to prevent the tissue drying out.
5. Flip the coverslip over a depression slide, such that the embryo side is down, and seal the four corners of the coverslip with nail polish (*see* **Note 20**).

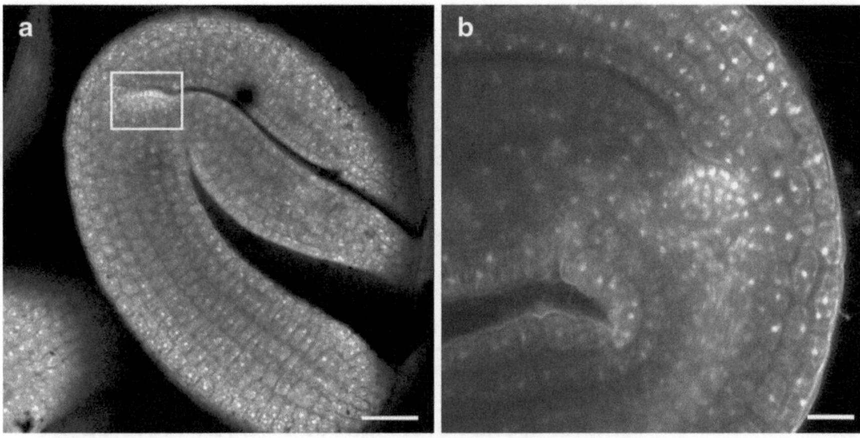

Fig. 1 Confocal laser scanning micrographs of mature Arabidopsis embryo and embryonic shoot apical meristem. (**a**) Optical longitudinal section of a wild-type Landsberg *erecta* (L*er*) mature embryo. The shoot apical meristem is *boxed*. (**b**) Optical longitudinal section of an embryonic shoot apical meristem from a mature wild-type L*er* embryo. The apical meristem cells stain very brightly with propidium iodide. Scale bars: 50 μm in (**a**) and 20 μm in (**b**)

6. Visualize each slide using a Confocal Laser Scanning Microscope (Fig. 1). Propidium iodide can be excited by a 514 nm argon laser beam and emits between 580 and 610 nm.

3.2 Histological Sectioning of the Vegetative Meristem

Plant materials for histological analysis should be grown on slanted agar plates. Depending on the growing conditions, 4–10-day-old seedlings are appropriate for vegetative SAM observation.

3.2.1 Tissue Dissection and Fixation

1. Remove the roots of each seedling at the base of the hypocotyl using clean forceps or small scissors.
2. Immediately place each dissected sample into a glass scintillation vial containing FAA fixation solution in a fume hood (*see* **Note 16**).
3. Loosen the caps of the scintillation vials and place them in a vacuum chamber.
4. Pull the vacuum slowly to 25 psi and let the samples sit for 20–30 min. This step removes air bubbles from the samples to allow the penetration of the fixative into the tissue. The samples will begin to sink. Slowly release the vacuum to return the samples to air.
5. Repeat the above step once to permit all the samples to sink to the bottom of the vials (*see* **Note 21**).
6. Slowly release the vacuum and remove the vials from chamber. Keep the sample vials in the fume hood overnight to complete the fixation process.

3.2.2 Tissue Dehydration and Infiltration

1. The next day, remove the FAA fixation solution using a Pasteur pipet and replace with 50 % ethanol. Incubate for 30 min at room temperature.
2. Dehydrate the samples through a graded ethanol series (60, 70, 80, 90, and 95 %), leaving the tissues in each solution for 30 min.
3. Wash the samples twice with 100 % (v/v) ethanol for 1 h each.
4. Transfer the dehydrated samples into 50 % (v/v) Technovit 7100 resin. Keep the samples at room temperature for 30 min.
5. Repeat the above step with 70 % (v/v) and 90 % (v/v) resin.
6. Replace with 100 % resin (v/v) twice for 1 h each. Store the samples at least overnight at 4 °C (*see* **Note 22**).

3.2.3 Tissue Staining and Embedding

1. Place a drop of neutral red solution onto a slide and place the sample in it for 3–4 s (*see* **Note 23**).
2. Add together Technovit 7100 Hardener II to Hardener I at a ratio of 1:15, mix the solution and pour it into the embedding mold. Place the neutral red-stained samples into the mold (*see* **Note 24**).
3. Allow polymerization for 1 h.
4. Place Histobloc on the top of each mold in an upside-down position.
5. Store the samples at least overnight at room temperature to allow full polymerization.

3.2.4 Tissue Sectioning

1. After removing the sample from the mold, place it on a rotary microtome at an 8° knife inclination angle. Use a very sharp microtome knife.
2. Section the tissue at 1–4 μm thickness. The sections will be released by the microtome one by one, unattached to each other.
3. Check continuously for sections that contain the SAM tissue (*see* **Note 25**).
4. Remove each SAM tissue section from the microtome with tweezers and float it in a petri dish filled with distilled water (*see* **Note 26**).
5. Put one section on a coverslip and dry it on a slide warmer at 42 °C for 10–15 min (*see* **Note 27**).

3.2.5 Toluidine Blue Staining

1. Soak the coverslip with ribbon in Toluidine blue solution for 1–2 min.
2. Soak the coverslip with ribbon in distilled water for 3–4 min to remove the excess stain. Check the sections under the microscope.

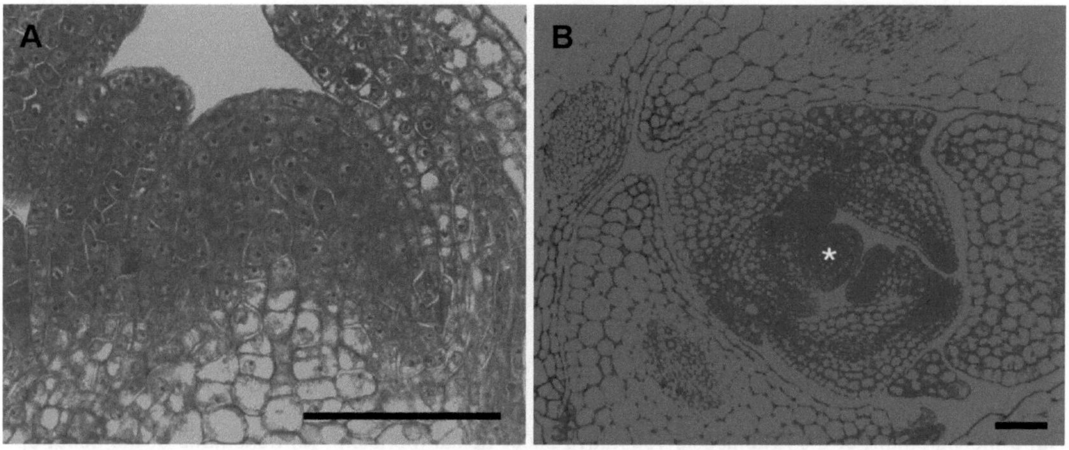

Fig. 2 Histological sections of Arabidopsis vegetative shoot apical meristems. (**a**) Longitudinal section. (**b**) Transverse section. The shoot apical meristem is marked (*asterisk*). Ten-day-old seedlings were fixed in FAA solution and infiltrated with Technovit 7100 resin before sectioning at 2 μm thickness. Scale bars: 50 μm

Repeat **step 1** if the sections are too lightly stained, and repeat **step 2** if they are too strongly stained (*see* **Note 28**).

3. Place the coverslip with the ribbon on the slide warmer at 42 °C for 10–15 min.

3.2.6 Mounting and Visualization

1. Place a drop of Permount onto a glass slide (*see* **Note 29**).
2. Place the coverslip with the completely dried ribbon upside-down on top of the Permount, ribbon side down (*see* **Note 30**).
3. Place the glass slide on the slide warmer overnight to allow complete fixation of the coverslip to the glass slide.
4. Visualize the stained sections using a stereomicroscope (Fig. 2).

3.3 Meristem Size Measurement

1. Launch the Image J software and open a meristem image file.
2. Set the scale by choosing the "straight line selections" tool from the tool bar below the main menu and drawing a straight line along the scale bar in the raw image (*see* **Note 31**). Go to the Analyze scroll down menu and select "set scale". In the window that appears, fill in the "known distance" and "unit of length" customizable space. After entering the correct values click "ok" to close the calibration window.
3. Draw lines on the image in order to make the desired measurements. Two parameters can be measured on longitudinal shoot apical meristem images: width and height (Fig. 3). The width is measured by drawing a horizontal line between the interior boundaries of the visible ridges that will form the next primordia. The height is measured by drawing a perpendicular line from the point equidistant along the width up to the tip of the meristem.

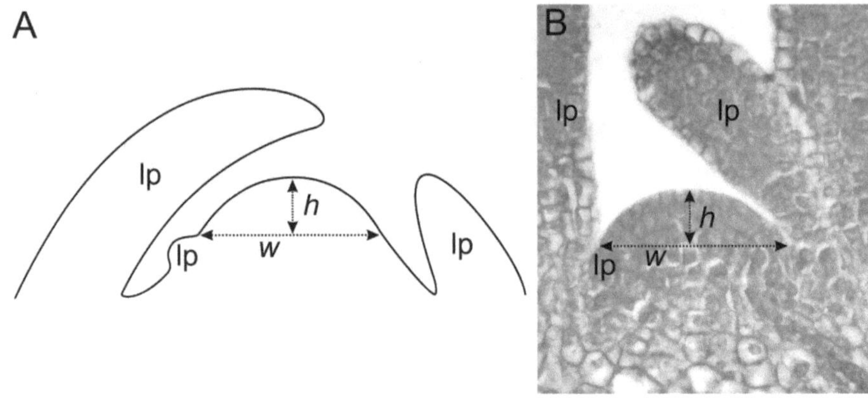

Fig. 3 Measurement of the Arabidopsis vegetative shoot apical meristem. (**a**) Schematic of the shoot apex showing the meristem as a dome of cells between the developing leaf primordia (lp). (**b**) Longitudinal section through a 7-day-old shoot apex. The meristem width (*w*) and height (*h*) measurements are made as shown

4. Once an accurate line is drawn, use the Ctrl + M keyboard combination to open a results window where all the measurements will be saved.
5. To make a line permanent on the image (useful for example to measure SAM height in longitudinal sections), click the "Edit" scroll down menu in the menu bar and select "draw".
6. Open the next image file and repeat **steps 2–4**. The new measurement will be saved in the same results window, which can be saved later as a Microsoft Excel table (*see* **Note 32**).
7. Perform statistical analysis using a program such as Microsoft Excel or OpenOffice SpreadSheet.

3.4 Vegetative Meristem RNA In Situ Hybridization

3.4.1 Fixation of Tissue Sections

For section preparation, the seedling tissues must be embedded in paraffin. To preserve morphology the tissues should first be fixed. Next they are dehydrated and stained with a dye to facilitate sectioning. For tissue embedding, an organic solvent gradually replaces the ethanol present in the tissue. The non-toxic solvent Histo-Clear is typically used in paraffin-based embedding procedures. Finally, paraffin is slowly introduced into the solvent solution until it reaches 100 % concentration. Numerous solution changes are needed in order for the paraffin to fully penetrate the tissue.

Fixation is one of the most critical steps for successful in situ hybridization, for which a compromise must be found between tissue morphology preservation and good probe penetration. A 4 % paraformaldehyde fixation solution penetrates 2 mm of tissue in about 1 h at room temperature. Because of the structure formed by the cotyledons, a bubble of air is usually trapped on the top of the shoot apex, preventing full penetration of the fixative

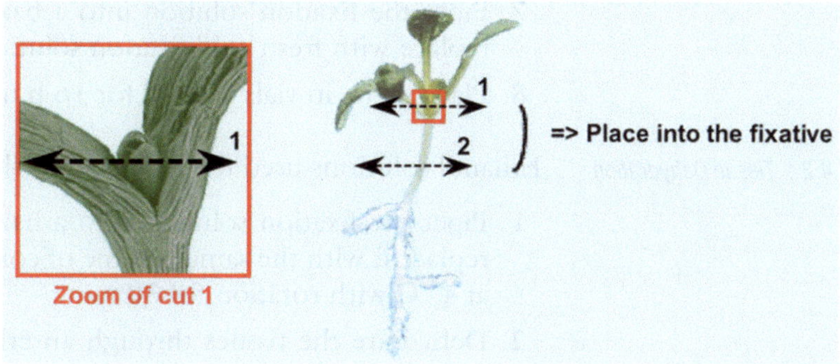

Fig. 4 Dissection of the Arabidopsis vegetative shoot apex for in situ hybridization. Lay a seedling sideways on a slide under a dissecting microscope. Holding the base of the hypocotyl firmly, make a sharp cut to remove the cotyledons and the oldest leaf primordia (*arrow 1*). Make another cut midway along the hypocotyl to remove the root system (*arrow 2*). Immediately place the dissected tissue in cold fixation solution

into the SAM tissue. Moreover, the stacking of leaf primordia over the SAM creates a layer of tissue for the fixative to pass through before reaching the meristematic cells. Chopping the tip of the seedling greatly enhances the SAM fixation process and allows all tissues to be cross-linked at the same pace.

1. Prepare the 4 % paraformaldehyde fixation solution fresh. Aliquot ~10 mL of the 4 % paraformaldehyde fixation solution into each scintillation vial kept on ice (*see* **Note 16**).
2. Pull a seedling from its culture plate (*see* **Note 33**). Lay the seedling sideways on a slide under a dissecting microscope. Using a sharp single-edge razor blade trim the cotyledons and oldest leaf primordia, and then cut midway along the hypocotyls to remove the roots (Fig. 4).
3. Immediately place the dissected tissue in the fixative and swirl. The dissecting process should be prompt to preserve tissue integrity. If the sample begins to lose moisture during dissection, add a few drops of fixative while trimming to avoid drying of the sample (work under a chemical hood!). Up to 50 dissected seedlings can be placed in a single scintillation vial.
4. Loosen the caps of the scintillation vials and place them on ice in the vacuum apparatus.
5. Pull the vacuum slowly to 25 psi and let the samples sit for 10 min. This step removes air bubbles from the samples to allow the penetration of the fixative into the tissue. The samples will begin to sink. Slowly release the vacuum to return the samples to air.
6. Repeat the above step once to permit all the samples to sink to the bottom of the vials.

7. Pipet the fixation solution into a hazardous waste bottle and replace with fresh cold fixation solution.
8. Place tissue in vials at 4 °C for 16 h under constant rotation.

3.4.2 Tissue Dehydration

Ethanol solutions used for dehydration should be kept at 4 °C.

1. Pipet the fixation solution into a hazardous waste bottle and replace it with the same volume of cold 15 % ethanol. Incubate at 4 °C with rotation for 30 min.
2. Dehydrate the tissues through an ethanol series (30, 50, 75, and 95 %), incubating at 4 °C with rotation for 30 min each.
3. Pour off the 95 % ethanol solution and replace with 100 % ethanol. Incubate at 4 °C with rotation for 30 min.
4. Pour off the 100 % ethanol solution and replace with 100 % ethanol containing 0.1 % Eosin Y to stain the tissue (*see* **Note 34**). Let the vials rotate overnight at 4 °C.

3.4.3 Tissue Embedding in Paraffin

1. Remove the vials from 4 °C and allow the solution to come to room temperature.
2. Remove the 100 % ethanol solution, replace with room temperature 100 % ethanol and incubate for 30 min.
3. Remove 1/4 of the volume of ethanol and replace it with the same volume of Histo-Clear, incubate for 30 min (*see* **Note 35**). Repeat this step four times.
4. Pour off the remaining solution and replace with 100 % Histo-Clear. Repeat this step once.
5. Pour off the solution and fill the vial 1/3 full with 100 % Histo-Clear. Add an equal volume of Paraplast X-tra paraffin chips. Incubate overnight at room temperature.
6. Prepare a 500 mL beaker of Paraplast chips and leave it in a 56 °C oven for the chips to melt overnight.
7. In the morning, place vials at 42 °C to melt the Paraplast chips. Keep adding Paraplast chips and allow them to melt at 42 °C. Repeat until the vials are full. The remainder of the chips should melt within an hour or 2.
8. When all chips have melted, transfer the vials to a 56 °C oven.
9. Remove the Histo-Clear–paraffin mixture, replace with pure molten Paraplast from the 500 mL beaker and incubate at 56 °C.
10. Replace the molten Paraplast every 8–10 h. Make at least six changes of molten Paraplast.
11. On a benchtop microscope slide warming table, pour the Paraplast solution containing the tissues into weighing boats. Split the contents of each vial into two weighing boats. Align and orient the samples using a pipet or needle (*see* **Note 36**). The embedded tissues can be stored at 4 °C for several months.

3.4.4 Tissue Sectioning and Mounting

1. Cut into the paraffin bed with a sharp single-edge razor blade in order to isolate each embedded seedling in a block, leaving at least 4 mm of paraffin on each side of the tissue.
2. Mount the block on the microtome sample holder (*see* **Note 37**).
3. Trim the block further to a rectangular or trapezoid shape, leaving about 2 mm of paraffin around the tissue. The long edges must be parallel to one another and to the hypocotyl of the sample.
4. Insert the sample holder into the microtome.
5. Align the hypocotyl of the tissue sample parallel to the knife blade, by moving the sample holder as needed (*see* **Note 38**).
6. Section the tissue at an 8° knife inclination angle for 8 μm-thick sections. Manipulate the paraffin ribbons containing the sectioned tissue onto a clean microscope slide using fine paintbrushes.
7. Screen the ribbons under a dissecting microscope for the serial sections containing SAM tissue.
8. Trim the ribbons of interest with a razor blade to remove unwanted sections.
9. Transfer the ribbons, shiny side down, onto an adhesion slide using a fine paintbrush. Orient them on the slide in parallel rows. Place the ribbons such that each is a few millimeters away from the edge of the slide.
10. Add RNase-free water beneath the ribbons and transfer the slide onto the slide warmer set at 42 °C.
11. Add more water so as to cover nearly all of the slide surface and let the ribbons expand for 10 min. The ribbon should expand approximately 50 % of its initial length.
12. Once the ribbons are fully expanded, slightly incline each slide and wick away the water using a kimwipe, always avoiding touching the ribbons.
13. Leave on a slide warmer overnight at 42 °C to dry. Cover with a lid to prevent dust falling on the slides. Once fully dry, sections can be stored in a clean microscope slide box at 4 °C for a couple of months.

*3.4.5 Riboprobe Preparation (**See Note 39**)*

In situ hybridization techniques were initially established using radioactively labeled probes. Improvements in non-radioactive labeling methods have led to equivalent sensitivity in signal detection. The most commonly used DIG indirect labeling approach uses a hapten coupled to UTP. Riboprobe is made by run-off transcription from a linearized plasmid template. This requires cloning of the cDNA of interest into a suitable vector (such as pBSKII) that allows linearization with restriction enzymes leaving 5′ overhangs, because 3′ overhangs or blunt ends can lead to RNA synthesis artifacts. The probe synthesis protocol is adapted from the Roche DIG RNA labeling mix manual. RNA is labeled to a density of 1 digoxigenin for every 20–25 nucleotides.

1. Linearize 10 μg of plasmid DNA in a restriction digest with a total volume of 100 μL. Incubate the digestion reaction at 37 °C for at least 2 h. Check for completion of the digestion by running a few μL of the digestion on a 1 % agarose minigel.
2. Add 100 μL of phenol–chloroform–isoamyl alcohol (25:24:1) and vortex.
3. Incubate for 5 min at room temperature and spin at 17,900 × *g* for 5 min.
4. Transfer the aqueous phase containing the DNA to a new tube and add 100 μL chloroform–isoamyl alcohol (24:1). Vortex briefly.
5. Incubate for 5 min at room temperature and spin at 17,900 × *g* for 5 min.
6. Transfer the aqueous phase containing the DNA to a new tube and precipitate the linearized DNA by adding 1/10 volume (10 μL) of 3 M NaOAc and 3 volumes (300 μL) of 95 % ethanol. Vortex briefly.
7. Incubate for 20 min at −20 °C and spin at 17,900 × *g* for 10 min.
8. Discard the supernatant and wash the pellet with 80 % ethanol. Spin at 17,900 × *g* for 5 min.
9. Discard the supernatant, air-dry the pellet, and resuspend in 50 μL of RNase-free MilliQ water.
10. Prepare the transcription mix by combining in a 1.5 mL eppendorf tube 10 μL linearized template (2 μg), 2 μL DIG RNA labeling mix, 4 μL 5× transcription buffer, 1 μL RNaseOUT and add RNase-free MilliQ distilled water to a total volume of 19 μL. Add 1 μL of RNA polymerase (SP6, T7 or T3). Incubate at 37 °C for 1 h.
11. Load 1 μL of the transcription reaction on a 1 % agarose minigel to check the RNA synthesis.
12. Add 1 μL DNaseI and incubate at 37 °C for 15 min.
13. Precipitate the riboprobe by adding 1/10 volume (2 μL) of 3 M NaOAc and 3 volumes (60 μL) of 95 % ethanol. Vortex briefly.
14. Incubate at −20 °C for 20 min and spin at 17,900 × *g* for 10 min.
15. Wash the pellet with 80 % ethanol and air-dry.
16. Resuspend the pellet in 50 μL RNase-free MilliQ distilled water.

17. Run 3 μL of the riboprobe on a 1 % agarose minigel to check probe synthesis and verify DNA absence. Multiple RNA bands can be observed because the riboprobe may adopt secondary structures. Save 3 μL of the probe for the gel in the next step (*see* **Note 40**).
18. If the riboprobe is longer than 1,000 nucleotides, it may be beneficial to hydrolyze it for better tissue penetration (*see* **Note 41**). Long probes tend to stick randomly to the sample and give unspecific background. Add 50 μL of 2×Probe hydrolysis solution to the 50 μL of riboprobe. Incubate at 60 °C for time t where $t = (L_i - L_f)/(k \times L_i \times L_f)$, where L_i is the initial length of the RNA probe in bp, L_f is the final length of the RNA probe in bp (optimal at 200–250 bp), K is the rate constant ($K = 110$ bp/min).
19. Stop the hydrolysis with 1/20 volume (5 μL) of 10 % glacial acetic acid. Load 6 μL on a 1 % agarose minigel, along with the 3 μL retained from the pre-hydrolysis riboprobe. The post-hydrolysis probe should give a fuzzy band of smaller molecular weight.
20. Precipitate the riboprobe by adding 1 μL of 10 mg/mL yeast tRNA, 1/10 volume (10 μL) of 3 M NaOAc, and 3 volume (300 μL) of 95 % ethanol.
21. Incubate at −20 °C for 2 h and spin at 17,900×g at 4 °C for 30 min.
22. Wash the pellet with 80 % ethanol and air-dry.
23. Resuspend the probe in 40 μL of 50 % formamide and freeze at −20 °C. This should yield enough probe for 40 slides (*see* **Note 42**).

3.4.6 Slide Pre-hybridization and Hybridization Treatments

Slides can be placed back to back in the rack slots to maximize space. Make sure to pull them apart at each wash/incubation step to remove any solution trapped between them from the prior step. An 18-slide experiment, corresponding to one rack, will require about 6 L of RNase-free Milli-Q water.

After rehydration of the sectioned tissue samples, incubation with a protease is conducted as a pretreatment to permeabilize the tissue and thereby facilitate penetration of the probe. Proteinase K is an effective protease, the concentration of which is critical because too extensive protein degradation results in loss of tissue morphology whereas too limited protein degradation reduces probe penetration and produces a weaker signal. To enhance tissue permeabilization a gentle acid hydrolysis step is used prior to the proteinase K treatment, which helps break down the cell walls and partially solubilizes highly cross-linked basic nuclear proteins.

Temperature, buffer composition and pH, blocking reagents and probe concentration are interdependent parameters that affect the ability of the riboprobe and tissue RNA to form duplexes during hybridization. Because the target RNA is embedded in the tissue sections, classical hybridization kinetics and Tm (melting point temperature, the temperature at which 50 % of the probe is dissociated) cannot be calculated but must be determined empirically. Optimal hybridization temperatures vary from 45 to 60 °C. Formamide, which destabilizes the hydrogen bonds between probe and target sequences, is the organic solvent of choice to reduce the melting temperature of RNA–riboprobe duplexes, thereby permitting the hybridization to take place at lower temperatures. Hybridization stringency is also determined by the concentration of monovalent cations: under high Na^+ concentrations (high stringency), only sequences with high degree of homology form stable duplexes. Yeast tRNA is used as blocking reagent to reduce non-specific binding of the riboprobe. Increased probe concentration leads to increased signal until a saturation point is reached. Further concentration increases result in non-specific binding that masks the true RNA localization.

1. Dewax the tissue sections in 100 % Histo-Clear twice for 10 min each with gentle shaking. Wash twice with 100 % ethanol for 2 min each.
2. Rehydrate by dipping the slides sequentially into a freshly made graded series of ethanol dilutions (95, 90, 80, 70, 50, and 30 %) with gentle shaking, for 2 min each step. Wash twice in RNase-free MilliQ distilled water for 2 min each.
3. Incubate the slides in 0.2 M HCl for 20 min.
4. Prepare 50 mL of proteinase K buffer and preheat to 37 °C in the oven.
5. Wash the slides in RNase-free Milli-Q distilled water for 5 min.
6. Wash the slides in 1× PBS for 5 min to neutralize the remaining acid.
7. Wash the slides in RNase-free Milli-Q distilled water for 5 min.
8. Add 10 μL of proteinase K stock to 50 mL of proteinase K buffer, mix and incubate the slides at 37 °C for 30 min.
9. Wash the slides in 1× PBS for 5 min.
10. Stop the digestion by washing the slides in 2 mg/mL glycine in 1× PBS for 2 min.
11. Wash the slides in 1× PBS for 30 s.
12. Incubate the slides in 1× PBS containing 3.7 % formaldehyde (mix 5 mL of 37 % formaldehyde with 45 mL of 1× PBS) for 20 min (*see* **Note 43**).
13. Wash the slides in 1× PBS for 5 min.
14. Dehydrate the tissues in graded ethanol series in the reverse of **step 2** (from RNase-free Milli-Q water to 100 % ethanol).

15. Pour off all ethanol and air-dry the slides. Under humid conditions, a vacuum treatment may be needed to eliminate any remaining ethanol.
16. Prehybridize the tissues by placing the slides on 55 °C slide warmer while adding 200 μL of prewarmed hybridization solution per slide. Cover with a coverslip and incubate at 55 °C for at least 1 h in a microscope slide box humidified with paper towels soaked in water.
17. The amount of riboprobe to use per slide ranges from 10 to 50 ng/kb. For riboprobe hybridization to an RNA target with an expected identical sequence, it is recommended to start the hybridization with a 50 % formamide solution containing 1 M Na^+ and 40 ng of probe (which should correspond to 0.5–5 μL of the 40 μL probe), at 55 °C. Use a total volume of 100 μL hybridization buffer per slide.
18. Per slide (multiply by the number of slides to be hybridized with the same riboprobe): aliquot 1 μL of DIG-labeled riboprobe, add 5 μL of 50 % formamide and heat to 75 °C for 2 min to disrupt the secondary structure.
19. Add 100 μL of warm hybridization solution per slide (multiply by the number of slides) and pipet onto the slide from one edge. Overlay tissue sections by gently applying a coverslip from one edge of the slide to the other, avoiding air bubbles. Incubate slides overnight at 55 °C in a microscope slide box humidified with paper towels soaked in water.
20. Prepare the wash solutions for the next day. Prewarm 2 L of 0.2× SSC, 0.1 % SDS washing solution at 55 °C overnight.

3.4.7 Post-hybridization Washes

Post-hybridization washes with a high stringency buffer remove unbound riboprobe and separate mismatched duplexes. An RNAse treatment between the washes degrades the non-hybridized or washed ssRNA probes. Riboprobe degradation products will then be eliminated with subsequent washes. For efficient washes, it is crucial to lay the slides flat in a glass dish and immerse them in 200 mL of washing solution under gentle agitation.

1. Dip the slides in 0.2× SSC, 0.1 % SDS that has been prewarmed to 55 °C. The slide/coverslip duplexes will separate. Remove the coverslips.
2. Lay the slides flat at the bottom of a glass dish and wash with 0.2× SSC, 0.1 % SDS that has been prewarmed to 55 °C on a shaker for 10 min at 55 °C. Repeat this step once.
3. During the washes, prepare 400 mL of 2 × SSC and warm it at 37 °C for the RNase treatment in **step 5**.
4. Incubate the slides in 2× SSC for 2 min at room temperature.

5. Prepare 10 μg/mL RNase solution in 2× SSC that has been prewarmed to 37 °C and incubate the slides for 30 min at 37 °C, under gentle agitation (*see* **Note 44**).
6. Wash the slides in 2× SSC for 2 min at room temperature.
7. Perform a high stringency wash with 0.2× SSC, 0.1 % SDS that has been prewarmed to 55 °C on a shaker, for 10 min at 55 °C. Repeat this step once.

3.4.8 Detection

The detection process starts with immunological detection of the DIG-labeled riboprobes. Best results are obtained using an anti-DIG antibody coupled to alkaline phosphatase (AP). The AP reaction with NBT and BCIP produces formazan, a stable blue/purple dye with bright reflective properties. Formazan does not precipitate, allowing long incubation times of greater than 24 h. Because formazan is soluble in organic solvents, it is crucial to embed the slides in aqueous mounting medium after detection. Mount the slides in a 50 % glycerol solution, which gives good resolution for microscope observation and imaging as well as allowing unlimited storage of the stained sections. Process all washes and buffer incubations in glass dishes, with the slides laying flat at the bottom. Gentle rotation should be applied during the wash steps.

1. Rinse the slides in 1× TBS for 5 min at room temperature.
2. Incubate the slides in 0.5 % Boehringer Blocking reagent in 1× TBS for 1 h at room temperature.
3. Rinse the slides in 1× TBS containing 0.5 % Bovine Serum Albumin (BSA) and 0.1 % Triton for 30 min at room temperature.
4. Replace the solution in **step 3** with 1× TBS containing 0.5 % BSA and incubate at room temperature for 5 min.
5. Remove the slides from the glass dish and add anti-DIG AP conjugate (*see* **Note 45**) in 1× TBS containing 0.5 % BSA. Use 100 μL per slide and gently apply a coverslip, avoiding air bubbles.
6. Place the slides flat in a microscope slide box humidified with paper towels soaked in water, and incubate 2 h at room temperature.
7. Dip the slide/coverslip duplexes in 1× TBS, 0.5 % BSA, 0.1 % Triton-X100. Remove the coverslips.
8. Lay the slides flat in the glass dish and rinse them in 1× TBS, 0.5 % BSA, 0.1 % Triton X-100 for 10 min at room temperature on a shaker. Repeat this step three times.
9. Rinse the slides in detection buffer for 15 min at room temperature with shaking. Save a few mL of this buffer for the next step.

10. Apply 100 μL of detection buffer per slide, add coverslip, and place the slides in a microscope slide box humidified with paper towels soaked in water. Incubate in the dark for 4–36 h (*see* **Note 46**).
11. When a satisfactory signal is observed, stop the reaction by dipping the slides in MilliQ water.
12. Mount the slides by adding 80 μL of 50 % glycerol and overlay with a coverslip, being careful that glycerol is evenly distributed across the slide.
13. Visualize the slides using a stereomicroscope and troubleshoot as necessary (*see* **Note 47**).

3.5 Confocal Laser Scanning Microscopy of the Inflorescence Meristem

3.5.1 Tissue Fixation

1. Prepare the fixation solution and chill it on ice (*see* **Note 1**). Dispense ~10 mL fixation solution into each glass scintillation vial or similar container (*see* **Note 16**).
2. Clip off the young inflorescences when the bolting stem is a few cm tall. Remove the older flowers and leave 5–7 visible flower buds. Retain several cm of stem because it will facilitate later manipulations.
3. Immediately place the tissue into the fixation solution. The tissue will float on the surface so gently swirl the vial to completely cover the tissue with fixation solution.
4. Loosen the caps of the scintillation vials and place them on ice in a vacuum chamber.
5. Pull the vacuum slowly to 25 psi and let the samples sit for 15 min. This step removes air bubbles from the samples to allow the penetration of the fixative into the tissue. The samples will begin to sink. Slowly release the vacuum to return the samples to air.
6. Swirl the vials to redistribute the fixation solution over the tissues.
7. Apply the vacuum for another 10 min, after which the tissues should sink.
8. Remove the fixation solution and add a fresh cold aliquot (*see* **Note 48**). Place the vials at 4 °C and incubate overnight.

3.5.2 Tissue Dehydration

Ethanol solutions should be kept at 4 °C.

1. Pour off fixation solution and replace it with cold 50 % ethanol. Incubate for 1 h.
2. Dehydrate the tissue through an ethanol series (70, 85, 95, and 100 %), leaving the tissues in each solution for 1 h. Ethanol solutions should be cold and samples should be kept at 4 °C.
3. Remove the 100 % ethanol and replace it with a fresh aliquot. Leave it overnight at 4 °C to remove the remaining chlorophyll and complete the fixation process. The following morning the

tissue should be white. If some chlorophyll remains in the tissue, continue replacing the 100 % ethanol at 1 h intervals until the tissue is completely white.

3.5.3 Tissue Staining and Rinsing

1. Rehydrate the samples through a decreasing graded ethanol series (95, 85, 70, 50, 30, and 15 %, distilled water), leaving the tissues in each solution for 1 h at room temperature.
2. Wash twice briefly with distilled water.
3. Prepare a stock solution of propidium iodide (*see* **Note 2**).
4. Add 500 μL of the 5 μg/mL propidium iodide staining solution per apex to each vial and incubate at room temperature for 24 h. The samples should be completely submerged in the staining solution. After the staining incubation period, the tissue appears pale orange.
5. Replace the staining solution with 5 mL of L-arginine buffer rinsing solution. Leave in L-arginine buffer at 4 °C for 4 days, changing the rinsing solution once every day.

3.5.4 Tissue Clearing

1. Dehydrate the samples through a graded ethanol series (15, 30, 50, 70, 85, 95, and 100 %), leaving the tissues in each solution for 1 h at room temperature.
2. Wash twice more with 100 % ethanol for 15–20 min each.
3. Incubate the tissues in a Histo-Clear series (75:25 Ethanol–Histo-Clear, followed by 50:50 and 25:75) to 100 % Histo-Clear. Leave the samples in each solution for at least 2 h to completely clear the tissue.
4. Wash with 100 % Histo-Clear three times for 2 h each and leave the tissues overnight in the last change.

3.5.5 Tissue Dissection

It is necessary to remove the older, larger flower buds in order to view the floral meristems. Dissecting in Histo-Clear is perfectly safe as this reagent is non-toxic. Nevertheless it has a strong odor and breathing its vapors for a long time can be unpleasant. For this reason it is preferable to dissect in immersion oil, which is also used as the mounting medium.

1. Place four individual drops of immersion oil on a microscope slide. Take one sample by the stem with fine forceps and put it in the first oil drop. Begin removing the older flower buds under the dissecting microscope (*see* **Note 49**).
2. When the first drop of oil has become too dirty to continue dissecting, grip the sample with forceps by the stem and move it into the second drop of oil. Continue dissecting, increasing the magnification if necessary. Repeat this step as many times as needed to remove most of the visible flower buds.
3. Once the sample has been dissected, cut off as much of the stem as possible.

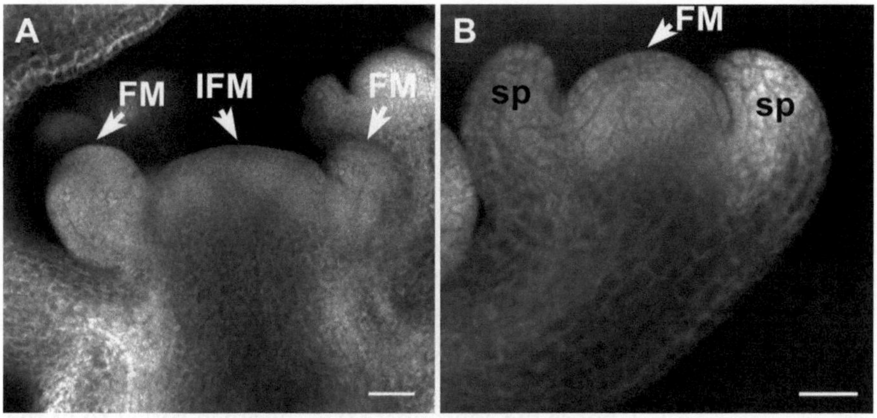

Fig. 5 Confocal laser scanning micrographs of an Arabidopsis inflorescence meristem and a floral meristem. (**a**) Optical longitudinal section of a wild-type Landsberg *erecta* (L*er*) inflorescence meristem (IFM) producing floral meristems (FM) on the flanks. (**b**) Optical longitudinal section of a wild-type L*er* stage 3 flower with sepal primordia (sp) arising from the flanks of the floral meristem (FM). Scale bars: 30 μm in (**a**) and 20 μm in (**b**)

3.5.6 Mounting and Imaging

1. Transfer the sample into the center of a coverslip. Ensure there is enough immersion oil to keep the sample in place. The top of the inflorescence meristem should lay flat against the coverslip (*see* **Note 50**). If the samples are sufficiently small, several can be transferred to a single coverslip leaving space between them to obtain clear images.
2. Flip the coverslip over atop a depression slide and seal the four corners of the slide with nail polish (*see* **Note 20**).
3. Visualize each slide using a Confocal Laser Scanning Microscope (Fig. 5). Propidium iodide can be excited by a 514 nm argon laser beam and emits between 580 and 610 nm.

3.6 Scanning Electron Microscopy of the Inflorescence Meristem

3.6.1 Tissue Fixation

1. Aliquot ~8 mL of glutaraldehyde fixation solution into each glass scintillation vial or similar container (*see* **Note 16**).
2. Using forceps or sharp scissors, gently clip off each inflorescence apex or single flower with 1 cm of the stem remaining and immediately place it into a scintillation vial. The tissue will float on the surface so gently swirl the vial to completely cover the tissue with fixation solution.
3. Incubate overnight at room temperature under constant rotation.
4. Remove the fixation solution into a hazardous waste bottle using a Pasteur pipet.
5. Optional: perform a secondary osmium tetroxide (OsO_4) coating step (*see* **Note 51**).

3.6.2 Tissue Rinsing and Dehydration

1. Rinse the tissues three times with 25 mM PB wash solution each. Empty the first two washes into a hazardous waste bottle using a Pasteur pipet.

2. Dehydrate the samples through a graded ethanol series (30, 50, 65, 75, 89, 95, and 100 %), leaving the tissues in each solution for 15–30 min.
3. Wash the tissues three times with 100 % ethanol for 15–30 min each and leave them in the third change overnight at room temperature.
4. The next day, repeat the 100 % ethanol wash twice for 15–30 min each.
5. Store the samples in 100 % ethanol until ready to dry them (*see* **Note 52**).

3.6.3 Critical Point Drying

1. Choose the appropriate size specimen basket to fit the samples.
2. Remove the basket lid and place the basket in a petri dish.
3. Cut small pieces of paper to fit in the specimen basket and write the genotype or sample name with a pencil. Using the forceps transfer each piece of paper into a separate chamber of the basket.
4. Fill the bottom of the petri dish with 100 % ethanol.
5. Quickly pour the samples from one scintillation vial into the petri dish. Use forceps to gently transfer the samples into the corresponding chamber(s), minimizing their exposure to air. Repeat this step for each of the scintillation vials.
6. Once all the samples have been transferred close the lid over the specimen basket.
7. Dry the samples in the critical point dryer, following the manufacturer's instructions (*see* **Note 53**).
8. Use forceps to transfer the dried samples to clean, dry scintillation vials for storage.

3.6.4 Tissue Mounting

1. Place two mounting bases on the dissecting microscope base.
2. On top of one mounting base, place a white index card that has been folded in the middle. On the other mounting base, place a mounting stub.
3. Use forceps to lift a conductive sticker by the edge and place it over the mounting stub (*see* **Note 54**). Place the two mounting bases side by side.
4. Carefully tip a specimen from the first vial onto the folded white index card.
5. View the specimen under the microscope. For an inflorescence meristem, grip it by the stem with one pair of forceps. With the other pair of forceps gently break off each older flower by pushing it very carefully down away from the stem (*see* **Note 55**), until the inflorescence apex and floral primordia are exposed. For a single flower, hold it by the stem and break off two sepals

Fig. 6 Scanning electron micrographs of an Arabidopsis inflorescence apex and a developing flower. (**a**) Wild-type Landsberg *erecta* (L*er*) inflorescence meristem producing floral meristems in a spiral phyllotaxy. Stage 1 through stage 5 floral meristems are shown. (**b**) Wild-type L*er* developing flower with all four sepals removed to reveal the petal, stamen, and carpel morphology. Scale bars: 50 μm

and petals from the tip downward in order to be able to view the internal floral organs.

6. Once the dissection is finished, use forceps to transfer the specimen onto the mounting stub covered by the sticker. For an inflorescence, carefully affix it by the base of the stem such that it sits perpendicular to the mounting stub. For a flower, affix the side that still contains the sepals and petals to the mounting stub (*see* **Note 56**).
7. Coat the samples using a sputter coating apparatus (*see* **Note 57**) and visualize them with a scanning electron microscope (Fig. 6), following instructions specific to the apparatus.

3.7 Floral Organ Number Counting

The first ten flowers from each of ten plants per genotype are sequentially removed over a period of days and all organs are counted and recorded.

3.7.1 Flower Dissection and Counting

1. Use small scissors or fine forceps to remove the first ten flowers on the first plant (*see* **Note 58**).
2. Grip the first flower by the pedicel with forceps and move it under a dissecting microscope.
3. With the other hand, use another set of forceps to remove each sepal sequentially by drawing the organ down and away from the pedicel and then pulling gently. Score and record the total number of sepals (*see* **Note 59**).
4. Repeat **step 3** with the petals and then the stamens.

a

Plant #	Flower #	Sepals	Petals	Stamens	Carpels
1	1	5	6	7	2
	2	5	5	6	2
	3	6	5	6	2
	4	5	6	6	2
	5	6	6	6	2
	6	6	4	6	2
	7	5	5	7	2
	8	4	5	6	2
	9	5	6	7	2
	10	4	6	6	2
2	1	4	5	8	3
	2	4	6	7	4
	3	4	6	7	3
	4	4	7	6	3
	5	4	6	7	3
	6	5	5	6	2
	7	6	6	5	3
	8	4	5	6	4
	9	4	4	6	2
	10	4	4	7	2

b

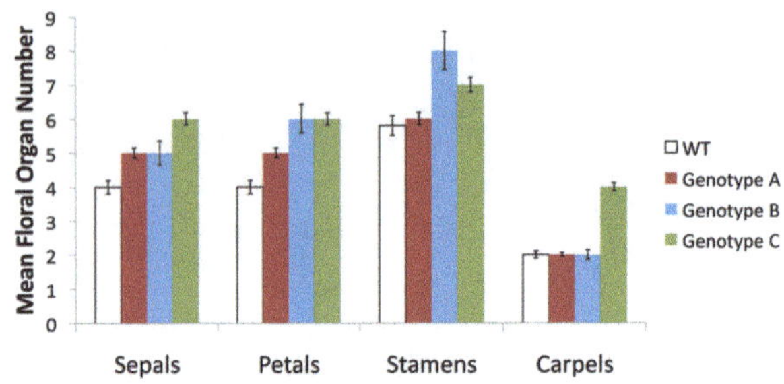

Fig. 7 Arabidopsis floral organ counts. (**a**) Sample tabulation of floral organ counting raw data in an Excel file format. (**b**) Sample bar graph generated in Excel showing mean floral organ number for the sepals, petals, stamens and carpels from wild-type plants and those of three additional genotypes (A, B, and C). The *error bars* represent standard error (SE)

5. For the carpels that remain, make a cross section through the intact gynoecium using a sharp blade. Count and record the number of individual carpels revealed in the cross section (*see* **Note 60**).
6. Repeat **steps 1–5** with the first ten flowers on the next plant.

3.7.2 Data Analysis

1. Enter the data into a statistical program (e.g. Microsoft Excel or OpenOffice Spreadsheet) for analysis (Fig. 7a). Arrange the data according to each genotype. For each genotype, the first ten flowers of ten plants are used for analysis, for a total of 100 flowers per genotype.
2. Calculate the mean and standard error values for each data set. A chi-square test can be performed to determine the statistical significance of values that differ between two genotypes.
3. Generate a bar graph showing the mean number of sepals, petals, stamens, and carpels for each genotype, including wild type (Fig. 7b). Mean floral organ number is placed along the *Y* axis and each floral organ type is placed along the *X* axis. Include error bars representing the standard error of measurement (SE) for each data set.

4 Notes

1. Formaldehyde and formaldehyde-containing solutions are toxic and should be handled wearing gloves under a fume hood. Treat all materials (pipets, tubes, etc.) that touch the solutions as hazardous waste.
2. Propidium iodide is a nucleic acid intercalating agent and should always be handled wearing gloves and suitable protecting clothing. Treat all materials (pipets, tubes, etc.) that touch the solution as hazardous waste.

3. Technovit 7100 resin is useful for working with plastic sections at room temperature. This resin hardens easily at room temperature, so it should be stored at 4 °C.
4. Toluidine blue is a general purpose stain/dye. This dye stains certain cellular components with different colors: i.e., Lignin/phenol is stained green/blue-green, pectins stain pink/reddish purple, and DNA stains green/blue-purplish.
5. The free software can be downloaded at http://rsbweb.nih.gov/ij/. Linux, Windows, or MacOSX versions are available.
6. The Image J program can open, process, and save images in any format (TIFF, JPEG, PNG, GIF, BMP, and raw data). For a complete list of supported data types, refer to the software documentation Web page.
7. Glassware (cylinders, beakers, flasks, bottles, slide racks, or Copeland jars) must be baked for 4 h at 180 °C to inactivate any trace of RNAse. It is not necessary to treat the water with DEPC, but simply use freshly autoclaved Milli-Q water stored in clean, baked glassware.
8. The Roche RNA DIG labeling and detection system is excellent for preparation and in situ detection of riboprobes. The system contains T7/T3/SP6 RNA Polymerases, RNaseOUT, DNaseI, Blocking reagent, NBT, and BCIP. Alternative reagents are available from other commercial sources and give good results (e.g., RNA Polymerases and RNasin from Promega). To prepare the 5× blocking reagent solution, the 1× TBS buffer must be warmed before adding the blocking reagent powder. The solution will look cloudy.
9. Paraformaldehyde powder and solution is toxic and should be handled wearing gloves under a fume hood. Treat all materials (pipets, tubes, etc.) that touch the powder or solution as a hazardous waste. Because paraformaldehyde vapors are also toxic, verify that the vacuum system used for tissue infiltration does not exhaust into the laboratory.
10. Plant tissues have a cuticle and thus float on the surface of the fixative, preventing proper infiltration. The combination of the Triton X-100 detergent and DMSO solvent enhances the penetration of the fixative while reducing the disrupting effect of the vacuum on the tissue morphology.
11. Formamide is highly corrosive in contact with skin and eyes, so the hybridization buffer should be handled wearing gloves under a fume hood. Treat all materials (pipets, tubes, etc.) that touch the solution as hazardous waste.
12. For inflorescences and single flowers, use a 10 mm × 5 mm specimen mount and corresponding mounting bases and conductive stickers.

13. Glutaraldehyde is highly toxic so it should always be handled wearing gloves under a fume hood. Treat all materials (pipets, tubes, etc.) that touch the solution as hazardous waste.
14. Imbibing seeds at cold temperature prevents embryo development during this step.
15. Isolating intact embryos is a key step for high quality imaging and morphology studies. The technique is difficult to master at the beginning; it requires patience and practice so include extra seeds when trying to dissect embryos for the first time because inevitably some samples will be damaged. Sit comfortably with your forearms resting on the bench so as to have steady hands.
16. One scintillation vial should be used per genotype or experimental batch. Take care not to pack the samples tightly into the vial; there should be room between them as they float on the surface of the solution. Write the name of the sample on the outside of the glass using a marker and cover it with a piece of clear tape so it does not wash off during the ethanol steps.
17. Fixation is not necessary to prepare embryos for confocal laser scanning microscopy. To avoid tissue loss or damage, remove solutions by carefully pipetting off the liquid with Pasteur pipets or a micropipet.
18. Cut off the very ends of the pipet tips with a razor blade. This will widen the tip end to avoid any damage to the sample during the pipetting process.
19. Do not try to grasp the embryos directly with the forceps. Almost completely close the forceps around the embryo to be transferred and lift the forceps. Some liquid will be trapped within the ends of the forceps and with it the embryo, which can now be safely placed in another drop of immersion oil.
20. Sealing with nail polish is preferable to other methods because it can easily be dissolved with acetone in the event that samples are dropped or need to be re-oriented for imaging.
21. The fixed samples should remain at the bottom of the vials once the vacuum has been released. If the samples rise to the surface of the fixation solution, reapply the vacuum for another 5–10 min.
22. Resin-infiltrated samples can be kept for at least 6 months at 4 °C.
23. This step will make the sample handling easier at later steps.
24. The resin will harden in 10–15 min, so it is important to maintain the correct position of the sample continuously until the hardening process is complete.
25. The SAM should appear as a dome between the leaf primordia for a longitudinal section (Fig. 2a), or as a circle in the center

of the primordia for a transverse section (Fig. 2b). After reaching the SAM tissue, section very carefully so as not to lose the ribbons that contain the central meristematic sections.

26. Perform this step quickly so that the section does not have time to roll up.
27. Use sharp tweezers to unfold the ribbon in distilled water. After putting the ribbon on the cover glass, check under the microscope again to determine whether the ribbon is completely unfolded and make any needed corrections using sharp forceps.
28. For optimal viewing the tissues should be stained light to medium blue. If the staining is too light the tissues will not be visible against the background, and if the staining is too dark the individual cells and layers will not be distinguishable. Note that the addition of Permount during the mounting step tends to lighten the tissue stain a few shades.
29. Permount is very harmful if inhaled, so work in the fume hood.
30. Before placing the cover glass carrying the ribbon on the Permount, carefully blow the dust off the ribbon to avoid permanently fixing debris onto the slide.
31. Include a scale bar in each raw image to serve as a reference for the Image J measurements.
32. Measure at least ten individual meristems to obtain statistically significant data.
33. For observation of the true vegetative shoot apex the seedling should not be more than 7 days old if grown in continuous light, to ensure that the SAM has not gone through reproductive transition.
34. The Eosin Y staining step is crucial for later tissue sectioning as it helps visualize and orient the tissue.
35. These solution replacements avoid the need to prepare a series of multiple dilutions because pure Histo-Clear is slowly added to the previous mixed solution, gradually bringing the content of the solution to 100 % Histo-Clear.
36. The subsequent step of mounting the embedded tissue is made easier by aligning the samples as the paraffin hardens around them. Orient the inflorescence apices on their sides with the stems pointing in the same direction, and align them in straight rows of 12–14 inflorescences each. Leave ~1 cm of paraffin between each sample.
37. Scoop a small piece of paraffin onto the tip of a metal spatula and melt it in a flame. Transfer the melted paraffin onto the top of the sample holder and affix it to the bottom of the paraffin block.

Hold the two together for ~1 min until the melted paraffin seals around the block. The paraffin block should be mounted such that the microtome knife will strike the longest side.

38. This step will ensure a clean longitudinal section through the shoot apical meristem in the majority of samples.

39. For expression analysis of a gene of unknown pattern, it is important to prepare a sense probe that will serve as negative control for specific hybridization of the antisense probe.

40. The yield of the transcription reaction can be estimated by running an aliquot of the probe on a 1.5 % mini agarose gel, next to an RNA standard of known concentration (e.g., RNA Molecular Weight Marker III, 0.3–1.5 kb, Roche). For testing the DIG labeling reaction yield, 1 μL of the probe can be deposited on a piece of filter, UV-cross-linked and incubated with 5 mL of detection buffer containing 5.5 μL of NBT and 4 μL of BCIP. A dark blue spot should become visible in the place where the probe was pipetted.

41. The necessity for riboprobe hydrolysis is controversial. Some researchers hydrolyze any riboprobe greater than 1 kb in length, whereas others find that it is not required in order to obtain a good signal. If a riboprobe greater than 1 kb in length gives a weak signal, then hydrolysis is recommended.

42. When using side-by-side sense and antisense probes, it is crucial to first load them on a 1.5 % mini agarose gel to compare their concentrations in order to use the same amount of riboprobe per slide.

43. This step helps to re-fix the tissues after the destabilizing proteinase K treatment.

44. This step can be essential to eliminate any background signals. The washing temperature can also be raised to 65 °C to help reduce background.

45. Dilute 1:1,000–1:500 for low abundance transcripts. Alternatively, the antibody can be diluted to 1:3,000 for an overnight incubation at 4 °C.

46. Most probes require an overnight incubation. Signal from very rare transcripts can be better observed after 48 h; in this case, add fresh detection buffer plus substrate after ~24 h and continue the incubation.

47. If the signal is brown, the pH of the detection solution is probably incorrect. Make sure that the pH is 9.6 and increase the washing time in the detection solution before adding the NBT and BCIP. If the signal appears as a purple haze of ubiquitous staining, there may be unspecific hybridization or antibody recognition problems. Several modifications to the protocol can

be tried, such as decreasing the amount of probe or antibody, increasing the hybridization temperature, and/or increasing the duration and temperature of the post-hybridization washes. If this does not solve the problem, try a probe designed from another region of the gene of interest. The absence of signal may be due to a variety of causes, such as transcript abundance below the threshold of detection, poorly labeled probe, excessively stringent post-hybridization washes, deficient anti-AP antibody, and/or old NBT/BCIP substrates. Add positive controls to troubleshoot this situation. When using a probe for the first time, it is very informative to run the hybridization experiment side by side with a DIG-labeled riboprobe that is already known to work.

48. Pipette the liquid off with a Pasteur pipet or a micropipet, trying not to touch the samples so as to avoid damaging them.
49. To remove the flower buds use two sets of fine forceps. With one set hold the inflorescence by the stem. With the other set touch the top of the bud and apply gentle pressure down and outwards away from the inflorescence stem. When the flower pedicel is bent outwards the flower bud can be safely pulled off the stem without damage to the surrounding tissues.
50. It is important to let the inflorescence meristem touch the coverslip. The surface of the meristem should lay flat against the coverslip for best imaging of the flower meristems.
51. Osmium tetroxide (OsO_4) may be used as a secondary fixative if necessary to add additional density and contrast to the tissue [18]. Prepare 4 mL of a 1 % OsO_4 solution: 1 mL 4 % OsO_4, 1 mL 0.1 M PB, 2 mL distilled water for each sample vial and add it to the vial using a Pasteur pipet. Note that OsO_4 is highly poisonous, even at low exposure levels, so it should always be handled under a fume hood while wearing gloves. Samples are incubated from overnight to several days, and the tissue should turn black. Usually overnight is sufficient for both inflorescences and single flowers. If the samples are left too long in the solution, the OsO_4 may begin to sediment and leave a grainy black residue on the sample surfaces. After incubation, empty the 1 % OsO_4 fixation solution into a hazardous waste bottle using a Pasteur pipet and replace it with 25 mM PB. Treat all materials (pipets, tubes, etc.) that touch the OsO_4 solution as hazardous waste.
52. For long periods, samples should be stored in 70 % ethanol rather than in 100 % ethanol.
53. Use of safety glasses is advised while operating the critical point dryer.

54. When placing the adhesive sticker on the stub, take care to lay the sticker flat, so that the surface of the stub is as smooth as possible. This will minimize topographical background when viewing the mounted sample in the electron microscope.
55. The dried tissues are brittle and prone to damage unless handled extremely carefully [18].
56. Place 3–4 inflorescences, or as many as five single flowers horizontally aligned, on a single mounting stub.
57. This step coats the samples with a conductive metal to prevent the buildup of high-voltage charges on the surface during the microscopy process [18]. Generally a 15–40 nm coating thickness is adequate, but use the minimum coating thickness possible. Over-coating obscures the surface detail and prevents high-resolution imaging, whereas under-coating can lead to charge buildup in the electron microscope. Samples may be re-coated if charging occurs. When first performing this protocol, use a few wild-type samples to empirically determine the optimal coating thickness needed to obtain satisfactory data.
58. This procedure generally takes several days to complete. Begin when the first arising flowers on the plant(s) of interest have opened, but before the floral organs begin to senesce and drop off. Remove each open flower for analysis. Because organ number in unopened flower buds is difficult to accurately quantify, when all the open flowers have been analyzed stop and continue the analysis the next day, once more buds have opened.
59. Some genotypes may produce flowers with fused or mosaic organs consisting of two types of tissue. If an abnormal floral organ is observed, add a new category to the results table and note the frequency of its occurrence [19]. The cellular composition of mosaic floral organs may be investigated using scanning electron microscopy.
60. To prevent excessive damage to the gynoecium, use a sharp cutting motion rather than a sawing motion.

Acknowledgements

We thank George Chuck, Harley Smith, Sabine Zachgo, Elliot Meyerowitz, Beth Krizek, Joshua Levin, and Mark Running for sharing protocols and suggesting improvements, and Elisa Fiume, Helena Pires, Jinsun Kim, Cristel Carles, Chanman Ha, and JiHyung Jun for helpful comments. This work is supported by grants from the NSF (IOS-1052050) and USDA (CRIS 5335-21000-029-00D).

References

1. Steeves TA, Sussex IM (1989) Patterns in plant development. Cambridge University Press, New York
2. Barton MK (2010) Twenty years on: the inner workings of the shoot apical meristem, a developmental dynamo. Dev Biol 341:95–113
3. Bhalla P, Singh MB (2006) Molecular control of stem cell maintenance in shoot apical meristem. Plant Cell Rep 25:249–256
4. Williams L, Fletcher JC (2005) Stem cell regulation in the *Arabidopsis* shoot apical meristem. Curr Opin Plant Biol 8:582–586
5. Running MP, Clark SE, Meyerowitz EM (1995) Confocal microscopy of the shoot apex. Methods Cell Biol 49:217–229
6. Clark SE, Running MP, Meyerowitz EM (1995) *CLAVATA3* is a specific regulator of shoot and floral meristem development affecting the same processes as *CLAVATA1*. Development 121(7):2057–2067
7. Fletcher JC (2001) The *ULTRAPETALA* gene controls shoot and floral meristem size in Arabidopsis. Development 128:1323–1333
8. Jin L, Lloyd RV (1997) In situ hybridization: methods and applications. J Clin Lab Anal 11:2–9
9. Gall JG, Pardue ML (1969) Formation and detection of RNA-DNA hybrid molecules in cytological preparations. Proc Natl Acad Sci U S A 63:378–383
10. Houben A, Orford SJ, Timmis JN (2006) In situ hybridization to plant tissues and chromosomes. Meth Mol Biol 326:203–218
11. Jackson D (1992) In situ hybridization in plants. In: Bowles DJ, Gurr SJ, McPherson R (eds) Molecular plant pathology: a practical approach. Oxford University Press, Oxford, pp 163–174
12. Ambrose BA, Lerner DR, Ciceri P, Padilla CM, Yanofsky MF, Schmidt RJ (2000) Molecular and genetic analyses of the *Silky1* gene reveal conservation in floral organ specification between eudicots and monocots. Mol Cell 5: 569–579
13. Carles CC, Lertpiriyapong K, Reville K, Fletcher JC (2004) The *ULTRAPETALA1* gene functions early in Arabidopsis development to restrict shoot apical meristem activity, and acts through *WUSCHEL* to regulate floral meristem determinacy. Genetics 167: 1893–1903
14. Chuck G, Muszynski M, Kellogg EA, Hake S, Schmidt RJ (2002) The control of splikelet meristem identity by the *branched silkless1* gene in maize. Science 298:1238–1241
15. Bowman JL, Smyth DR, Meyerowitz EM (1991) Genetic interactions among floral homeotic genes of Arabidopsis. Development 112(1):1–20
16. Zhao Y, Medrano L, Ohashi K, Fletcher JC, Yu H, Sakai H, Meyerowitz EM (2004) HANABA TARANU is a GATA transcription factor that regulates shoot apical meristem and flower development in Arabidopsis. Plant Cell 16: 2586–2600
17. Murashige T, Skoog F (1962) A revised medium for rapid growth and bioassays with tobacco tissue cultures. Physiol Plant 15: 473–497
18. Bozzola JJ, Russell LD (1999) Electron microscopy: principles and techniques for biologists, 2nd edn. Jones and Bartlett, Sudbury, MA
19. Levin JZ, Meyerowitz EM (1995) UFO: an Arabidopsis gene involved in both floral meristem and floral organ development. Plant Cell 7(5):529–548

Chapter 8

Genetic and Phenotypic Analyses of Petal Development in *Arabidopsis*

Judit Szécsi, Barbara Wippermann, and Mohammed Bendahmane

Abstract

The link between gene regulation/function and organ shape (morphogenesis) is poorly understood and remains one of the major issues in developmental biology. Petals are attractive model organs for studying organogenesis mainly because they have a simple laminar structure with a small number of cell types. Moreover, because petals are dispensable for plant growth and reproduction, one can experimentally manipulate petal development and dissect the genetic mechanisms behind the changes without serious effects on plant viability. Here, we describe the methods used to study petal development at the molecular, cytological, and genetic level, and more specifically mitotic and post-mitotic growth control during petal morphogenesis.

Key words Petal, Morphogenesis, Molecular, Genetic, Mitotic growth, Post-mitotic growth

1 Introduction

Studying organ formation addresses some of the most fundamental questions in plant development. All plant organs are formed from founder cells in meristems. The meristematic cells are morphologically homogeneous; however, they divide, expand, and differentiate to give rise to different cells types. These cells associate with each other to form distinct groups, or tissues, which further interact to form a functional organ. In order to understand the processes by which complex organs arise from their simple progenitors, one must dissect the mechanisms underlying these processes by isolating and characterizing the molecular components. The physical dimensions of leaves, roots, and stem can vary considerably within a given species depending on the amount of light, the growth temperature or soil conditions. In contrast, flower organ size is remarkably constant, indicative of strong endogenous control during morphogenesis. Therefore, petals can represent an attractive model for studying organ formation. Moreover, petals have a simple laminar structure with a small number of cell types (*see* Figs. 1 and 2) and are dispensable for plant growth and reproduction.

José Luis Riechmann and Frank Wellmer (eds.), *Flower Development: Methods and Protocols*, Methods in Molecular Biology, vol. 1110, DOI 10.1007/978-1-4614-9408-9_8, © Springer Science+Business Media New York 2014

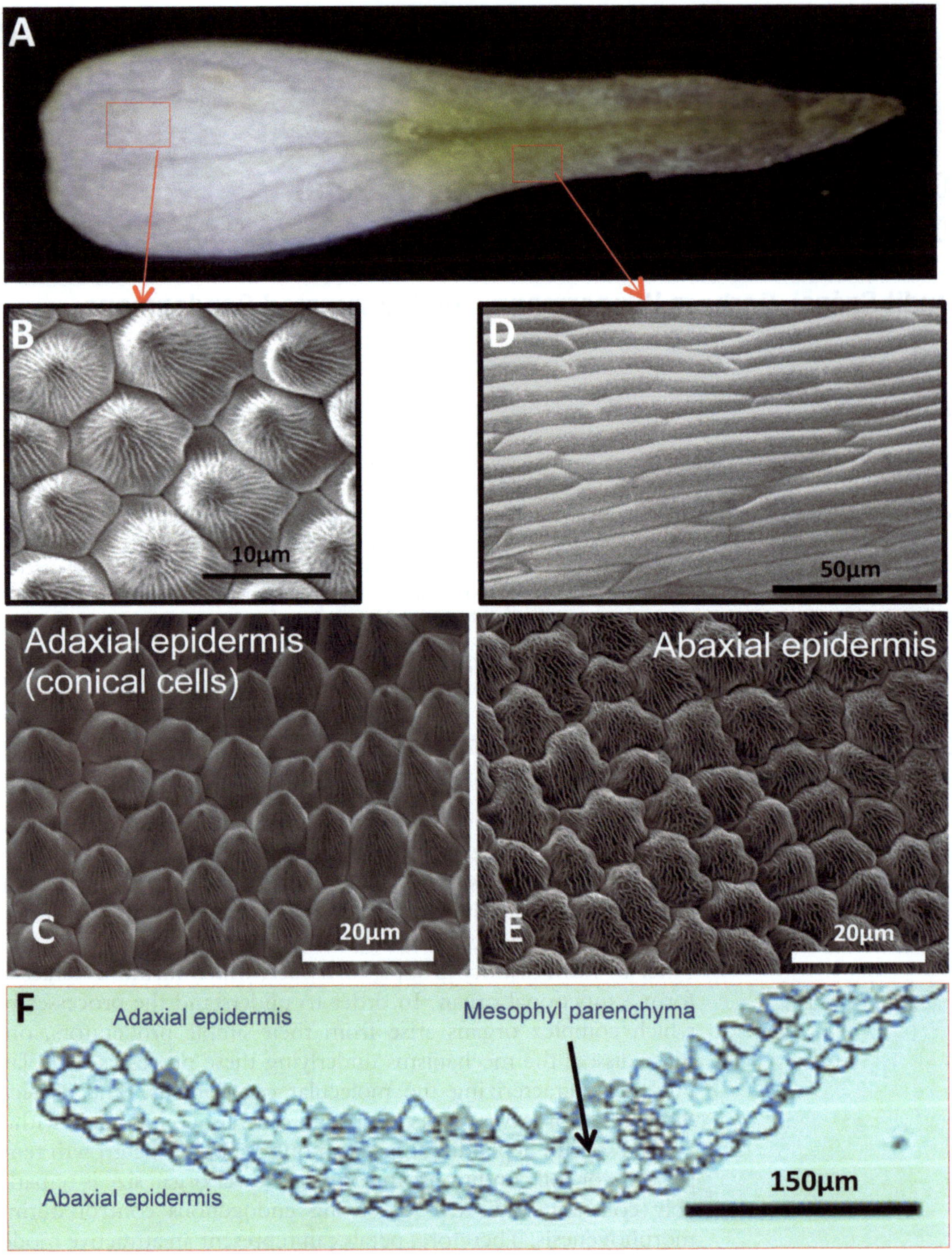

Fig. 1 The *Arabidopsis* petal. (**a**) Fully expanded *Arabidopsis* petals at flower development stage 14 [1]. (**b**, **c**) SEM views of the distal adaxial surface of petals showing the typical conical cells. (**d**) SEM view of the proximal adaxial surface of petals. (**e**) SEM view of the distal abaxial surface of petals. (**f**) Transverse section of a distal part from *Arabidopsis* mature petal at development stage 14

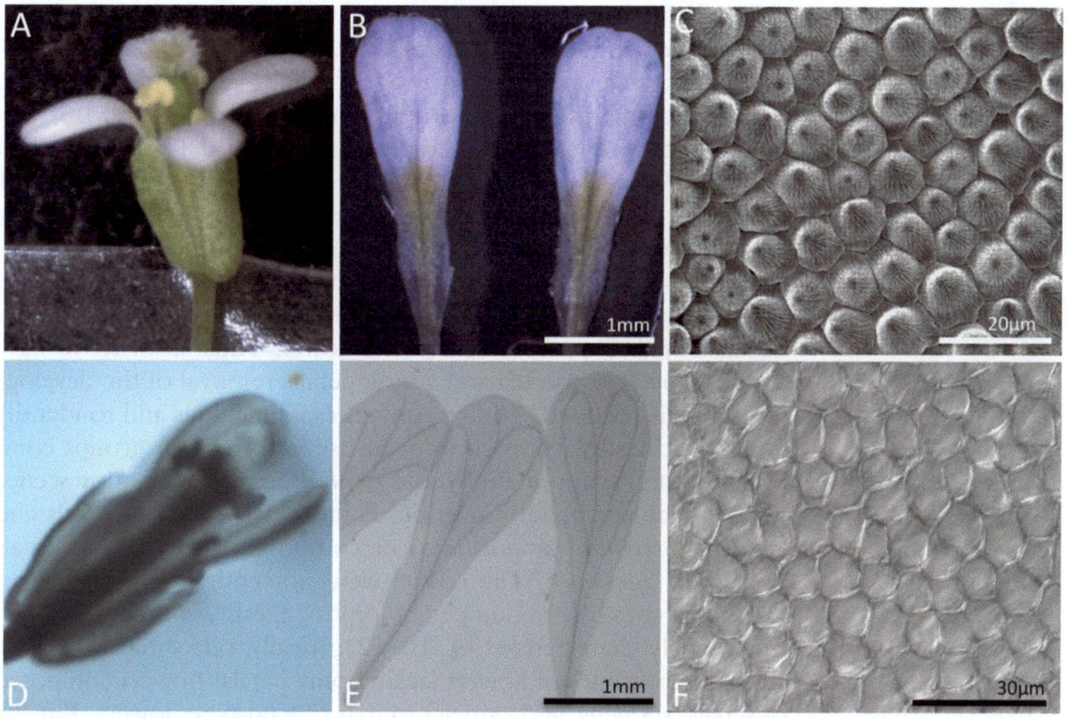

Fig. 2 Microscopy analyses of *Arabidopsis* petals. (**a**) Flower of wild-type *Arabidopsis* at development stage 14 [1]. (**b**) Fully expanded petals at flower development stage 14. (**c**) SEM view of the distal adaxial surface of petals. (**d**) Cleared wild-type Arabidopsis flower at development stage 14. (**e**) Cleared detached fully expanded petals. (**f**) Light microscopy image of the distal adaxial surface of cleared petals

These advantages make the petal a suitable organ for dissecting and analyzing developmental processes and the data from these studies can serve as a paradigm for analyzing other, more complex organs and tissues of the plant. Furthermore, the form, color, and scent of petals are important horticultural traits. Maintaining correct growth patterns of petals is important for the positioning of the sexual organs to ensure successful pollination and/or pollen dispersal. Therefore, studying petal development is of great interest for the horticultural and agricultural sector.

The morphological events that occur during petal development have been well described in *Arabidopsis thaliana* [1]. Petal primordia emerge at floral stage 5. Before stage 8, petal growth depends on cell division. After stage 8, petals grow by cell expansion and expanding petals will burst the bud open [1–3]. The mature petal has a bifacial laminar structure, with a sheet of epidermal cells covering an internal layer of mesophyll with limited vasculature (*see* Fig. 1). The mesophyll cells are usually uniform in size and generally rounded. The epidermis at the adaxial side is composed of specialized conical-papillate cells with thickened cell walls and cuticles (*see* Fig. 1b, c). The cells of the epidermis at the abaxial side of the petal have a cobblestone-like shape with irregular epicuticular ridges (*see* Fig. 1e) [4, 5].

In *Arabidopsis*, genes involved in petal identity determination and development have been studied for over two decades. It is well established that the transcription factors PISTILLATA (PI), APETALA3 (AP3), SEPALLATA1-4 (SEP1-SEP4), and APETALA1 (AP1) are responsible for the specification of petal identity [6–9]. The discovery of genes controlling petal morphogenesis downstream of these floral organ identity factors has been hampered by the fact that petals at early development stages are difficult to study because of their position within the flower. Thus, in order to have access to petal primordia and young petals, removal of the developing sepals is required. To overcome these difficulties and to identify genes specifically expressed in petals, several research groups compared the transcriptome in whole *Arabidopsis* flowers/inflorescence of wild-type and floral homeotic mutants either lacking petals and stamens (*apetala3* or *pistillata*) or composed of petals and sepals only (i.e., *agamous*). The results were not entirely satisfactory, as only few genes involved in petal morphogenesis were identified and functionally characterized [4, 10–12]. Ideally, one should compare the transcriptomes of flowers lacking only petals (flowers composed of sepals, stamens, and carpels) to that of wild-type flowers. But the nature of the ABC model is such that mutation in any of its components affects two adjacent flower whorls, making the ablation of only the petal whorl a difficult task. However, in *Arabidopsis*, a fragment of the *AP3* promoter (*D6*; see below) [13] was shown to drive expression only in the second whorl of the flower. Expressing the diphtheria toxin A (DTA) under the control of this AP3 promoter fragment (*D6:DTA*) [13] produces flowers completely lacking petals, but the other floral organs are intact. Therefore, we believe that transcriptome comparisons (microarrays and/or RNA-Seq approaches) between *D6:DTA* flowers (lacking only petals) and wild-type flowers performed at different developmental stages [1] may help to identify genes involved in petal morphogenesis.

Despite the difficulties listed above, few genes have been discovered and their functional analyses have contributed in part to the understanding of the molecular and genetic mechanisms that underlie petal development [10–12, 14–19]. Based on current literature, several genes control petal growth by affecting cell proliferation and/or cell expansion in an organ-specific manner [20, 21]. Some of these genes (e.g., *JAGGED*, *AINTEGUMENTA*, *ARGOS*) affect petal growth by positively regulating cell proliferation [22–24], whereas other genes (*BPEp*, *OPR3*, *ARF8*, *BIG BROTHER*, *KLU*, and *DA1*) control final organ size by negatively regulating the duration/period of cell proliferation [18, 25–27] or cell expansion [10, 14, 18]. However, these studies were mainly limited to linking specific gene expression/function to a given phenotype, and knowledge on comprehensive networks/pathways during organ (petal) morphogenesis is far from complete.

Several research groups, including our own, are focusing on the discovery and characterization of genes and regulatory pathways that mediate the development of petals. In this chapter, we describe selected methods, which have proven to be useful for this work.

2 Materials

1. Dexamethasone stock solution: 10 mM dexamethasone in ethanol. Dissolve 20 mg dexamethasone powder in 5 mL of absolute ethanol. Make aliquots of 1 mL and keep them at −20 °C in the dark. This stock solution is 1,000-fold concentrated.
2. Fixation/discoloration solution: 86 % ethanol, 14 % acetic acid. Mix 86 mL of absolute ethanol with 14 mL acetic acid for 100 mL of solution. This solution can be kept at room temperature (RT) in an airtight bottle for several months.
3. 86 % EtOH: Mix 86 mL absolute ethanol with 14 mL distilled water for 100 mL of solution. This solution can be kept at RT in an airtight bottle for several months.
4. Tweezers. Use high-quality straight or curved thin tip tweezers (http://www.dumonttweezers.com).
5. Eppendorf reaction tubes.
6. Small petri dishes.
7. Binocular microscope with attached camera.
8. Optical light microscope with 40× objective and attached camera.
9. Scanning electron microscope (SEM), Hirox SH-3000.

3 Methods

3.1 Dissection of Petals at Early Developmental Stages

1. Remove all open flowers from an inflorescence as well as the following 5–8 closed buds (these floral buds correspond to development stage 12 and thus contain fully developed petals) (*see* **Note 1**).
2. Detach the next floral bud and put it on a microscope slide with double scotch tape.
3. Under a binocular microscope, slightly press on the side of two opposite sepals of the bud with tweezers, which should create small gap at the top of the bud.
4. Slide your tweezers in the gap and delicately bend back the top sepal until it detaches. The petal is visible underneath.

5. Carefully detach the petal and place it on a microscope slide with a drop of distilled water. Repeat until all four petals are detached.
6. Cover the petals with a coverslip and observe petal shape and cell composition using light microscopy for changes in overall shape and in size and for changes in both abaxial and adaxial morphology (*see* Fig. 1a–e).

3.2 Morphological and Cellular Analysis of Fully Grown Petal Using Live Tissues: SEM and Image Analysis

1. Fully open flowers are used for these studies, but petals of early developmental stages can also be used after dissection (see above and *see* **Note 1**). You need a binocular microscope near your SEM for dissection.
2. Cool SEM stage to −10 °C (*see* **Note 2**).
3. Put a carbon conductive double-sided adhesive tape on the SEM sample holder.
4. Take out a fully open flower (*see* Fig. 2a) with tweezers and dissect petals under the binocular microscope.
5. Place all four petals flat on the SEM sample holder, adaxial side up (*see* Fig. 2b).
6. Place the sample holder immediately on the SEM stage, close the drawer and start the vacuum. At the same time, start freezing sample by decreasing the temperature to −50 °C (*see* **Note 3**).
7. Observe petals first at 5 kV at the smallest magnification (70× in our case). This analysis will provide information about petal shape and size (*see* **Note 4**).
8. For petal cell count and cell size determination, scan the distal part of the petal with conical cells at higher magnifications (500× or 700×; *see* Fig. 2c).
9. We take pictures of the conical cells at the distal part of the adaxial side of the petals, between veins.
10. Images are then analyzed with the ImageJ software (http://rsb.info.nih.gov/ij/docs/intro.html). First, set the scale using a picture of a micrometer taken at the same magnification as the one used for the imaging. Then determine the distance–pixel ratio on this picture using the "Set scale" command. For petal size measures, the edges of each petal are drawn manually with the "Polygon selections" or "Free hand selections" tools of the software. The surface area is then determined by the "Measure" command.
11. The number of cells on each picture can be counted.
12. All petal surface and cell number measures are downloaded in a Microsoft Excel table and analyzed by Excel statistical tools (average surface or cell number, standard deviation, *t*-test). These data are then used to calculate the cell density (the cell number divided by the surface area of the pictures that served for the cell counts) or the average cell size (the surface area

divided by the cell number), and to extrapolate the cell number on the petal surface to get a theoretical number of surface cells per organ.

13. Examine petal surface to monitor changes in the shape of specific cell types, such as the petal blade conical cells (*see* Fig. 1b, c), the elongated cells toward the base of the petal (*see* Fig. 1d) and vasculature (*see* Fig. 1a).
14. At higher magnification (1,500× or 2,000×), SEM enables to detect changes in specific cell types in further detail. Examples include the observation of the specialized conical cells in the adaxial side of the petal blade or cuticle structures (*see* Fig. 1b).

3.3 Measure Petal Size and Petal Cell Number Using Fixed Material

1. Prepare fixation/discoloration solution (86 % ethanol and 14 % acetic acid).
2. Put 1.5 mL of fixation/discoloration solution into a 2 mL Eppendorf tube.
3. Harvest fully open flowers with a pair of tweezers and put them immediately in the fixation/discoloration solution. Place a maximum of ten flowers in a tube.
4. Close the tube and invert it several times. Make sure that all flowers are immersed in the solution.
5. Keep tubes at RT for 1 h or at 4 °C overnight.
6. Discard fixation/discoloration solution, and rinse samples twice with 86 % ethanol to eliminate acetic acid.
7. Keep samples in 86 % ethanol at 4 °C until observation (*see* **Notes 5** and **6**).
8. We observe the fixed and discolored flowers under a binocular microscope (*see* Fig. 2d) (*see* **Note** 7). Pour flowers and ethanol from the Eppendorf tube to in a small petri dish. Make sure that all flowers are covered with 86 % ethanol. Add more 86 % ethanol if needed.
9. Put a few drops of 86 % ethanol on a microscope slide. Take one fixed flower and detach all petals under the binocular microscope, using tweezers. Flatten petals with adaxial face up and arrange them in a row.
10. Cover samples with cover slide.
11. Take pictures of each petal under the binocular microscope (*see* Fig. 2e). We usually use 20× magnification.
12. The samples on the microscope slide are then observed under a light microscope. Using 400× total magnification, take pictures of the adaxial size of each petal for cell number analysis (*see* Fig. 2f). We take pictures of the conical cells at the distal part of the adaxial side of the petals, between veins.
13. Pictures are analyzed for petal surface measure and petal cell count using ImageJ program, as described above.

3.4 Petal Development Studies Using Gene Knockdown or Overexpression Approaches

1. To study the effect of a gene in petal development, a commonly applied strategy is to generate transgenic plants in which the expression of this gene can be manipulated. Phenotypic changes induced by such alteration of gene expression are then observed and analyzed.
2. Generate constructs for gene knockdown or for gene overexpression under the control of the constitutive 35S promoter or under control of the *AP3* 5D6.5 promoter fragment to drive gene mis-expression specifically in petals (*see* **Note 8**) [13]. Alternatively, use inducible promoter systems in which gene mis-expression can be controlled in temporal and/or spatial-specific manner, to achieve a gene knockdown or overexpression at a specific petal developmental stage. Dexamethasone (DEX)-inducible and several other inducible promoter systems have been described [28, 29]. Several Gateway-compatible cloning vectors are available to create DEX-inducible promoter constructs [30].
3. Using *Agrobacterium*-dependent floral-dip transformation [31], generate transgenic *Arabidopsis* plants for the constructs.
4. After isolating the transgenic plants, they are first grown under standard long day conditions (18 h light at 20 °C–6 h dark at 18 °C) until inflorescences develop.
5. When DEX-inducible promoter systems are used, remove all open flowers from an inflorescence.
6. Prepare a 10 μM DEX working solution (stock solution diluted 1,000× in milliQ water) containing 0.01 % (v/v) Tween-20.
7. Dip whole inflorescence into DEX working solution for 10 s.
8. Repeat this treatment once daily for a 5-day period with freshly prepared DEX working solution.
9. Flowers of control plants are treated with a solution containing 0.01 % ethanol and 0.01 % (v/v) Tween-20.
10. On day 5, collect the youngest fully open flowers for morphological and cellular analysis of the petals as described above (Subheadings 3.2 and 3.3).

3.5 Petal Development Studies Using the Cre-Lox Recombination System

1. This two components system allows the generation of mosaic wild-type and mutant tissues in the same organ using the Cre/loxP recombination system [32–34] (*see* **Note 9**).
2. Generate a construct harboring the wild-type sequence corresponding to a given gene that is flanked by the loxP recombination sites. Expression of this construct in plants mutated for the corresponding gene should lead to complementation of the gene loss of function-associated phenotype and generates phenotypically wild-type plants.

3. A second plant expressing the Cre site-specific recombinase from bacteriophage P1 is generated. The expression of Cre can be driven by a constitutive or tissue specific promoters to achieve recombination ubiquitously in the whole plant or in selected tissues/organs. Moreover, for temporal control of the recombination, the expression of the Cre recombinase can be placed under the control of an inducible promoter [29].
4. Once the two above mentioned plants are crossed, the loxP flanked wild-type gene sequence and the Cre recombinase will be expressed in the same cells. The Cre recombinase will excise the loxP flanked wild-type transgene, resulting in the conversion of wild-type to a mutant phenotype in the targeted cells. Depending on the promoter (*see* **Notes 8**) that drives the Cre recombinase expression, the mutant phenotype will appear in the whole plant (constitutive promoter) or in specific tissues (tissue-specific promoter). When using inducible promoters, the wild-type phenotype is converted into a mutant phenotype upon induction.
5. Morphological and cellular analysis of petals containing mosaic wild-type and mutant tissues can be performed as described above (Subheadings 3.2 and 3.3).

4 Notes

1. At early flower developmental stages, petals are inaccessible because of their position within the flower, and closed flower buds require removal of the sepals in order to have access to petal primordia and young developing petals. Petal primordia become recognizable at stage 5 as lateral outgrowths of the flank of the floral meristem [1]. Petals remain difficult to access up to stages 9. Starting at stage 9, petal primordia rapidly lengthen mainly through cell expansion and become more easily accessible, although sepals still have to be removed [1]. At early developmental stages, petals are small and fragile, and dissection has to be done quickly and carefully in order to preserve tissue integrity.
2. Scanning electron microscopy (SEM) is widely used in flower morphology studies. Sample preparation for SEM typically requires sample fixation and numerous dehydration steps. Moreover, samples have to be coated with metal particles to achieve better contrast during observation. However, recent progress in SEM technology has made it possible to image the surface of tissues at high resolution without fixation. Moreover, some SEM instruments can now fit on a bench top and are equipped with user-friendly computer software that facilitates

image acquisition. Thus, SEM is now an easily accessible technique to study organ surface morphology without long and laborious sample preparation that can interfere with tissue integrity. In our laboratory, we use a Hirox SH-3000 SEM equipped with a Coolstage MK3 (Deben UK), to freeze the samples.

3. For SEM observation, it is important to work quickly. After dissection, samples have to be mounted immediately on the pre-cooled stage and then cooled to −50 °C. For SEM observation, one can use higher temperatures (up to −10 °C) but the petals will dry out more quickly. One can put several petals on the sample holder at the same time.
4. We usually use 5 kV settings for SEM up to 1,000× magnification. If higher magnification is needed to observe sub-cellular structures on the surface, increase electron intensity up to 30 kV.
5. For fixed samples, make sure that samples are always covered with ethanol. Do not leave samples in the air as they will dry out very quickly.
6. Fixed flowers become fragile after the discoloration step. Manipulate them quickly but carefully.
7. It is easier to observe petals that are arranged in a row (*see* Fig. 2b, e), but it is not absolutely necessary.
8. The most widely used promoter for driving gene expression in plants is the 35S promoter. This promoter results in high levels of expression in most plant tissues and can be used to study the effect of ectopic expression of petal-expressed genes of interest in all plant organs (e.g., in leaves). On the other hand, constitutive expression of the genes of interest in all organs may severely interfere with plant development. In such case, one can use petal-specific promoters, such as the *AP3* promoter, to drive gene miss-expression only in petals. However, the full-length promoter of *AP3* drives gene expression not only in petals but also in stamens [35]. To circumvent this issue, versions of the *AP3* promoter were developed to allow expression only in petals [13]. Based on this study, the *AP3* promoter region −285 (*AP3* 5D6.5) was shown to drive petal-specific expression from early developmental stages onward.
9. In very elegant studies using the Cre/loxP recombination system [32–34] wild-type and mutant tissues were tagged by different fluorescent molecules to unambiguously distinguish between different tissue types [32, 34]. Even though it can achieve visually satisfying results, the Cre/loxP system is not widely used in *Arabidopsis* research, due to variable level of the Cre expression, low level of excision efficiency, and problems in reproducibility [36].

Acknowledgements

We thank Dr. Dali Ma for helpful discussion and for critical reading of this review. Work in our group is funded by the Agence Nationale de Recherche (ANR-09-BLAN-0006), by the Biology Department of the French National Institute for Agronomic Research (INRA), and by the Ecole Normale Supérieure de Lyon.

References

1. Smyth DR, Bowman JL, Meyerowitz EM (1990) Early flower development in Arabidopsis. Plant Cell 2(8):755–767
2. Pyke KA, Page AM (1998) Plastid ontogeny during petal development in Arabidopsis. Plant Physiol 116(2):797–803
3. Hill JP, Lord EM (1989) Floral development in *Arabidopsis thaliana*: a comparison of the wild type and the homeotic pistillata mutant. Can J Bot 67(10):2922–2936
4. Irish VF (1999) Petal and stamen development. Curr Top Dev Biol 41:133–161
5. Noda K, Glover BJ, Linstead P, Martin C (1994) Flower colour intensity depends on specialized cell shape controlled by a Myb-related transcription factor. Nature 369(6482):661–664
6. Honma T, Goto K (2001) Complexes of MADS-box proteins are sufficient to convert leaves into floral organs. Nature 409(6819): 525–529
7. Irish VF (2010) The flowering of Arabidopsis flower development. Plant J 61(6):1014–1028
8. Krizek BA, Fletcher JC (2005) Molecular mechanisms of flower development: an armchair guide. Nat Rev Genet 6(9):688–698
9. Theissen G, Saedler H (2001) Plant biology. Floral quartets. Nature 409(6819):469–471
10. Szecsi J, Joly C, Bordji K, Varaud E, Cock JM, Dumas C, Bendahmane M (2006) BIGPETALp, a bHLH transcription factor is involved in the control of Arabidopsis petal size. EMBO J 25(16):3912–3920
11. Wellmer F, Riechmann JL, Alves-Ferreira M, Meyerowitz EM (2004) Genome-wide analysis of spatial gene expression in Arabidopsis flowers. Plant Cell 16(5):1314–1326
12. Zik M, Irish VF (2003) Global identification of target genes regulated by APETALA3 and PISTILLATA floral homeotic gene action. Plant Cell 15(1):207–222
13. Hill TA, Day CD, Zondlo SC, Thackeray AG, Irish VF (1998) Discrete spatial and temporal cis-acting elements regulate transcription of the Arabidopsis floral homeotic gene APETALA3. Development 125(9):1711–1721
14. Brioudes F, Joly C, Szecsi J, Varaud E, Leroux J, Bellvert F, Bertrand C, Bendahmane M (2009) Jasmonate controls late development stages of petal growth in *Arabidopsis thaliana*. Plant J 60(6):1070–1080
15. Brioudes F, Thierry AM, Chambrier P, Mollereau B, Bendahmane M (2010) Translationally controlled tumor protein is a conserved mitotic growth integrator in animals and plants. Proc Natl Acad Sci USA 107(37):16384–16389
16. Nakayama N, Arroyo JM, Simorowski J, May B, Martienssen R, Irish VF (2005) Gene trap lines define domains of gene regulation in Arabidopsis petals and stamens. Plant Cell 17(9):2486–2506
17. Schmid M, Davison TS, Henz SR, Pape UJ, Demar M, Vingron M, Scholkopf B, Weigel D, Lohmann JU (2005) A gene expression map of *Arabidopsis thaliana* development. Nat Genet 37(5):501–506
18. Varaud E, Brioudes F, Szecsi J, Leroux J, Brown S, Perrot-Rechenmann C, Bendahmane M (2011) AUXIN RESPONSE FACTOR8 regulates Arabidopsis petal growth by interacting with the bHLH transcription factor BIGPETALp. Plant Cell 23(3):973–983
19. Zimmermann P, Hirsch-Hoffmann M, Hennig L, Gruissem W (2004) GENEVESTIGATOR. Arabidopsis microarray database and analysis toolbox. Plant Physiol 136(1):2621–2632
20. Breuninger H, Lenhard M (2010) Control of tissue and organ growth in plants. Curr Top Dev Biol 91:185–220
21. Irish VF (2008) The Arabidopsis petal: a model for plant organogenesis. Trends Plant Sci 13(8):430–436
22. Dinneny JR, Yadegari R, Fischer RL, Yanofsky MF, Weigel D (2004) The role of JAGGED in shaping lateral organs. Development 131(5):1101–1110
23. Hu Y, Xie Q, Chua NH (2003) The Arabidopsis auxin-inducible gene ARGOS controls lateral organ size. Plant Cell 15(9):1951–1961
24. Mizukami Y, Fischer RL (2000) Plant organ size control: AINTEGUMENTA regulates

growth and cell numbers during organogenesis. Proc Natl Acad Sci USA 97(2):942–947
25. Anastasiou E, Kenz S, Gerstung M, MacLean D, Timmer J, Fleck C, Lenhard M (2007) Control of plant organ size by KLUH/CYP78A5-dependent intercellular signaling. Dev Cell 13(6):843–856
26. Disch S, Anastasiou E, Sharma VK, Laux T, Fletcher JC, Lenhard M (2006) The E3 ubiquitin ligase BIG BROTHER controls arabidopsis organ size in a dosage-dependent manner. Curr Biol 16(3):272–279
27. Li Y, Zheng L, Corke F, Smith C, Bevan MW (2008) Control of final seed and organ size by the DA1 gene family in *Arabidopsis thaliana*. Genes Dev 22(10):1331–1336
28. Samalova M, Brzobohaty B, Moore I (2005) pOp6/LhGR: a stringently regulated and highly responsive dexamethasone-inducible gene expression system for tobacco. Plant J 41(6):919–935
29. Moore I, Samalova M, Kurup S (2006) Transactivated and chemically inducible gene expression in plants. Plant J 45(4):651–683
30. Karimi M, Bleys A, Vanderhaeghen R, Hilson P (2007) Building blocks for plant gene assembly. Plant Physiol 145(4):1183–1191
31. Logemann E, Birkenbihl RP, Ulker B, Somssich IE (2006) An improved method for preparing Agrobacterium cells that simplifies the Arabidopsis transformation protocol. Plant Methods 2:16
32. Eriksson S, Stransfeld L, Adamski NM, Breuninger H, Lenhard M (2010) KLUH/CYP78A5-dependent growth signaling coordinates floral organ growth in Arabidopsis. Curr Biol 20(6):527–532
33. Sieburth LE, Drews GN, Meyerowitz EM (1998) Non-autonomy of AGAMOUS function in flower development: use of a Cre/loxP method for mosaic analysis in Arabidopsis. Development 125(21):4303–4312
34. Stransfeld L, Eriksson S, Adamski NM, Breuninger H, Lenhard M (2010) KLUH/CYP78A5 promotes organ growth without affecting the size of the early primordium. Plant Signal Behav 5(8):982–984
35. Irish VF, Yamamoto YT (1995) Conservation of floral homeotic gene function between Arabidopsis and antirrhinum. Plant Cell 7(10):1635–1644
36. Marjanac G, De Paepe A, Peck I, Jacobs A, De Buck S, Depicker A (2008) Evaluation of CRE-mediated excision approaches in *Arabidopsis thaliana*. Transgenic Res 17(2):239–250

Chapter 9

Cell Biological Analyses of Anther Morphogenesis and Pollen Viability in *Arabidopsis* and Rice

Fang Chang, Zaibao Zhang, Yue Jin, and Hong Ma

Abstract

Major advances have been made in recent years in our understanding of anther development through a combination of genetic studies, cell biological technologies, biochemical analysis, microarray and high-throughput sequencing-based approaches. In this chapter, we summarize the widely used protocols for pollen viability staining; the investigation of anther morphogenesis by light microscopy of semi-thin sections; TUNEL assay for programmed tapetum cell death; and laser microdissection procedures to obtain specialized cells or cell layers for carrying out transcriptomics.

Key words Anther anatomy, Semi-thin section, Pollen viability, Programmed cell death, TUNEL assay, Laser microdissection, Pollen starch test, Callose staining

1 Introduction

Male reproductive development of flowering plants begins with the initiation of stamens in the flower and involves the differentiation of five specialized cell layers in each of four anther lobes. At the center of each anther lobe are the pollen mother (meiotic) cells, which are surrounded by four well-organized somatic cell layers: the tapetum; the middle cell layer; the endothecium; and epidermal cell layer from the interior to the surface. The somatic cell layers support the male meiosis and subsequent microspore and pollen mitoses, resulting in the production of the mature pollen grains [1, 2].

One of the most important aspects of male fertility is the viability of pollen. Three methods widely used to assess the pollen viability are as follows: Alexander staining, FDA-PI staining, and I_2/KI (iodine) staining for starch. All these protocols are very useful, simple to handle, and do not require expensive equipment. After staining, normal pollen can be easily distinguished from the abnormal ones by color. Specifically, the Alexander's staining method assesses pollen viability by yielding a magenta-red color in

José Luis Riechmann and Frank Wellmer (eds.), *Flower Development: Methods and Protocols*, Methods in Molecular Biology, vol. 1110, DOI 10.1007/978-1-4614-9408-9_9, © Springer Science+Business Media New York 2014

normal pollen grains and blue-green in aborted pollen grains [3]; the combination of fluorescein diacetate (FDA) and propidium iodide (PI) stains living cells green and dead cells orange [4–6]; and in the I_2/-KI staining method, iodine reacts with starch in normal pollen grains, resulting in a black staining [7]. The I_2/-KI staining method is also used for detecting starch accumulation in other anther tissues [7–9].

One unique feature of male meiosis in flowering plants is the formation and dissolution of a callose-containing cell wall surrounding the meiocytes during and after meiosis [10–12]. The callose wall is composed mainly of β-1,3-glucan, which can be visualized by aniline blue staining. Whereas free aniline blue is colorless, β-1,3-glucan bound with aniline blue exhibits yellowish green fluorescence, revealing the presence of callose [13, 14].

One important process required for normal pollen development is the programmed cell death (PCD) of the tapetum [15–19]. In *Arabidopsis* and rice, tapetum PCD likely commences at the tetrad stage and terminates after the second pollen mitosis [19, 20]. Because PCD is characterized by DNA breaks, TUNEL (terminal deoxynucleotidyl transferase-mediated dUTP nick end labeling) assay is increasingly applied to identify and quantify cell death in many tissues and cell types by catalytically incorporating at the 3′-OH DNA ends labeled-16-dUTP, which can then be visualized by microscopy. In this chapter, we describe the biotin-labeled 16-dUTP based TUNEL method, which is widely used in plants.

In addition to methods used for phenotypic studies of anther development, laser capture microdissection (LCM) can be applied to isolate individual cells from tissue sections. The harvested cells can then provide DNA, RNA, and protein for the profiling of genomic characteristics, gene expression, and the proteome of individual cell types. LCM combined with high-throughput technologies such as microarray analysis and next-generation sequencing makes it an extremely powerful tool for understanding the molecular events in different cell and tissue types [21–26]. In this chapter we describe a protocol for preparing samples for LCM of anthers and other floral tissues for high-throughput transcriptomics.

2 Materials

2.1 Alexander Red Staining

1. A Leica microscope (Leica Microsystems Ltd., Wetzlar, Germany) with a Spot Insight digital camera (Diagnostic instruments, Inc. Sterling Heights, MI, USA).
2. Carnoy's fixative: 6:3:1 (v/v) ethanol–chloroform–acetic acid.
3. Alexander's staining solution: ethanol, malachite green, glycerol, phenol, chloral hydrate, acid fuchsin, and glacial acetic acid. Mix the components in the order given below to obtain a total volume of 100 mL of Alexander's staining solution (*see* **Note 1**).

95 % ethanol	10 mL
Malachite green (1 % solution in 95 % ethanol)	1 mL
Distilled water	52.5 mL
Glycerol	25 mL
Phenol	5 mL
Chloral hydrate	0.5 g
Acid fuchsin [1 % (w/v) solution in water]	5 mL
Orange G [1 % (w/v) solution in water]	0.5 mL
Glacial acetic acid	1 mL

2.2 FDA Staining and Imaging

1. FDA-PI staining solution: 10 μg/mL PI and 0.5 μg/mL FDA in distilled water.

2.3 Iodine Pollen Starch Test

1. I_2/KI solution: 0.2 % (w/v) potassium iodide and 1 % iodine in distilled water. *See* **Note 2** for how to prepare the solution.

2.4 Anther Anatomy Using Semi-thin Sections

1. Glass slides and glass covers.
2. 42 °C heat plate.
3. A rotary microtome (e.g., RM 2265 from Leica Microsystems).
4. A fluorescence microscope with CCD camera.
5. FAA fixative: formaldehyde, acetic acid, and ethanol. Mix 1 mL of 38 % formaldehyde, 1 mL of acetic acid, and 18 mL of 50 % ethanol to obtain a total volume of 20 mL of FAA fixative.
6. Technovit 7100 kit including Technovit 7100, liquid, Hardener I, and Hardener II (Heraeus Kulzer Technik, Wehrheim, Germany).
7. 10× Phosphate Buffered Saline (PBS, pH 7.4): NaCl, KCl, Na_2HPO_4, KH_2PO_4. Dissolve 80 g NaCl, 2 g KCl, 14.2 g Na_2HPO_4, and 2.7 g KH_2PO_4 in 900 mL distilled water. Adjust the PH of the buffer to 7.4 using 10 N HCl. Then add distilled water to the buffer to obtain a total volume of 1 L of 10× PBS buffer.
8. Toluidine Blue O staining solution: 0.005 % (w/v) Toluidine Blue O in 0.1 M PBS, pH 7.4 (*see* **Note 3**).

2.5 Callose Staining

1. Resin-embedded inflorescence tissues.
2. 0.067 M phosphate buffer (pH 6.8): $Na_2HPO_4 \cdot 12H_2O$, KH_2PO_4. Dissolve 11.81 g $Na_2HPO_4 \cdot 12H_2O$ and 4.5 g KH_2PO_4 in 1 L distilled water.
3. Aniline blue staining solution: 0.05 % (w/v) aniline blue in 0.067 M phosphate buffer (pH 6.8).

2.6 Detection of Programmed Cell Death Using TUNEL Assay

1. Paraplast X-tra (Sigma Chemical Company, St. Louis, MO, USA).
2. 10× TdT buffer: 0.3 M Tris–HCl (pH 7.2), 1.4 M sodium cacodylate ($C_2H_6AsNaO_2$), 1 mM DTT (DL-Dithiothreitol, $C_4H_{10}O_2S_2$).
3. TdT incubation buffer: $CoCl_2$, Bio-16-dUTP, and TdT in 1× TdT buffer. Mix 41.5 μL of deionized water, 5 μL of 10× TdT buffer, 2 μL of 25 mM $CoCl_2$, 1 μL of 1 mM Bio-16-dUTP, and 0.5 μL of 25 U/μL TdT in order to get total volume of 50 μL of TdT incubation buffer.
4. 10× Phosphate Buffered Saline (PBS, pH 7.4): NaCl, KCl, Na_2HPO_4, KH_2PO_4. Dissolve 80 g NaCl, 2 g KCl, 14.2 g Na_2HPO_4, and 2.7 g KH_2PO_4 in 900 mL distilled water. Adjust the pH of the buffer to 7.4 using 10 N HCl. Then add distilled water to the buffer to obtain a total volume of 1 L of 10× PBS buffer.
5. ExtrAvidin-peroxidase (Sigma): 1:50 diluted in 1× PBS buffer (pH 7.4), 1 % (w/v) Bovine Serum Albumen (BSA), 0.5 % (v/v) Tween-20.
6. Terminating buffer: 0.3 M NaCl, 30 mM sodium citrate ($C_6H_5Na_3O_7$).
7. AEC: 3-amino-9-ethylcarbazole (Sigma).

2.7 Sample Preparation for Laser Capture Microdissection

1. A laboratory microwave oven (e.g., BP-111RS; Microwave Research & Application, Inc, Laurel MD, USA).
2. A rotary microtome (e.g., RM 2135; Leica Microsystems Ltd.).
3. Positively (+) charged slides (e.g., from Thermo Fisher Scientific).
4. Acetone and xylene.
5. Paraffin.

3 Methods

3.1 Alexander Staining and Photography

This Alexander staining procedure is used to distinguish normal pollen grains from aborted pollen grains by color (*see* **Note 4**).

1. Collect stage 12 flower buds, fix them in a microfuge tube with 1 mL of Carnoy's fixative for at least 2 h.
2. Transfer the flower buds to a glass slide.
3. Remove the fixative from the plant material with absorbent paper.
4. Dissect the buds under a dissecting microscope to obtain individual anthers.

5. Place individual anthers into a microtube with an appropriate volume of Alexander's staining solution to submerge the anthers completely.
6. Incubate at 25 °C for 30 min.
7. Place the anthers on a slide, cover it with a coverslip, and apply even pressure on the coverslip to ensure that all anthers converge to one plane. The coverslip can be sealed using nail polish or wax.
8. Visualize the anthers using a Leica microscope. Take micrographs using a Spot Insight digital camera (Diagnostic instruments, Inc. Sterling Heights, MI, USA). Viable pollen grains should be stained magenta-red and aborted pollen grains stained blue-green (*see* Fig. 1a, b).

3.2 Iodine Pollen Starch Test

1. Collect stage 12–13 flowers.
2. Dissect out anthers and prefixed in 70 % ethanol.
3. Place the flowers on a glass slide with one drop of water.
4. Slice anthers open with forceps until they release pollen grains.
5. Add one or two drops (~5–10 μL) of 1 % I_2/KI solution.
6. Incubate for 5 min to stain the pollen grains.
7. Cover the pollen with a coverslip, observe, and photograph. The normal mature pollen grains containing starch granules stain black; however, the immature pollen grains appear orange to red (*see* Fig. 1c).

3.3 FDA Staining and Imaging

1. Collect pollen grains from plants and place them into the FDA-PI staining solution for 5 min at room temperature.
2. Observe and photograph using a fluorescence filter block for blue excitation and a long-pass emission filter to detect the red fluorescence from PI and the green fluorescence from fluorescein. This method stains living cells green and dead cells orange (*see* Fig. 1d).

3.4 Anther Anatomy Using Semi-thin Sections

1. Collect the entire flower inflorescence, quickly place it into the FAA fixative.

   Make sure to use at least three times as much FAA fixative as the volume of the tissue.
2. Keep the container with tissues under vacuum for about 15 min.
3. Release the vacuum slowly and the tissue is likely to sink down.
4. Apply vacuum a second time to improve tissue preservation.
5. Change the fixative with fresh fixative and incubate overnight.

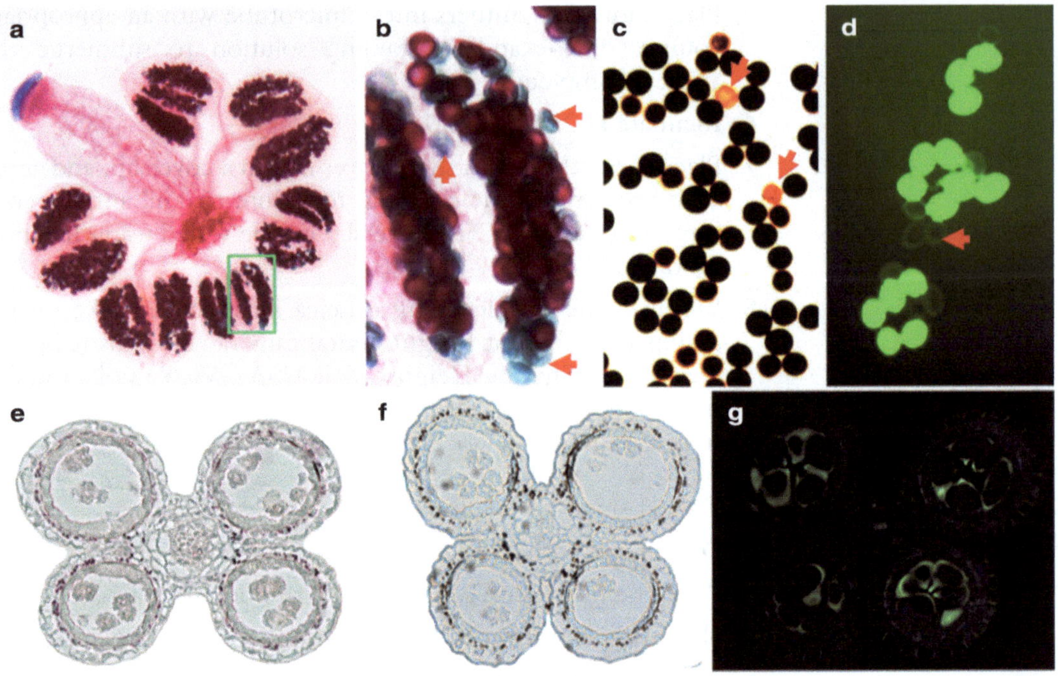

Fig. 1 Detection of pollen viability and anther starch accumulation. (**a–b**) Anthers and pollen grains stained with Alexander's staining solution. (**a**) An *Arabidopsis* stage-12 flower with six anthers. The area marked with a *green box* is shown in higher magnification in (**b**). (**d**) Green florescence from FDA-stained *Arabidopsis* pollen grains. **c**, **e**, and **f** present starch accumulation in rice pollen (**c**) and anther before (**e**) and after (**f**) drought stress by I_2/KI staining. (**g**) The callose structure surrounding the microsporocytes in an anther of rice after aniline blue staining. *Red arrowheads* indicate aborted pollen grains

6. Dehydrate for 30 min in each of the following ethanol concentrations:

   70 % ethanol.
   85 % ethanol.
   95 % ethanol.

   Subsequently, dehydrate twice in 100 % ethanol for 2 h. Add 0.1 % (w/v) Eosin Y to the second 100 % ethanol wash.
7. After dehydration, discard the 100 % ethanol and add a mixture of 100 % ethanol and an equal volume of Base liquid Technovit 7100 resin. Mix gently and incubate for 1–2 h.
8. Prepare the infiltration solution: add 1 g of Hardener I to 100 mL of Base liquid Technovit 7100 and mix. It takes approximately 5 min to dissolve the hardener thoroughly. The prepared solution remains stable for approximate 4 weeks at 4 °C. We usually prepare fresh before use.
9. Transfer the inflorescence to the infiltration solution and incubate for 3 h to overnight.
10. Add 1/15 volume of Hardener II with a pipette, mix gently and thoroughly.

11. Add 200 μL of the solution into the Histoform S/Q, then place the infiltrated inflorescence into the Histoform and adjust its position as required. We suggest aligning the vertical axis of the inflorescence to the Histoform if you want to prepare transverse sections.
12. Incubate at 40 °C in an oven for more than 1 day for solidification. Then perform the following sectioning procedure.
13. Adjust a heat plate to 42 °C.
14. Trim the molds to remove the embedding material without the tissue, to form a rectangular block with surfaces as close to the tissue as possible.
15. Prepare sections with a rotary microtome. We usually use sections of 0.5 μm in thickness.
16. Add several drops of water on the glass slides and transfer the sections to the top of the water to remove wrinkles.
17. Label the slides and place them on the 42 °C heat plate until the sections are completely dry. It usually takes 30 min to dry the sections and to affix them onto the glass slides.
18. Stain the sections using Toluidine Blue O staining solution for 1 min.
19. Wash the sections twice in water.
20. Dry the sections.
21. Observe and photograph. Figure 2 shows images from Arabidopsis anther treated according to this method.

3.5 Callose Staining

1. Collect floral buds, fix them, and then embed them with Technovit 7100 resin following **steps 1–12** of Subheading 3.4.
2. Prepare sections and dry the sections following **steps 13–17** of Subheading 3.4.
3. Stain the sections with 0.05 % (w/v) aniline blue staining solution for 5–10 min.
4. Wash three times with 0.067 M phosphate buffer (pH 6.8).
5. Observe the fluorescence under the microscope with a 390–440 nm excitation filter and a 478 nm blocking filter.

3.6 Detection of Programmed Cell Death Using TUNEL Assay

See Fig. 3 for a protocol overview of this assay.

1. Collect inflorescences with floral buds and fix them in glass vials following the **steps 1–5** of Subheading 3.4.
2. Keep Paraplast X-tra at 55 °C.
3. Dehydrate sample through 45 min washes in each of the following solutions: 50 % ethanol, 60 % ethanol, 70 % ethanol, 80 % ethanol, 90 % ethanol, and 95 % ethanol with 0.1 % (w/v) Eosin Y.

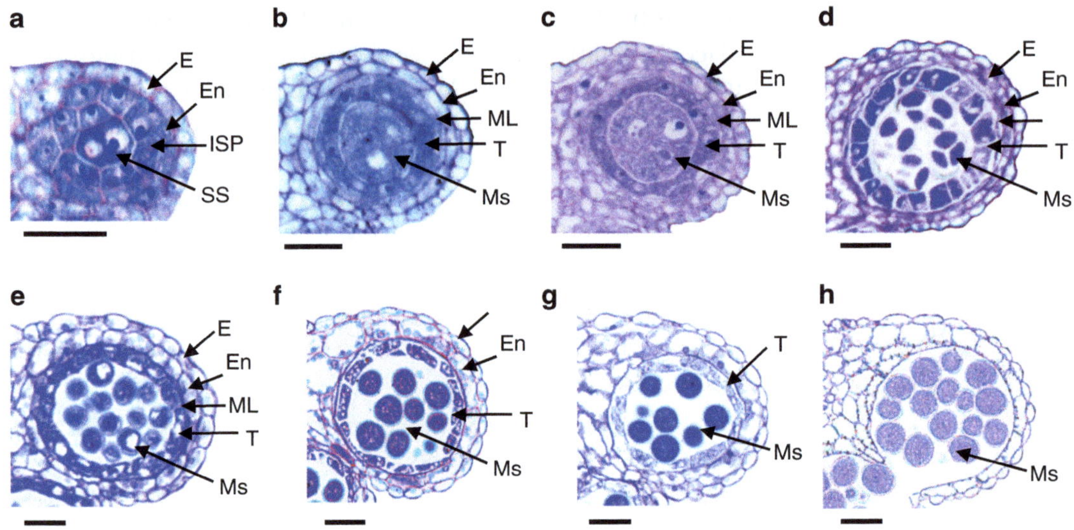

Fig. 2 The transverse view of rice anther development stages. Stage 1a (*A*), Stage 2a (*B*), stage 2–3 (*C*), stage 3a (*D*), stage 4 (*E*), stage 5a (*F*), stage 5–6 (*G*), stage 6 (*H*), stage 7 (*I*), stage 8 (*J*), stage 9 (*K*), stage 10 (*L*), stage 11 (*M*), stage 12 (*N*), stage 13 (*O*), stage 14 (*P*). Sections were generated and stained using the protocol described in Subheading 3.4. Scale bar: 50 μm

4. Incubate the floral buds in 95 % ethanol with 0.1 % (w/v) Eosin Y for 2 h.
5. Replace the 100 % ethanol by Histo-Clear by incubating samples for 30 min. in the following solutions:

   25 % Histo-Clear, 75 % ethanol.
   50 % Histo-Clear, 50 % ethanol.
   75 % Histo-Clear, 25 % ethanol.
   100 % Histo-Clear.
   Repeat with fresh Histo-Clear.
6. Place about 20 chips of Paraplast into the glass vials (1/4 volume) and incubate overnight at room temperature.
7. Place the vial at 42 °C till the chips melt into the solution.
8. Move the vial to 55 °C and incubate for 12 h.
9. Change the melted Paraplast every 12 h for total of at least six changes.
10. Turn on the heat plate and adjust to 70 °C. Place pre-labeled molds, a pair of forceps, a needle, and a 5 mL tip in the heat plate to preheat them.
11. Bring the vials of floral buds and quickly transfer the material into the molds.
12. Using the preheated forceps and needle to orient the tissue quickly, then pour more wax to fill up the mold.

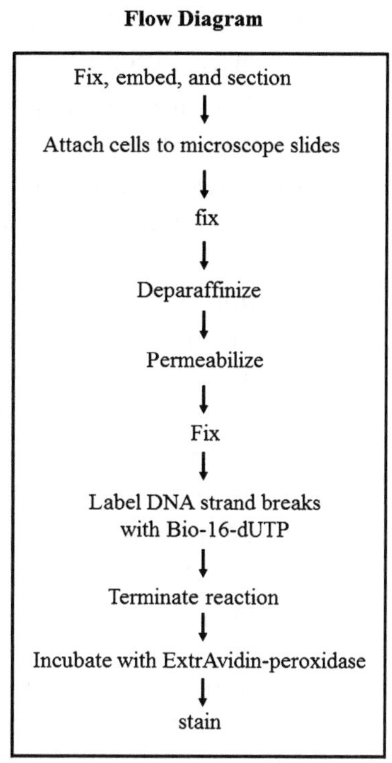

Fig. 3 Protocol overview for the TUNEL assay

13. Keep mold undisturbed to allow solidification of the wax. Remove the molds and store the blocks at 4 °C.
14. Adjust the heat plate to 42 °C.
15. Label the slides, put them on the heat plate and add several drops of water on each slide.
16. Trim the molds to remove excess embedding material. Cut as close to the tissue as possible and generate rectangular blocks.
17. Place the mold into a rotary microtome and cut 8–12 μm sections.
18. Put several drops of water onto the glass slides and transfer the sections onto the water surface.
19. Incubate for several minutes to smoothen the sections.
20. Carefully remove the water with absorbent paper and avoid air bubbles.
21. Place the slides back on the heat plate and bake them overnight at 42 °C. The sections will adhere to the slides after this step.
22. Fix the inflorescence sections by immersing the slides in 4 % methanol-free formaldehyde solution in 1× PBS buffer in a Coplin jar for 15 min at room temperature.

23. Wash the slides by immersing them into an appreciate volume of 1× PBS buffer, and incubate at 70 °C for 10 min.
24. Transfer the slides into fresh xylene for 5 min at room temperature. Repeat once for a total of two xylene washes.
25. Wash by immersing the slides in 96 % ethanol for 3 min at room temperature. Repeat once for a total of two 96 % ethanol washes.
26. Rehydrate the samples by sequentially immersing the slides through graded ethanol washes: 96 %, 90 %, and 80 % ethanol for 3 min each at room temperature. Repeat once for each concentration of ethanol.
27. Wash the sample by immersing the slides in deionized water for 3 min at room temperature.
28. Wash the sample by immersing the slides in 10 mM Tris–HCl (pH 8.0) for 5 min at room temperature.
29. Drain off the excess water and add 100 μL of 20 μg/mL proteinase K to each slide so that the sections are covered by solution. Place the slides on a flat surface and incubate for 15 min at room temperature.
30. Rinse the slides in deionized water for 2 min at room temperature. Repeat three times for a total of four washes.
31. Fix the inflorescence sections again as descript in **step 22**.
32. Wash the samples by immersing the slides in deionized water for 2 min at room temperature. Repeat three times for a total of four washes.
33. Remove excess liquid and add 100 μL of fresh prepared TdT incubation buffer to each slide to cover the sections.
34. Put the slides in a dark humidified chamber, incubate the box in 37 °C for 60 min. Do not let the sections to dry out and avoid exposure to light.
35. Wash the slides three times in deionized water in the Coplin jar at room temperature.
36. Terminate the reaction by immersing the slides in the terminating buffer for 15 min at room temperature.
37. Wash the sections by immersing the slides in 1× PBS buffer (pH 7.4) for 5 min at room temperature.
38. Transfer the sections into 1× PBS buffer (pH 7.4) containing 2 % (w/v) BSA and incubate for 10 min at room temperature.
39. Wash the sections by immersing the slides in 1× PBS buffer (pH 7.4) for 5 min at room temperature.
40. Incubate the samples in ExtrAvidin-peroxidase solution for 15 min.

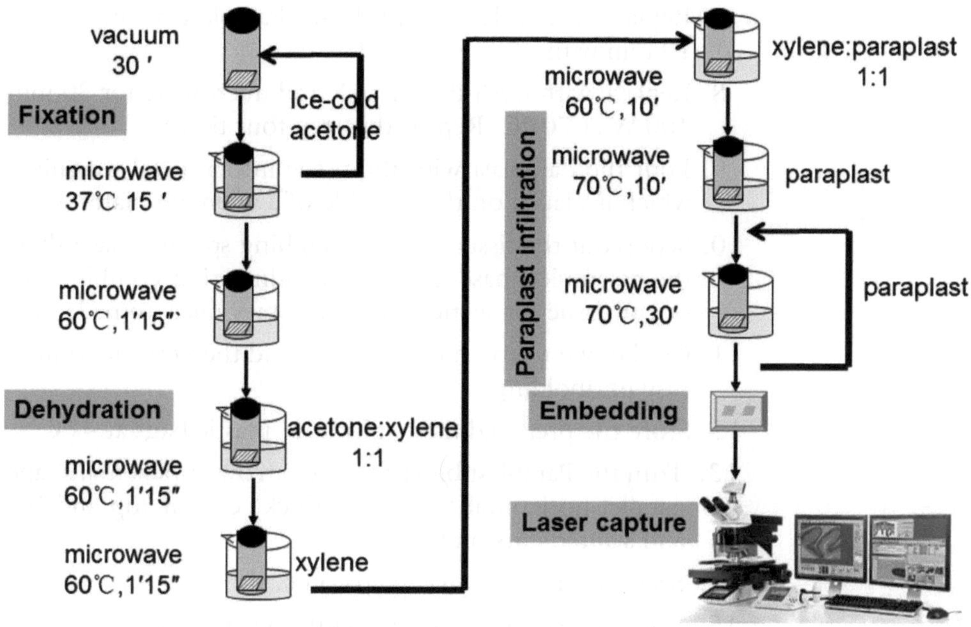

Fig. 4 Flow diagram of sample preparation for LCM

41. Incubate the samples in 1× PBS buffer (pH 7.4) containing 2 % (w/v) BSA for 10 min.
42. Wash the sections by immersing the slides in 1× PBS buffer (pH 7.4) for 5 min. Repeat three times for a total of four washes.
43. Stain with AEC staining solution for 30 min in 37 °C.
44. Rinse in deionized water three times.
45. Drain off excess water.
46. Observation and photograph under microscope. Red color indicates signals for broken DNA ends.

3.7 Sample Preparation for Laser Capture Microdissection

See Fig. 4 for a flow diagram of this assay.

1. Collect inflorescences with floral buds and immediately place them in pure ice-cold acetone.
2. Heat for 15 min with 400 W at 37 °C in a microwave oven.
3. Vacuum (400 mmHg) at room temperature for 30 min to help the infiltration.
4. Transfer floral buds to fresh acetone and microwave for 15 min with 400 W at 37 °C. Repeat twice.
5. Transfer floral buds to an acetone/xylene (1:1) solution. Microwave for 75 s with 500 W at 60 °C.
6. Transfer floral buds to pure xylene. Microwave for 75 s with 500 W at 60 °C.

7. Replace the xylene with fresh Paraplast-X, microwave for 10 min with 400 W at 70 °C.
8. Replace with fresh Paraplast-X and microwave for 30 min with 400 W at 70 °C. Repeat this step four times.
9. Pour the Paraplast with the tissue into a metal weighing dish, which is placed on the hot side of a warming plate.
10. Scoop out the tissues with a weighing spatula, place them into the assembled base mold/embedding ring combination, and orient tissue for sectioning on a rotary microtome later.
11. Cool down to room temperature and then transfer onto ice for easy un-molding.
12. Store the prepared tissue blocks in plastic bags at 4 °C.
13. Trim the Paraplast blocks into a narrow trapezoidal shape with parallel horizontal (5–10 μm thick) cuts at top and bottom, and slanted cuts on the sides.
14. Section using a rotary microtome.
15. Place ribbons of sections gently onto the surface of diethylpyrocarbonate (DEPC)-treated water on positively (+) charged slides, which were prewarmed to 40 °C on the heat plate; incubate for 5 min or more.
16. Remove the water by tipping the slide onto absorptive paper towels while holding one end of the ribbon with a fine paintbrush. Wick off the residual water with tissue paper.
17. Store the dry slides at 4 °C until PALM procedure.
18. Warm up the slides to room temperature.
19. Deparaffinize the sections in two changes of xylene (5 min each xylene change).
20. Wash the sections three times with DEPC-treated water.
21. Place on the microscope stage to mark tissue regions to be collected.
22. Catapult tissues into inverted tube-caps (500 μL microcentrifuge tubes), which were filled with about 50 μL RNA extraction buffer (e.g., XB from the PicoPure RNA Isolation kit). After collecting sufficient amount of tissue, extract RNA according to an RNA extraction protocol.

4 Notes

1. The final stain solution should be prepared by adding the components in the order shown. Store solution in the dark.
2. Dissolve 0.2 g potassium iodide in a small amount of distilled water. Add 1 g iodine crystal while stirring. Once dissolved, add water to bring up the volume to 100 mL. Keep the solution in a light stopped brown glasses bottle.

3. Other laboratories use 0.01 % (w/v) Toluidine Blue O, but in our hands, concentrations of 0.0025–0.005 % (w/v) work better, especially for young flower tissues. The prepared Toluidine Blue O staining solution should be stored in dark.
4. This method is used to analyze whether the pollen is well developed or not. In contrast, it cannot be used for the detection of the real time effects caused by a certain treatment to pollen vitality.

References

1. Ma H (2005) Molecular genetic analyses of microsporogenesis and microgametogenesis in flowering plants. Annu Rev Plant Biol 56: 393–434
2. Singh MB, Bhalla PL (2007) Control of male germ-cell development in flowering plants. Bioessays 29:1124–1132
3. Alexander MP (1969) Differential staining of aborted and nonaborted pollen. Stain Technol 44:117–122
4. Jones KH, Senft JA (1985) An improved method to determine cell viability by simultaneous staining with fluorescein diacetate-propidium iodide. J Histochem Cytochem 33: 77–79
5. Zhang H, Zdolsek JM, Brunk UT (1991) Effects of alloxan and reducing agents on macrophages in culture. APMIS 99:1038–1048
6. Smith DJ, Oliver CE, Caton JS et al (2005) Effect of sodium [36Cl]chlorate dose on total radioactive residues and residues of parent chlorate in beef cattle. J Agric Food Chem 53:7352–7360
7. Zhu QH, Ramm K, Shivakkumar R et al (2004) The ANTHER INDEHISCENCE1 gene encoding a single MYB domain protein is involved in anther development in rice. Plant Physiol 135:1514–1525
8. Pressman E, Peet MM, Pharr DM (2002) The effect of heat stress on tomato pollen characteristics is associated with changes in carbohydrate concentration in the developing anthers. Ann Bot 90:631–636
9. Mamun EA, Alfred S, Cantrill LC et al (2006) Effects of chilling on male gametophyte development in rice. Cell Biol Int 30:583–591
10. Rhee SY, Somerville CR (1998) Tetrad pollen formation in quartet mutants of *Arabidopsis thaliana* is associated with persistence of pectic polysaccharides of the pollen mother cell wall. Plant J 15:79–88
11. Worrall D, Hird DL, Hodge R et al (1992) Premature dissolution of the microsporocyte callose wall causes male sterility in transgenic tobacco. Plant Cell 4:759–771
12. Yeung EC, Oinam GS, Yeung SS et al (2011) Anther, pollen and tapetum development in safflower, *Carthamus tinctorius* L. Sex Plant Reprod 24:307–317
13. Dong X, Hong Z, Sivaramakrishnan M et al (2005) Callose synthase (CalS5) is required for exine formation during microgametogenesis and for pollen viability in Arabidopsis. Plant J 42:315–328
14. Zhang ZB, Zhu J, Gao JF et al (2007) Transcription factor AtMYB103 is required for anther development by regulating tapetum development, callose dissolution and exine formation in Arabidopsis. Plant J 52:528–538
15. Phan HA, Iacuone S, Li SF et al (2011) The MYB80 transcription factor is required for pollen development and the regulation of tapetal programmed cell death in *Arabidopsis thaliana*. Plant Cell 23:2209–2224
16. Li H, Yuan Z, Vizcay-Barrena G et al (2011) PERSISTENT TAPETAL CELL1 encodes a PHD-finger protein that is required for tapetal cell death and pollen development in rice. Plant Physiol 156:615–630
17. Rogers HJ (2006) Programmed cell death in floral organs: how and why do flowers die? Ann Bot 97:309–315
18. Li X, Gao X, Wei Y et al (2011) Rice APOPTOSIS INHIBITOR5 coupled with two DEAD-box adenosine 5′-triphosphate-dependent RNA helicases regulates tapetum degeneration. Plant Cell 23:1416–1434
19. Varnier AL, Mazeyrat-Gourbeyre F, Sangwan RS et al (2005) Programmed cell death progressively models the development of anther sporophytic tissues from the tapetum and is triggered in pollen grains during maturation. J Struct Biol 152:118–128
20. Zhang C, Guinel FC, Moffatt BA (2002) A comparative ultrastructural study of pollen development in *Arabidopsis thaliana* ecotype

Columbia and male-sterile mutant aptl-3. Protoplasma 219:9–71

21. Takahashi H, Kamakura H, Sato Y et al (2010) A method for obtaining high quality RNA from paraffin sections of plant tissues by laser microdissection. J Plant Res 123:807–813
22. Barbazuk WB, Emrich SJ, Chen HD et al (2007) SNP discovery via 454 transcriptome sequencing. Plant J 51:910–918
23. Ohtsu K, Takahashi H, Schnable PS et al (2007) Cell type-specific gene expression profiling in plants by using a combination of laser microdissection and high-throughput technologies. Plant Cell Physiol 48:3–7
24. Taylor TB, Nambiar PR, Raja R et al (2004) Microgenomics: Identification of new expression profiles via small and single-cell sample analyses. Cytometry A 59:254–261
25. Zheng Z, Andersson AF, Ye W et al (2011) A method for metagenomics of Helicobacter pylori from archived formalin-fixed gastric biopsies permitting longitudinal studies of carcinogenic risk. PLoS One 6:e26442
26. Murphy SJ, Cheville JC, Zarei S et al (2012) Mate pair sequencing of whole-genome-amplified DNA following laser capture microdissection of prostate cancer. DNA Res 19: 395–406

Chapter 10

Molecular Cell Biology of Male Meiotic Chromosomes and Isolation of Male Meiocytes in *Arabidopsis thaliana*

Yingxiang Wang, Zhihao Cheng, Pingli Lu, Ljudmilla Timofejeva, and Hong Ma

Abstract

Plants typically produce numerous flowers whose meiotic chromosomes are relatively easy to observe, making them excellent structures for studying the cellular processes underlying meiosis. In recent years, breakthroughs in light and electron microscopic technologies for small chromosomes, combined with molecular genetic methods, have resulted in major advances in the understanding of meiosis in the model plant *Arabidopsis thaliana*. In this chapter, we summarize protocols for basic cytology, fluorescence in situ hybridization, immunofluorescence, electron microscopy, and isolation of male meiocytes for the analysis of *Arabidopsis* meiosis.

Key words Meiosis, Chromosome spread, Electron microscopy, FISH, Immunofluorescence, Meiocytes

1 Introduction

Meiosis is essential for eukaryotic sexual reproduction and redistributes genetic diversity among individual progeny. In flowering plants, male and female meioses occur in anthers and in ovules, respectively [1]. Because male meiocytes are formed in larger numbers than female meiocytes and relatively easy to obtain, they are studied much more often than female meiocytes. Traditionally, early studies were mainly limited to plants with relatively large chromosomes, such as lily and maize, because of the ease with which chromosome behavior could be observed using light and electron microscopy [2–4]. With the improvements of light and electron microscopic technologies, cytological procedures have been developed even for small chromosomes. Additionally, *Arabidopsis thaliana* offers many advantages of a model system for plant biology, such as a short life cycle of about 6 weeks, a small stature, a small genome size of approximately

José Luis Riechmann and Frank Wellmer (eds.), *Flower Development: Methods and Protocols*, Methods in Molecular Biology, vol. 1110, DOI 10.1007/978-1-4614-9408-9_10, © Springer Science+Business Media New York 2014

125 Mb distributed over five chromosomes ($2n = 10$), and a great deal of available genetic resources, such as those at the Arabidopsis Biological Resource Center (ABRC) [5]. Such advantages have made *Arabidopsis* a powerful model organism for studying the cellular and molecular biology of flowering plants as well as for investigating the fundamental processes in meiosis [6–8].

Meiosis entails one round of DNA replication and two successive rounds of nuclear division, meiosis I and meiosis II. One of the most fundamental aspects in prophase I is the establishment of the homologous chromosome association, which assures the accurate segregation of chromosomes during meiosis I. Structures associated with chromosome synapsis and recombination such as synaptonemal complexes (SCs) and recombination nodules (RNs) can be seen in various organisms by electron microscopy (EM) [9], which have also been used to observe SCs and RNs successfully in *Arabidopsis* [10–13].

Fluorescence in situ hybridization (FISH) is a powerful method to analyze homologous chromosome interactions, including pairing and synapsis, using specific DNA sequences as probes [14]. *Arabidopsis* provides an attractive system for such studies because of the small number of chromosomes and the availability of a large number of bacterial artificial chromosome (BAC) clones covering the whole genome. The FISH method described in this chapter is generally applicable to either single-copy DNA sequences, such as BAC clones, or repetitive DNA sequences.

Another often-used method for meiotic studies is immunocytology, for protein localization in meiotic cells, such as proteins of the axial/lateral elements and/or transverse filaments of synaptonemal complexes, centromere-specific histones, and recombination-associated proteins. However, unlike the Carnoy's fixation for chromosome spread with staining by 4′-6-diamidino-2-phenylindole (DAPI), the protocol for immunocytology usually utilizes formaldehyde fixation and gently squashed chromosomes, making it difficult to obtain good images of chromosome spreads. Although applications of this method have led to great advances in the understanding of meiotic processes in *Arabidopsis*, rice, and maize [14–16], previous procedures usually generated relatively blurry images of chromosomes under UV light. Here, we present a method for immunofluorescence of *Arabidopsis* meiotic chromosomes with the improved images under UV light.

With the development of next-generation DNA sequencing technologies, tissue- or cell-specific transcriptomes have been analyzed in many species [17]. However, the *Arabidopsis* male or female meiocytes constitute only a tiny portion of reproductive organs and are surrounded by somatic cells in the anther or the ovule [18, 19], making it challenging to collect pure meiocytes. Consequently, the transcriptional profile of pure meiocytes can greatly facilitate the identification of meiotic genes. Here, we describe a relatively simple method to collect the *Arabidopsis* male meiocytes using a micromanipulator system [20].

2 Materials

2.1 *Arabidopsis* Plants

Seed sterilization: *Arabidopsis* seeds; 1.5 mL microfuge tubes; 20 % bleach. Rinse the seeds using autoclaved ddH_2O four times in the laminar flow hoods. Plant germination: Seeds were placed on petri dishes with 0.5× MS medium. The plates were kept in the dark for 2–3 days at 4 °C and then were placed vertically in a growth chamber with long-day conditions (16/8-h light/dark cycle) at 18–22 °C temperature for seed germination. Growth in soil: Plastic trays with small pots for further growth after 7–10 days on plates. Seedlings were transplanted into the small pots with soil and grown until use for experiments.

2.2 Light Microscopy of Tetrads

1. Carnoy's fixative solution: 3 volumes of 100 % ethanol, 1 volume of glacial acetic acid.
2. Basic fuchsine: Basic fuchsine, ethyl alcohol, aqueous phenol. 1 volumes of stock A, 3 volumes of stock B. Stock A: Dissolve 3.0 g basic fuchsine in 100 mL 95 % ethyl alcohol; stock B: make 5 % aqueous phenol by adding 6 mL 88 % liquid phenol to 94 mL water (H_2O).

2.3 Chromosome Spread or Squash

1. 10× PBS: NaCl, KCl, Na_2HPO_4, KH_2PO_4. Prepare 1 L of 10× PBS by dissolving 80 g NaCl, 2 g KCl, 14.2 g Na_2HPO_4, and 2.7 g KH_2PO_4 in 900 mL deionized water. Use 10 N HCl to adjust the pH to 7.4.
2. Fixative: 2 % paraformaldehyde, 0.1 % (v/v) Triton X-100 in 1× PBS buffer pH 7.4.
3. 10 mM citrate buffer (pH 4.5): Sodium citrate, citric acid. Mix 4.45 mL 0.1 M sodium citrate with 5.55 mL 0.1 M citric acid and make up to 100 mL.
4. Digestion cocktail for the detection of chromosomes with DAPI and FISH: 0.5 % cytohelicase, 0.3 % cellulase, and 0.3 % pectolyase in 10 mM citrate buffer (pH 4.5).
5. Digestion cocktail for immunolocalization: 0.5 % (w/v) cytohelicase, 3 % (w/v) cellulase, 3 % (w/v) pectinase in 0.01 M citrate buffer pH 4.5.
6. 60 % acetic acid: 6 volumes of glacial acetic acid with 4 volumes of sterile deionized water.
7. Counterstaining solution: Vectashield mounting medium with DAPI (Vector Laboratories).

2.4 Fluorescent In Situ Hybridization

2.4.1 Templates for Synthesis of Probes

1. A template for probing repetitive centromere DNA: A plasmid derived from the pUC18 vector, carrying an insert of a 180 bp repetitive sequence of Arabidopsis centromeres.
2. Templates for probing single-copy DNA: *Arabidopsis* BAC clones F1N21 and F9K16 (ordered from ABRC).

2.4.2 Labeled Oligonucleotide Probe

Telomere: CY3-$(CCCTAAA)_4$ modified at the 5′ terminus with the CY3 fluorescence marker.

2.4.3 DIG-Nick Translation

1. Digoxigenin (DIG)-Nick Translation Mix (Roche, Diagnostics, IN, USA): 5× conc. stabilized reaction buffer in 50 % (v/v) glycerol, *E. coli* DNA polymerase I, DNase I, 0.25 mM dATP, 0.25 mM dCTP, 0.25 mM dGTP, 0.17 mM dTTP, and 0.08 mM DIG-11-dUTP 160 μL (40 labeling reactions).
2. 0.5 M ethylenediaminetetraacetic acid (EDTA).
3. Nucleotide purification spin column of ProbeQuant G-50 Microcolumns (Roche, Diagnostics, IN, USA).
4. Agarose gel (1 % agarose, 50 mL volume): Place 0.5 g agarose into a 250 mL conical flask; add 50 mL of 0.5× TBE buffer (Tris base, boric acid, EDTA) to mix by swirling; place flask in a microwave oven, heat for about 1 min to melt the agarose, and then cool the solution to about 60 °C. Add 1 μL GoldenView (a dye to visualize nucleic acid, similar to ethidium bromide) (10 mg/mL, Shanghai Yangguang, China), and pour the gel mix slowly into a gel-casting tray.
5. Loading buffer: Prepare 10 mL loading buffer by dissolving 25 mg bromophenol blue, xylene cyanol, and 4 g sucrose in 10 mL H_2O. Store as 1 mL aliquots at room temperature.

2.4.4 Hybridization, Washing, and Immunodetection

1. Hybridization cocktail: 45 % formamide, 10 % dextran sulfate, 250 μg/mL sheared and denatured DNA (from fish).
2. 20× SSC buffer: NaCl, sodium citrate ($C_6H_5O_7Na_3{\cdot}2H_2O$). Prepare 1 L 20× SSC buffer by dissolving 175.3 g NaCl and 88.2 g sodium citrate ($C_6H_5O_7Na_3{\cdot}2H_2O$) in 900 mL of deionized water. Use 10 N HCl to adjust the pH to 7.0.
3. 10× PBS: NaCl, KCl, Na_2HPO_4, KH_2PO_4. Prepare 1 L of 10× PBS by dissolving 80 g NaCl, 2 g KCl, 14.2 g Na_2HPO_4, and 2.7 g KH_2PO_4 in 900 mL deionized water. Use 10 N HCl to adjust the pH to 7.4.
4. Wash buffer I: 1× PBS buffer, 1 % (v/v) Triton X-100.
5. Wash buffer II: 1× PBS buffer, 0.1 % (v/v) Tween 20.
6. Blocking buffer: 1× PBS buffer, 5 % (w/v) bovine serum albumin (BSA), 0.1 % (v/v) Tween-20, 1 mM EDTA.

7. Detection solution with antibody: Mix 80 μL deionized water with 20 μL buffer II and 1 μL antibody for each slide. Prepare fresh just before use.
8. Anti-DIG–rhodamine Fab fragment (Roche, Diagnostics, IN, USA).

2.5 Transmission Electron Microscopy

1. Fixative: 2.8 % glutaraldehyde in 0.1 M HEPES buffer + 0.1 % Triton X-100 (pH 7.2).
2. Post-fixative: 1 % (w/v) OsO_4 in 0.1 M HEPES buffer.
3. Acetone.
4. Spurr's embedding kit (Electron Microscopy Sciences, Hatfield, PA, USA).
5. 0.1 % toluidine blue: 0.1 g toluidine blue + 100 mL 1 % borax.
6. 3 % uranyl acetate: 0.3 g uranyl acetate + 10 mL double-distilled water.
7. Lead citrate solution: $Pb(NO_3)_2$, $Na_3C_6H_5O_7{\cdot}2H_2O$. For 50 mL, resuspend 1.33 g $Pb(NO_3)_2$ and 1.76 g $Na_3C_6H_5O_7{\cdot}2H_2O$ in 40 mL of ddH_2O. Use about 8 mL of 1 M NaOH to adjust the pH of the solution to approximately 12, and then add ddH_2O to 50 mL.

2.6 Isolation of Male Meiocytes

1. Petri dish (100 × 15 mm).
2. Laminar flow hood.
3. Syringes (BD Tuberculin 1 mL).
4. Needles (BD 27 G × ½).
5. Watchmaker forceps (sharp-tip).
6. Micro slides (25 × 75 mm, 1.0 mm thick).
7. Depression slides (VWR).
8. A dissection microscope (Nikon or similar).
9. An inverted microscope (e.g., Zeiss).
10. Capillary tubes (glass, 1.5 mm outer diameter, 10 cm length).
11. Glass capillary pipette puller (two-step; e.g., Narishige).
12. Manipulator (three-axis coarse and fine micromanipulator; e.g., Narishige).
13. Micropipettors (10 and 200 μL).

3 Methods

3.1 Isolation and Observation of Tetrads

1. Fix inflorescences with unopened floral buds in Carnoy's fixative overnight on a bench at room temperature.
2. Wash the fixed inflorescence three times with water, and discard the older buds on the periphery of the inflorescence,

which contain yellow anthers visible without removing sepals (*see* **Note 1**).

3. Place the inflorescence with several remaining immature buds on a glass slide, separate the immature anthers from other floral organs using a fine needle, and add two drops of water.
4. Break the anthers by pinching several times using a pair of sharp forceps.
5. Dry the slides on air, and visualize tetrads by staining with 0.01 % basic fuchsine for 2–3 min.
6. Cover with a cover slip for microscopic observation of the tetrads, which will show a red appearance under a light microscope.

3.2 Chromosome Spread for Detection of Meiotic Chromosomes with DAPI

1. Fix the inflorescences with unopened floral buds in Carnoy's fixative for 4 h at room temperature (*see* **Note 2**).
2. Wash the fixed buds three times for 10 min each with 10 mM citrate buffer.
3. Discard the older buds with yellow anthers visible from the exterior of the bud before digestion.
4. Digest the buds using digestion cocktail for 80 min at 37 °C.
5. Wash the buds with 10 mM citrate buffer and store at 4 °C for up to 1 week.
6. Place five buds on one slide with a Pasteur pipette, and use a needle to cut and push the anthers away from other floral organs.
7. Add a drop of 60 % acetic acid on the anthers, and break the anthers by pinching several times using a pair of sharp forceps.
8. Place the slides on a heat block at 45 °C for 30 s and add another drop of 60 % acetic acid (*see* **Note 3**).
9. Spread the meiotic chromosomes by adding one drop (~20 μL) of Carnoy's fixative that was pre-chilled at −20 °C to the center of the samples (*see* **Note 4**).
10. Dry the slides in air for approximately 5 min.
11. Add 5 μL of mounting medium containing DAPI, cover with a cover slip, and seal the edge of cover slip with nail polish. The slide is now ready for microscopy (Fig. 1).

3.3 Fluorescent In Situ Hybridization

3.3.1 Probe Labeling

We describe FISH with two kinds of labeled probes in this chapter. One is a directly labeled probe when synthesizing an oligonucleotide for repetitive nucleic acid sequences, using a modification with fluorescence markers CY3, CY5, or FITC at the 5′ terminus. The example provided in Subheading 2.4.2 is the probe for telomeres: CY3-$(CCCTAAA)_4$ modified at the 5′ terminus.

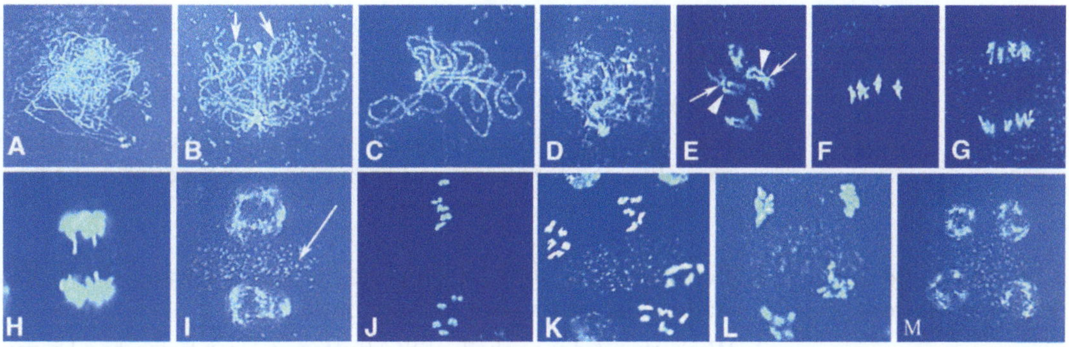

Fig. 1 Male meiosis in *Arabidopsis*. (**a**) Leptotene. (**b**) Zygotene; *arrows* indicate regions of pairing between homologs. (**c**) Pachytene. (**d**) Diplotene. (**e**) Diakinesis; *arrows* point to the position of chiasmata, and *arrowheads* indicate the centromeric regions. (**f**) Metaphase I. (**g**) Anaphase I. (**h**) Telophase I. (**i**) Prophase II; the *arrow* points to the organelle band between the two groups of chromosomes. (**j**) Metaphase II. (**k**) Anaphase II. (**l**) Telophase II. (**m**) Four newly formed nuclei (modified from ref. [7])

The second kind of probes are plasmids containing templates for probe labeling using a standard nick translation procedure, described as follows:

1. Add 1 μg template DNA to ddH_2O with a combined volume of 16 μL.
2. Add 4 μL DIG-Nick translation mix to the above DNA solution and centrifuge briefly for a few seconds.
3. Incubate at 15 °C for 90 min.
4. Mix a 3 μL aliquot for each 20 μL reaction with 7 μL loading buffer, denature the DNA by heating at 95 °C for 3 min, and place the mix on ice for 3 min and then the sample onto an agarose gel.
5. Visualize the probe to determine its length, which should range between 200 and 500 nucleotides (*see* **Note 5**).
6. Stop the reaction by adding 1 μL 0.5 M EDTA (pH 8.0) and incubate at 65 °C for 10 min.

3.3.2 Preparation of Chromosome Spread for FISH

1. Prepare the chromosome spreads as described in Subheading 3.2, **steps 1–9**.
2. Apply 100 μL of 70 % formamide in 2× SSC buffer to the dry slides, and cover each of the slides with a piece of parafilm. Incubate slides in an oven at 80 °C for 5 min.
3. Dehydrate the slides with a precooled (at –20°C) ethanol series (70, 80, 90, and 100 %) for 5 min each. Dry the slides on air.

3.3.3 Hybridization with a Probe

1. Prepare 10 μL hybridization mixture for each slide by mixing 2 μL of the labeled probe with the hybridization cocktail.

2. Denature the mixture by incubating at 85 °C for 5 min and then immediately chill on ice.
3. Apply 10 μL mixture to each slide and cover with a cover slip.
4. Place wet paper towels on the bottom of a moisture box that is specially used for slides, and then place the slides into the box; incubate at 37 °C overnight.

3.3.4 Washing and Detection of the Signals

1. Remove cover slips by dipping the slides up and down in a staining jar containing 2× SSC buffer.
2. Wash the slides with 2× SSC buffer at least twice for 15 min each (*see* **Note 6**).
3. Incubate the slides in 1× PBS buffer for 5 min.
4. Detect the DIG-labeled probe with 100 μL for each slide of the detection solution containing the anti-DIG–rhodamine antibody (1:1,000 dilution) and incubate at 37 °C for approximately 1 h.
5. Wash the slides with 1× PBS buffer three times for 5 min each.
6. Dry the slides in air, pipette 7 μL counterstain with DAPI on each slide, and cover with a cover slip.
7. Observe the signals under a fluorescence microscope (Fig. 2).

3.4 Immunolocalization

1. Fix the inflorescences with unopened floral buds (around 0.2–0.4 mm) in fixative and place under vacuum with the pressure of 0.01 MPa for 15 min at room temperature (*see* **Note 7**).
2. Wash the fixed inflorescences with the citrate buffer twice for 5 min each.
3. Discard the buds on the periphery of the inflorescence with yellow anther visible from the outside.
4. Remove the appropriately staged anthers quickly from buds and digest them with a digestion medium at 37 °C for 120 min.

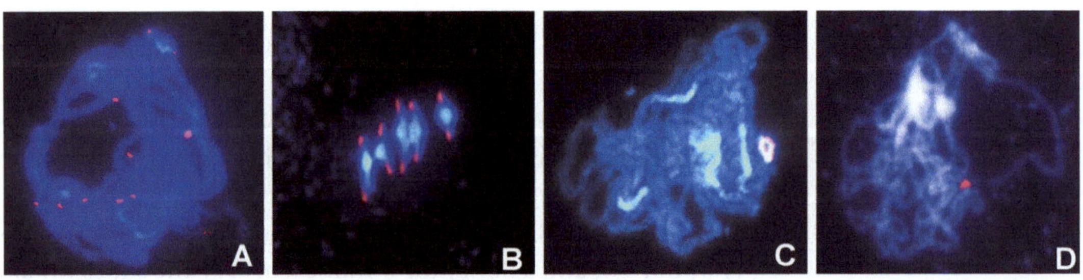

Fig. 2 Analysis of wild-type male meiosis by FISH. (**a**) A pachytene nucleus with the telomeric probe labeled at the 5′ terminus with the CY3. (**b**) Metaphase I bivalents with the centromeric probe labeled at the 5′ terminus with the CY3. (**c**) Pachytene chromosomes hybridized with the 45S rDNA probe labeled by DIG. (**d**) Pachytene chromosomes hybridized with the F1N21 BAC labeled by DIG. *Red dots* indicate FISH signals

5. Wash the anthers with ddH_2O twice for 5 min each.
6. Pipette the anthers onto the slides with ddH_2O, use a needle to cut and push the anthers away from other floral organs, and cover with a cover slip.
7. Apply pressure to the cover slips gently using a stirring rod (*see* **Note 8**).
8. Immerse the slides in liquid nitrogen for 1 min, and quickly pry the cover slip away from the slide with a razor blade. Dry the slides in air.
9. Circle the region containing the anther materials with a marker pen.
10. Immerse the slides in wash buffer I twice for 30 min each at room temperature.
11. Block nonspecific binding by adding 100 μL blocking buffer to one slide. Cover with a parafilm, and incubate at 37 °C for 30 min.
12. Dilute the primary antibody with blocking buffer with appropriate concentrations (determine by trial for individual antibody), add 100 μL to each slide, cover it with a parafilm, and incubate overnight at 4 °C in a moisture chamber.
13. Wash the slides with washing buffer II three times for 15 min each, and put the slides in the blocking buffer at 37 °C for 30 min.
14. Dilute the secondary antibody 1:500–1:1,000-fold in blocking buffer, add 100 μL to each slide, cover them with parafilm, and incubate in humidified atmosphere at 37 °C for 60 min in the dark.

    Perform the following steps in the dark.
15. Wash the slides with washing buffer II three times for 15 min each and mount in vectashield antifade medium with 2 mg/mL DAPI.
16. Observe the signals under a fluorescence microscope (Fig. 3).

3.5 Observation of Synaptonemal Complex by Transmission Electron Microscopy

1. Fix the inflorescences with unopened floral buds in fixative, place in vacuum with the pressure of 0.01 MPa for 2 h in ice, and then store at 4 °C overnight.
2. Wash the buds with 0.1 M HEPES buffer twice for 15 min each and then postfix with post-fixative overnight at 4 °C.
3. Wash the inflorescences with water twice for 15 min each and dehydrate in 30, 50, 70, 80, 90, and 95 % ethanol for 15 min each; incubate in an ethanol/acetone (1:1) mix and then in 100 % acetone twice for 15 min each at 4 °C.

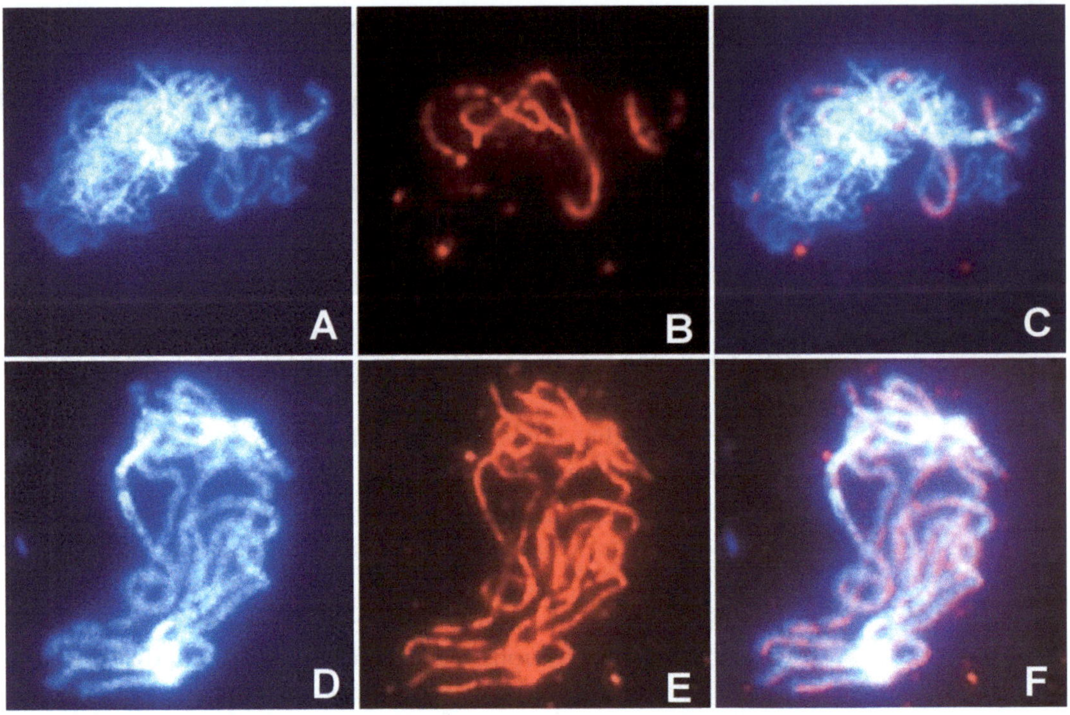

Fig. 3 Localization of AtZYP1 in wild-type male meiocytes. (**a**–**c**) Zygotene. (**d**–**f**) Pachytene. In each row of three panels, the *left panel* shows *blue*-colored chromosomes stained with DAPI; the *middle panel* shows *red-colored* signals for AtZYP1; and the *right panel* shows the merged image of DAPI and the AtZYP1 signal

4. Incubate the samples in a graded series of acetone/Spurr's resin solutions (2–4 h each) at room temperature: 2:1 acetone to Spurr's resin, 1:1 acetone to Spurr's resin, 1:2 acetone to Spurr's resin, 100 % Spurr's resin, again in fresh 100 % Spurr's resin overnight; then polymerize for 2 days at 65 °C in an oven.
5. Determine the prophase stage of male meiotic cells by examining semi-thin sections of the embedded inflorescences stained with 0.1 % toluidine blue.
6. After identifying the appropriate specimen, prepare ultrathin sections for TEM.
7. Select the sections with the grid and allow to dry.
8. Stain the sections with 0.5 % uranyl acetate for 15–30 min and then wash with ddH_2O.
9. Stain with lead citrate for 5–10 min and then wash with dH_2O.
10. Observe the SC structure by the examination of the grids with an electron microscope (Fig. 4).

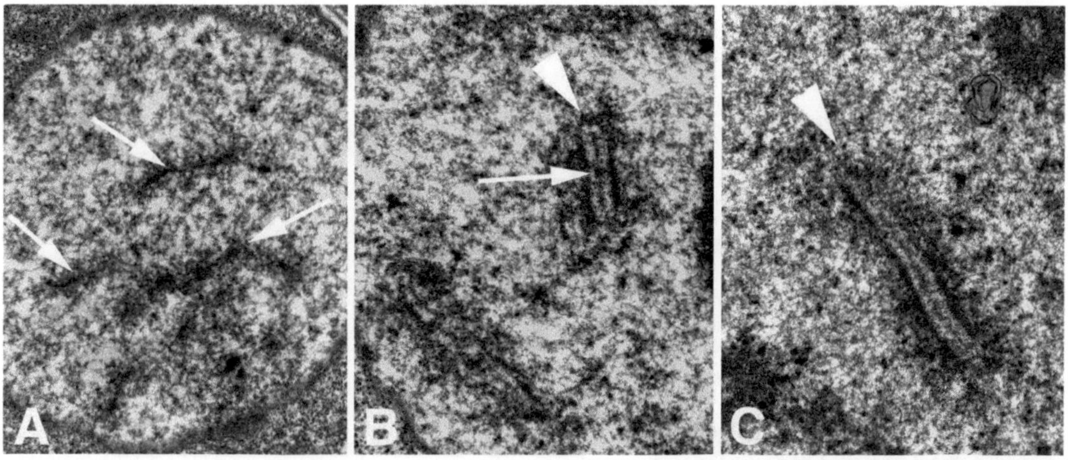

Fig. 4 Transmission electron microscopy of wild-type *Arabidopsis* male meiotic prophase I. (**a**) A section of a leptotene nucleus showing axial elements (indicated by *arrows*). (**b**) A section of a zygotene nucleus showing a short fragment of SC (indicated by *arrowhead*) with a recombination nodule (indicated by an *arrow*). (**c**) A section of a pachytene nucleus showing a long SC fragment (indicated by an *arrowhead*). Figure adopted from ref. [7]

3.6 Isolation of *Arabidopsis* Male Meiocytes

3.6.1 Micromanipulation Platform Setup

1. Use a two-step puller to pull each glass capillary into two micropipettes with a tip inner diameter of approximately 40–50 μM.
2. Store the micropipettes mounted on a clay strip in a covered dish for further use.
3. Insert a micropipette into a glass syringe connected with a long plastic tube and tightly seal them.
4. Tie the glass syringe with the micropipette and plastic tube onto a micromanipulator holder (Fig. 5).

3.6.2 Isolation of Male Meiocytes

1. Use a pair of watchmaker forceps to collect an inflorescence with young floral buds from a healthy plant.
2. Place the floral buds onto a glass slide and add one drop of liquid MS medium to keep the buds moist and healthy.
3. Use two needles to gently open flowers under a dissection microscope to gather stage 5–7 anthers.
4. Immediately transfer the dissected anthers to a micro-chamber of the depression slide with some liquid MS medium. Put the slides on ice.
5. After gathering 20–40 stage 5–7 anthers, move the micro-chamber under a dissection microscope and cut each anther open by moving two needles touching each other in opposite directions in a way similar to a pair of scissors.

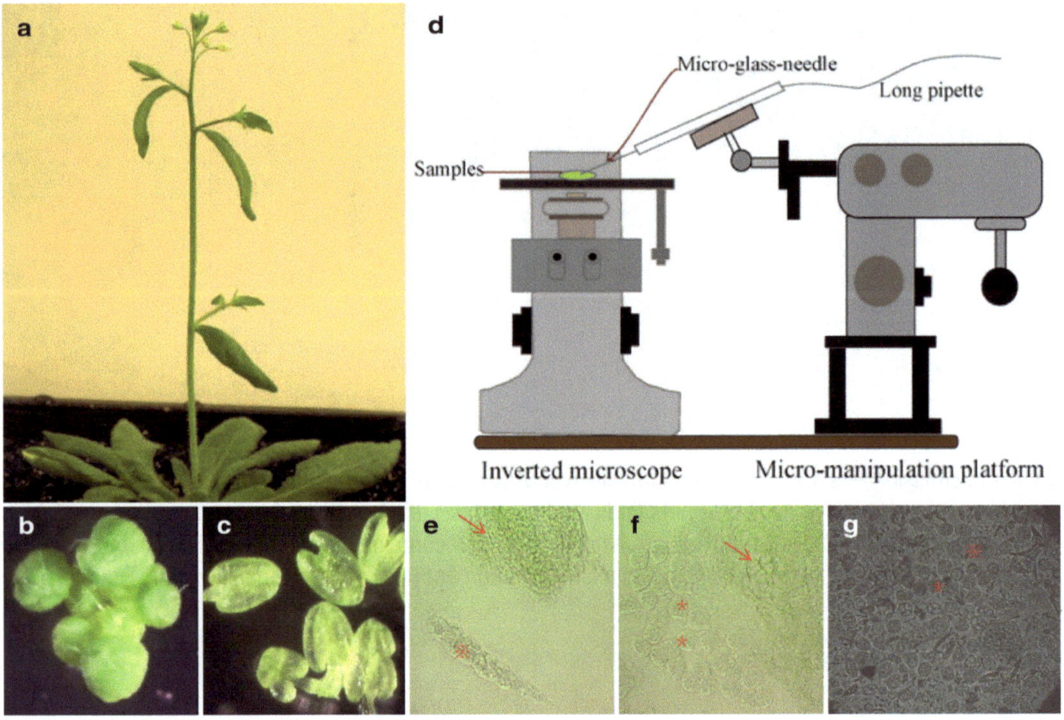

Fig. 5 An illustration of the procedure for isolating *Arabidopsis* male meiocytes. (**a**) A healthy *Arabidopsis* plant at a stage suitable for collecting meiocytes. (**b**) A cluster of young floral buds from the plant shown in (**a**). (**c**) Dissected anther from (**b**) at appropriate stages. (**d**) A setup of the system of micro-capture for released meiocytes from (**c**). (**e** and **f**) Released "worms" (with *asterisks*) and individual meiocytes (*arrow*). (**g**) Collected meiocytes in both types from (**e**) and (**f**). ⋇ in (**e**) and (**g**) indicate "worms." Figure adopted from ref. [20]

6. Release individual meiocytes or meiocyte cluster (nicknamed "worm" as they form a wormlike shape) out of an anther locule by gently pressing the uncut end of the anther (Fig. 5).
7. Place the depression slide on an inverted microscope, and add more MS medium in the chamber to suspend the cells.
8. Adjust and move the micropipette tip using the micromanipulator to come to close contact with the meiocytes or the whole "worm."
9. Absorb cells into the micropipette by gently sucking on the mouthpiece at the far end of the plastic tube.
10. Immediately transfer the collected cells into a 1.5 μL microfuge tube. For the isolation of total RNA, add 100 μL TRIzol into the tube and directly place cells in the TRIzol reagent. Keep the tube on ice.
11. After obtaining an appropriate number of cells, label the date of collection on the top of the tube, and quickly freeze them in liquid nitrogen.

12. Store the isolated meiocytes in a −80 °C freezer.
13. Repeat the above steps to obtain additional *Arabidopsis* male meiocytes.

4 Notes

1. Normally, the yellow color of anthers under a microscope indicates the presence of pollen grains and that the developmental stage is well after that during which meiosis occurs. The identification of buds containing tetrads will require some practice and experience.
2. The fixation time is very flexible. If one wants to obtain the results quickly, fixation for 3 h is sufficient. If one wants to store the samples for a long time (up to half a year) prior to analysis, the samples can be fixed overnight, then transferred to fresh fixative with 70 % ethanol, and stored at 4 °C.
3. Pollen mother cells should be treated with 60 % acetic acid for less than 3 min. Longer treatments lead to abnormal and artificial chromosome morphology.
4. Proper application of Carnoy's fixative is very important for chromosome spread. After adding a drop of pre-chilled Carnoy's fixative on slides, the fixative will gradually spread out the samples. Well-spread chromosomes will form a cycle (of approximately the size of a nickel) that can be easily visualized by eyes on the slide.
5. If the synthesized probes do not meet the standard, it is necessary to reincubate the reaction at 15 °C and check the fragment size again. When the desired probe length is achieved, stop the reaction by adding 1 μL 0.5 M EDTA (pH 8.0) per 20 μL reaction volume and heating at 65 °C for 10 min.
6. If the labeled oligonucleotide probe is used, after the step of washing with 2× SSC buffer, the slides can be directly mounted using the vectashield antifade medium with 2 mg/mL DAPI for microscopy. Be sure not to let the slides dry too much, as overdrying will cause high backgrounds.
7. For the immunofluorescence procedure, it is very important to determine the optimal time to fix samples. Normally, 15 min is appropriate for *Arabidopsis* plants of the Columbia accession, while prolonged fixation is required for Landsberg *erecta*. In our hands, higher fixative concentration and longer fixation times will result in higher fluorescence backgrounds. In general, the optimal time for fixation is when the samples have just sunken to the bottom instead of still floating in the fixative.
8. Unlike the DAPI and FISH using chromosome spread, chromosome squash is applied for immunofluorescence.

To avoid slippage between the cover slip and the slide, we hold two corners of the cover slip of a diagonal line on the slide with the left thumb and middle finger, and then press the cover slip with a thin stirring rod vertically. We then repeat this action by holding the cover slip at the corners of the other diagonal line and squashing on the remaining area. The optimal way should be determined by conducting several trials to ensure that the chromosomes are thoroughly scattered from the cytoplasm but not broken.

Acknowledgment

We thank Dr. Zaibao Zhang and Jiyue Huang for helpful suggestions and comments. This work was supported by grants from the Ministry of Sciences and Technology of China (2011CB944600).

References

1. Ma H (2005) Molecular genetic analyses of microsporogenesis and microgametogenesis in flowering plants. Ann Rev Plant Biol 56:393–434
2. Stack S (1973) The synaptonemal complex and the achiasmatic condition. J Cell Sci 13:83–95
3. Anderson LK, Stack SM, Fox MH et al (1985) The relationship between genome size and synaptonemal complex length in higher plants. Exp Cell Res 156:367–378
4. Golubovskaya IN (1979) Genetic control of meiosis. Int Rev Cytol 58:247–290
5. Initiative AG (2000) Analysis of the genome sequence of the flowering plant *Arabidopsis thaliana*. Nature 408:796–815
6. Mercier R, Grelon M (2008) Meiosis in plants: ten years of gene discovery. Cytogenet Genome Res 120:281–290
7. Ma H (2006) A molecular portrait of Arabidopsis meiosis. Arabidopsis Book 4:1–39
8. Osman K, Higgins JD, Sanchez-Moran E et al (2011) Pathways to meiotic recombination in *Arabidopsis thaliana*. New Phytol 190: 523–544
9. Zickler D, Kleckner N (1999) Meiotic chromosomes: integrating structure and function. Ann Rev Genet 33:603–754
10. Wijeratne AJ, Chen C, Zhang W et al (2006) The *Arabidopsis thaliana PARTING DANCERS* gene encoding a novel protein is required for normal meiotic homologous recombination. Mol Biol Cell 17:1331–1343
11. Li W, Yang X, Lin Z et al (2005) The *AtRAD51C* gene is required for normal meiotic chromosome synapsis and double-stranded break repair in Arabidopsis. Plant Physiol 138:965–976
12. Li W, Chen C, Markmann-Mulisch U et al (2004) The *Arabidopsis AtRAD51* gene is dispensable for vegetative development but required for meiosis. Proc Natl Acad Sci U S A 101:10596–10601
13. Chen C, Zhang W, Timofejeva L et al (2005) The *Arabidopsis ROCK-N-ROLLERS* gene encodes a homolog of the yeast ATP-dependent DNA helicase *MER3* and is required for normal meiotic crossover formation. Plant J 43:321–334
14. Armstrong SJ, Sanchez-Moran E, Franklin FC (2009) Cytological analysis of *Arabidopsis thaliana* meiotic chromosomes. Methods Mol Biol 558:131–145
15. Wang M, Wang K, Tang D et al (2010) The central element protein ZEP1 of the synaptonemal complex regulates the number of crossovers during meiosis in rice. Plant Cell 22:417–430
16. Hamant O, Ma H, Cande WZ (2006) Genetics of meiotic prophase I in plants. Ann Rev Plant Biol 57:267–302
17. Martin JA, Wang Z (2011) Next-generation transcriptome assembly. Nat Rev Genet 12: 671–682
18. Yang WC, Shi DQ, Chen YH (2010) Female gametophyte development in flowering plants. Ann Rev Plant Biol 61:89–108
19. Scott RJ, Spielman M, Dickinson HG (2004) Stamen structure and function. Plant Cell 16:S46–S60
20. Yang HX, Lu PL, Wang YX et al (2011) The transcriptome landscape of *Arabidopsis* male meiocytes from high-throughput sequencing: the complexity and evolution of the meiotic process. Plant J 65:503–516

Chapter 11

Genetic and Phenotypic Analyses of Carpel Development in Arabidopsis

Vicente Balanzà, Patricia Ballester, Monica Colombo, Chloé Fourquin, Irene Martínez-Fernández, and Cristina Ferrándiz

Abstract

Carpels are the female reproductive organs of the flower, organized in a gynoecium, which is arguably the most complex organ of a plant. The gynoecium provides protection for the ovules, helps to discriminate between male gametophytes, and facilitates successful pollination. After fertilization, it develops into a fruit, a specialized organ for seed protection and dispersal. To carry out all these functions, coordinated patterning and tissue specification within the developing gynoecium have to be achieved. In this chapter, we describe different methods to characterize defects in carpel patterning and morphogenesis associated with developmental mutations as well as a list of reporter lines that can be used to facilitate genetic analyses.

Key words Carpel, Gynoecium, Cloral hydrate, Vascular clearing, Pollination, Lignin

1 Introduction

Carpels are the female reproductive organs of the flower. In Arabidopsis, two congenitally fused carpels form a single pistil or syncarpous gynoecium. At maturity (anthesis), several parts are readily visible in an Arabidopsis pistil (*see* Fig. 1). At the base of the gynoecium, a short internode called the *gynophore* connects the pistil to the base of the flower. Above the gynophore is the *ovary*, which comprises most of the length of the gynoecium and contains between 50 and 80 ovules. The ovary is divided longitudinally into two chambers or locules by a *septum*, which is formed postgenitally. The two ovary wall regions of the pistil are termed the *valves*, and the external part of the septum is termed the *replum*. At the apical end of the ovary are the *style* and the *stigma*. The stigma consists of a single layer of specialized epidermal cells termed stigmatic papillae, and it is in charge of receiving and inducing the germination of pollen grains. The stigma constitutes the initial portion of the *transmitting tract*, a polysaccharide-rich tissue

José Luis Riechmann and Frank Wellmer (eds.), *Flower Development: Methods and Protocols*, Methods in Molecular Biology, vol. 1110, DOI 10.1007/978-1-4614-9408-9_11, © Springer Science+Business Media New York 2014

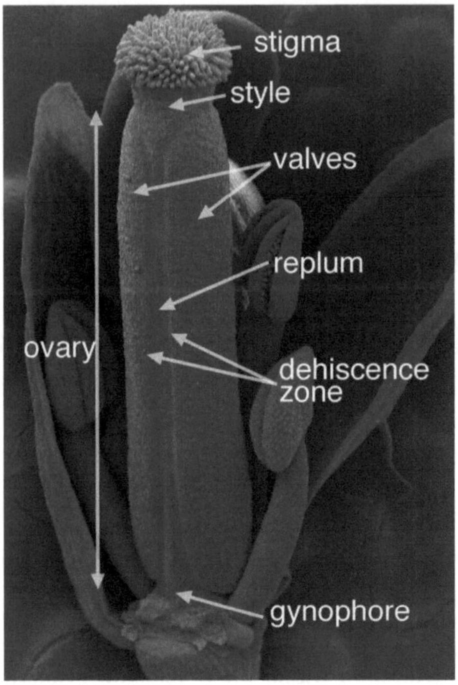

Fig. 1 The Arabidopsis mature gynoecium. Scanning electron micrograph of an Arabidopsis gynoecium at anthesis (stage 13 of flower development; *see* ref. 12). The different parts are indicated

specialized in directing pollen tube growth. Basal to the stigma, the style appears as a short, solid cylinder with distinctive epidermal morphology. The central core of the style is composed of transmitting tract tissue, surrounded by vascular tissue. The tract continues through the central part of the septum, guiding the pollen tubes towards the ovules. After fertilization, the ovules develop into seeds and the *Arabidopsis* gynoecium is transformed into an elongated bilocular fruit called silique. This structure opens at maturity to release its seeds along four *dehiscence zones* defined by longitudinal furrows of smaller cells on either side of the replum. The lignification of specific cells in these zones contributes to the dehiscence process (see below) (*see* Fig. 1, and [1, 2]).

Developmental mutants affect the identity and distribution of the different parts of the pistil [1, 2], and these morphological defects are best observed with standard scanning electron microscopy procedures (*see* Chapters 12 and 13). In addition to determining changes in overall morphology, a number of other specific techniques can help to visualize frequent defects associated with mutations affecting carpel and fruit development. In this chapter we outline methods for studying these defects.

First, we describe how to observe pollen tube growth by aniline blue staining. Aniline blue reveals callose deposits associated to pollen tube formation giving a brilliant yellow fluorescence with essentially no background signal. Mutants affected in transmitting tissue development show poor pollen tube growth [3]. Transmitting tissue is also easily revealed by alcian blue staining, a protocol that is also described in this chapter.

Second, we describe a method to optically clear plant tissues. We use this method to observe vascular development in the gynoecium using darkfield microscopy, where lignified xylem elements appear white. Vascular strands are frequently affected by mutations in genes involved in gynoecium morphogenesis and especially by altered auxin signaling [4, 5]. Main differences are observed in the position of the medial bundle bifurcation and the extent of vascular fans in the style. Tissue clearing is also a good option to visualize ovule development (*see* Chapter 12) and to obtain detailed GUS reporter activity when combined with differential interference contrast (DIC) optics microscopy (*see* Chapter 15).

Wild-type Arabidopsis gynoecium development is sensitive to auxin transport inhibition by 1-*N*-naphthylphthalamic acid (NPA), which affects overall fruit morphology and vascular development. Several classical Arabidopsis gynoecium mutants are hyper- or hyposensitive to this treatment [5–7], and, thus, the characterization of NPA sensitivity in a new mutant can provide clues on potential genetic interactions with other factors of known pathways directing gynoecium development.

Finally, we describe several alternative methods to reveal lignin deposition patterns. Lignin is one of the most abundant organic polymers and an integral part of the secondary wall of plants, to which it confers mechanical strength. Lignin is produced in high quantities in the secondary walls of vascular tissue cells. In Arabidopsis fruits, lignification is associated to vascular bundles, the endocarp*b* (end*b*, the subepidermal internal cell layer), and two or three cell rows adjacent to the separation layer of the dehiscence zone, where it contributes to the establishment of tensions which help to pod shatter [8–10]. In a temporal sequence, lignin becomes initially detectable in the vascular tissues around developmental stage 15–16 (for a definition of developmental stages, *see* refs. 11, 12), while lignification of the end*b* and the dehiscence zone appears later, between late stage 17A and stage 17B [11]. Lignin staining of fruits is often very useful to detect whether the specification of the dehiscence zone is defective. The methods described in this chapter can readily be applied for the characterization of lignin deposition in fruits from other species with only minor optimization.

In addition to these methods for phenotypic analyses, we include a table of published reporter lines that can be useful for genetic analyses. They correspond to markers for specific tissues within the pistil or for hormone distribution as well as to expression reporters for key factors involved in carpel morphogenesis. For details on the role of these factors, *see* refs. 1, 2.

2 Materials

2.1 Aniline Blue Staining of Arabidopsis Pollen Tubes

1. Fine forceps (tweezers; 5 s is a suitable grade).
2. Binocular microscope (or magnifying glasses).
3. Microcentrifuge tubes.
4. Pasteur pipettes or micropipettes.
5. Microscope slides.
6. Cover slips.
7. Microscope with UV light and DAPI filter set.
8. Fixing solution: Absolute ethanol:acetic acid (3:1).
9. Softening solution: 8 M NaOH.
10. Aniline blue solution: Prepare 0.1 % (w/v) aniline blue in 0.1 M K_2HPO_4–KOH buffer, pH 11 (*see* **Note 1**).
11. 50 % glycerol.

2.2 Cleared Tissue for Observation of Vascular Development

1. Fine forceps (tweezers; 5 s is a suitable grade).
2. Glass scintillation vials or similar containers.
3. Pasteur pipettes or micropipettes.
4. Microscope concavity slides (*see* **Note 2**).
5. Cover slips (20 × 20 mm).
6. Surgical blades and needles (*see* **Note 3**).
7. Nail polish (*see* **Note 4**).
8. White paper towels.
9. Microscope with darkfield optics.
10. Fixing solution: Absolute ethanol:acetic acid (6:1).
11. Clearing solution: Chloral hydrate:glycerol:H_2O (8 g:1 mL: 2 mL) (*see* **Note 5**).

2.3 NPA Treatment

1. Fine forceps.
2. Plastic spray bottles (0.5–1 L volume).
3. Clear humidity domes for trays (*see* **Note 6**).
4. Binocular dissecting scope.
5. NPA solution: 100 μM NPA, 0.01 % (v/v) Silwet L77.
6. Mock solution: 0.01 % (v/v) Silwet L77 (*see* **Note 7**).

2.4 Lignin Staining

1. Fine forceps (5 s or similar) and needles.
2. Glass scintillation vials or similar containers.
3. Vacuum bell.
4. Oven at 60 °C.

5. Microscope polylysine-pretreated slides.
6. Cover slips (24 × 60 mm).
7. Histoprep metal base molds and plastic embedding cassettes.
8. 37–55 °C microscope slide warming table.
9. FAA fixation solution: 3.7 % formaldehyde, 50 % ethanol, 5 % acetic acid. Prepare a fresh solution (*see* **Note 8**).
10. Eosin solution: 2 % Eosin Y in 95 % ethanol.
11. Histoclear (*see* **Note 9**).
12. Graded ethanol series in distilled water (30, 50, 70, 95, and 100 %).
13. Paraplast X-tra paraffin chips.
14. Microtome.
15. Tissue flotation bath.
16. Fine brushes.
17. Phloroglucinol solution: Prepare 2 % phloroglucinol solution in 96 % ethanol (*see* **Note 10**).
18. 50 % HCl (*see* **Note 11**).
19. Stock alcian blue: Prepare 1 % alcian blue 8G (also called Ingrain blue 1) solution in 50 % ethanol (*see* **Note 12**).
20. Stock safranin: Prepare 1 % safranin-O solution in 50 % ethanol (*see* **Note 13**).
21. Acetate buffer: 0.1 M NaOAc–HOAc, pH 5.0.
22. Toluidine blue solution: Prepare 0.1 % toluidine blue solution in distilled water (*see* **Note 14**).
23. Mounting medium (Entellan, Merkoglass).

2.5 Genetic Analyses

See Table 1 for seed stocks.

3 Methods

3.1 Aniline Blue Staining of Arabidopsis Pollen Tubes

3.1.1 Material Collection

To compare the general performance and extent of pollen tube growth, Arabidopsis pistils can be collected from flowers 1 or 2 days after anthesis (flowers are wide open). Remove sepals, petals, and stamens from around the pistils.

For in vivo pollen tube guidance experiments, flowers at the developmental stage 12 must be emasculated and pollinated manually.

1. Remove any siliques, open flowers, open buds (it is possible to see the stigma poking out through the top of the bud), meristem, and smaller buds from the inflorescence.

Table 1
Reporter lines that are frequently used in genetic analyses of carpel development in Arabidopsis

Reporter line	Use	Reference	ABRC/NASC
Hormone reporter lines			
pDR5rev::3× VENUS	Synthetic promoter responsive to auxin accumulation	[14]	N799364
TCS::GFP	Synthetic promoter responsive to cytokinin accumulation	[15, 16]	CS23900
Key developmental gene reporter lines			
SHP1::GUS	3.5 kb of SHP1 genomic sequence upstream of the translational start to GUS	[17]	
SHP2::GUS	2.1 kb of SHP2 genomic sequence upstream of the translational start fused to GUS	[18]	
IND::GUS	2.7 kb of IND genomic sequence upstream of the translational start and 0.5 kb of the coding region fused to GUS	[19, 20]	
FUL::GUS	*ful-1* bears en enhancer trap inserted in the 5′UTR, rendering a null allele. GUS activity mimics faithfully FUL expression patterns and can be used as reporter for FUL in ful-1 heterozygous backgrounds	[21]	N3759
	2.3 kb of FUL promoter sequence fused to GUS	[22]	N8847
STY1::GUS	2 kb of STY1 promoter sequence fused to GUS	[23]	
STY2::GUS	2.1 kb of STY2 promoter sequence fused to GUS	[23]	
NGA3::GUS	2.7 kb of NGA3 genomic sequence upstream of the translational start codon fused to GUS	[24]	
HEC1::GUS	2,972 bp of HEC1 genomic sequence upstream of the translational start and the coding region fused to GUS	[25]	
HEC2::GUS	3,058 bp of HEC2 genomic sequence upstream of the translational start codon fused to GUS		

HEC3::GUS	2,979 bp of HEC3 genomic sequence upstream of the translational start codon fused to GUS		
CRC::GUS	8 kb of CRC genomic sequence upstream of the translational start and the coding region fused to GUS	[17]	
SPT::GUS	6.5 kb of SPT genomic sequence upstream of the translational start and 313 bp of the coding region fused to GUS	[26]	
ETT::GUS (ARF3::ARF3:GUS)	Reporter line with a translational fusion of ETTIN/ARF3 to GUS driven by ETT promoter	[27]	CS66480
Tissue specific reporter lines			
SLG::GUS	Stigma Promoter region of the S locus glycoprotein (SLG) gene from *Brassica oleracea* fused to GUS. GUS expression in the gynoecium is limited to stigmatic papillar cells	[28, 29]	
ASA1::GUS	Style Promoter for the anthranilate synthase 1 (*ASA1*) *gene* fused to GUS. GUS expression in the gynoecium is limited to inner style tissues in pollinated wild-type gynoecia	[30]	
GT140	Dehiscence zone Enhancer trap inserted in the promoter region of IND. GUS is detected in the dehiscence zone during fruit development	[31, 32]	N26931

2. While working at the binocular microscope it is best to immobilize the inflorescence, for instance using a restraint from a thin strip of paper held with sticky tape.
3. Emasculate stage 12 flowers: First carefully remove (or displace) sepals and petals, and then remove the immature anthers. Flowers can be protected from undesired pollination by covering the inflorescence.
4. Perform pollination after 24 h by touching the stigma with a suitable anther from a mature flower (maximal pollination). Minimal pollination (pollination using minimal amount of pollen) can be used to observe the growth morphology and growth pattern of individual pollen tubes. In this case squash a mature anther on a microscope slide, and, under the microscope, use a fine hair to pick up 1–5 pollen grains and transfer them to the stigma. Flowers can be labeled with the female and male accession and the date of pollination.
5. Pollinations are allowed to progress for 24 h, unless specified otherwise (for example, to observe pollen tube progression, time series of 2, 4, 6, 12, 24, and 48 h are suggested). After the desired amount of time post pollination, the remaining sepals and petals are removed from around the pistil (*see* **Note 15**).

3.1.2 Tissue Fixation

1. Place the excised pistils in a 1.5 mL microcentrifuge tube containing the fixing solution for at least 1 h at room temperature. If necessary, tissue can be left in the fixing solution overnight or indefinitely.

3.1.3 Pistil Softening

1. Replace the fixing solution with the softening solution. Leave the pistils in the softening solution overnight at room temperature.
2. Carefully remove the softening solution by pipetting. The tissue is now extremely fragile and can be easily damaged.
3. Gently wash the pistils three to five times with distilled water.

3.1.4 Pistil Staining

1. Replace the water with the aniline blue solution. Leave for at least 2 h at room temperature under dark condition (for instance using a piece of aluminum foil).

3.1.5 Pistil Mounting and Visualization

1. Place the stained pistils in a drop of 50 % glycerol on a microscope slide and cover with a cover slip. Put the cover slip on it carefully, starting from the tip of the pistils, in order to avoid bubbles.
2. Observe with a standard fluorescence microscope under UV irradiation condition and broadband DAPI filter set (excitation filter 390 nm; dichroic mirror 420 nm; emission filter 460 nm) to view the fluorescent signal from the tissue (*see* Fig. 2).

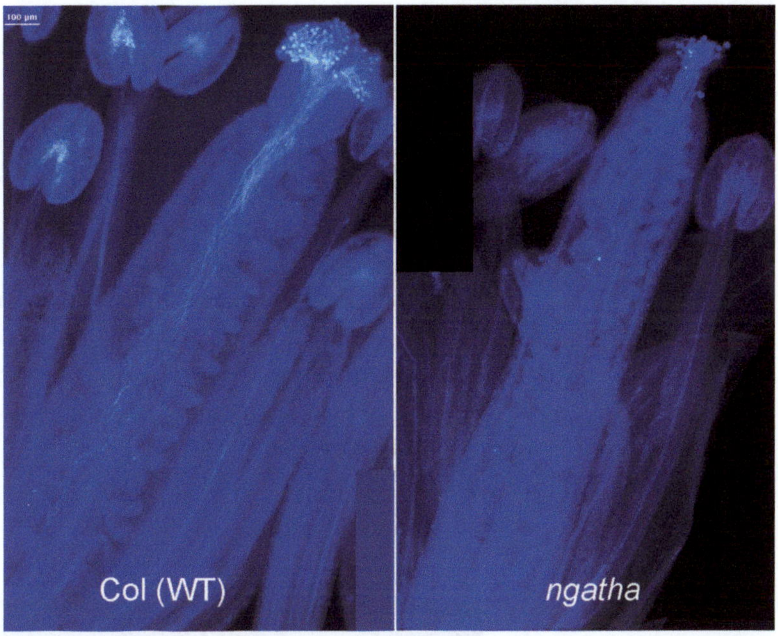

Fig. 2 Pollen tube growth staining. Pollen tube growth of one-day postanthesis pistils from wild-type Columbia (*left*) and *ngatha* quadruple mutant (*right*), stained with *aniline blue*. *ngatha* mutants are strongly defective in stigma and style development [6, 25] and therefore show poor pollen tube growth when compared to wild type

3.2 Cleared Tissue for Observation of Vascular Development

3.2.1 Material Collection

1. Anthesis (stage 13 of flower development, *see* ref. 12) is the preferred reference stage to compare vascular development in the gynoecium. This stage is easily recognized because the flower is fully open and anthers are dehiscent and leveled with the stigma of the pistil. When collecting the flowers, leave a long pedicel to facilitate manipulation without harming the pistil.

3.2.2 Tissue Fixation

1. Place anthesis flowers in vials containing enough fixing solution so as to completely cover plant tissue (*see* **Note 16**). Fix for at least 6 h, or leave overnight at room temperature.

3.2.3 Tissue Clearing

1. Replace fixative with absolute ethanol and incubate for 30 min at room temperature. Repeat twice.
2. Replace ethanol with clearing solution. Allow tissue to clear for at least 48 h at room temperature in darkness. Longer clearing times (up to 1 week) can improve transparency, but the tissue becomes very fragile and it is more difficult to manipulate.

3.2.4 Pistil Mounting and Visualization

1. Mount the samples under a dissection scope. We like to use concavity slides to avoid squashing the pistil. To properly orient the pistil, place the flower on a drop of clearing solution on

the flat surface of the concavity slide, close to the depression, with the replum facing upwards. Use fine forceps or needles to remove additional floral organs that may interfere with pistil visualization. Then, gently push the pistil to the center of the depression filled with clearing solution trying to maintain the proper orientation.

2. If needed, pipet enough clearing solution to completely cover the sample. Be careful not to touch the sample with the tip. Wipe away any excess solution with a white paper towel.
3. Place the cover slip over the depression of the concavity slide. Avoid making bubbles as much as possible.
4. Seal the cover slip with nail polish, and let it dry in the chemical fume hood.
5. Observe with a standard microscope under darkfield illumination (*see* Fig. 3).

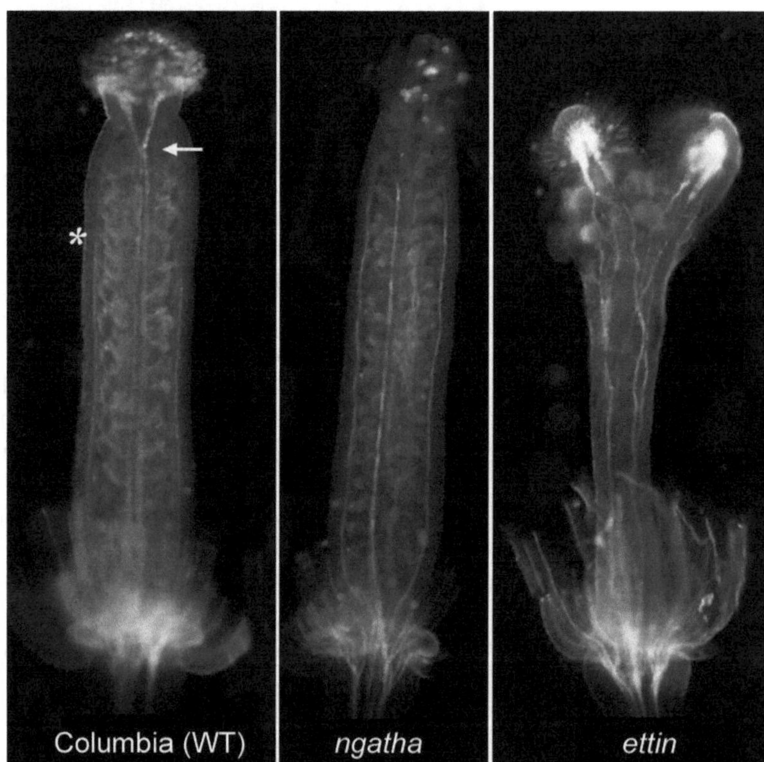

Fig. 3 Cleared pistils observed under darkfield to reveal vascular development. Vascular patterning in wild-type (*left*), *ngatha* quadruple mutant (*center*), and *ettin* mutant [30] (*right*) pistils at anthesis. Medial vein bifurcation in the wild type is visible at the boundary of style and ovary (*arrow*), while lateral veins do not reach the apical end of the ovary (*asterisk*). In the *ngatha* mutant, vascular development is very reduced, and medial veins stop prematurely and do not bifurcate. In the *ettin* mutant, veins bifurcate and overproliferate extensively

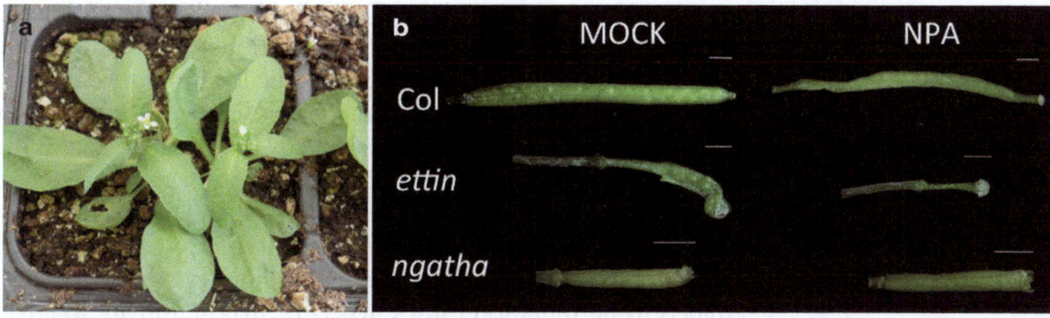

Fig. 4 Typical fruit phenotypes observed after NPA treatment. (**a**) Optimal stage for plant treatment. (**b**) On the *left*, typical morphology of fruits of wild type, *ettin* mutant, and *ngatha* quadruple mutant when untreated or treated with mock solution. On the *right*, examples of fruit morphologies for these three genotypes observed after NPA treatment. Wild-type fruits frequently show elongated styles and gynophores and shorter ovaries. *ettin* mutants are hypersensitive to NPA treatment, and most fruits are extremely reduced, valveless, and with extensive stigma proliferation. *ngatha* mutants are insensitive to treatment, displaying the same morphology to that observed in mock-treated plants

3.3 NPA Treatment

3.3.1 Plant Growth and Preparation

1. Arabidopsis plants are grown on soil under standard greenhouse conditions (*see* **Note 17**). Prepare two duplicate batches of plants in separate trays including wild-type controls and mutants of interest. Usually, 10–20 individual plants for each genotype are sufficient. One batch will be treated with NPA and another with mock solution.
2. Grow plants until bolting is apparent and the first open flowers are visible (*see* Fig. 4a).

3.3.2 Plant Treatment

1. Spray NPA solution plentifully to one batch of plants and mock solution to the duplicate.
2. Cover the trays with clear domes or clear plastic bags to maintain humidity (*see* **Note 6**). Leave plants covered for 8–12 h.
3. Repeat NPA treatments two more times, spaced by 8–12 h each time.
4. After the three treatments have been performed, wash plants by generously spraying them with water. Grow plants uncovered under standard greenhouse conditions for 3–4 more weeks.

3.3.3 Phenotype Visualization

1. 3–4 weeks post treatment, developing fruits should show visible phenotypes. Collect all fruits from the main inflorescence of each plant. Fruits of stage 16 and onwards [11] can be directly observed under the dissecting scope, while young fruits (stages 13–15) should be fixed for scanning electron microscopy analyses (*see* Chapter 13).
2. Score the number of fruits affected by NPA treatment and the severity of their phenotypes, comparing among the different genotypes and with the mock-treated plants. Establish phenotypic categories and assign fruits to each one. Fruit phenotypes usually

vary within the same plant. Typically, in wild-type Columbia accession, NPA treatment causes a reduction in the length of the valves, with a concomitant increase in the length of the style and the gynophore. A useful set of categories has been described for young fruits [13]: *mild*, when valves cover more than half of the length of the fruit; *medium*, when valves are very reduced and cover less than half of the length of the fruit; and *strong*, for valveless fruits. By scoring the proportion of affected fruits in each category for wild type and the mutant of interest, it can be determined if the mutation causes hyper- or hyposensitivity to auxin transport inhibition (*see* Fig. 4b).

3.4 Lignin Staining

3.4.1 Whole-Mount Phloroglucinol Staining

Tissue Fixation

1. Collect whole fruits at late stage 17B or beginning of stage 18 (fruit yellowing but closed).
2. Place fruits in glass vials filled with FAA solution.
3. Loosen the caps of the scintillation vials, and place the vials in a vacuum chamber.
4. Pull the vacuum slowly for 20 min. This step removes trapped air bubbles from the fruits and improves penetration of the fixative (*see* **Note 18**).
5. Release the vacuum slowly until vacuum bell can be opened. Keep the samples for 2 h at room temperature and then overnight at 4 °C.
6. The next day, remove the FAA fixation solution and replace with 70 % ethanol. Leave in 70 % ethanol for at least 30 min (*see* **Note 19**).

Fruit Staining

1. Remove 70 % ethanol from the vial and replace with phloroglucinol solution. Incubate for 2–5 min (time can be increased up to 30 min if staining is weak).

Fruit Lignin Visualization

1. Remove phloroglucinol solution, and replace it with 50 % HCl. Leave for 30 s to 2 min.
2. Immediately examine and photograph the fruits since the staining is only clearly visible for less than 30 min. For this, place the fruits on a Petri dish or a slide, uncovered, and preferably over a dark background (*see* Fig. 5a).

3.4.2 Section Staining

Tissue Fixation

Follow the steps described in Subheading 3.4.1.1.

Tissue Dehydration and Paraplast Embedding

1. Pour off the 70 % ethanol from vials and replace with 95 % ethanol. Incubate for 30 min.
2. Pour off the 95 % ethanol and replace with eosin solution (*see* **Note 20**). Incubate for 2 h at room temperature and then overnight at 4 °C.

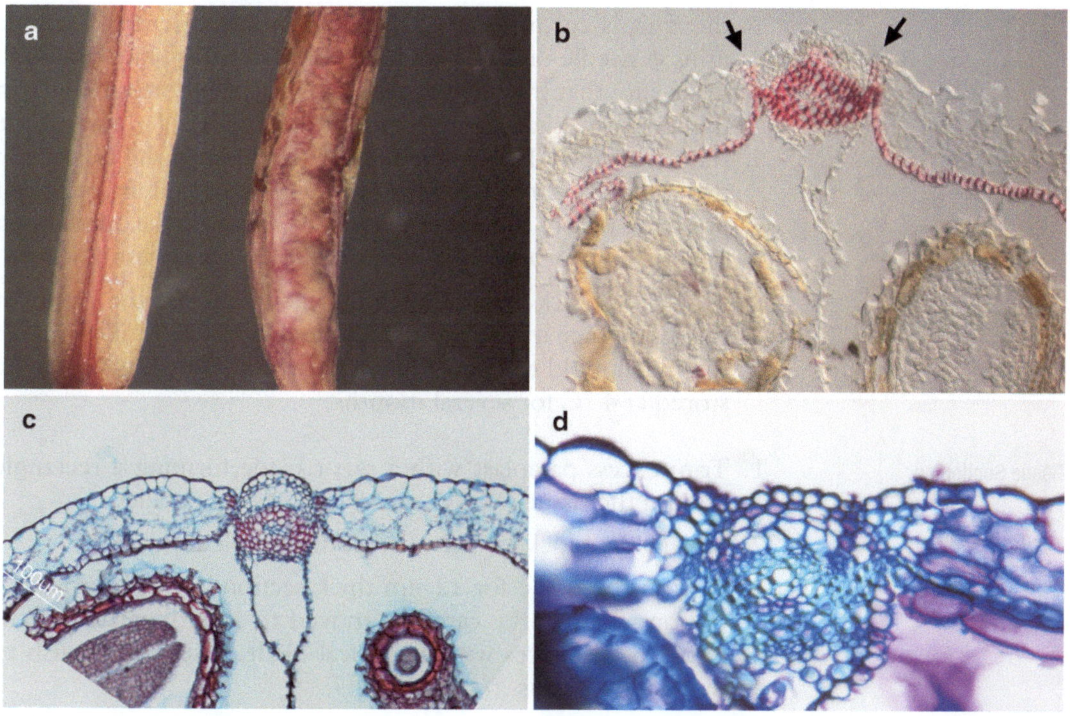

Fig. 5 Examples of fruit lignin staining obtained with the different described protocols. (**a**) Whole-mount phloroglucinol staining of wild type (*left*) and *fruitful* (*right*), a mutant with overlignified fruits [8, 21]. (**b**) Transversal section of an Arabidopsis mature fruit stained with phloroglucinol. The region corresponding to the valve–replum boundaries is shown. Lignin appears *bright pink* in the medial vascular vein, the subepidermal inner cell layer, and two small patches on the dehiscence zones (*arrows*). (**c**) Transversal section of an Arabidopsis mature fruit stained with *alcian blue* and safranin-O. Lignin appears in *red*. Mucilage in the seed is also stained in *bright red*. (**d**) Transversal section of an Arabidopsis mature fruit stained with *toluidine blue*. Lignin appears *turquoise*

3. Incubate the samples in the following solutions for 1 h each: 100 % ethanol (twice), 25 % Histoclear/75 % ethanol, 50 % Histoclear/50 % ethanol, 75 % Histoclear/25 % ethanol, and 100 % Histoclear (three times).
4. Pour off half of the final Histoclear wash, and add an approximately equal volume of paraplast X-tra paraffin chips. Incubate overnight at 60 °C in the oven.
5. Prepare a beaker full of paraplast chips, and leave it also at 60 °C to melt overnight.
6. The next morning, remove the Histoclear/paraffin mixture and replace with pure melted paraplast from the beaker. Incubate at 60 °C for 3–4 h. Repeat the step replacing molten paraplast at least three times (*see* **Note 21**).
7. On a warming table (55 °C) pour the paraplast solution containing the tissue into metal molds. Place one fruit per mold orienting the sample, so the length of the fruit remains perpendicular to the bottom of the mold. This is usually easier if

the fruit is cut into two halves. Align and orient the sample using a needle or a fine forceps. If half fruit is being oriented, make sure that the section plane is facing the bottom. Once in this position, slide carefully the mold to a cooler area of the warming table. This will make the paraplast in the bottom of the mold to harden slightly, trapping the fruit in the right position. Before the upper paraplast sets, place the embedding cassette on top, adding some more molten paraplast if needed.

8. Move the block off the warming table, and let the paraplast fully set. Once it is solid, it can be moved to 4 °C. This makes the metal mold to pop off easily. The embedded tissue can be stored at 4 °C for several months.

Tissue Sectioning

1. Trim excess paraplast with a razor blade forming a rectangle around the tissue. Make sure that edges are parallel.
2. Mount the block on the microtome sample holder.
3. Section the tissue for 12 μm thick sections. It will form long ribbons that can be carefully moved to a cardboard, and cut into 1.5 cm pieces with a surgical blade. Use a fine brush to move the ribbons.
4. Float the ribbons on 42 °C water for more than 1 min. This expands the tissue and reduces wrinkles.
5. Fish out the ribbons from the bath with a polylysine-pretreated microscope slide. Use a blunt wooden stick to help orient the ribbons in the slide.
6. Place the slides on a slide set at 40–45 °C. Let them dry overnight (*see* **Note 22**).

Tissue Staining

Three different staining procedures can be used to reveal lignin deposition.

Phloroglucinol stain for lignin

1. Dewax the sections by placing the slides for 10 min in Histoclear. Repeat this step with clear Histoclear. Then, wash the slides twice in 100 % ethanol for 2 min.
2. Stain slides with phloroglucinol solution for 2 min (up to 30 min).
3. Place the slides in 50 % HCl for 30 s.
4. Place a cover slip on each slide, and wipe the edges of the slide to prevent acid from damaging the microscope.
5. Observe with a standard microscope preferentially equipped with DIC optics. Examine and photograph the slide immediately as the staining lasts only around 30 min (*see* Fig. 5b).

Safranin and alcian blue staining for lignin

1. Prepare the staining solution: Add 5 mL of alcian blue stock solution and 2 mL of safranin stock solution to 200 mL of acetate buffer. The color should be dark purple.
2. Dewax the tissue sections in Histoclear twice for 10 min each. Then, wash twice in 100 % ethanol for 2 min each.
3. Rehydrate the tissue by sequential washes in a graded series of ethanol (90, 70, 50, and 30 %) for 2 min each. Wash the slides in distilled water for 2 min.
4. Place the slides in the staining solution for 30 min.
5. Wash with distilled water.
6. Let the slides dry for 1 h.
7. Apply a mounting medium, and protect the slides with a cover slip (*see* **Note 23**).
8. Observe with a standard microscope under bright field (*see* Fig. 5c).
9. Alternatively, lignin stained with safranin-O can also be observed with a fluorescence microscope, exciting at 492 nm and using a B-2A filter at 520 nm. Lignin will then fluoresce yellow to red (*see* **Note 24**).

Toluidine blue staining

1. Dewax the tissue sections in Histoclear twice for 10 min each. Then, wash twice in 100 % ethanol for 2 min each.
2. Rehydrate the tissue by sequential washes in a graded series of ethanol (90, 70, 50, and 30 %) for 2 min each. Wash the slides in distilled water for 2 min.
3. Place the slides in a solution of 0.02 % of toluidine blue for 5 min.
4. Wash generously with distilled water, until water comes out almost clear.
5. Let the slides dry for 1 h.
6. Apply a mounting medium, and protect the slides with a cover slip (*see* **Note 23**).
7. Observe with a standard microscope under bright field (*see* Fig. 5d).

3.5 Genetic Analyses

For genetic analyses, combination of mutants, overexpression lines, and/or reporter lines should be generated by crossing and genotyping of segregating populations following standard procedures. Several reporter lines can be useful to characterize a carpel mutant. Here, we provide a table with some of the most widely used.

4 Notes

1. Aniline blue contains a fluorochrome (Sirofluor), with excitation wavelength of 370 nm and emission wavelength of 509 nm, that specifically binds to ß-1,3-glucan, a major component of the pollen tube wall. The water-soluble aniline blue (Sigma) is decolorized in aqueous phosphate (0.067–0.1 M K_3PO_4, or K_2HPO_4, or a mixture of both). The solution, initially dark blue or purple, will finally turn pale yellow as it decolorizes. The pH of the phosphate buffer used varies in the literature, but in general the higher the pH (up to 11.5), the more intense the fluorescence [33]. After preparation, store the aniline blue solution in the dark at 4 °C.
2. Normal microscope slides can also be used instead of concavity slides. For this, place transparent sticky tape on both sides of the sample to avoid squashing it with the cover slip.
3. It is important to use disposable materials as much as possible, because the chloral hydrate can damage them.
4. Sealing of cover slips with nail polish prevents damage to the microscope lens.
5. Chloral hydrate is toxic to tissues of the mucous membranes and upper respiratory tract. It may be harmful by inhalation, ingestion, or skin absorption. Wear appropriate gloves, safety glasses, and mask, and use it in a chemical fume hood. Once prepared, clearing solution can be kept in darkness at room temperature for several months.
6. Clear domes fitting your standard growth trays can be used to keep humidity if they are tall enough to accommodate the bolting plants. Alternatively, use large plastic bags, placing the trays inside and sealing loosely with tape.
7. For a standard growth tray fitting 48 cell inserts or aracons, 200 mL of NPA and mock solution are sufficient.
8. Formaldehyde is toxic and volatile. It should be handled wearing gloves and under a fume hood.
9. Histoclear is moderately toxic, and it has a strong lemon scent. Use preferably with gloves and under a fume hood.
10. Phloroglucinol solution can be stored in darkness for several days at room temperature.
11. Hydrochloric acid (HCl) is toxic and volatile. It should be used with great care under a fume hood, with gloves and safety glasses.
12. Alcian blue 8G is one of the most widely used cationic dyes for the demonstration of glycosaminoglycans and mucopolysaccharides. It will stain in blue the non-lignified cell walls. In

addition, because transmitting tract cells secrete a complex extracellular matrix (ECM) very rich in acidic glycoproteins such as arabinogalactans, alcian blue is also widely used to reveal ECM by very intense blue staining. When used to stain ECM, counterstaining with neutral red or a similar dye is recommended.

13. Safranin-O stains in red the lignified walls, nuclei, and chloroplasts.
14. Toluidine blue is a general-purpose dye; it stains the cell walls in blue and lignin in turquoise blue.
15. In wild-type carpels, pollen tubes are able to pass through the style in 4 h after pollination (hap) and reach the base of the ovary at 10–12 hap [3, 34].
16. We prefer not to use microcentrifuge tubes, since it is much easier to damage the pistil, which remains soft, when changing solutions.
17. In our case, plants are usually grown in cabinets at 21 °C under long-day (16-h light) conditions, illuminated by cool-white fluorescent lamps (150 μE m^{-2} s^{-1}), in a 1:1:1 mixture of sphagnum:perlite:vermiculite.
18. Be careful not to pull too strong vacuum. You should be able to observe tiny air bubbles forming, but it is important to avoid boiling of the fixation solution.
19. At this point, the samples can be stored at 4 °C for weeks or months.
20. The eosin staining will help to visualize and orient the tissue when sectioning, but it will not give any color after sectioning, dewaxing, and specific staining are performed.
21. If needed, you can leave the samples overnight in molten paraplast in the oven and continue with the remainder paraplast changes the next morning.
22. Once dry, the slides can be stored in a dry place, covered to avoid dust, for several months.
23. The mounted slides can be stored for a long time.
24. Due to changes in fluorescence emission, safranin can differentiate regions of high and low lignin content more accurately than phloroglucinol; regions of high lignin fluoresce red/orange, and regions with low lignin fluoresce yellow [35].

References

1. Balanzá V, Navarrete M, Trigueros M, Ferrándiz C (2006) Patterning the female side of Arabidopsis: the importance of hormones. J Exp Bot 57(13):3457–3469
2. Ferrandiz C, Fourquin C, Prunet N, Scutt C, Sundberg E, Trehin C, Vialette-Guiraud A (2010) Carpel Development. Adv Bot Res 55:1–74

3. Crawford BCW, Ditta G, Yanofsky MF (2007) The NTT gene is required for transmitting-tract development in carpels of Arabidopsis thaliana. Curr Biol 17(13):1101–1108
4. Alvarez J, Smyth D (2002) CRABS CLAW and SPATULA Genes Regulate Growth and Pattern Formation during Gynoecium Development in Arabidopsis thaliana. Int J Plant Sci 163:17–41
5. Nemhauser JL, Feldman LJ, Zambryski PC (2000) Auxin and ETTIN in Arabidopsis gynoecium morphogenesis. Development 127:3877–3888
6. Alvarez JP, Goldshmidt A, Efroni I, Bowman JL, Eshed Y (2009) The NGATHA distal organ development genes are essential for style specification in Arabidopsis. Plant Cell 21:1373–1393
7. Staldal V, Sohlberg JJ, Eklund DM, Ljung K, Sundberg E (2008) Auxin can act independently of CRC, LUG, SEU, SPT and STY1 in style development but not apical-basal patterning of the Arabidopsis gynoecium. New Phytol 180:798–808
8. Ferrándiz C, Liljegren SJ, Yanofsky MF (2000) Negative regulation of the SHATTERPROOF genes by FRUITFULL during Arabidopsis fruit development. Science 289:436–438
9. Liljegren SJ, Roeder AHK, Kempin SA, Gremski K, Østergaard L, Guimil S, Reyes DK, Yanofsky MF (2004) Control of fruit patterning in Arabidopsis by INDEHISCENT. Cell 116:843–853
10. Spence J, Vercher Y, Gates P, Harris N (1996) "Pod shatter" in *Arabidopsis thaliana Brassica napus* and *B. juncea*. J Microsc 181:195–203
11. Roeder AHK, Yanofsky MF (2005) Fruit Development in Arabidopsis. Arabidopsis Book 52(1):1–50
12. Smyth DR, Bowman JL, Meyerowitz EM (1990) Early flower development in *Arabidopsis*. Plant Cell 2:755–767
13. Sohlberg JJ, Myrenas M, Kuusk S, Lagercrantz U, Kowalczyk M, Sandberg G, Sundberg E (2006) STY1 regulates auxin homeostasis and affects apical-basal patterning of the Arabidopsis gynoecium. Plant J 47(1):112–123
14. Heisler MG, Ohno C, Das P, Sieber P, Reddy GV, Long JA, Meyerowitz EM (2005) Patterns of auxin transport and gene expression during primordium development revealed by live imaging of the Arabidopsis inflorescence meristem. Curr Biol 15(21):1899–1911
15. Marsch-Martinez N, Ramos-Cruz D, Irepan Reyes-Olalde J, Lozano-Sotomayor P, Zuniga-Mayo VM, de Folter S (2012) The role of cytokinin during Arabidopsis gynoecia and fruit morphogenesis and patterning. Plant J 72(2):222–234
16. Müller B, Sheen J (2008) Cytokinin and auxin interaction in root stem-cell specification during early embryogenesis. Nature 453:1094–1097
17. Baum SF, Eshed Y, Bowman JL (2001) The Arabidopsis nectary is an ABC-independent floral structure. Development 128(22):4657–4667
18. Savidge B, Rounsley SD, Yanofsky MF (1995) Temporal relationship between the transcription of two Arabidopsis MADS box genes and the floral organ identity genes. Plant Cell 7:721–733
19. Girin T, Paicu T, Stephenson P, Fuentes S, Körner E, O'Brien M, Sorefan K, Wood TA, Balanzá V, Ferrándiz C, Smyth DR, Ostergaard L (2011) INDEHISCENT and SPATULA Interact to Specify Carpel and Valve Margin Tissue and Thus Promote Seed Dispersal in Arabidopsis. Plant Cell 23(10):3641–3653
20. Sorefan K, Girin T, Liljegren SJ, Ljung K, Robles P, Galván-Ampudia CS, Offringa R, Friml J, Yanofsky MF, Ostergaard L (2009) A regulated auxin minimum is required for seed dispersal in Arabidopsis. Nature 459(7246):583–586
21. Gu Q, Ferrandiz C, Yanofsky MF, Martienssen R (1998) The FRUITFULL MADS-box gene mediates cell differentiation during Arabidopsis fruit development. Development 125(8):1509–1517
22. Hempel FD, Weigel D, Mandel MA, Ditta G, Zambryski PC, Feldman LJ, Yanofsky MF (1997) Floral determination and expression of floral regulatory genes in Arabidopsis. Development 124(19):3845–3853
23. Kuusk S, Sohlberg JJ, Long JA, Fridborg I, Sundberg E (2002) STY1 and STY2 promote the formation of apical tissues during Arabidopsis gynoecium development. Development 129(20):4707–4717
24. Trigueros M, Navarrete-Gomez M, Sato S, Christensen SK, Pelaz S, Weigel D, Yanofsky MF, Ferrandiz C (2009) The NGATHA genes direct style development in the Arabidopsis gynoecium. Plant Cell 21(5):1394–1409
25. Gremski K, Ditta G, Yanofsky MF (2007) The HECATE genes regulate female reproductive tract development in Arabidopsis thaliana. Development 134(20):3593–3601
26. Groszmann M, Bylstra Y, Lampugnani ER, Smyth DR (2010) Regulation of tissue-specific expression of SPATULA, a bHLH gene involved in carpel development, seedling germination, and lateral organ growth in Arabidopsis. J Exp Bot 61:1495–1508
27. Fahlgren N, Montgomery TA, Howell MD, Allen E, Dvorak SK, Alexander AL, Carrington JC (2006) Regulation of AUXIN RESPONSE

FACTOR3 by TAS3 ta-siRNA affects developmental timing and patterning in Arabidopsis. Curr Biol 16(9):939–944
28. Roe JL, Nemhauser JL, Zambryski PC (1997) *TOUSLED* participates in apical tissue formation during gynoecium development in Arabidopsis. Plant Cell 9(3):335–353
29. Toriyama K, Thorsness MK, Nasrallah JB, Nasrallah ME (1991) A Brassica S locus gene promoter directs sporophytic expression in the anther tapetum of transgenic Arabidopsis. Dev Biol 143(2):427–431
30. Sessions RA, Zambryski PC (1995) Arabidopsis gynoecium structure in the wild and in ettin mutants. Development 121(5):1519–1532
31. Liljegren SJ, Ditta GS, Eshed Y, Savidge B, Bowman JL, Yanofsky MF (2000) SHATTERPROOF MADS-box genes control seed dispersal in Arabidopsis. Nature 404(6779):766–770
32. Sundaresan V, Springer P, Volpe T, Haward S, Jones JD, Dean C, Ma H, Martienssen R (1995) Patterns of gene action in plant development revealed by enhancer trap and gene trap transposable elements. Genes Dev 9(14):1797–1810
33. Johnson-Brousseau SA, McCormick S (2004) A compendium of methods useful for characterizing Arabidopsis pollen mutants and gametophytically-expressed genes. Plant J 39(5):761–775
34. Jiang L, Yang SL, Xie LF, Puah CS, Zhang XQ, Yang WC, Sundaresan V, Ye D (2005) VANGUARD1 encodes a pectin methylesterase that enhances pollen tube growth in the Arabidopsis style and transmitting tract. Plant Cell 17(2):584–596
35. Bond J, Donaldson L, Hill S, Hitchcock K (2008) Safranine fluorescent staining of wood cell walls. Biotech Histochem 83(3):161–171

Part III

Microscopy and Histology

Chapter 12

Microscopic Analysis of Arabidopsis Ovules

Balaji Enugutti and Kay Schneitz

Abstract

Ovules are the major female reproductive organs in higher plants. Furthermore, ovules of *Arabidopsis thaliana* are successfully used as model system to study plant organogenesis. Here we describe two microscopic techniques to analyze ovule development in Arabidopsis. Both methods involve fixed specimens and allow rapid, easy, and reproducible morphological comparisons between wild-type and mutant ovule development.

Key words Arabidopsis, Confocal laser scanning microscopy, Light microscopy, Microscopy, Ovule, Ovule mutants, Pseudo-schiff propidium iodide

1 Introduction

The ovule is the major female plant organ involved in sexual reproduction. It is the organ where the egg cell originates, fertilization takes place, and embryo develops during seed formation. Therefore, the genetic, molecular, and evolutionary basis of ovule development has been the focus of intense study [1–5]. The remarkable success in this endeavor has been made possible by a number of technical advantages that render the Arabidopsis ovule an excellent model to study plant organogenesis. An individual plant can produce several hundreds to thousands of ovules. As their development is independent, a researcher can quickly analyze large numbers of ovules at all developmental stages from a single mutant plant. Importantly, Arabidopsis ovules undergo a stereotypic mode of development [6]. This feature allows easy detection of developmental aberrations in mutant analysis. Moreover, ovules within a gynoecium develop in a largely synchronous fashion, except during the short phase of meiosis. Finally, ovules are devoid of chlorophyll and consist of six cell layers at the most, including the embryo sac, before fertilization. Thus, morphological inspection is rapidly achieved using either standard light microscopy on fixed and cleared

José Luis Riechmann and Frank Wellmer (eds.), *Flower Development: Methods and Protocols*, Methods in Molecular Biology, vol. 1110, DOI 10.1007/978-1-4614-9408-9_12, © Springer Science+Business Media New York 2014

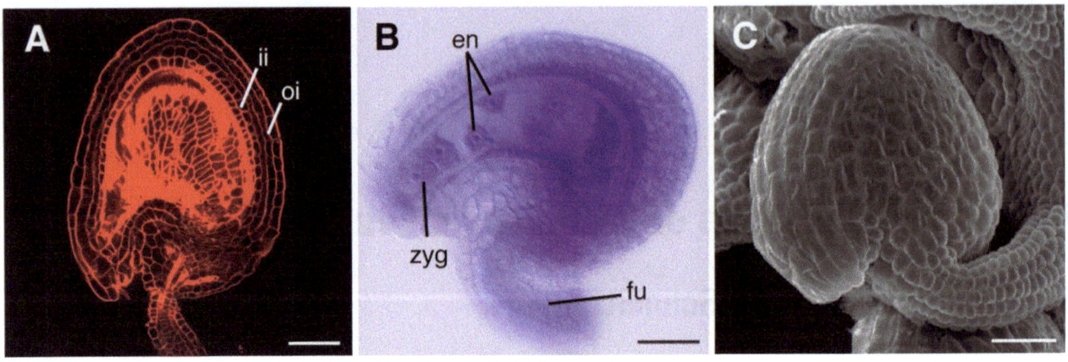

Fig. 1 Micrographs of mature Arabidopsis ovules. (**a**) Mid-optical laser confocal scanning micrograph of a wild-type stage 4-I ovule stained with mPS-PI. (**b**) Mid-optical section of a cleared wild-type stage 4-V ovule stained with Mayer's hemalum. Endosperm nuclei are visible. (**c**) Scanning electron micrograph of a wild-type stage 3-VI/4-I ovule. *en* endosperm nucleus, *fu* funiculus, *ii* inner integument, *oi* outer integument, *zyg* zygote. Size bars: 20 μm

specimens or confocal microscopy. Except for ultrastructural studies there is no need to embed and section ovules to analyze their morphology in Arabidopsis.

Here, we describe two simple microscopic techniques to study Arabidopsis ovule development. The first method uses fixed specimens stained by a modified pseudo-Schiff propidium iodide (mPS-PI) technique followed by confocal laser scanning microscopy (CLSM) (*see* Fig. 1a). This method enables a straightforward analysis of integument development at the cellular level. It is less well suited to study the ontogenesis of the embryo sac. It has been taken from published accounts [7, 8] with only minor ovule-specific adaptions. Alternative CLSM-based protocols can also be used successfully [9, 10]. Another simple protocol involving cleared whole-mount specimens and conventional light microscopy is also available [6, 11] (*see* Fig. 1b). The second technique described in this chapter is based on scanning electron microscopy (SEM) and provides a more global view of ovule development [12] (*see* Fig. 1c).

2 Materials

2.1 mPS-PI Staining of Fixed Ovules

1. Methanol/acetic acid fixative: 50 % (v/v) methanol, 10 % (v/v) acetic acid in deionized water.
2. 80 % ethanol.
3. SDS stock solution: 10 % (w/v) sodium dodecyl sulphate. Dissolve 10 g sodium dodecyl sulphate in 80 mL deionized water and make up to 100 mL with deionized water.
4. 0.2 N NaOH solution: 0.2 N NaOH. Dissolve 4 g of NaOH in 80 mL of deionized water and make up to 100 mL with deionized water to yield 1 N NaOH. Dilute to 0.2 N with deionized water.

5. Bleach solution: Mix one volume of sodium hypochlorite (12 % NaOCl) with four volumes of deionized water resulting in 2.5 % active Cl^- ions.
6. Periodic acid solution: 1 % (w/v) periodic acid. Dissolve 1 g of periodic acid in 80 mL deionized water and make up to 100 mL with deionized water. Solution is stable for 1 year at room temperature.
7. Schiff's reagent: 100 mM sodium metabisulfite, 0.15 N HCl. Dissolve 190 mg of sodium metabisulfite in 8 mL of deionized water, add slowly 125 μL of 37 % HCl (12 N), and make it up to 10 mL with deionized water. Prepare the Schiff's reagent always fresh.
8. Propidium iodide (PI): Add 100 μg PI per 1 mL of Schiff's reagent.
9. Activated charcoal.
10. Chloral hydrate solution: 4 g chloral hydrate, 1 mL glycerol, and 2 mL water.
11. Immersion oil.
12. U-100 insulin syringes, 0.5 mL/27 G.
13. Forceps, class 5 Dumont INOX.
14. Double-sticky tape.
15. Tungsten needle: Cut a 5 cm piece of tungsten wire (0.25 mm in diameter), and introduce it into the tip of a 25 G needle. Push it up towards the rear end of the needle, so that 1–2 cm stick out of the rear end of the needle. Cut the tip of the needle including a small piece of tungsten wire with a pair of scissors (this will prevent the needle from closing up at the tip due to deformation). Introduce a small kink on the rear end side of the tungsten wire, and push the wire back through the tip. Pull the wire in this direction until it is tightly fixed in the needle. Connect the needle to a power supply (using a clamp and wire). Dip the tungsten needle into a 10 % $NaNO_2$ (sodium nitrite) solution in H_2O, which is connected to the other pole of the power supply. Pass enough current through the needle to let it glow. Slowly retract the needle, and repeat until the desired shape of the tip has been created. To remove the oxidation debris created in the process, reduce the current.
16. Razor blade.
17. Kimwipe.
18. Microscope slides 76 × 26 × 1 mm.
19. Microscope cover glasses 22 × 22 mm.

2.2 Scanning Electron Microscopy of Ovules

1. Acetone.
2. Glutaraldehyde.
3. 0.5 M Sodium cacodylate trihydrate (pH 7.0): Dissolve 21.4 g of sodium cacodylate trihydrate in 100 mL of distilled water, adjust the pH to 7.0 using HCl, and make the volume up to 200 mL.
4. Osmium tetroxide (EM grade).
5. Aluminum specimen stubs, 0.5″.
6. Leit adhesive carbon discs, 12 mm.
7. SEM pin stub specimen mount gripper tweezers.
8. SEM pin stub specimen storage box.
9. Cotton.
10. Parafilm.
11. Desiccator.
12. Glass tubes No. 2775/1 (Hecht-Assistent, Germany).
13. Glass scribe.
14. U-100 insulin syringes, 0.5 mL/27 G.
15. Forceps, class 5 Dumont INOX.
16. Tungsten needle (*see* Subheading 2.1, **step 15**).
17. Double-sticky tape.

3 Methods

Detailed staging information of Arabidopsis ovule development, also relative to flower development, is available [6]. Ovules within a pistil develop in a quite synchronous fashion. In wild-type plants of accession Landsberg *erecta* (L*er*) each pistil carries about 50 ovules. The first-open flower (floral stage 13 [13]) contains ovules that have just been fertilized and show zygotes and early endosperm development. Starting with the last or the oldest closed flower bud (floral stage 12) and counting "inwards," the next 4–6 consecutively younger flowers contain ovules at various stages of development. The 2–3 oldest closed flowers carry stage 3 ovules (integument and embryo sac development). The next two buds carry stage 2 ovules (megaspore mother cell formation, meiosis, integument initiation, and early outgrowth), while the youngest buds carry stage 1 ovules (fingerlike ovule protrusions). For stage 13 flowers and the next 2–3 younger flowers it is feasible to dissect individual pistils. For younger flowers, just slightly press open the flower bud, if possible poke small holes in the pistil with a fine needle, and process the flower as a whole. Before observation, dissect out ovules on a microscope slide (mPS-PI technique) or on aluminum specimen stubs covered with adhesive carbon discs (SEM).

3.1 mPS-PI Staining of Fixed Ovules

Use 1.5 mL reaction tubes filled with 1 mL of each solution.

1. Dissect pistils, and remove ovary walls under a dissecting microscope using fine needles or tungsten needles (work carefully without damaging the ovules but also reasonably fast as ovules rapidly dry out in the strong light of the dissecting scope) (*see* **Notes 1–3**).
2. Fix in freshly prepared methanol/acetic acid fixative for at least 2 h at room temperature or overnight at 4 °C. Tissue can be stored for extended periods of time (weeks or months) at 4 °C (*see* **Notes 4–6**).
3. Exchange the methanol/acetic acid fixative with 80 % ethanol and incubate at 80 °C for 2 min.
4. Replace the 80 % ethanol with 1 % SDS and 0.2 N NaOH solution, and leave it at room temperature overnight (*see* **Note 7**).
5. Rinse with H_2O.
6. Incubate the tissue for 2 min in bleach solution containing 2.5 % active Cl^- ions (*see* **Note 8**).
7. Rinse with H_2O.
8. Put tissue into 1 % periodic acid and incubate for 40 min at room temperature (*see* **Note 9**).
9. Rinse with H_2O.
10. Put tissue into Schiff's reagent containing propidium iodide (100 μg/mL) and incubate for 2 h (*see* **Notes 10** and **11**).
11. Exchange the Schiff's PI solution with chloral hydrate solution and incubate overnight at room temperature (*see* **Notes 12** and **13**).
12. Prepare slide for observation: Fix one layer of sticky tape onto the slide, and cut out a rectangular section using a razor blade to create a small (about 1–1.5 cm on the bottom side) "empty" cavity on the slide (to prevent squashing the ovules) (*see* **Note 14**).
13. Place a small drop of chloral hydrate solution into the empty cavity, and disperse it with a pipette tip or a needle.
14. Add the tissue in a drop of chloral hydrate solution on the slide (between the tape layers), dissect the ovules, carefully cover with a cover slip, and avoid bubbles by carefully placing the left edge of the cover slip on the tape, fixing its position with forceps and slowly lowering the right side of the cover slip using a second pair of tweezers or a needle (*see* **Note 2**).
15. If the chloral hydrate solution does not reach the edge of the cover slip, add some by carefully pipetting a small amount of chloral hydrate solution beneath one edge of the cover slip. If there is too much chloral hydrate solution and the cover slip floats, try to remove some solution using a Kimwipe.

16. Observe with 20× (air) or 40× (oil) objectives using a CLSM with an Olympus FV1000 setup and FluoView software after excitation at 488 nm with a multi-line argon laser, and collect the propidium iodide fluorescence (580–630 nm slit width). One-way scan images (scan rate 12.5 μs/pixel, 512×512 pixels, Kahlman frame, average of four scans) are obtained using an Olympus UApo 40× oil objective. Alternatively, use differential interference contrast (DIC or Nomarski) objectives on a good research light microscope, such as the Olympus BX61/63, Leica DM6000 B, or Nikon Eclipse 80i (*see* **Note 15**) (*see* Fig. 1a).

3.2 Scanning Electron Microscopy of Ovules

Use 1.5 mL reaction tubes filled with 1 mL of each solution. Unless mentioned, all steps are carried out at room temperature under a fume hood.

3.2.1 Fixation I

1. Fix dissected carpels or whole flower buds in freshly prepared 2 % glutaraldehyde and 69 % acetone fixative in H_2O for 6 h to overnight (*see* **Notes 1**, **6**, and **16**).
2. Wash four times in 70 % acetone for 15 min each.
3. Wash six times in 70 % acetone for 30 min each. If necessary, tissue can be stored in refrigerator at this point.

3.2.2 Fixation II

1. Wash for 15 min with 50 % acetone in 50 mM sodium cacodylate buffer (pH 7.0) (*see* **Note 17**).
2. Wash for 10 min in 25 % acetone/50 mM cacodylate buffer (pH 7.0).
3. Wash for 10 min in 10 % acetone/50 mM cacodylate buffer (pH 7.0).
4. Wash for 5 min in 50 mM cacodylate buffer (pH 7.0).
5. Fix for 2 h with 2 % osmium tetroxide in 50 mM cacodylate buffer (pH 7.0) (*see* **Notes 17** and **18**).
6. Wash twice in 50 mM cacodylate buffer (pH 7.0) for 15 min each.
7. Repeat **step 3**.
8. Wash for 10 min in 10 % acetone.
9. Dehydrate with an acetone series (p.a. 20, 40, 60, and 70 %) for 30 min each. Tissue can be stored for extended periods of time (weeks or months) at 4 °C.
10. Wash in 100 % acetone three or four times with 2-h intervals in between each wash before critical point drying (C.P.D.).

3.2.3 Critical Point Drying

1. Label the glass tubes by engraving using a glass scribe.
2. Transfer the tissue to the labeled glass tubes in 100 % acetone (*see* **Note 19**).

3. Without introducing any air bubbles plug the neck of the vial with very loose cotton to avoid the sample mix-up if you are drying different specimens.
4. Submerge all the sample vials in acetone tank.
5. To completely dry the sample while maintaining the morphological features, the C.P.D. procedure is carried out (using a critical point dryer) for 2–3 h using liquid CO2.
6. Seal the vials with parafilm and place in the desiccator until mounting of samples.

3.2.4 Specimen Preparation and Scanning

1. Stick the Leit adhesive carbon disc on the aluminum specimen stub by holding it with SEM specimen mount gripper tweezers on a reaction tube rack.
2. Pistils of all stages are mounted in desired orientation on the carbon discs and dissected with fine tungsten needles or fine 0.5 mL/27 G needles under a dissecting microscope (see Notes 1 and 3).
3. Do the gold sputtering of specimens for 60 s, according to the instructions on the sputter instrument (such as Leica EM SCD 005).
4. Scan the specimens between 5 and 15 kV voltage levels (monitoring the optimum resolution and less charge-up of the sample) at the desired magnification on a good SEM, such as the JEOL JSM-5900L scanning electron microscope (*see* Fig. 1c).

4 Notes

1. Prepare a slide covered with a double-stick tape. Remove the pistil from the flower, and place it on the sticky tape. For further dissection, use a conventional dissecting scope with 5× to 40× magnification lenses. If possible use a ring-light source as this facilitates the observation of the specimen. Orient the pistil with the replum pointing upwards (i.e., the valves directed sideways). Cut off the pedicel and the bottom first millimeter of the pistil, and with a fine needle poke a small hole next to the replum and at the top of each valve. Then use the needle to slice along the edges of the valves (next to the replum), and remove the valves. Use forceps to pick up the pistil at the stigma, and quickly place the pistil with the attached ovules in the fixative.
2. For younger ovules, dissect free the pistil by removing much of the other floral organs with a forceps. Then either dissect out the ovules or just cut the pistil into many pieces, and look for appropriately oriented ovules later in the microscope. Collect pistils from several flowers. For older pistils, make sure

that you cut off and remove the stigmas as well. Stigmas are often too hard and bulky and can create problems when putting a cover slip. Be patient; the dissecting steps require experience and may not work the first time.

3. Avoid damaging the samples. Pistils of stage 8–9 flowers (carrying stage 1 ovules) are very difficult to dissect and require practice. Carefully try to break open small pieces of the carpel wall generating small holes through which the ovules can be observed. Despite careful handling one often observes depressions on the ovule primordia generated by the contact with the needles.
4. Methanol is poisonous, and ingestion can cause blindness or death. Avoid an exposure to the vapors. Wear appropriate gloves and safety glasses, and use in a chemical fume hood.
5. The addition of acetic acid results in a better morphology.
6. Make sure that the tissue is always submerged, also for consecutive steps.
7. Sodium dodecyl sulphate (SDS) is toxic, an irritant to eyes. Avoid inhaling SDS powder; wear a mask over nose and mouth.
8. Bleach (NaOCl) is poisonous and corrosive and can burn skin and cause eye damage; hence, wear gloves and safety glasses, and work under the chemical fume hood.
9. Periodic acid is a strong oxidizer and corrosive. Wear gloves and goggles.
10. Hydrochloric acid (HCl) is volatile and may be fatal if inhaled. The vapors are destructive to mucous membranes and respiratory tract. Wear gloves and safety glasses, and use in a chemical fume hood.
11. Propidium iodide is a mutagen. Solutions containing propidium iodide should be poured through activated charcoal before disposal. Then incinerate the charcoal to destroy the dye.
12. Chloral hydrate is destructive to mucous membranes and respiratory tract. Wear appropriate gloves and safety glasses, and use in a chemical fume hood.
13. Use a yellow pipet tip that has its bottom third cleaved off while pipetting chloral hydrate solution.
14. Make sure that the sticky tape is tightly pressed on the upper surface of the microscope slide (particularly in the alignment next to the cavity; no air bubbles; otherwise, poke in the air bubble with a fine needle).
15. Be aware of the working distance of an objective. Be careful not to squeeze the ovules when using a 40× objective.

16. It is important to use glutaraldehyde that was bottled under N_2. Upon opening, dispense the content of the bottle quickly into 1 mL aliquots and immediately store at −20 °C.
17. Wear gloves, and take care when handling cacodylic acid and osmium tetroxide. Avoid inhalation and contact with skin. Both chemicals are poisonous.
18. Osmium is a heavy metal and has to be disposed after neutralizing with twice the volume of vegetable oils containing high percentage of unsaturated bonds, such as corn oil, to the one volume of 2 % osmium for several hours. Neutralization is monitored with a piece of filter paper by dipping into the osmium tetroxide/corn oil mix. Upon complete neutralization the filter paper keeps its color; otherwise, it turns black.
19. C.P.D. can be done in many sorts of glass vials/tubes, even in standard reaction tubes.

Acknowledgment

Our work is supported by grants from the Deutsche Forschungsgemeinschaft (German Research Council) to Kay Schneitz.

References

1. Chevalier D, Sieber P, Schneitz K (2002) The genetic and molecular control of ovule development. In: O' Neill SD, Roberts JA (eds) Plant Reproduction. Sheffield Academic Press, Sheffield
2. Colombo L, Battaglia R, Kater MM (2008) Arabidopsis ovule development and its evolutionary conservation. Trends Plant Sci 13: 444–450
3. Kelley DR, Gasser CS (2009) Ovule development: genetic trends and evolutionary considerations. Sex Plant Reprod 22:229–234
4. Shi D, Yang W (2011) Ovule development in Arabidopsis: progress and challenge. Curr Opin Plant Biol 14:74–80
5. Enugutti B, Kirchhelle C, Schneitz K (2012) On the genetic control of planar growth during tissue morphogenesis in plants. Protoplasma 250:651–661
6. Schneitz K, Hülskamp M, Pruitt RE (1995) Wild-type ovule development in *Arabidopsis thaliana*: a light microscope study of cleared whole-mount tissue. Plant J 7:731–749
7. Haseloff J (2003) Old botanical techniques for new microscopes. Biotechniques 34: 1174–1182
8. Truernit E, Bauby H, Dubreucq B, Grandjean O, Runions J, Barthelemy J, Palauqui JC (2008) High-resolution whole-mount imaging of three-dimensional tissue organization and gene expression enables the study of phloem development and structure in *Arabidopsis*. Plant Cell 20:1494–1503
9. Christensen CA, King EJ, Jordan JR, Drews GN (1997) Megagametogenesis in *Arabidopsis* wild type and the *Gf* mutant. Sex Plant Reprod 10:49–64
10. Running MP, Clark SE, Meyerowitz EM (1995) Confocal microscopy of the shoot apex. Methods Cell Biol 49:217–229
11. Enugutti B, Oelschner M, Schneitz K (2013) Microscopic analysis of ovule development in *Arabidopsis thaliana*. Methods Mol Biol 959:127–136
12. Schneitz K, Hülskamp M, Kopczak SD, Pruitt RE (1997) Dissection of sexual organ ontogenesis: a genetic analysis of ovule development in *Arabidopsis thaliana*. Development 124:1367–1376
13. Smyth DR, Bowman JL, Meyerowitz EM (1990) Early flower development in *Arabidopsis*. Plant Cell 2:755–767

Chapter 13

Scanning Electron Microscopy Analysis of Floral Development

Robert G. Franks

Abstract

Scanning Electron Microscopy (SEM) allows the morphological characterization of the surface features of floral and inflorescence structures in a manner that retains the topography or three-dimensional appearance of the structure. Even at relatively low magnification levels it is possible to characterize early developmental stages. Using medium to high power magnification at later stages of development, cell surface morphology can be visualized allowing the identification of specific epidermal cell types. The analysis of the altered developmental progressions of mutant plants can provide insight into the developmental processes that are disrupted in that mutant background.

Key words Development, Morphology, Primordia, Flower, Inflorescence, Critical point drying, Sputter coating

1 Introduction

Scanning Electron Microscopy (SEM) is a useful tool for the characterization of floral development. Many morphological aspects of floral development in both wild-type and mutant plants are best characterized with SEM. With respect to the model dicot plant species *Arabidopsis thaliana*, a classic compendium containing results from SEM analysis is John Bowman's *Arabidopsis: An Atlas of Morphology and Development* [1]. This volume contains SEM and histological analysis of both wild-type Arabidopsis floral development and the analysis of the developmental progressions of several Arabidopsis mutants. In addition to Arabidopsis [1, 2], SEM has also been used in the analysis of many other flowering plants, including *Zea mays* [3], *Antirrhinum majus* [4], and many others not listed here. In many species, a system to define and standardize the developmental stages of floral development depends on morphological landmarks that are best identified with SEM (e.g., [5]).

José Luis Riechmann and Frank Wellmer (eds.), *Flower Development: Methods and Protocols*, Methods in Molecular Biology, vol. 1110, DOI 10.1007/978-1-4614-9408-9_13, © Springer Science+Business Media New York 2014

The analysis of the developmental progression of floral mutants by SEM allows the characterization of the earliest stages of floral development when relevant structures are difficult to visualize with light microscopy. Additionally, the sepals, petals, stamens, and carpels can be differentiated by the characteristic surface morphologies of the epidermal cells. This can be useful in the analysis of homeotic mutant phenotypes in which one floral organ is replaced by the likeness of another floral organ [6–8]. As these homeotic transformations are often incomplete and generate chimeric organs of mixed identity, the analysis of cell surface morphology by SEM can help to resolve the identities of these organs. SEM has also been used to examine alterations in organ polarity or organ patterning defects. As the adaxial and abaxial surfaces of the floral organs are typically distinguishable by SEM, it is possible to use SEM to describe abaxial and adaxial patterning events during floral organ development [9–14]. Similar approaches have been used to describe alterations along the apical to basal axis of the developing floral organs [15, 16]. Although SEM analysis is relatively time-intensive and requires sophisticated equipment, the ability to visualize the three-dimensional topography of developing structures makes SEM an excellent complement to histological and differential interference contrast (Nomarski optical) analysis of cleared tissues [17].

2 Materials

1. Glutaraldehyde (25 % solution, EM Grade).
2. Osmium tetroxide (crystalline) capsules (Ted Pella Inc.).
3. 20 mL glass scintillation vials.
4. Specimen holder and plastic box (Ted Pella Inc.).
5. Dumont Mount gripper tweezers (Ted Pella Inc.).
6. Dumont #5 Forceps-Inox Biologie (Fine Science Tools).
7. Carbon conductive tabs (Ted Pella Inc.).
8. Standard pin stub SEM mount (Ted Pella Inc.).
9. Glass thin wall capillary tubes (Sutter Instrument).
10. Screen basket for critical point drying (Ted Pella Inc.).
11. Osmium tetroxide spill kit (*see* **Note 1**).
12. Sodium phosphate buffer: 0.1 M sodium phosphate (pH 7.0). For 200 mL, mix 39 mL of 0.2 M NaH_2PO_4 with 61 mL of 0.2 M Na_2HPO_4 and add 100 mL of deionized water.
13. Glutaraldehyde fixative: 3 % glutaraldehyde, 25 mM sodium phosphate buffer (pH 7.0). For 100 mL, add 12 mL of 25 % glutaraldehyde, 25 mL of 0.1 M sodium phosphate buffer (pH 7.0), and 63 mL deionized water. Prepare fresh fixative on the day of tissue fixation.

14. Osmium tetroxide stock solution: 2 % (w/v) OsO_4. Place 25 mL of distilled water in a bottle. Working in a fume hood, carefully put a 0.5 g capsule of OsO_4 in the bottle under the water and break it open with a metal spatula. Let solution sit at room temperature to dissolve fully, swirling occasionally. OsO_4 crystals will be slow to dissolve. The stock solution can be stored in a tightly closed bottle, wrapped in aluminum foil at 4 °C for at least 1 month. As OsO_4 is very volatile, it is best to keep the bottle tightly wrapped with Parafilm and then place in another secondary sealed container. The OsO_4 solution should be straw colored. If the color has changed toward purple it is no longer good. OsO_4 is hazardous and should be disposed of in accordance with your local hazardous waste procedures.
15. Osmium tetroxide fixative: 1 % (w/v) OsO_4, 25 mM sodium phosphate buffer (pH 7.0). Add two parts of osmium tetroxide stock solution, one part of 0.1 M sodium phosphate buffer, and one part of deionized water.

3 Methods

The standard procedure for analysis of plant material by SEM includes chemical fixation, dehydration to 100 % ethanol, critical point drying (CPD), mounting on a stub, dissection of material, coating with a metal, and then imaging in the SEM chamber. These steps are each covered in detail below. The methods presented here are those that we have used for the analysis of floral and inflorescence development in Arabidopsis [18–21] and several *Cornus* species [22] and should be generally applicable to a large number of flowering plant species.

Please note that glutaraldehyde and osmium tetroxide (OsO_4) are toxic substances and should be used in a fume hood and with protection of a lab coat, gloves and goggles. Osmium tetroxide is extremely hazardous because it is a strong oxidizer and is very volatile. If used improperly, fumes from the OsO_4 solution can fix the corneal and nasal epithelial cells of the user. This solution must be used in a certified chemical fume hood. Also *see* **Note 1** for additional procedures for OsO_4 use and for a simple and effective kit to quickly neutralize OsO_4 spills.

3.1 Fixation of the Tissue

Plant tissues must be fixed and dehydrated to preserve the structure of the tissue. The subsequent coating and imaging steps of SEM analysis will subject the tissue to high vacuum conditions that would distort or disrupt the morphology of unfixed tissues. Additionally, the use of OsO_4 as a post-fix increases the electrical conductivity of the tissue reducing electrostatic charging artifacts (*see* **Note 2**).

1. Working in the fume hood, place 15–20 mL of the glutaraldehyde fixative into a 20 mL glass scintillation vial and set on ice.
2. Cut off inflorescence or floral tissue with scissors or forceps and place tissue into the glutaraldehyde fixative. The tissue will often float in the fixative. It is helpful to subject the tissue to a vacuum to help remove the air from the intercellular spaces within the tissue and allow a better penetration of the fixative. When pulling a vacuum, loosen the caps of the scintillation vials and place them on ice in a bell jar or vacuum oven in a fume hood. Slowly pull a vacuum until the vacuum reaches to 20–25 mmHg. Incubate the tissue on ice for 10–15 min under this vacuum, and then release the vacuum slowly to prevent disruption of the tissue morphology. Finally, allow the tissue to sit in fixative on ice at atmospheric pressure for 10–15 min. This cycle of pulling and releasing a vacuum should be repeated two or three times if your tissue is difficult to fix or has many intercellular air spaces.
3. Incubate at 4 °C for 12–24 h. Overnight is convenient.
4. Pour off the glutaraldehyde fixative (treat glutaraldehyde as organic waste and dispose of properly).
5. Wash in 50 mM sodium phosphate buffer; two washes of 15 min each.
6. Post-fix the tissue in 10–20 mL of the osmium tetroxide fixative, using enough to cover the tissue. Because of the volatility of the OsO4, we do not pull a vacuum on during the OsO4 fixation step.
7. Incubate at 4 °C for 2 h to overnight depending on the thickness of your sample. The osmium tetroxide solution and the tissue will darken considerably.

3.2 Dehydration of the Tissue

1. Pour off the osmium tetroxide fixative and rinse three times with 50 mM sodium phosphate buffer. Treat the first wash as osmium tetroxide waste and dispose of it properly.
2. Put tissue through an ethanol series (30, 50, 70, 85, and 95 %, followed by three incubations in 100 % ethanol) to dehydrate with 30 min to 1 h incubation in the ethanol solution for each step. Do the three incubations in 100 % ethanol using the highest quality (200-proof) ethanol. This ensures that all water is removed from the tissue before progressing to the critical point drying. Note that it is important that the tissue remains immersed in the appropriate liquid at each step. Do not allow the tissue to air-dry and minimize the amount of time that the tissue is exposed to air. This will reduce morphological artifacts that can be caused by shrinking of the tissue. For long-term storage of the tissue, a 70 % ethanol solution is best.
3. Incubate the tissue in fresh 100 % ethanol overnight to fully dehydrate before continuing on to critical point drying.

3.3 Critical Point Drying (CPD)

In this step you are replacing the ethanol with liquid CO_2 and then allowing the liquid CO_2 to evaporate under controlled conditions so that the tissue comes out dried without damaging the tissue morphology. There are many different critical point drying (CPD) apparatuses that you might find within a electron microscopy (EM) facility. Most of these are now relatively automated. As the steps for each type of critical point drying apparatus are different, we will cover the general steps, but suggest that you refer to the manufacturer's recommendations or the instructions from the staff at your EM facility. In the Center for Electron Microscopy at North Carolina State University, where we process our samples, we typically use a Tousimis SAMDRI-795 critical point drying apparatus (Tousimis Research Corp., Rockville, MD, USA).

1. Turn on CPD equipment and allow it to warm up if necessary.
2. Place metal screen basket or other sample holder in a beaker filled with enough 200-proof ethanol to completely cover the basket.
3. Transfer the fixed tissue from the glass scintillation vials into the screen basket sample holders. Sometimes it is convenient to have one basket that is divided into four sections to allow the critical point drying of several samples/genotypes in one CPD run. Always keep the tissue submerged in 200-proof ethanol during this process.
4. Place a small volume of 200-proof ethanol into the chamber of the CPD apparatus.
5. Place the screen sample holder into the CPD apparatus, ensuring that the tissue is covered by ethanol.
6. Seal the chamber of the CPD apparatus.
7. Open the main valve on the CO_2 tank.
8. Follow manufacturer-specific instructions to have apparatus carry out cycles of filling the chamber with liquid CO_2 and then purging the chamber. This replaces all of the ethanol with liquid CO_2. The amount of time required to purge the ethanol from your samples depends upon the size of your samples; larger samples require longer purge steps. For a batch of samples containing several Arabidopsis inflorescences, 10–15 min of purging is sufficient.
9. Allow equipment to cycle through its heat and bleed or vent cycles. This allows the liquid CO_2 to evaporate without damaging the morphology of the tissue.
10. Remove the tissue from the screen baskets after the CPD run has been completed. The tissue will be quite brittle.
11. At this point the tissue can be stored for later mounting or dissection by maintaining the tissue in a desiccator. It is critical that the tissue not be allowed to reabsorb moisture from the air to ensure the highest quality images.

3.4 Mounting and Dissecting Samples

Often it is desirable to dissect the inflorescences or flowers before sputter coating and imaging. For example, one might want to remove older flowers from an inflorescence to allow imaging of young floral stages. Alternatively, one might need to remove perianth organs in order to visualize the more internally positioned reproductive organs. One convenient manner to dissect the tissue is to attach the inflorescence or flower to the metal SEM pin stub with a carbon conductive tab that is sticky on both surfaces.

1. Place the sticky tab onto the stub and then place your tissue onto the sticky tape. Ensure that the sample is well attached to the tape because a good contact between the sample and the conductive tape helps to ensure good electrical grounding.
2. One can use dissecting needles or forceps to knock off undesired portions of the sample. We often generate home-made dissecting needles by pulling borosilicate glass thin wall capillary tubes (B100-75-10 or similar, Sutter Instrument) on a Flaming/Brown-Type Micropipet Puller (Model P-87, Sutter Instrument) [23]. These glass needles can be used to break off the sepals or other undesired portions of the sample. Others report using eyelashes that have been affixed to toothpicks as effective home-made dissection tools [24].

Note that the CPD dried sample will be brittle. If the sample starts to soften because of high humidity in the room, it is best to replace the sample in the desiccator for a while before continuing on to sputter coating. Once you have your specimens mounted onto the metal stubs and have dissected them as you desire, you are ready to coat the samples.

3.5 Coating the Sample

The purpose of coating your tissue with metal (typically a gold–palladium alloy) is to make the tissue electro-conductive, thus allowing proper imaging under the SEM and reducing image artifacts due to the unwanted buildup of electrostatic charge. Again there are many different possibilities with respect to coating your sample before imaging in the SEM. We recommend that you follow the manufacturer's recommendations for your specific apparatus. However, we outline the basic steps below. We typically are using an Anatech Hummer 6.2 Sputter System (Anatech, Union City, CA, USA). The sputter coater works by using a high voltage differential and a vacuum chamber filled with argon gas to produce ionized gas plasma. The plasma causes metal atoms to be ejected (or sputter off) from a target of gold–palladium alloy. These metal atoms will then be deposited on your sample in a thin layer.

1. Open up the chamber and place samples into chamber on the stage using Dumont mount gripper tweezers to hold your stub mounts.
2. Replace bell jar or chamber cover.
3. Turn on the sputter coater and pull a vacuum in the chamber.

4. Flush the chamber with argon gas.
5. Adjust the pressure of the argon-filled chamber. We typically use a pressure of approximately 40 mTorr.
6. Set the voltage and adjust the flow of argon to generate a current of 10 mA.
7. Using a digital thickness monitor, coat the sample to the desired thickness. We typically coat to 20–30 nm.
8. We use a stage that is tilted such that the tissue sits at a 45° angle relative to the gold–palladium target. This allows a better coating of recessed surfaces. After application of the first coat, we rotate the samples 180° and recoat to ensure that all surfaces are properly coated. For topographically complex samples, four rotations of 90° may be necessary.
9. When coating is complete, turn off equipment and vent chamber.
10. Remove samples and store them in a desiccator or directly proceed to imaging in the SEM.

3.6 Imaging the Sample

As there are many different SEM machines, you will need to refer to your SEM lab technician for detailed operating instructions. Here we discuss the various steps required to capture an image and give guidelines for various parameters that we have found to be appropriate for our samples. At NCSU, we use a JEOL JSM-5900LV SEM (JEOL, Peabody, MA, USA).

When you go to image your sample, your SEM facility may have a technician who will set up the machine, saturate the filament and adjust the electron beam. If not, you will need to first adjust the electron beam. This is typically done using a test grid.

1. To use the test grid, vent and open the SEM chamber, place the test grid onto the stage, close and evacuate the chamber. Follow the manufacturer's instructions to properly saturate the filament, adjust the alignment tilt of the beam, center the objective aperture, and correct the astigmatism.
2. To view your sample, vent the chamber, remove the specimen holder from the stage, and place your sample stub mount onto specimen holder. Then place the specimen holder back onto the stage, and reactivate the vacuum to evacuate the chamber.
3. Select a desired accelerating voltage (10–20 kV is appropriate for many applications).
4. Set the spot size (for magnification less than 1,000×, use 50–30 as spot size; for magnification between 1,000× and 20,000×, use 30–20 as spot size).
5. Set the working distance—for low magnification images (i.e., 50× to 500×) it is beneficial to use a relatively long working distance (greater than 20 mm). Using a long working distance

will give you a greater depth of field and allow you to capture an image in which the entire height of the flower is in focus. For higher magnification images (i.e., greater than 2,000×), it is important to reduce the working distance (8 mm or less). This will reduce the depth of field, but will ensure that you have good resolution at the higher magnifications. Note that it is important to readjust the astigmatism correction after changing the working distance, the spot size, or the alignment tilt.

6. Set the required magnification and find the area of interest.
7. Focus up and down on the area of interest. If there is "smearing" of the image, you will need to adjust the astigmatism correction. It is best to use a higher power magnification to focus and correct the astigmatism and then reduce the magnification to obtain the image you desire (i.e., if you desire a 5,000× image, check the focus and astigmatism at 10,000×).
8. Adjust brightness and contrast, if necessary.
9. Capture and save your image using the software that is available on your system.

After imaging your sample in the SEM it is possible to further dissect the sample and then recoat. In this way one can image the overall structure and subsequently image more internal structures from the same sample. You will need to recoat in the sputter coater after each dissection.

Although beyond the scope of this article, replica techniques have been applied to SEM analysis. This approach involves using a thin coating of a polymer or resin to make an impression of the plant organ surface that can then be analyzed with the SEM [26, 27]. This approach has the advantage of preserving the structure of epicuticular waxes [28], as well as being non-destructive and thus enabling the sequential analysis of living material [29].

For additional technical details regarding SEM analysis, or alternative techniques not covered here including freeze-drying, environmental SEM, and low-temperature SEM, please refer to the following references [30–32].

4 Notes

1. As indicated above, OsO_4 must be used in a fume hood to prevent contact of the volatile fumes with eyes or nasal passages. Proper safety equipment such as disposable gloves, lab coat, and eye protection is also required. It is best to make up as little of this solution in advance as possible and thus reduce the chance of spills. In the case of a spill, OsO_4 can be absorbed and neutralized by a mixture of kitty litter and corn oil [25]. The double bonds in the corn oil quickly neutralize the OsO_4.

To prepare an OsO_4 emergency spill kit, mix 300 g of kitty litter with 150 mL of corn oil and place in a tightly sealed plastic bag. This amount of the mixture is sufficient to neutralize 50 mL of 2 % OsO_4 solution. This kit should be prepared in advance and be available in case of a spill. If it is not possible to very quickly neutralize the spill, evacuate the area before exposure of eyes and nasal passages to OsO_4 fumes.

2. One common problem with SEM imaging is the buildup of excess electrostatic charge on the sample due to poor grounding of the tissue. This leads to overly bright portions of the SEM image. This is mitigated if you have a good contact between the stub/carbon tab and your tissue. The addition of a very small amount of liquid silver paint (Ted Pella Inc.) to the base of the sample can help to ground the tissue; however, care must be taken not to allow the silver paint to wick up onto the portions of your sample that you hope to image. Alternatively, reducing the energy of the electron beam (reducing kV), increasing the amount of time in the osmium tetroxide fix, or applying a thicker metal coating with the sputter coater may also help to reduce charging artifacts.

   Another issue can be the undesired collapse of cells. Some cell types are more prone to collapse during the critical point drying or sputter coating steps and thus can be difficult to image properly. Better images can sometimes be obtained by increasing the time in the fixative or by pulling a vacuum during the fixation to allow better penetration of the fixative into the tissue. Tightly opposed bud scales or other woody externally located organs can significantly reduce the penetration of fixative to the interior portions of a sample. In these cases, it is often critical to remove the external structures before fixation to allow the fixative to penetrate. If your tissue does not sink during the fixation step, a small amount of nonionic detergent (0.01 % Tween-20 or NP-40) can be used to reduce surface tension and help the tissue to sink into the fixative.

Acknowledgment

I thank John Mackenzie Jr. and Valerie K. Lapham from the Center for Electron Microscopy at NCSU for assistance and advice on our SEM analyses. I also thank April Wynn, Mia Chunmiao Feng, John Mackenzie Jr., and Valerie K. Lapham for comments and suggestions on the manuscript. I thank Mia Chunmiao Feng for providing the SEM images used in Fig.1

I apologize to those whose work is not cited due to space limitations.

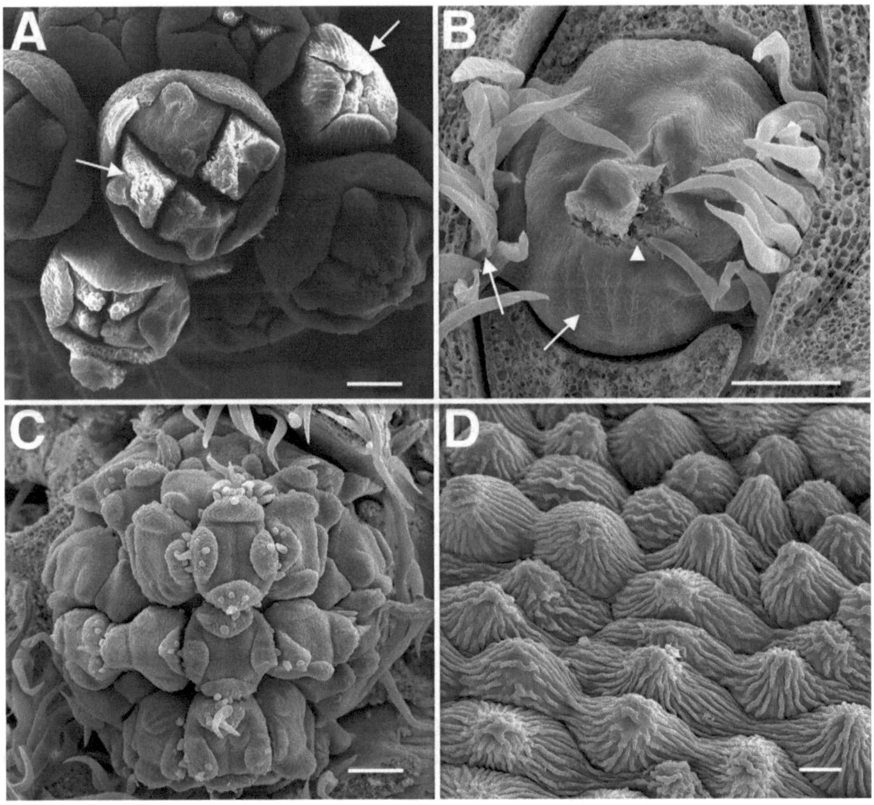

Fig. 1 Examples of scanning electron microscopy micrographs. (**a**) Excessive buildup of electrostatic charge results in "overexposed" sections of the image (*arrows*). This is often caused by poor electrical grounding of the tissue. (**b**) Collapsed cells are evident (*arrows*). This is often caused by incomplete penetration of the fixative. The *arrowhead* indicates a section of the tissue that was inadvertently damaged during the dissection required to remove the external structures of this sample. (**c**) A relatively low-magnification image (taken at 130×) of a *Cornus sanguinea* inflorescence. (**d**) Higher magnification image of *C. sanguinea* petal surface (taken at 1,000×) showing a more-detailed epidermal cell surface morphology

References

1. Bowman JL (1994) Arabidopsis: an atlas of morphology and development. Springer, New York, NY
2. Bowman JL, Smyth DR, Meyerowitz EM (1989) Genes directing flower development in *Arabidopsis*. Plant Cell 1(1):37–52
3. Cheng PC, Greyson RI, Walden DB (1983) Organ initiation and the development of unisexual flowers in the tassel and Ear of Zea mays. Am J Bot 70(3):450–462
4. Sommer H, Beltran JP, Huijser P, Pape H, Lonnig WE, Saedler H, Schwarzsommer Z (1990) Deficiens, a homeotic gene involved in the control of flower morphogenesis in antirrhinum-majus: the protein shows homology to transcription factors. Embo Journal 9(3):605–613
5. Smyth DR, Bowman JL, Meyerowitz EM (1990) Early flower development in *Arabidopsis*. Plant Cell 2(8):755–767
6. Bateson W (1894) Materials for the study of variation. Macmillian, London
7. Leavitt RG (1909) A vegetative mutant, and the principle of homoeosis in plants. Bot Gaz 47:30–68
8. Sattler R (1988) Homeosis in plants. Am J Bot 75(10):1606–1617
9. Siegfried KR, Eshed Y, Baum SF, Otsuga D, Drews GN, Bowman JL (1999) Members of the *YABBY* gene family specify abaxial cell fate in *Arabidopsis*. Development 126(18):4117–4128
10. Sawa S, Watanabe K, Goto K, Kanaya E, Morita EH, Okada K (1999) FILAMENTOUS

FLOWER, a meristem and organ identity gene of Arabidopsis, encodes a protein with a zinc finger and HMG-related domains. Genes Dev 13(9):1079–1088
11. McConnell JR, Barton MK (1998) Leaf polarity and meristem formation in *Arabidopsis*. Development 125(15):2935–2942
12. Kerstetter RA, Bollman K, Taylor RA, Bomblies K, Poethig RS (2001) *KANADI* regulates organ polarity in *Arabidopsis*. Nature 411(6838):706–709
13. Eshed Y, Baum SF, Perea JV, Bowman JL (2001) Establishment of polarity in lateral organs of plants. Curr Biol 11(16):1251–1260
14. Waites R, Hudson A (1995) *phantastica*: a gene required for dorsoventrality of leaves in *Antirrhinum majus*. Development 121(7):2143–2154
15. Sessions RA, Zambryski PC (1995) *Arabidopsis* gynoecium structure in the wild and in *ettin* mutants. Development 121(5):1519–1532
16. Sessions A, Nemhauser JL, McColl A, Roe JL, Feldmann KA, Zambryski PC (1997) *ETTIN* patterns the *Arabidopsis* floral meristem and reproductive organs. Development 124(22):4481–4491
17. Berleth T, Jurgens G (1993) The role of the monopteros gene in organising the basal body region of the Arabidopsis embryo. Development 118(2):575–587
18. Franks RG, Liu Z, Fischer RL (2006) SEUSS and LEUNIG regulate cell proliferation, vascular development and organ polarity in Arabidopsis petals. Planta 224(4):801–811
19. Franks RG, Wang C, Levin JZ, Liu Z (2002) *SEUSS*, a member of a novel family of plant regulatory proteins, represses floral homeotic gene expression with *LEUNIG*. Development 129(1):253–263
20. Azhakanandam S, Nole-Wilson S, Bao F, Franks RG (2008) SEUSS and AINTEGUMENTA mediate patterning and ovule initiation during gynoecium medial domain development. Plant Physiol 146(3):1165–1181
21. Bao F, Azhakanandam S, Franks RG (2010) SEUSS and SEUSS-LIKE transcriptional adaptors regulate floral and embryonic development in Arabidopsis. Plant Physiol 152(2):821–836
22. Feng C-M, Xiang Q-YJ, Franks RG (2011) Phylogeny-based developmental analyses illuminate evolution of inflorescence architectures in dogwoods (Cornus s. l., Cornaceae). New Phytol 191(3):850–869
23. Dean DA, Gasiorowski JZ (2011) Preparing injection pipettes on a flaming/brown pipette puller. Cold Spring Harb Protoc 2011(3):prot5586. doi:10.1101/pdb.prot5586
24. Bomblies K, Shukla V, Graham C (2008) Scanning electron microscopy (SEM) of plant tissues. Cold Spring Harb Protoc 2008(4):prot4933. doi:10.1101/pdb.prot4933
25. Cooper K (1980) Neutralization of osmium tetroxide in case of accidental spillage and for disposal. Bulletin Micros Society Canada 8(3):24–28
26. Marlow P, Presland AEB, Wield DV (1970) Some replica techniques for the scanning electron microscope. Micron 2(2):139–147
27. Sampson J (1961) Method of replicating Dry or moist surfaces for examination by light microscopy. Nature 191(479):932–933
28. Williams MH, Vesk M, Mullins MG (1987) Tissue preparation for scanning electron microscopy of fruit surfaces: comparison of fresh and cryopreserved specimens and replicas of banana peel. Microsc Acta 18(1):27–31
29. Williams MH, Green PB (1988) Sequential scanning electron microscopy of a growing plant meristem. Protoplasma 147(1):77–79
30. Hayat MA (1974) Principles and techniques of scanning electron microscopy. Biological applications, vol 1. Van Nostrand Reinhold Company, New York
31. Pathan AK, Bond J, Gaskin RE (2008) Sample preparation for scanning electron microscopy of plant surfaces: horses for courses. Micron 39(8):1049–1061
32. Bozzola JJ, Russell LD (1999) Electron microscopy: principles and techniques for biologists, 2nd edn. Jones and Bartlett, Sudbury, MA

Chapter 14

Detection of mRNA Expression Patterns by Nonradioactive In Situ Hybridization on Histological Sections of Floral Tissue

Anna Medzihradszky, Kay Schneitz, and Jan U. Lohmann

Abstract

Analysis of gene activity with high spatial resolution is a prerequisite for deciphering regulatory networks which underlie developmental programs. Over many years, in situ hybridization has become the gold standard for the identification of in vivo expression patterns of endogenous mRNAs. Nonetheless, the method has several limitations, and the detection of lowly expressed transcripts is still a challenge. Here, we present a robust protocol for sensitive analysis of expression patterns in inflorescence tissue of *Arabidopsis thaliana*. We describe how the samples are fixed, embedded, and sectioned in preparation for in situ hybridization, how RNA probes are prepared, and how hybridization and detection is carried out. While the described protocol is optimized for inflorescence meristems, it can possibly be used for other tissues as well.

Key words Arabidopsis, DIG-labelling, Flower, In situ hybridization, Light microscopy, Microscopy

1 Introduction

The advent of genomic technologies has fundamentally changed the way biologists approach the complex questions of gene regulation and function. While just 20 years ago, analyzing single genes was challenging, today we are able to reliably measure the activity of all genes of an organism, sample differential splicing, and even discover new transcribed units in a single RNAseq experiment. However, a major problem remains with multicellular organisms, because genome-wide approaches are usually unable to resolve expression patterns at the cellular level. Since many developmental regulators have tissue or even cell-type specific roles, it is imperative to reliably record their expression patterns with maximum resolution in space and time (*see* Fig. 1 for an example). RNA-based in situ hybridization has for many years been the method of choice to meet this experimental need and despite the fact that the method involves a substantial number of fairly challenging steps,

José Luis Riechmann and Frank Wellmer (eds.), *Flower Development: Methods and Protocols*, Methods in Molecular Biology, vol. 1110, DOI 10.1007/978-1-4614-9408-9_14, © Springer Science+Business Media New York 2014

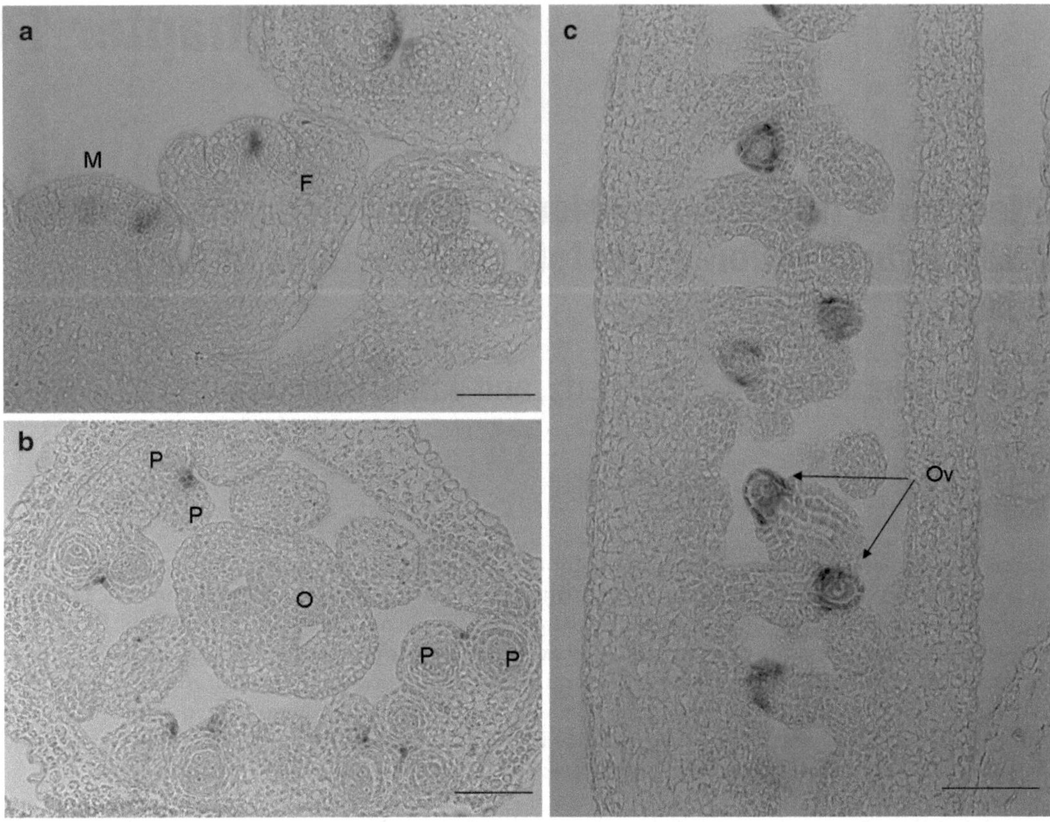

Fig. 1 Expression pattern of the *WUSCHEL* (*WUS*) gene [13] in young floral meristems, anthers and ovules. (**a**) Longitudinal section of a shoot apical meristem with young floral buds. Note the signal in the subepidermal cell layers. *M* shoot meristem, *F* late stage 3 flower; (**b**) cross section of a stage 8 flower. *WUS*-derived signal in anthers is restricted to the stomium region. *P* pollen sacs, *O* ovary; (**c**) longitudinal section of a stage 11 carpel displaying stage 2-III ovules. Signal can be detected in the nucellus. *Ov* ovules. Scale bars: 50 μm

has remained a favorite among plant biologists. In comparison to alternative strategies, such as reporter gene analysis, it offers a number of advantages, which are still unmatched. Firstly, RNA in situ hybridization records the localization of endogenous mRNA, which remains the closest approximation to recording endogenous protein distribution. In contrast to reporter gene analysis, which might (or might not) faithfully visualize the activity of a promoter, analysis of endogenous mRNA in addition also includes the effects of RNA turnover and influence of regulatory RNAs, such as miRNAs. Secondly, since endogenous mRNA is the target, there is no need to generate transgenic plants, greatly facilitating the quick analysis of multiple different genotypes without tedious transformation or introgression of existing marker lines in mutant backgrounds. No method comes without drawbacks and RNA in situ hybridization on plant tissue suffers from a fair number of those: sensitive RNA in situ hybridization on shoot tissue so far only works reliably on tissue sections and consequently access to histological equipment is required. Preparing histological sections also is fairly time consuming and requires training, making in situ hybridization experiments a matter of days rather than hours. As any RNA based method, in situ hybridization requires a strict RNase-free experimental regime, further complicating matters. While these experimental challenges can be met, one of the major limitations of in situ hybridization lies in the cell biology of plants and thus cannot be avoided. Since mRNA mostly accumulates in the cytoplasm, detection sensitivity of the hybridization varies greatly depending on the target cell type. While small cytoplasm-rich cells of the meristems lend themselves well to analysis, large vacuolated cells in differentiated tissues respond much poorer to the method. And finally, since static, fixed tissue sections are used, it is impossible to trace expression patterns over time.

Taken together, RNA-based in situ hybridization, despite all its drawbacks, still is the gold standard for the characterization of gene expression patterns with cellular resolution. In the following chapter we describe a detailed protocol for robust and sensitive non-radioactive in situ hybridization, which has evolved in multiple labs over several years.

This protocol essentially goes back to an early plant in situ hybridization protocol [1] that was modified by several labs [2–6]. We eventually took advantage of an advanced version of the protocol [7], made a few adaptations [8] and modified it to include the use of a Leica ASP 200S instrument for fixation and paraffin embedding of floral tissue [9]. By and large it can also be used with tissues of other model systems, such as Antirrhinum and corn. A similar protocol has also been used with ovules of, for example, Gnetum [10]. Interpreting tissue sections requires intimate knowledge of the different stages of floral development. Detailed staging information of Arabidopsis flower and ovule development is available from the literature [11, 12].

2 Materials

2.1 General Components and Solutions (See Note 1*)*

1. Ethanol (denatured and p.a.).
2. 10× PBS: 1.3 M NaCl, 30 mM NaH_2PO_4, 70 mM Na_2HPO_4.
3. 1 M $MgCl_2$.
4. 5 M NaCl.
5. 0.5 M EDTA pH 8.0.
6. Formamide (deionized).

2.2 Components for Fixation and Tissue Infiltration

1. FAA (**F**ormalin–**A**cetic acid–**A**lcohol Fixative): 50 % ethanol (Abs. Sigma for sample collection, manual infiltration and postfixation on the slides, denatured for automated infiltration),

5 % Glacial acetic acid (Ph.Eur), 3.7 % Formaldehyde (histological grade, 10–15 % methanol stabilized).

2. Sample vials of glass, falcon tubes, or eppendorf tubes.

2.2.1 Components for Tissue Infiltration with Automated Vacuum Tissue Processor Leica ASP200S

1. Xylol (Isomer, Roth).
2. Paraplast (Leica).
3. Eosin Y.
4. Embedding cassettes (Leica).

2.2.2 Components for Manual Tissue Infiltration

1. 4 % Formaldehyde solution from paraformaldehyde (PFA): set the pH of 1× PBS to 11 with NaOH and heat up to 60 °C. Pour into a bottle containing the paraformaldehyde (Sigma) and mix until dissolved. Place on ice. When cooled down, adjust pH to 7.0 with H_2SO_4 (not HCl, as this will release carcinogenic vapors). Add 0.1 % (v/v) Triton X-100 for fixation.

   The fixative should be fresh or stored at -20 °C in aliquots, but only thawed once.
2. Histo-Clear (Roth: Roti-Histol).
3. Eosin Y.
4. Paraplast (Leica or Sigma, we tested both).
5. Vacuum desiccator and vacuum pump.

2.3 Components for Tissue Embedding

1. Embedding station (e.g., Leica EG1160).
2. Embedding cassettes (Leica).
3. Metal molds (Leica).
4. Forceps and preparation needles for handling the samples.

2.4 Components for Sectioning

1. Microtome (e.g., Leica), microtome blade (R35 Feather).
2. Scalpel or razor blade.
3. Brush or preparation needle to handle ribbons.
4. Heated water bath.
5. Heating plate.
6. Pretreated microscope slides (StarFrost from Knittel, Menzel-Gläser, or Poly-L-Lysine from Sigma).

2.5 Components for Probe Synthesis

1. DIG RNA labelling kit from Roche or individual components (Roche: SP6 RNA Polymerase, T7 RNA Polymerase, T3 RNA Polymerase, RNase Inhibitor, DIG RNA labelling mix).
2. RNase-free DNase (Roche DNaseI or Promega).
3. 2× Hydrolysis Carbonate Buffer: 120 mM Na_2CO_3, 80 mM $NaHCO_3$, aliquot and store at -20 °C.

4. 4 M LiCl.
5. Glycogen (Fermentas).
6. 10 % acetic acid.

2.6 Components for Hybridization

1. Proteinase-K buffer: 100 mM Tris pH 7.5, 50 mM EDTA pH 8.0.
2. Proteinase K (Solution: 18 +/− 4 mg/ml, Roche).
3. Fixative: see tissue fixation.
4. Glycine solution: 2 mg/mL glycine dissolved in 1× PBS.
5. 10× Salts: 3 M NaCl, 100 mM Tris pH 8.0, 50 mM EDTA, 53.7 mM NaH_2PO_4, 46.3 mM Na_2HPO_4. Aliquot and store at -20 °C.
6. 50× Denhardt's solution: 1 % Ficoll, 1 % Polividon 25, 1 % BSA. Aliquot and store at -20 °C.
7. 50 % Dextran sulfate dissolved in water. Heat to dissolve (needs a few hours at 70 °C), aliquot and store at -20 °C.
8. tRNA (Type XXI, Strain W, lyophilized powder, Sigma).
9. Glass staining Coplin jar or dishes with removable racks (baked, 25–30 pieces are needed) (*see* **Note 20**).
10. Membrane forceps (bent tip, Sartorius).
11. Heating block, incubator (37 and 55 °C), large box with a lid as a humid chamber, plastic foil.

2.7 Components for Washing, Antibody Binding and Detection

1. 10× TBS: 1 M Tris pH 7.5, 1.5 M NaCl.
2. 20× SSC: 3 M NaCl, 0.3 M Na-citrate.
3. Triton X-100.
4. Blocking solution: 1 % Blocking reagent (Roche), 0.3 % Triton-X 100 in 1× TBS. Heat to 60 °C to dissolve, cool to room temperature before use. Make sure that the blocking reagent is fully dissolved!
5. Antibody buffer: 1 % BSA (Albumin fraction V, ≥98 %, molecular biology grade), 0.3 % Triton-X 100 in 1× TBS (*see* **Note 28**).
6. Anti-DIG solution: anti-Digoxigenin-AP Fab fragments (Roche) diluted in antibody buffer, 1:1,250, 100 μL/slide.
7. Detection buffer: 100 mM Tris pH 9.5, 50 mM $MgCl_2$, 100 mM NaCl.
8. Detection solution: NBT-BCIP (Roche) in detection buffer 1:50, 100 μL/slide.
9. 50 % glycerol.
10. Membrane forceps, shaker, humid chamber.

3 Methods

3.1 Fixation and Tissue Infiltration

1. Harvest inflorescences and transfer them directly to the sample vials filled with ice-cold fixative (FAA or PFA), and keep on ice. Choose vials depending on sample size and number. For manual fixation 15 mL falcon tubes or eppendorf tubes are recommended. Fixative should be in at least 5× excess to the sample volume (*see* **Notes 2** and **3**).

3.1.1 Automated Vacuum Tissue Processor Leica ASP200S

1. Transfer the plant material into sample cassettes, close lids and label them with a pencil, since other markers will be washed off. Transfer cassettes into the cassette holder and put it in the retort of the ASP200S (*see* **Note 4**).
2. Start the program outlined in Table 1, making sure that the functions "stirrer," "recirculation," and "pressure-vacuum" are on. Set the program to stop when samples are in the last wax step until you manually press: "drain retort".
3. After taking the samples out, immediately store them at 65 °C in a tissue embedding station, such as the Leica EG1160 or a 55 °C incubator (*see* **Note 5**).
4. Clean the retort with the routine cleaning cycle of ASP200S.

Table 1
Program for the Leica automated vacuum tissue processor ASP200S

Station	Reagent	Duration (h)	Temperature (°C)	Drain (s)
1	FAA-Fixative	4		140
2	Ethanol 70 % denatured	1		80
3	Ethanol 90 % denatured	1		80
4	Ethanol 90 % denatured	1		80
5	Ethanol 100 % denatured + Eosin Y	1		80
6	Ethanol 100 % denatured	1		80
7	Ethanol 100 % absolute	1		140
8	Xylol	1		80
9	Xylol	1		80
10	Xylol	1.25		140
wax1	Paraplast	1	62	140
wax2	Paraplast	1	62	140
wax3	Paraplast	3	62	140

3.1.2 Manual Tissue Infiltration

Using an automated tissue processor produces more reliable and reproducible infiltration of plant tissues and, because the processing time is dramatically reduced, usually allows for better sensitivity. Nonetheless, because the equipment required is expensive, manual fixation and tissue infiltration might be needed. The time frame for the entire protocol using manual tissue infiltration is at least 7 days.

Use 4 % PFA for manual fixation (*see* **Note 6**). The samples should be on ice at all times during vacuum infiltration, fixation, and dehydration.

1. Fill sample collection vessels with fixative and make sure that the harvested tissue pieces are covered by fixative. After sample collection place the collection vessels into a vacuum desiccator. Use inverted pipette tips to submerge the tissue pieces and leave them in the collection vessels to keep the samples weighed down and submerged during infiltration. Apply vacuum until bubbles start to emerge from the fixative. Leave the samples under vacuum for 5 min and then release the vacuum very slowly. Releasing the vacuum too fast can damage the samples. Repeat the above steps until the samples turn dark, translucent green and sink down. Exchange the fixative at least once during vacuum infiltration and also after it, leaving the samples in fresh solution for the fixation on ice with gentle shaking. The time of fixation differs for tissue types, for inflorescences 4–6 h is ideal. Samples that do not sink and thus clearly are not penetrated by fixative should be removed (*see* **Note 7**).
2. Dehydrate the fixed samples by incubating them in the following solutions for the specified times. Ethanol solutions below 95 % should be diluted with 0.75 % NaCl instead of water. Keep the samples on ice and use gentle shaking. Make sure that the tissue does not dry out while solutions are changed. The samples can be left overnight in any of the solutions after the 50 % ethanol step if needed (*see* **Note 8**).

1× PBS	30 min
1× PBS	30 min
30 % Ethanol	1 h
40 % Ethanol	1 h
50 % Ethanol	1 h
60 % Ethanol	1 h
70 % Ethanol	1 h
85 % Ethanol	1 h
95 % Ethanol	overnight

3. Exchange ethanol for Histo-Clear using the following steps. These should be done at room temperature with gentle shaking (*see* **Note 9**). (Falcon tubes are not compatible with Histo-Clear. If these were used for sample fixation, they should be exchanged for glass vials).

100 % Ethanol + Eosin Y	30 min
100 % Ethanol	30 min
100 % Ethanol	1 h
100 % Ethanol	1 h
25 % Histo-Clear, 75 % Ethanol	1 h
50 % Histo-Clear, 50 % Ethanol	1 h
75 % Histo-Clear, 25 % Ethanol	1 h
100 % Histo-Clear	1 h
100 % Histo-Clear	1 h

100 % Histo-Clear (~5 mL in the glass vials) with 1/4 volume Paraplast chips overnight (no shaking). Place Paraplast into a 60 °C oven to melt for the next day.

4. Place samples to 42 °C. Wait until the Paraplast in the sample has fully melted. Saturate the solution with Paraplast by adding a few chips and allowing them to melt. Once the chips are completely melted add a few more and so on, until the chips will not melt any more. It usually takes 1–2 h for the chips to melt.
5. Transfer the samples to 60 °C for several hours (at least 2). Replace half of the wax–Histo-Clear mixture with freshly melted wax and leave overnight at 60 °C.
6. Remove Histo-Clear from the samples by repeatedly exchanging half the volume of the Paraplast–Histo-Clear mix while covering the samples with freshly melted Paraplast. Change the Paraplast six times leaving at least 6 h between changes. Gently shake the samples from time to time. Leave the glass vials open for the last day so that the rest of the Histo-Clear can evaporate from the samples.

3.2 Embedding

Heat all your equipment (forceps, metal molds, etc.) and the Paraplast to 65 °C before starting.

1. Label the plastic molds with pencil.
2. Place the metal mold on the heating plate, fill it with wax and quickly transfer the tissue into the mold. Orient the tissue into

the middle of the mold using the preheated forceps or needle. Cool down the bottom part of the metal mold until the wax surrounding the sample solidifies, but the top is still melted. Place the labelled, prewarmed plastic mold on top and, if necessary, fill up with wax.

3. Carefully move the mold to the -4 °C area (Leica EG 1160), or to ice and leave there until the wax fully solidifies. When the block is cold and solid, remove the metal mold using a metal spatula and store the wax blocks at +4 °C.

3.3 Sectioning

1. Cut the paraffin blocks to a trapezoid shape (overhead view), using a scalpel or razor blade, as close to the tissue as possible. The longer of the two parallel sides should be at the bottom (*see* **Note 10**).
2. Place the plastic mold into the microtome and cut sections. Use 20–30 μm steps for trimming and 8 μm steps for proper sections (*see* **Note 11**).
3. Gently lift the ribbons containing the needed sections (*see* **Note 12**) using either a small brush or needle and place in a 40 °C water bath to remove compressions. Fish out the stretched ribbons with a microscope slide. Remove as much of the water as possible from under the sections by gently taping the edge of the slide on a paper towel (*see* **Note 13**). Place the slide on a 42 °C heating plate immediately and let it dry for at least 4 h (*see* **Note 14**).

3.4 Probe Synthesis

1. Insert the transcribed sequence of the gene of interest into a vector with a RNA polymerase promoter (T7, T3 or SP6) and linearize the plasmid with an enzyme that generates 5′ overhangs (*see* **Note 15**).
2. Run the linearized plasmid on a 1 % agarose gel and estimate the concentration by loading 1 μg of lambda DNA, or similar, as a reference. Purify the band using a standard gel purification kit (i.e., Fermentas).
3. Synthesize the RNA transcript by assembling the following reaction using about 1 μg of the linearized plasmid and incubating at 37 °C for 2 h:

10× transcription buffer	2.0 μL
RNase inhibitor	1.0 μL (20 Units)
10× Dig NTP mix	2.0 μL
Enzyme	2.0 μL (40 Units)
Linearized plasmid	13 μL
Total	20 μL

4. Add 2 µL of RNase-free DNase (20 Units) and incubate for 15 min at 37 °C. Stop the reaction by adding 2 µL of 0.2 M EDTA (pH 8.0).
5. Precipitate the RNA by adding 2.5 µL of 4 M LiCl (final concentration 400 mM) and 75 µL of 100 % ethanol p.a. (threefold volume). Mix well by turning the tubes and incubate at -20 °C for at least 2–3 h. Centrifuge in a microcentrifuge at 4 °C for 30 min with maximum speed and wash the pellet with ice-cold 80 % ethanol p.a. (100–150 µL). Air-dry the pellet and resuspend in 100 µL of sterile, RNase-free water (*see* **Note 16**).
6. Hydrolyze the probe by adding 100 µL of 2× hydrolysis carbonate buffer to the transcript. Mix and incubate at 60 °C. The time needed for incubation can be calculated by the following formula: time = $Li–Lf/0.11 \times Li \times Lf$ (*Li*: Initial length of the probe in kb; *Lf*: Final length of the probe in kb, about 0.1–0.2). Neutralize the reaction after incubation by adding 20 µL of 10 % acetic acid. It is important not to hydrolyze longer, or the probe fragments will be too small, resulting in poor hybridization results (*see* **Note 17**).
7. Precipitate the probe by adding 1 µL of 1 M $MgCl_2$, 1 µL glycogen, and 600 µL of 100 % ethanol p.a. Mix well and incubate at -20 °C overnight. Centrifuge in a microcentrifuge for at least 30 min at 4 °C with maximum speed and wash with 750 µL ice-cold 80 % ethanol p.a. Air-dry the pellet and resuspend in 50 µL of sterile, RNase-free water. Dilute with 50 µL of (deionized) formamide and store at -20 °C (*see* **Note 18**).
8. Determine the amount of probe needed by pilot experiments (*see* **Note 19**).

3.5 Hybridization

Before starting, heat a water bath to 37 °C, prepare the proteinase K buffer and heat it to 37 °C. Prepare fixative and glycine solutions.

1. Dewax and rehydrate the sections before proteinase K treatment by placing the slides in a staining dish with rack and incubate them in the following solutions for the specified times. All steps are to be carried out at room temperature except when noted otherwise. Gentle shaking and work under a fume hood is required for the Histo-Clear washes. No shaking is needed for the other steps (*see* **Note 20**).

100 % Histo-Clear	10 min
100 % Histo-Clear	10 min
100 % Ethanol	5 min (no shaking from this point on)
100 % Ethanol	2 min

95 % Ethanol/water	1 min
90 % Ethanol/water	1 min
80 % Ethanol/water	1 min
60 % Ethanol/0,75 % NaCl	1 min
30 % Ethanol/0,75 % NaCl	1 min
0,75 % NaCl	2 min
1× PBS	2 min

2. Add proteinase K to the warm proteinase K buffer to a final concentration of 1 μg/mL. Place the slides into the proteinase K solution and digest for 30 min at 37 °C. Stop the reaction by placing the slides into room-temperature glycine solution for 2 min. Wash the slides in 1× PBS for 2 min (*see* **Note 21**).
3. Postfix the samples in fixative (FAA or PFA) for 2–5 min under the fume hood. Wash two times for 5 min with 1× PBS (*see* **Note 22**).
4. Dehydrate the sections before applying the probe by incubating them in the following solutions for the specified times.

0.75 % NaCl	2 min
30 % Ethanol/0.75 % NaCl	1 min
60 % Ethanol/0.75 % NaCl	1 min
80 % Ethanol/water	1 min
90 % Ethanol/water	1 min
95 % Ethanol/water	1 min
100 % Ethanol	2 min
100 % Ethanol (p.a.)	Until the probes are ready to apply

5. Prepare a humid chamber by putting two sheets of kitchen paper on the bottom of a large plastic box. Fix plastic pipettes or metal holders on the bottom to keep the slides elevated and soak the kitchen paper with a solution of 50 % formamide in 2× SSC.
6. Pipette the volume of probe needed for one slide into an eppendorf tube and denature at 80 °C for 5–10 min. Place directly on ice (*see* **Note 23**).

7. Prepare the hybridization mix of 50 % (deionized) formamide, 10 % dextran sulfate, 1× in situ Salts, 1× Denhardt's solution, 0.5 mg/mL tRNA. Prepare 100 μL per slide and keep at 50 °C (*see* **Note 24**).
8. Add the hybridization mix to the cold probe, mix with pipette tip and put the tube in the heating block. Take the slide out of the ethanol and place it on the heating block to warm up and dry (the sections turn white when they are dry). Remove the slide from the block and pipette the probe on it. Gently lower a coverslip on the sections using it to spread the probe solution over the entire slide, making sure that no air bubbles are trapped. Place the slide in the humid chamber and move on to the next slide (*see* **Note 25**).
9. Seal the humid chamber with plastic foil before placing it into the incubator.
10. Incubate the slides at 55 °C overnight (at least 14 h) (*see* **Note 26**).
11. Prepare 200 mL 2× SSC and 1 L 0.2× SSC for the next day and place them (together with a vertical glass jar and a dish with rack for the slides) into the incubator to warm up.

3.6 Washing, Antibody Binding and Detection

1. Remove the slides from the humid chamber and put them into a vertical glass jar filled with 2× SSC at the hybridization temperature. Wait until the solution seeps under the coverslips (1–2 min). Lift up one slide, so that the coverslip can slide down. Remove the coverslip from the solution and place the slide back into the jar (*see* **Note 27**).
2. When all the slides are ready, transfer them to a staining dish with rack filled with warm 0.2× SSC. Place them into the incubator at the hybridization temperature for 30 min. During the washing steps gentle shaking can be applied, but it is not essential. Exchange the solution every 30 min for a total duration of 2 h (four washes).
3. Wash the slides with 0.2× SSC for 5 min at 37 °C then transfer them with the rack to a new staining dish with a fresh solution of 0.2× SSC at 37 °C and leave it on the bench to cool down for about 10 min.
4. Wash the sections with 1× PBS for 5 min. This and all the following steps should be carried out at room temperature.
5. Place the slides in blocking solution for 30 min with gentle shaking.
6. Equilibrate in antibody buffer for 15 min with gentle shaking (*see* **Note 28**).
7. Prepare 100 μL of anti-DIG solution per slide and prepare a humid chamber, as done for the hybridization, but soak the

paper with water. Gently dry the bottom and edges of the slides, add the anti-DIG solution, and cover with a coverslip. Place the slides into the humid chamber for 90 min (*see* **Note 29**).

8. Remove coverslips by transferring the slides into a vertical jar filled with antibody buffer.
9. Transfer the slides to a dish with rack filled with fresh antibody buffer and wash with gentle shaking for 1 h changing the solution twice.
10. For the detection place the slides into detection buffer for 5–10 min, without shaking. Prepare the NBT-BCIP solution. Gently dry the bottom and edges of the slides, add NBT-BCIP solution and cover with a coverslip. Place the slides back into the humid chamber, cover them with aluminum foil to protect them from light and wait for the signal to develop. The signal can be checked under a microscope before stopping the reaction, but exposure to intense light will stop or slow the reaction (*see* **Note 30**).

    For most probes a signal can be detected after color reaction overnight, but some probes might require several days of incubation.
11. To stop the reaction place the slides into a vertical staining jar filled with distilled water and wait until the water seeps under the coverslips. This can take between 2 and 30 min. Remove the coverslips and leave the slides in the water for a few additional minutes (*see* **Note 31**).
12. Gently dry the bottom and edges of the slides, add 50 % glycerol (80 μL/slide) and cover with a coverslip. Keep the slides in a humid chamber (few days), or seal them with nail polish and store at 4 °C (up to 1 year) (*see* **Note 32**).

4 Notes

1. It is not important to DEPC treat the solutions, but bake all the glassware, metals (spoons for chemicals, magnetic stirrers, forceps) and keep them separate and RNase-free as much as possible. Use ultrapure water for all solutions and autoclave them if possible. Always use clean gloves and use filter tips.
2. Samples should be kept on ice during sample collection until fixation is complete or the samples are transferred to the ASP200S. Do not put too many samples in the same vial or cassette and use the smallest samples possible. Large samples are harder to infiltrate and might result in incomplete fixation and problems with embedding and sectioning. The samples can stay in the fixative on ice for an hour, to help collect all the samples, but it is not good to leave them longer without vacuum infiltration.

3. At the end of the fixation, samples should be completely soaked with the fixative and the air removed from the intercellular space. Air frequently gets trapped in young flowers and siliques. In these and other cases tissue infiltration can be facilitated by introducing cuts into the tissue, for example by dissecting the ends of a silique.
4. Make sure that the sample size is bigger than the pore size of the cassettes. If they are smaller, line the inside of the cassette with nylon mesh.
5. After removing the samples from the machine, they should not dry out, immediately place them into a box or petri dish, if not continuing with embedding right away.
6. Alternative: FAA can also be used for fixation as for the automated embedding. In case of using FAA the ethanol series for dehydration should start at 50 %. The rest of the steps are the same as for PFA.
7. The problem is usually not too short fixation, but overfixation of the tissue, which reduces the hybridization signal strength.
8. It can be useful to degas the ethanol solutions with vacuum before use to make sure that no air is introduced to the samples, which can hinder further steps.
9. Staining the tissue with 0.1 % eosin is optional. It is easier to see the tissue during embedding and sectioning when it is stained. The eosin will be washed out during the rehydra -tion steps.
10. The top and the bottom of the paraffin block have to be parallel, relative to the blade, to get straight ribbons of sections. The block should not be smaller than 3 × 3 mm. If the block is too small it does not offer enough resistance to the blade and the sample will be squashed during sectioning, thereby deforming the sample.
11. To get nice ribbons, the angle of the blade should be set to around 4°. The blade needs to be clean and the paraffin blocks need to be at the correct temperature for sectioning (in case of Sigma Paraplast ice cold, for Leica Paraplast at room temperature or higher). If the temperature of the block is not room temperature and consecutive sections are needed, then do not stop cutting in the middle of the sample, as the block temperature changes fast.
12. The sections can be examined and manipulated under a stereomicroscope. To select the sections of interest place the ribbon on a dry, room-temperature slide before putting it into the water bath. Alternatively, the ribbons can be fished out from the water bath keeping a big drop of water under them using a prewarmed slide. (Once the sections are placed on a slide

without water under them, they stick onto the slide and cannot be removed without damage.) After selecting the sections, the ribbon can be put back on the top of the water and fished out again with a slide collecting the sections for the in situ hybridization.

13. It is very important to leave as little water as possible below the sections so they can melt onto the slide. Sections might be lost during the following steps if excess water remains.

    Alternative: instead of a water bath the sections can be put directly on a room-temperature slide, room-temperature water added below (~1 mL) and the slides transferred to the hot-plate, so that the water can warm up and the sections expand. After expansion very carefully remove as much water as possible from bellow the sections with a pipette and tap the edge of the slide on a paper towel to remove the rest. For this method it is useful to degas the water with vacuum before use so no bubbles form below the sections.

14. Leaving the sections on the hotplate at 42 °C overnight ensures that the sections melt onto the slide. Using pretreated slides (e.g., StarFrost from Knittel, Menzel-Gläser, or Poly-l-Lysine from Sigma) or treating the slides before use greatly helps to ensure that the sections stick to the slides. Label slides with a pencil to make sure that the markings are not washed off by the subsequent washing steps in alcohol and Histo-Clear! The slides can be stored in a box at 4 °C for an extended period of time (up to 1 year).

15. Take care to work clean and RNase-free (use gloves) already during the template preparation. The template can also be generated by a PCR reaction using primers containing one of the RNA polymerase promoter sequences. For the sense probe the promoter should reside in the forward primer and for anti-sense probes in the reverse primer. In this case about 200 ng of PCR product should be used in the transcription reaction.

16. The RNA transcripts can be analyzed on a 1 % agarose gel before or after precipitation (0.5–1 μL before, or 2–3 μL after precipitation). Double bands on the gel are normal for the transcription, but it is a good idea to load a bit from the templates next to the transcript to see if the DNase treatment was efficient.

17. Probe hydrolysis is not necessary when the transcript is smaller than 500 nt. The hydrolyzed RNA probe can also be checked on a 1.2 % agarose gel (it will appear as a smear between 100 and 400 bp) or in a dot blot (before adding the formamide).

18. If larger dilutions are needed, the probe can also be diluted up to 10× with Hybridization Buffer instead of formamide.

19. How much probe to use in a reaction: The best way to determine the amount of probe needed for hybridization is by doing pilot experiments. From the 100 μL probe/formaldehyde mixture 0.5–2 μL per slide usually works well. It is important to note that if the target gene expression is too low for detection, raising the amount of probe will only increase the background. Also, if the target gene expression is low, the vector sequence in the probe might also give a strong background, which can be excluded by using PCR products containing no vector sequences as the template for probe synthesis.

    The amount of labelled RNA can be estimated by performing a *dot blot* according to the following protocol: spot 0.5–1 μL of RNA dilutions (undiluted, 1:10, 1:100) on a Hybond-N + membrane. Do not forget a positive/negative control. Cross-link the probes by drying the membrane at 55 °C for 30 min, or by UV illumination. Wet the membrane with 1× TBS and block it for 30 min with 0.5 % Blocking buffer (with shaking). Incubate with the anti-DIG antibody diluted 1:5,000 in fresh Blocking buffer for 30 min with gentle shaking and wash 2–3 times for 5 min with 1× TBS. Wash once for 2 min with Substrate buffer and incubate with NBT-BCIP in Substrate buffer in darkness until the signal is visible (about 10–20 min). Wash the membrane with water and let it dry. It is important to note that the dot blot can only be used to obtain a rough quantitative estimate of labelled RNA and has no correlation to the efficiency of the hybridization of the probe and thus the acquired signal in the in situ hybridization experiment.

20. Use glass staining jars or dishes for the washing steps, cleaned and baked for each one. (There are different ones: from small, vertical Coplin jars for 8 slides holding 70 mL solution to large staining dishes for 20–24 slides with removable glass racks, holding 200–250 mL solution. These are interchangeable, but the text states which one is the most convenient for each step.) The sections on the slides should not dry out at any time except directly before the hybridization step. To ensure this prepare all solutions in separate containers and move the slides, preferably with a rack, from one container to another. Move them fast but carefully, so as to not lose any sections.

    The ethanol series can be made by simply diluting ethanol with 0.75 % NaCl, even though salt concentration will vary from step to step. The same ethanol series can be used for the dehydration and can be used for 3–4 rounds of experiments.

21. The proteinase K treatment is necessary to facilitate the penetration of the probe into the tissue. Proteinase K concentration and length of treatment is critical, but can be changed

depending on the thickness of the sections. The above conditions are for 8 μm sections.

22. Important: the fixatives must be the same during initial fixation and postfixation of the slides.
23. If you use the same probe for multiple slides, it is possible to denature the probe for up to three slides in the same tube, but not for more, otherwise it is harder to mix afterwards. Probes can also be mixed with the hybridization mix, denatured together and kept on ice until use. In this case the mixture does not need to be warmed up before applying it to the slides and the slides can be dried at room temperature for 30 min.
24. Some protocols advise to store the final hybridization mix at 4 °C, however it is preferable to make it fresh for each experiment. Before assembling the hybridization mix warm up the required aliquots of ingredients to 50 °C, except for the tRNA which should stay on ice. Cut the pipette tip before transferring dextran sulfate as it is very viscous. After mixing the ingredients, keep the mixture at 50 °C (in a heating block) until use.
25. Important: Vortex the hybridization mix well from time to time before putting it on the probes as the dextran sulfate might sink to the bottom. Try to apply the hybridization mix to the slides quickly, so the slides can go to the incubator as soon as possible, but take care to avoid cross-contamination between the probes. A small amount of probe is enough to get false results.

    The best way to cover the samples with the hybridization solution and the coverslip is to pipette the required amount close to one end of the slide. Place the short edge of the coverslip on the slide where the solution has been pipetted, while holding the far edge of the coverslip with forceps. Slowly lower the far edge allowing the solution to spread under the coverslip and over the sections, until all the sections are covered and the coverslip rests fully on the slide. (Make sure that the tips of the forceps are even, otherwise it is very easy to break the coverslip).

    Alternatives: Parafilm is also suitable to cover the slides, however, better results are obtained with coverslips. It is also important that Parafilm shrinks at high temperature, and therefore it is not suitable for hybridization over 55 °C.
26. Too much background can be solved with reduced amount of probe or changing the hybridization temperature (to 60–62 °C). In this case the washing steps need to be adjusted to the hybridization temperature.
27. Important: Take care to lift the slides vertically each time when the coverslip needs to be removed. This way it will not damage the sections as it slides off. This also needs to be done quickly, so that the samples do not dry out. If there is enough solution

on the slides, such as after incubation with anti-DIG antibody, the coverslip might slip off even without immersing the slide, and thus turning the slide to the vertical position should be done with one decisive movement.

28. The blocking solution can also be used for all steps instead of antibody buffer (BSA). This slightly reduces the background, but the signal is also weaker.
29. Wiping the bottom and edges of the slide is important to avoid spilling (drawing off) of the solution and the consecutive drying of the sections. Care must be taken to wipe the slides very quickly so that the sections themselves do not dry out.
30. In case of weak expression levels, the addition of 10 % PVA (polyvinyl alcohol) to the NBT-BCIP solution is possible. This enhances the signal, but makes it very hard to stop the reaction and remove the coverslip afterwards. It is of course possible to take pictures before stopping the reaction and removing the NBT-BCIP solution thereby making sure that the sections are not damaged.
31. After stopping the reaction the slides can stay in water for a few days.
32. Different mounting solutions are also available for longer storage.

References

1. Jackson D (1991) In situ hybridization in plants. In: Bowles DJ, Gurr SJ, McPherson MJ (eds) Molecular plant pathology. A practical approach. Oxford University Press, Oxford, UK
2. Coen ES, Romero JM, Doyle S, Elliott R, Murphy G, Carpenter R (1990) *floricaula*: a homeotic gene required for flower development in Antirrhinum majus. Cell 63:1311–1322
3. Drews GN, Bowman JL, Meyerowitz EM (1991) Negative regulation of the Arabidopsis homeotic gene *AGAMOUS* by the *APETALA2* product. Cell 65:991–1002
4. Lincoln C, Long J, Yamaguchi J, Serikawa K, Hake S (1994) A *knotted1*-like homeobox gene in Arabidopsis is expressed in the vegetative meristem and dramatically alters leaf morphology when overexpressed in transgenic plants. Plant Cell 6:1859–1876
5. Long JA, Moan EI, Medford JI, Barton MK (1996) A member of the KNOTTED class of homeodomain proteins encoded by the *STM* gene of *Arabidopsis*. Nature 379:66–69
6. Vielle-Calzada JP, Thomas J, Spillane C, Coluccio A, Hoeppner MA, Grossniklaus U (1999) Maintenance of genomic imprinting at the *Arabidopsis medea* locus requires zygotic *DDM1* activity. Genes Dev 13: 2971–2982
7. Long JA, Barton MK (1998) The development of apical embryonic pattern in Arabidopsis. Development 125:3027–3035
8. Balasubramanian S, Schneitz K (2000) *NOZZLE* regulates proximal-distal pattern formation, cell proliferation and early sporogenesis in *Arabidopsis thaliana*. Development 127:4227–4238
9. Maier AT, Stehling-Sun S, Wollmann H, Demar M, Hong RL, Haubeiß S, Weigel D, Lohmann JU (2009) Dual roles of the bZIP transcription factor PERIANTHIA in the control of floral architecture and homeotic gene expression. Development 136: 1613–1620
10. Shindo S, Ito M, Ueda K, Kato M, Hasebe M (1999) Characterization of MADS genes in the gymnosperm *Gnetum parvifolium* and its

implication on the evolution of reproductive organs in seed plants. Evol Dev 1:180–190

11. Smyth DR, Bowman JL, Meyerowitz EM (1990) Early flower development in *Arabidopsis*. Plant Cell 2:755–767
12. Schneitz K, Hülskamp M, Pruitt RE (1995) Wild-type ovule development in *Arabidopsis thaliana*: a light microscope study of cleared whole-mount tissue. Plant J 7:731–749
13. Mayer KFX, Schoof H, Haecker A, Lenhard M, Jürergens G, Laux T (1998) Role of WUSCHEL in regulating stem cell fate in the Arabidopsis shoot meristem. Cell 95: 805–815

Chapter 15

The GUS Reporter System in Flower Development Studies

Janaki S. Mudunkothge and Beth A. Krizek

Abstract

The β-glucuronidase (GUS) reporter gene system is an important technique with versatile uses in the study of flower development. Transcriptional and translational GUS fusions are used to characterize gene and protein expression patterns, respectively, during reproductive development. Additionally, GUS reporters can be used to map cis-regulatory elements within promoter sequences and to investigate whether genes are regulated posttranscriptionally. Gene trap/enhancer trap GUS constructs can be used to identify novel genes involved in flower development and marker lines useful in mutant characterization. Flower development studies primarily have used the histochemical assay in which inflorescence tissue from transgenic plants containing GUS reporter genes are stained for GUS activity and examined as whole-mounts or subsequently embedded into wax and examined as tissue sections. In addition, quantitative GUS activity assays can be performed on either floral extracts or intact flowers using a fluorogenic GUS substrate.

Key words GUS, Reporter gene, Transcriptional reporter, Translational reporter, X-Gluc, Histochemical staining, Sections, MUG, Fluorometric assay

1 Introduction

A key step in investigating gene function is to know the precise spatial and temporal expression of a gene. Tissue-specific gene expression patterns can be determined directly at the mRNA level by in situ hybridization or at the protein level by immunolocalization. However, many plant genes are members of gene families and it may be difficult to obtain RNA probes or antibodies that are specific for a particular gene or protein. In addition, these procedures are complicated and time consuming. An alternative strategy is to use reporter gene constructs for such studies. Reporter genes encode proteins whose presence is easily assayed. The two most commonly used reporters in plants are β-glucuronidase (GUS) encoded by the *E. coli uidA* gene and green fluorescent protein (GFP) from the jellyfish *Aequorea victoria*. There are many different GFP derivatives now available and their use in flower development studies is covered in another chapter of this book. This chapter focuses on the use of the GUS reporter system.

José Luis Riechmann and Frank Wellmer (eds.), *Flower Development: Methods and Protocols*, Methods in Molecular Biology, vol. 1110, DOI 10.1007/978-1-4614-9408-9_15, © Springer Science+Business Media New York 2014

To report on gene expression patterns, two types of GUS gene fusion constructs are typically used. Transcriptional reporters consist of the promoter region of a gene of interest driving GUS expression and thus report on promoter activity. Translational reporters consist of an in-frame fusion of a gene of interest to GUS under the control of the promoter region. Translational reporters are often in a genomic context with 5' upstream sequence, exons and introns of the gene of interest, and 3' UTR sequence. Both traditional and Gateway compatible GUS plasmids are available from the Arabidopsis Biological Resource Center (ABRC) (https://abrc.osu.edu). In addition, plasmids containing GUSPlus, a new GUS reporter with higher sensitivity and greater tolerance to fixatives, are available from CAMBIA (http://www.cambia.org/daisy/cambialabs/3698.html).

For GUS reporter fusions to reflect endogenous gene expression patterns, the constructs must contain all of the required gene regulatory regions. Since in many cases this is not known, if possible one should confirm that the GUS expression pattern matches the gene expression pattern that is determined using a direct method (in situ hybridization or immunolocalization). Alternatively, one can determine whether a translational reporter fusion complements the corresponding mutant.

Besides providing information on gene expression patterns, GUS reporter constructs can also be used to investigate gene regulation and to identify new genes involved in flower development. Cis-acting regulatory elements can be identified by generating a series of transcriptional GUS fusions containing different lengths of 5' sequence and determining which fusions reproduce the endogenous gene expression pattern [1, 2]. Comparison of GUS expression in transcriptional and translational fusions can be used to determine if a gene is regulated primarily at the transcriptional level and whether cis-regulatory elements are present in introns, coding region or 3' UTR sequence [3]. GUS reporter lines can also be used in gene identification strategies either as gene traps/enhancer traps to identify genes that are expressed in a particular tissue or at a particular developmental stage [4, 5] or as material for forward genetic screens to identify genes required for expression of the reporter [6]. Gene trap strategies have identified cell and tissue specific marker lines useful in the characterization of flower development in mutant plants [7].

The GUS enzyme is a hydrolase that can cleave many different β-glucuronides. Different GUS substrates are commercially available for both qualitative histochemical assays and quantitative fluorometric assays. Flower development studies typically involve histochemical assays utilizing the substrate 5-bromo-chloro-3-indolyl glucuronide (X-gluc), which produces a blue precipitate at the site of GUS activity. Cleavage of X-gluc does not directly result in the indigo pigment; instead the hydrolysis product must undergo an oxidative dimerization to form the precipitate. 4-methyl umbelliferyl glucuronide (MUG) is a fluorogenic substrate that can be used to detect GUS activity in tissue extracts as well as intact tissue. Cleavage of MUG by GUS produces the fluorescent compound 4-methyl umbelliferone (MU), which is easily quantified using a fluorometer. The protocols described here are adapted from previously published protocols [8–11].

2 Materials

2.1 Histochemical GUS Assay

Use ultrapure water to prepare all solutions. Dispose of acetone, formaldehyde, and xylenes according to waste disposal regulations.

2.1.1 GUS Staining

1. 1 M Na_2HPO_4.
2. 1 M NaH_2PO_4.
3. 100 mM Potassium ferricyanide ($K_3Fe(CN)_6$) (*see* **Note 1**).
4. 100 mM Potassium ferrocyanide ($K_4Fe(CN)_6$) (*see* **Note 1**).
5. 100 mM X-Gluc (*see* **Note 2**).
6. 90 % acetone.
7. Rinse solution: 50 mM sodium phosphate buffer pH 7.2, 0.5 mM $K_3Fe(CN)_6$, 0.5 mM $K_4Fe(CN)_6$. For 100 mL, add 3.42 mL of 1 M Na_2HPO_4, 1.58 mL of 1 M NaH_2PO_4, 500 μL of 100 mM $K_3Fe(CN)_6$, and 500 μL of 100 mM $K_4Fe(CN)_6$ to a graduated cylinder and add water to 100 mL. Prepare fresh before use.
8. Stain solution: 50 mM sodium phosphate buffer pH 7.2, 0.5 mM $K_3Fe(CN)_6$, 0.5 mM $K_4Fe(CN)_6$, 2 mM X-gluc. To 10 mL of Rinse solution, add 200 μL of 100 mM X-Gluc stock. Prepare fresh before use.
9. 20 mL glass scintillation vials.
10. Wire mesh screen.
11. Vacuum oven.
12. Ethanol.

2.1.2 Embedding/Sectioning/Mounting

1. Xylenes.
2. Paraplast tissue embedding medium.
3. Microscope slide warming table (Eberbach Corporation).
4. Microtome.
5. Tissue floatation bath.
6. Slide warmer.
7. Superfrost plus glass slides.
8. Permount mounting medium.

2.2 Fluorometric GUS Assay

Use ultrapure water to prepare all solutions.

2.2.1 Protein Extracts

1. Micropestles that fit microcentrifuge tubes.
2. 1 M Na_2HPO_4.
3. 1 M NaH_2PO_4.
4. β-mercaptoethanol.
5. 0.5 M EDTA pH 8.0.
6. 10 % (w/v) SDS.
7. 10 % (v/v) Triton X-100.
8. GUS extraction buffer: 50 mM sodium phosphate buffer pH 7, 10 mM β-mercaptoethanol, 10 mM EDTA, 0.1 % SDS, 0.1 % Triton X-100. For 10 mL, add 288.5 μL 1 M Na_2HPO_4, 211.5 μL 1 M NaH_2PO_4, 7 μL β-mercaptoethanol, 200 μL 0.5 M EDTA, 100 μL SDS, and 100 μL Triton X-100 to a 10 mL graduated cylinder and add water to 10 mL (*see* **Note 3**).

2.2.2 Fluorometric Assay

1. 25 mM MUG (*see* **Note 4**).
2. GUS assay buffer: 50 mM sodium phosphate buffer pH 7, 10 mM β-mercaptoethanol, 10 mM EDTA, 0.1 % SDS, 0.1 % Triton X-100, 1 mM MUG. To 5 mL of GUS extraction buffer, add 200 μL of 25 mM MUG stock. Prepare fresh before use.
3. 1 mM MU (*see* **Note 5**).
4. Stop buffer: 0.2 M Na_2CO_3.
5. Fluorometer (*see* **Note 6**).
6. Fluorometer cuvettes.
7. 1 mg/mL BSA.
8. Bradford solution for protein concentration determination.
9. Spectrophotometer.

3 Methods

3.1 Histochemical GUS Assay

3.1.1 GUS Staining

1. Add 10 mL of cold 90 % acetone to scintillation vials and store vials on ice.
2. Harvest inflorescence tissue into acetone vials on ice. Incubate vials on ice for 15 min.
3. Remove acetone completely (*see* **Note 7**) and replace with approximately 10 mL of rinse solution. Swirl vials gently and incubate at room temperature for 5 min.
4. Remove Rinse solution completely and replace with Stain solution. Use 2 mL of Stain solution for every ten Arabidopsis

inflorescences collected. The tissue should be completely covered by the Stain solution.

5. Vacuum infiltrate the samples by putting the vials (with loose lids) into a vacuum oven and bring the pressure to between 10 and 15 inHg. After 10 min, slowly release the vacuum and swirl the tissue gently to get rid of air bubbles. Repeat the vacuum infiltration for a total of four times, 10 min each.
6. Close the lids of the vials and incubate at 37 °C for several hours to several days (*see* **Note 8**).
7. Remove the stain solution completely and process the samples through an ethanol series: 15, 30, 50, 70, 85, 95, 100 and 100 % for 30 min each at room temperature (*see* **Note 9**). At the end of the ethanol series, there may still be some chlorophyll in the tissue. By the next day, all of the chlorophyll should be gone and the blue precipitate visible.
8. Examine the tissue under a dissecting microscope (*see* **Note 10**). Fill a depression slide with 100 % ethanol, place the inflorescence tissue in the ethanol, orient and dissect the tissue as needed using a dissecting microscope. Take pictures with a digital camera connected to the dissecting microscope.
9. Better resolution of the GUS staining pattern can be achieved by examining tissue sections of the material. The procedure for embedding and sectioning tissue is described below.

3.1.2 Embedding

1. Remove the 100 % ethanol and process the tissue samples through an ethanol–xylenes series: 75 % ethanol–25 % xylenes, 50 % ethanol–50 % xylenes, 25 % ethanol–75 % xylenes for 30 min each at room temperature (*see* **Note 11**).
2. Replace the last ethanol–xylenes mixture with 100 % xylenes and leave for 1 h at room temperature. After 1 h, replace the 100 % xylenes with fresh 100 % xylenes and incubate for one more hour.
3. Remove the last xylenes and fill the vials halfway with 100 % xylenes. Fill the remaining space in the vial with Paraplast chips. Incubate the vials at 42 °C until all chips are in solution. Pour out the xylenes–Paraplast mix into a waste container and replace with 100 % molten Paraplast. Incubate the vials in an oven at 60 °C (*see* **Note 12**).
4. Incubate the vials overnight at 60 °C and replace the molten Paraplast a total of six times within two days (three to four times per day with 3–4 h between Paraplast changes) (*see* **Note 13**).
5. Heat up microscope slide warming table. Place plastic molds, weigh boats or paper boats made using 3 × 5 cards onto hot side of slide warming table.
6. Remove vial from 60 °C, swirl briefly and quickly pour tissue into boat on slide warming table. Add more molten Paraplast

as needed to cover all of the tissue. Distribute and orient inflorescences in boats using a wooden handled teasing needle and carefully slide the boat to a cooler region of slide warming table (*see* **Note 14**). Repeat tissue orientation and continue sliding boat to coolest side of slide warming table. Leave here for several minutes until Paraplast has solidified a bit around the tissue. Then move the boat to ambient temperature surface and leave for several hours until fully solidified. Store tissue blocks at 4 °C until ready to section.

3.1.3 Sectioning and Mounting

1. Cut individual inflorescences out of the tissue block with a razor blade. Trim away excess Paraplast and mount tissue on tissue holder with either inflorescence straight up and down (for longitudinal sections) or pointing directly at you (for transverse sections).
2. Cut sections through the entire inflorescence taking sections at 8–10 μm and forming a long ribbon. Transfer the ribbon to a piece of black paper using two paintbrushes, one at either end of the ribbon. Using a razor blade, cut the ribbon into pieces that will fit widthways on a microscope slide. Transfer each ribbon piece to a water bath using a wood applicator, maintaining the pieces in order (*see* **Note 15**).
3. Put a Superfrost Plus microscope slide into the waterbath at a 45° angle and push the batch of ribbon pieces onto the slide using a wood applicator.
4. Bake the slides on a slide warmer at 42 °C overnight. Store slides at 4 °C until ready to mount.
5. To remove the Paraplast, incubate slides in 100 % xylenes for 10 min. Repeat with a second batch of 100 % xylenes for 10 min.
6. Remove one slide at a time from xylenes and blot briefly on paper towels. Add two drops of Permount and cover with a coverslip, carefully pressing out air bubbles. Let the slides dry overnight in the hood.
7. Once slides are dry, wipe them clean with a tissue wipe soaked in xylenes to remove any excess mounting material.
8. Examine slides on compound microscope and take pictures. If GUS staining is intense, a blue color is apparent under bright-field illumination. Weaker GUS staining can be observed under dark-field illumination and appears pink.

3.2 Fluorometric GUS Assay

3.2.1 Preparation of Protein Extracts

1. Fill 1.5 mL microcentrifuge tube on ice with liquid N_2. Harvest tissue into tube and grind tissue with a chilled micropestle until it is a fine powder.
2. Add 150 μL of GUS extraction buffer to tube while tissue is still frozen. Leave sample on ice. Grind tissue further when buffer has thawed.

3. Spin tube in microcentrifuge at 4 °C.
4. Transfer supernatant to a new microcentrifuge tube and keep on ice (*see* **Note 16**).

3.2.2 MU Standard Curve

1. The fluorometer needs to have the following filter set: excitation 365 nm, emission 455 nm. Turn on fluorometer and let it warm up for about 30 min.
2. Prepare fresh MU dilutions in Stop buffer from 1 mM stock of MU. Typical concentrations are 1 μM, 500 nM, 150 nM, 50 nM, and 20 nM MU.
3. Calibrate the fluorometer using the highest MU standard and measure the fluorescence of the other MU standards to generate a MU standard curve.

3.2.3 MUG Assay on Plant Extracts

1. Aliquot 600 μL of GUS assay buffer to a microcentrifuge tube for each sample. Preheat tubes to 37 °C.
2. For each sample, prepare five 1.5 mL microcentrifuge tubes with 900 μL of Stop buffer.
3. Add 10 μL of plant protein extract to preheated Assay buffer. Vortex to mix.
4. Immediately remove 100 μL from the reaction and add to 900 μL of Stop buffer.
5. After 1 min, remove another 100 μL from the reaction and add to 900 μL of Stop buffer.
6. Repeat removals of 100 μL of reaction at 5 min, 10 min and 15 min (*see* **Note 17**).
7. Measure fluorescence of each sample in a fluorometer.
8. Plot the fluorescence of each sample versus minutes and fit the data to a line. The unit of slope is fluorescent units/minute.
9. Divide the fluorescent units/minute slope of each plant extract by the slope of the MU standard curve to calculate the amount of MU generated in each sample per unit of time (nmol MU/min).
10. To calculate GUS activity in nmol MU/min/mg, divide the nmol MU/min values by the total mg of protein used in each assay as determined in the Bradford Assay below.

3.2.4 Bradford Assay

1. Prepare dilutions of BSA stock solution: 0.2, 0.4, 0.6, 0.8 and 1.0 mg/mL.
2. Prepare diluted Bradford solution (one part dye reagent and four parts ultrapure water) and transfer 1 mL to microcentrifuge tubes. You need one tube for each BSA stock and each plant sample.
3. Add 20 μL of BSA stock solution or 20 μL of plant protein extract to a microcentrifuge tube and vortex to mix.

4. Incubate for 10 min at room temperature.
5. Measure absorbance at 595 nm in plastic cuvettes.
6. Calculate the total amount of protein in the plant samples using the BSA standard curve.

3.2.5 MUG Assay on Intact Inflorescences or Flowers

1. Prepare 96 well microtiter plates containing 100 μL GUS extraction buffer in each well.
2. Harvest tissue into microtiter plate and incubate at 37 °C (*see* **Note 18**).
3. Stop reaction by the addition of 50 μL Stop buffer.
4. Transfer 100 μL of each well to a fresh microtiter plate and measure fluorescence.
5. The GUS activity in the tissue can be normalized per tissue weight or per mg of protein if the tissue is subsequently assayed for total protein.

4 Notes

1. Make up 10 mL of these solutions at a time and store in the dark at 4 °C. The potassium ferrocyanide stock will oxidize quickly, do not keep it more than 1–2 months. Potassium ferricyanide and potassium ferrocyanide are used to oxidize the soluble hydrolysis product so that it does not diffuse away from the site of production prior to undergoing the oxidative dimerization. High concentrations of potassium ferricyanide and potassium ferrocyanide can inhibit GUS activity.
2. The X-Gluc stock solution is made up in dimethyl formamide. Work in a hood when using dimethyl formamide. The X-Gluc solution can be stored at −20 °C for several weeks but should be colorless. Don't use the solution if it has turned red.
3. Add β-mercaptoethanol to GUS extraction buffer right before use. Work in a hood when using β-mercaptoethanol.
4. The MUG stock solution is made up in GUS extraction buffer. Prepare fresh before use.
5. Dissolve 19.8 mg of MU in 100 mL of ultrapure H_2O and store in a dark bottle at 4 °C for up to a month.
6. If assay will be performed on intact tissue in microtiter well plates then a fluorometer with a microplate reader is required.
7. Remove the acetone by pouring it out of the vial while pressing a wire mesh screen against the vial lip to retain the tissue. Remove the last bit of acetone with a pipettor. For all subsequent steps involving removal of solutions from scintillation vials (except those involving Paraplast), use a wire mesh screen to keep the inflorescence tissue in the vial.

8. The staining period includes the time during the vacuum infiltration time and the incubation period at 37 °C. The exact time needed for staining depends on the individual line. Typically samples are incubated from several hours to overnight.
9. For better tissue preservation, samples can be fixed in FAA (50 % ethanol, 10 % glacial acetic acid, 3.7 % formaldehyde) in place of the 50 % ethanol. Then continue with the ethanol series.
10. There have been reports of GUS staining in pollen grains due to diffusion of the primary GUS reaction product from anthers [12]. To confirm GUS staining in pollen, isolate pollen grains from the anther before staining the tissue with GUS and stain them separately for GUS activity. In addition, some SAIL T-DNA insertion lines carry a *LAT52:GUS* marker which will result in GUS staining in pollen grains [13]. This construct could potentially result in gene silencing of other GUS reporters that might be introduced into the T-DNA line.
11. All work with xylenes should be performed in a fume hood. Exposure of tissues to xylenes must be minimized since xylenes can solubilize the blue precipitate.
12. The transition of tissue from 100 % ethanol to 100 % Paraplast should take place in a single day.
13. After removing the vial from the 60 °C oven, place it on your hand for a short time to cool the bottom before pouring out the molten Paraplast. A thin layer of partially solidified Paraplast will keep the tissue from coming out of the vial.
14. Don't leave tissue too long on hot part of warming table as bubbles will form around the tissue and interfere with getting good sections.
15. Place the ribbon on the paper such that the shiny side faces up. Wet the tip of the wood applicator in the water bath, then touch it to the ribbon piece to pick up the piece. Flip the wood applicator over so that the piece is on top and slowly put the wood applicator into the water bath. The ribbon piece should come off the wood applicator and float in the bath. Carefully push the ribbon piece up against the side of the water bath. It should stick there. Then transfer the second piece and push it up against the first piece and repeat with each additional piece. They will gently stick to each other and the order can be preserved.
16. Tissue extracts can be stored at −70 °C.
17. More than one sample can be processed at a time but the different samples will need to be staggered by appropriate periods of time.
18. The exact time needed for incubation in the MUG solution depends on the individual line. Weaker reporter lines may require up to 16 h [14].

References

1. Hill TA, Day CD, Zondlo SC, Thackeray AG, Irish VF (1998) Discrete spatial and temporal cis-acting elements regulate transcription of the Arabidopsis floral homeotic gene *APETALA3*. Development 125:1711–1721
2. Tilly JJ, Allen DW, Jack T (1998) The CArG boxes in the promoter of the Arabidopsis floral organ identity gene *APETALA3* mediate diverse regulatory effects. Development 125:1647–1657
3. Lee J-Y, Colinas J, Wang JY, Mace D, Ohler W, Benfey PN (2006) Transcriptional and posttranscriptional regulation of transcription factor expression in *Arabidopsis* roots. Proc Natl Acad Sci U S A 103:6055–6060
4. Springer PS (2000) Gene traps: tools for plant development and genomics. Plant Cell 12: 1007–1020
5. Sundaresan V, Springer P, Volpe T, Haward S, Jones JDG, Dean C, Ma H, Martienssen R (1995) Patterns of gene action in plant development revealed by enhancer trap and gene trap transposable elements. Genes Dev 9:1797–1810
6. Chiu W-H, Chandler JW, Cnops G, Van Lijsebettens M, Werr W (2007) Mutations in the *TORNADO2* gene affect cellular decisions in the peripheral zone of the shoot apical meristem of *Arabidopsis thaliana*. Plant Mol Biol 63:731–744
7. Liljegren SJ, Ditta GS, Eshed Y, Savidge B, Bowman J, Yanofsky MF (2000) *SHATTERPROOF* MADS-box genes control seed dispersal in Arabidopsis. Nature 404: 766–770
8. Blazquez M (2002) Quantitative GUS activity assays. In: Weigel D, Glazebrook J (eds) Arabidopsis: a laboratory manual. Cold Spring Harbor Laboratory Press, Cold Spring Harbor, New York, pp 249–252
9. Bomblies K (2002) Whole-mount GUS staining. In: Weigel D, Glazebrook J (eds) Arabidopsis: a laboratory manual. Cold Spring Harbor Laboratory Press, Cold Spring Harbor, New York, pp 243–248
10. Jefferson RA (1987) Assaying chimeric genes in plants: the GUS gene fusion system. Plant Mol Biol Rep 5:387–405
11. Jefferson RA, Kavanagh TA, Bevan MW (1987) GUS fusions: β-glucuronidase as a sensitive and versatile gene fusion marker in higher plants. EMBO J 6:3901–3907
12. Mascarenhas JP, Hamilton DA (1992) Artifacts in the localization of GUS activity in anthers of petunia transformed with a CaMV 35S-GUS construct. Plant J 2:405–408
13. Sessions A, Burke E, Presting G, Aux G, McElver J, Patton D, Dietrich B, Ho P, Bacwaden J, Ko C, Clarke JD, Cotton D, Bullis D, Snell J, Miguel T, Hutchison D, Kimmerly B, Mitzel T, Katagiri F, Glazebrook J, Law M, Goff SA (2002) A high throughput Arabidopsis reverse genetics system. Plant Cell 14:2985–2994
14. Blazquez MA, Soowal LN, Lee I, Weigel D (1997) *LEAFY* expression and flower initiation in *Arabidopsis*. Development 124: 3835–3844

Part IV

Experimental Systems

Chapter 16

A Floral Induction System for the Study of Early *Arabidopsis* Flower Development

Diarmuid Seosamh Ó'Maoiléidigh and Frank Wellmer

Abstract

Assessing the molecular changes that occur over the course of flower development is hampered by difficulties in isolating sufficient amounts of floral tissue at specific developmental stages. This is especially problematic when investigating molecular events at very early stages of *Arabidopsis* flower development, as the floral buds are minute and are initiated sequentially such that a single flower on an inflorescence is at a given developmental stage. Moreover, young floral buds are hidden by older buds, which present an additional challenge for dissection. To circumvent these issues, a floral induction system that allows the simultaneous induction of a large number of flowers on the inflorescence of a single plant was generated. To allow the plant community to avail of the full benefits of this system, we address some common problems that can be encountered when growing these plants and collecting floral buds for analysis.

Key words Floral induction system, Synchronous flowering, Stage-specific flower development, Tissue collection

1 Introduction

Over the past decade, a wealth of information has been gathered regarding the topology of the gene regulatory network underlying flower development using genomic technologies [1–9]. However, a recurrent difficulty that delayed progress was the isolation of sufficient amounts of floral material at distinct developmental stages due to the diminutive size of young flowers and the nature of their development from an inflorescence meristem [10]. To circumvent these problems, we PCR-amplified the entire genomic locus of the *APETALA1* (*AP1*) gene, which controls the onset of flower development, and translationally fused the final exon of this gene to the hormone binding domain of the rat glucocorticoid receptor (noted GR) [11]. The resulting $AP1_{PRO}$:AP1-GR transgene mediates the expression of the AP1-GR fusion protein in a domain that resembles that of the endogenous *AP1* gene. The fusion protein, when expressed in plants, remains in the cytoplasm due to an interaction

José Luis Riechmann and Frank Wellmer (eds.), *Flower Development: Methods and Protocols*, Methods in Molecular Biology, vol. 1110, DOI 10.1007/978-1-4614-9408-9_16, © Springer Science+Business Media New York 2014

Fig. 1 The response of $AP1_{PRO}$:AP1-GR *ap1-1 cal-1* plants to treatment with a dexamethasone-containing solution. (**a**, **b**) Inflorescence-like meristems 6 days after treatment with (**a**) a "mock" solution and (**b**) a solution containing 10 μM dexamethasone. (**c**, **d**) Flowers that developed after treatment of inflorescence-like meristems with (**c**) a "mock" solution and (**d**) a solution containing 10 μM dexamethasone

of the GR portion with a heat-shock protein. Treatment of plants with a synthetic steroid hormone such as dexamethasone leads to a release of the fusion protein from its cytoplasmic retention and subsequently to its nuclear import [3, 5]. For the construction of a floral induction system, we combined the $AP1_{PRO}$:AP1-GR transgene with plants that are doubly mutant for *AP1* and its paralog *CAULIFLOWER* (*CAL*) [11]. Flower formation in *ap1-1 cal-1* plants is blocked for a prolonged period of time and, as a result, these plants undergo a massive over-proliferation of inflorescence-like meristems (Fig. 1a) [12]. Treatment of $AP1_{PRO}$:AP1-GR *ap1-1 cal-1* plants with a dexamethasone-containing solution results in the transformation of these meristems into floral buds, which develop in a relatively synchronized manner (Fig. 1b), as was observed for a similar transgenic line that expressed the AP1-GR fusion under the control of the constitutive Cauliflower Mosaic Virus 35S promoter in the *ap1-1 cal-1* background [5]. Examination of the flowers after treatment with a dexamethasone-containing solution revealed that they closely resemble those of wild-type plants (Fig. 1d). In the absence of treatment, plants develop in an almost indistinguishable manner from non-transgenic *ap1-1 cal-1* plants (Fig. 1c).

This floral induction system provides the plant community with a valuable tool to study the molecular events that bring about the formation of a mature flower. It has been used successfully to assess the roles of several floral regulators during flower development using genomic as well as proteomic approaches [3, 5, 8, 13–16]. Below, we present a detailed description of how to grow and maintain these plants and how to collect tissue, which can be used for a variety of downstream applications.

2 Materials

2.1 Plant Lines and Growth

There are currently two $AP1_{PRO}$:AP1-GR *ap1-1 cal-1* plant lines available. The first transgenic plant line expresses a phosphinothricin acetyl transferase, which confers resistance to the herbicide glufosinate ("Basta"). The second transgenic plant line expresses an aminoglycoside 3′-phosphotransferase, which confers resistance to the antibiotic kanamycin.

1. Autoclaved soil–vermiculite–perlite (3:1:1) mixture (*see* **Note 1**).
2. Pots.

2.2 Reagents for Induction of Flower Formation

1. Dexamethasone.
2. Dexamethasone stock solution: 10 mM dexamethasone in 100 % ethanol. Store at −20 °C for up to several months.
3. Silwet L-77 surfactant.
4. Dexamethasone treatment solution: 10 μM dexamethasone, 0.015 % (v/v) Silwet L-77. Add 10 μL of dexamethasone stock solution and 1.5 μL of Silwet L-77 to 10 mL of distilled water (*see* **Note 2**).

2.3 Reagents for Agrobacterium-Mediated Transformation Using the Floral Dip Method

1. Vacuum pump.
2. Vacuum desiccator.
3. Silwet L-77 surfactant.
4. Sucrose.
5. Liquid LB medium.
6. Transformation solution: 5 % (w/v) sucrose, 0.025 % (v/v) Silwet L-77.

3 Methods

3.1 Plant Growth and Treatment

1. Sow $AP1_{PRO}$:AP1-GR *ap1-1 cal-1* seeds on soil (*see* **Note 3**) and grow at a temperature between 16 and 20 °C until the plant stems have grown to between 1 and 3 cm (*see* **Note 4**) (Fig. 2a).

Fig. 2 Procedure for treating the inflorescence-like meristems with the induction solution. (**a**) The stems of the plants should have grown to between 1 and 3 cm before treatment with the induction solution. (**b**, **c**) The inflorescence-like meristems should be directly treated with the induction solution until they turn dark-green as in (**c**), compared to (**b**) in the absence of treatment. (**d**, **e**) Removal of flower buds from a synchronously flowering inflorescence. (**d**) Inflorescence meristem 6 day after induction before tissue collection. (**e**) Floral tissue has been removed from the inflorescence shown in (**d**) by scraping the surface using jeweler's forceps

2. Liberally apply the dexamethasone treatment solution (*see* **Note 2**) onto the inflorescence-like meristems (Fig. 2b) using a Pasteur pipette, until the inflorescence-like meristems are well drenched and have turned dark green (Fig. 2c).
3. Allow flowers to develop up to the desired floral stage (*see* **Notes 5** and **6**) (e.g., Fig. 2d).

4. Collect floral tissue with a fine jeweler's forceps. Ensure that only floral tissue, and not the underlying vegetative tissue, is collected. To do this, scrape the top layer of tissue from the inflorescence-like meristem with the forceps as shown in Fig. 2e (*see* **Note 7**).
5. If the tissue will be processed for chromatin immunoprecipitation experiments for a transcription factor specifically expressed in flowers, it can be harvested more liberally. However, the harvest of non-meristematic or non-floral tissues may lead to a higher background in such assays.

3.2 Agrobacterium-Mediated Transformation Using the Floral Dip Method

1. Grow the plants until the inflorescence-like meristems have transitioned to flowering independently of dexamethasone-treatment (*see* **Note 8**).
2. Grow a dense culture of *Agrobacterium* in liquid LB (approximately for 24 h at 28 °C). Pellet the bacteria by centrifugation at ~4,500 × *g*. Decant the supernatant and resuspend in the transformation solution.
3. Place the transformation solution into the vacuum desiccator and dip the plants into the *Agrobacterium*-containing solution. Seal the vacuum desiccator with the lid and apply the vacuum (500 mbar for 5 min).
4. Remove the plants and shake off excess *Agrobacterium-containing* transformation solution. Lay the plants down in a tray, cover in a plastic wrap, and incubate at 4 °C overnight in the dark (*see* **Note 9**).
5. Remove the plastic wrap and move the plants back to the growth chamber. Wait for the plants to produce seeds and select transgenic plants using the appropriate selection protocols [17].

4 Notes

1. Growing plants on a mixture of sterile compost, perlite, and vermiculite in a ratio of 3:1:1 (or similar) is recommended, as plants of the floral induction system sown on other growth media often suffer from stunted growth. The addition of vermiculite and perlite aerates the compost and prevents soil compaction. Plants of the floral induction system appear to be more susceptible to pathogen infection, so sterilization of the growth medium is required to decrease the risk of disease. To sterilize the medium, we seal a moistened mixture of compost, perlite, and vermiculite in an autoclave bag and autoclave for 1 h at 121 °C. It is also important to let the soil dry slightly before watering, as over-watering impairs growth of these lines and favors fungus growth.

2. Preparation of the induction solution immediately before use is essential. A 10 mM dexamethasone stock solution (dissolved in 100 % ethanol) can be prepared in advance and stored at −20 °C. This stock solution can then be used to prepare the dexamethasone treatment solution (Subheading 2.2).
3. Do not overcrowd the plants. A minimum distance of ~2.5 cm should be kept between each seedling. We find it helpful to remove surplus seedlings on two separate occasions: the first round to thin the population of seedlings shortly after germination and the second round after identifying plants that are experiencing some growth difficulties (e.g., stunted growth). Increase the distance between plants if growth appears stunted or plants transition to flowering earlier than expected.
4. Temperature fluctuations can lead to early transitioning of the inflorescence-like meristems to flowering. If you encounter this problem, grow the plants between 16 and 18 °C and avoid temperature fluctuations. Furthermore, rotating pots and trays to homogenize airflows, light, and temperatures experienced by individual plants will decrease the frequency of asynchronous bolting.
5. After treatment with the induction solution, young floral buds can sometimes grow in an irregular manner (i.e., there can be areas of induced synchronous flower growth and areas of unresponsive inflorescence-like meristematic tissue on a single inflorescence-like meristem). To improve the response of the inflorescence-like meristems to dexamethasone treatment, grow the plants between 16 and 18 °C. When the plants have bolted to between 1 and 3 cm, transfer the plants to 20–22 °C and after approximately 24 h, treat the inflorescence-like meristems with the induction solution. After treatment with the dexamethasone-containing solution, continue growing the plants between 20–22 °C. If the inflorescence-like meristems turn to a pale green or a yellowish color, ensure that the Silwet L-77 is completely mixed into the induction solution. Otherwise, decrease the amount of Silwet L-77 used in the induction solution.
6. The number of days after dexamethasone treatment can be correlated with specific stages of flower development (*see* Table 1) [5, 8]. These correlations were made based on plants being grown in continuous cool-white light at 20 °C.
7. Tissue collection is routinely performed using a sharpened jeweler's forceps. When collecting tissue that will be processed and used for gene expression analysis, one must take care to remove only the top layer of tissue from the inflorescence-like meristems.

Table 1
Correlation of days after dexamethasone treatment with approximate stages of flower development (stages according to [10]). Correlations are based on plants of the floral induction system being grown at 20 °C in continuous light

Day	Stage
1	2
2	3
3	4
4	5–6
5	6–7
6	7–9
8	>9

8. Do not induce synchronous flowering if the plants are to be transformed with another transgene by floral dip, and allow the plants to transition to flowering independently of dexamethasone treatment. In this case, grow the plants at a temperature between 20 and 22 °C.
9. After transforming plants, do not incubate at 4 °C for longer than 20 h as this can negatively affect fertility levels.

Acknowledgements

Work in our laboratory is funded by grants from Science Foundation Ireland to F.W.

References

1. Gomez-Mena C et al (2005) Transcriptional program controlled by the floral homeotic gene *AGAMOUS* during early organogenesis. Development 132(3):429–438
2. Ito T et al (2004) The homeotic protein AGAMOUS controls microsporogenesis by regulation of *SPOROCYTELESS*. Nature 430(6997):356–360
3. Kaufmann K et al (2010) Orchestration of floral initiation by APETALA1. Science 328(5974):85–89
4. Schmid M et al (2005) A gene expression map of Arabidopsis thaliana development. Nat Genet 37(5):501–506
5. Wellmer F et al (2006) Genome-wide analysis of gene expression during early Arabidopsis flower development. PLoS Genet 2(7):e117
6. Wellmer F et al (2004) Genome-wide analysis of spatial gene expression in *Arabidopsis* flowers. Plant Cell 16(5):1314–1326
7. Winter CM et al (2011) LEAFY target genes reveal floral regulatory logic, *cis* motifs, and a link to biotic stimulus response. Dev Cell 20(4):430–443
8. Wuest SE et al (2012) Molecular basis for the specification of floral organs by APETALA3 and PISTILLATA. Proc Natl Acad Sci U S A 109:13452–13457

9. Yant L et al (2010) Orchestration of the floral transition and floral development in Arabidopsis by the bifunctional transcription factor APETALA2. Plant Cell 22(7): 2156–2170
10. Smyth DR et al (1990) Early flower development in *Arabidopsis*. Plant Cell 2(8):755–767
11. O'Maoileidigh DS et al (2013) Control of reproductive floral organ identity specification in Arabidopsis by the C function regulator AGAMOUS. Plant Cell 25(7)
12. Bowman JL et al (1993) Control of flower development in Arabidopsis thaliana by APETALA1 and interacting genes. Development 119(2):721–743
13. Smaczniak C et al (2012) Characterization of MADS-domain transcription factor complexes in Arabidopsis flower development. Proc Natl Acad Sci U S A 109(5):1560–1565
14. Sun B et al (2009) A timing mechanism for stem cell maintenance and differentiation in the Arabidopsis floral meristem. Genes Dev 23(15):1791–1804
15. Das P et al (2009) Floral stem cell termination involves the direct regulation of AGAMOUS by PERIANTHIA. Development 136(10):1605–1611
16. Jiao Y, Meyerowitz EM (2010) Cell-type specific analysis of translating RNAs in developing flowers reveals new levels of control. Mol Syst Biol 6:419
17. Zhang X et al (2006) Agrobacterium-mediated transformation of Arabidopsis thaliana using the floral dip method. Nat Protoc 1(2):641–646

Chapter 17

Fluorescence Activated Cell Sorting of Shoot Apical Meristem Cell Types

G. Venugopala Reddy

Abstract

Growing tips of plants harbor a set of stem cells in structures called shoot apical meristems (SAMs) which provide cells for development of aboveground biomass. Despite a periodic differentiation of stem cell progenitors into leaves, the stem cell pool remains constant over time. Genetic analysis has revealed molecular pathways involved in stem-cell specification, cell division patterns, and organ differentiation. Stem cells within SAMs are few in number, which imposes a limitation to the experimental approaches that can be used for deciphering the gene regulatory networks that underlie cell fate transitions. Here, I provide detailed experimental protocols for the protoplasting and subsequent purification through cell sorting of SAM cells, which allows genome-wide analyses of gene expression patterns at a single cell-type resolution.

Key words Stem cells, FACS, Protoplasts, Fluorescent reporter, Plant stem cells, Arabidopsis, Central zone, Peripheral zone, Rib-meristem, CLAVATA3, WUSCHEL

1 Introduction

SAMs harbor a set of 35–40 stem-cells and provide cells for the development of all the aboveground biomass of plants [1]. Most of the important pattern formation events such as maintenance of stem-cell identity, specification and differentiation of leaf/flower primordia, and the temporal control of the transition from vegetative to reproductive growth (flowering time) are determined in the SAMs [2]. Thus, the SAM represents a dynamic and interacting network of functionally distinct cell types. The cellular identities and functions are regulated by diverse signals. Genetic analysis has revealed a basic set of molecules and hormonal pathways involved in stem-cell maintenance, organ differentiation, and flowering time [3]. However, our knowledge is still limited as to how different pathways interact with each other to function as a network in specifying different cell types and their function. This is due to the lack of a complete description of the different cell types and of the gene expression programs associated with them.

José Luis Riechmann and Frank Wellmer (eds.), *Flower Development: Methods and Protocols*, Methods in Molecular Biology, vol. 1110, DOI 10.1007/978-1-4614-9408-9_17, © Springer Science+Business Media New York 2014

Traditionally, SAMs of higher plants have been divided into distinct domains of cells mostly based on their location within the SAMs and cytological criteria [2]. The central zone (CZ) is at the tip and harbors a set of stem cells. Progeny of stem cells enter into differentiation pathways when they enter the surrounding regions—the flanking peripheral zone (PZ). Within the PZ, cells differentiate as leaves or flowers in a specified spatiotemporal sequence followed by specific cellular behaviors and differentiation events lead to the development of boundary regions and separation of organs from the SAMs. Thus, stem cell daughters progress through a sequential differentiation process to acquire distinct cellular identities [2]. Besides this radial organization, the SAM of higher plants is a complex multilayered structure. In dicots, the tunica consists of the outer epidermal (L1 layer) and an inner sub-epidermal (L2 layer), whereas the corpus forms a multilayered structure located beneath the L2 layer [4, 5]. Cells of the corpus are also referred to as the Rib-meristem (RM), which differentiates to provide cells for the development of the stem and also provide cues to the overlying CZ to specify them as stem cells [6]. Thus, the SAM stem cell niche is a collection of distinct cell types, which express different genes and exhibit distinct cell behaviors both along the radial domain and across cell layers that are clonally distinct. Therefore, a tight regulation of gene expression dynamics by the cell–cell communication machinery is critical to ensure the timely transition from one cell type to another. The challenge is to understand how the interconnected network of cells interprets complex spatiotemporal signals to regulate gene expression patterns to bring about cell fate transitions. This requires an improved spatiotemporal resolution for gene expression profiling assays in the SAMs.

Fluorescence activated cell sorting (FACS) of protoplasts derived from three distinct cell types, followed by genome-wide expression profiling (microarray analyses) yielded a global map of gene expression within different cell types and through 15 different zones along the proximo-distal root developmental gradient [7]. This analysis showed that the gene expression patterns are continuous and traverse across the traditional demarcation of distinct cell types based on anatomical features. A later study has further refined the root gene expression map by utilizing additional cell-type specific markers, and revealed a complex spatiotemporal regulation of gene expression patterns in which expression fluctuations are observed along the developmental gradient [8].

In contrast to these studies on the *Arabidopsis* root, expression profiling studies of the SAM, until recently, used RNA isolated from the whole tissue. A recent study employed FACS mediated cell sorting to generate gene expression profiles of cells located within three broad spatial domains; the CZ, the PZ, and the RM of SAMs [9]. Here I describe a method for the Fluorescence Activated Cell Sorting (FACS)-based isolation of protoplasts from specific cell types of the Arabidopsis SAM.

2 Materials

2.1 Protoplasting and Cell Sorting

1. Plant material. 50–60 mg of *ap1-1;cal-1* SAM tissue (from about 200 shoots) expressing a fluorescent marker.
2. Solution A: 10 mM KCl, 2 mM $MgCl_2$, 2 mM $CaCl_2$, 0.1 % (w/v) Bovine Serum Albumin (BSA), 2 mM MES, 600 mM Mannitol. To prepare 25 mL of solution A, mix 250 μL of 1 M KCl, 50 μL of 1 M $MgCl_2$, 50 μL of 1 M $CaCl_2$, add 0.025 g of BSA, 0.0097 g of MES, and 2.75 g of Mannitol, and make up the volume to 25 mL with sterile water (*see* **Note 1** for manufacturers and catalog numbers).
3. Solution B: 1.5 % (w/v) Cellulase, 1 % (w/v) Pectolyase, 1 % (w/v) Hemicellulase in solution A. To prepare solution B, dissolve 300 mg of Cellulase, 200 mg of Pectolyase, and 200 mg of Hemicellulase in 20 mL of solution A (*see* **Note 1** for manufacturers and catalog numbers).
4. Incubator shaker (e.g., New Brunswick Excella E24).
5. 70 μm nylon cell strainer (BD Falcon).
6. 35 × 10 mm petri dish (BD Falcon).
7. Fine forceps and scalpels.
8. Temperature Controlled Centrifuge (e.g., Beckman Coulter Allegra X-15R).
9. FACS-Aria (Beckton Dickinson).

2.2 Isolation of Total RNA from Sorted Cells

1. RNA extraction kit (e.g., RNeasy Plant Mini Kit from QIAGEN).
2. Glycoblue.
3. 7.5 M ammonium acetate.
4. RNase-free water.
5. Agilent 6000 RNA Nano kit.
6. Tabletop microcentrifuge (e.g., Eppendorf Centrifuge 5415 D).
7. Agilent 2100 Bioanalyzer.
8. NanoDrop ND-1000 Spectrophotometer.

3 Methods

3.1 Protoplasting and Cell Sorting

This method has been developed based on two earlier studies [7, 10].

1. Grow *ap1-1;cal-1* plants carrying a fluorescent cell type marker (*see* **Note 2** for promoters and fluorescent tags) for 4 weeks.
2. Prepare 25 mL of solution A. Adjust the pH to 5.5 with 1 M Tris using HCl.

3. To prepare solution B, dissolve cell-wall-digesting enzymes cellulase (300 mg), pectolyase (200 mg), and hemicellulase (200 mg) in 20 mL of freshly prepared buffer A. Mix vigorously by pipette until the solution becomes transparent/clear (*see* **Note 3**).
4. In the meantime bring the temperature of the shaker to 22 °C.
5. Harvest 50–60 mg of *ap1;cal* SAMs as quickly as possible (*see* **Note 4** for instructions on tissue harvest).
6. Transfer the tissue to a 70 μm nylon cell strainer (BD Falcon) and place the strainer in a 35 × 10 mm petri dish (BD Falcon).
7. Add approximately 6–7 mL of solution B and place the petri dish on a shaker set at 120 rpm. maintained at 22 °C. At every 10 min interval, gently rinse the SAM surface with a jet of solution B using a 1 mL pipette (*see* **Note 5**).
8. Set the centrifuge at 4 °C.
9. If protoplasting is efficient, the protoplasting solution should turn turbid by 45 min into the procedure.
10. Pipetting and mixing of solution may result in partial loss of solution B. Supplement solution B at regular intervals (*see* **Note 5**).
11. Upon 1 h and 15 min of treatment, transfer the contents of petri dish to a 15 mL Falcon tube and centrifuge at 500 × *g* at 4 °C for 10 min. Remove the supernatant without disturbing the pellet.
12. Dissolve the pellet in 500–600 μL of solution A by pipetting gently. Once the pellet is dissolved completely, place the tube on ice until the protoplasts are loaded onto the FACS.
13. Separate mGFP5-ER or Ds-Red-N7 (*see* **Note 2** for promoters and fluorescent tags) expressing protoplasts by passing them through a FACS-Aria (Becton Dickinson) fitted with a 100 μm nozzle at a flow rate of 5,000–7,000 events per second and with a fluid pressure of 35 psi. Select GFP positive cells by their emission intensity in the green channel (~530 nm) and Ds-Red cells by their emission intensity in the red channel (~610 nm). Collect sorted protoplasts directly into the lysis buffer (Qiagen RLT buffer), mix, and freeze immediately at −80 °C. Forty-five minutes of sorting should yield about 125,000–150,000 CLV3 (*see* **Note 2** for promoters and fluorescent tags) positive protoplasts.

3.2 Isolation of Total RNA from Sorted Cells

1. Collect fluorescently labeled protoplasts from FACS flow in 500 μL RLT buffer (QIAGEN RNeasy kit) and adjust the volume of cell suspension up to (3× of original) 3 mL by adding RLT buffer.

2. Add 30 μL of *ß*-mercaptoethanol in it and one could either store it at −80 °C or proceed without freezing for RNA isolation.
3. Add 0.5 volume of chilled ethanol and mix gently.
4. Apply 700 μL of solution to an RNeasy column and centrifuge at 10,600 × *g* in a microcentrifuge for 15 s. Add the solution to same sample column and repeat the same step until all remaining amount of precipitated solution is loaded onto the column.
5. Add 350 μL of buffer RW1 to spin column and centrifuge at 10,600 × *g* in a microcentrifuge for 15 s.
6. To perform on column DNA digestion, add 80 μL of DNA digesting solution to sample column at RT for 15 min (10 μL DNase plus 70 μL RDD buffer).
7. Stop DNA digestion by adding 350 μL of buffer RW1 and centrifuge at 10,600 × *g* in a microcentrifuge for 15 s.
8. Apply 500 μL RPE buffer provided with RNeasy kit after supplementing it with ethanol.
9. Centrifuge at 10,600 × *g* in a microcentrifuge for 2 min.
10. Air-dry the sample column for 2 min.
11. Elute RNA from column by applying 50 μL RNase-free water, centrifuge at 17,900 × *g* for 1 min, and repeat the same step.
12. Precipitate RNA by adding 0.5× volumes of 7.5 M ammonium acetate and 2.5 volumes of ethanol, along with 0.5 μL of Glycoblue (15 mg/mL) overnight at −20 °C.
13. Centrifuge at 17,900 × *g* for 30 min, remove supernatant, and wash RNA pellet with cold 70 % ethanol.
14. Air-dry the pellet for 5 min and resuspend RNA pellet in 4–6 μL RNase-free water.
15. Determine RNA integrity using the Agilent 6000 RNA Nano kit on an Agilent 2100 Bioanalyzer according to manufacturer's protocol (*see* **Note 6** for expected RNA yield).

4 Notes

1. Manufacturer information and catalog numbers of chemicals, kits and enzymes used in this protocol: BSA and MES (Fisher), Mannitol (Sigma, USA), Hemicellulase (Sigma, USA Cat # H2125), Cellulase (Yakult, Japan Cat #203039), Pectolyase (Yakult, Japan Cat #202047), RNeasy Plant Mini Kit (# 74904 QIAGEN).
2. Promoters and fluorescent tags used for sorting SAM cell types have been described in an earlier study [9]. In brief, *pCLV3::mGFP5-ER* (*CLAVATA3* promoter driving the expression of endoplasmic reticulum localized mGFP5) is a

marker for stem cells/the CZ. *pWUS::mGFP5-ER* (*WUSCHEL* promoter driving the expression of endoplasmic reticulum localized mGFP5) is a marker for the RM and *pFIL::dsRED-N7* (*FILAMENTOUSFLOWER* promoter driving the expression of nuclear-localized dsRED) is a marker for differentiating cells of the organ primordia. In general, reporter lines with higher signal-to-noise ratio will reduce contamination during sorting and improve the purity of cell types. Therefore, using fluorescent proteins with tags that direct them to specific intracellular compartments is recommended.

3. Protoplasting solution (solution B) should be clear and without any sediments, to obtain optimum protoplast yield.
4. Do not chop SAM tissue into smaller bits as that would increase the debris, which interferes with the flow sorting process.
5. Applying a jet of protoplasting solution onto the SAM tissue at regular intervals (approximately every 10 min), by using 1 mL pipette, during the protoplasting process increases the protoplast yield. Pipetting and mixing of solution may result in partial loss of solution B. Supplement solution B at regular intervals.
6. 100,000 cell type-specific protoplasts will usually yield between 200 and 300 ng of total RNA, which is sufficient for microarray analysis with a two step amplification procedure.

Acknowledgements

Our genomics work is being carried out at the core facility of center for plant cell biology (CEPCEB) and institute of integrative genome biology (IIGB), University of California, Riverside. This work is funded by National Science Foundation grant (IOS-1147250) to G.V.R.

References

1. Reddy GV (2008) Live-imaging stem-cell homeostasis in the Arabidopsis shoot apex. Curr Opin Plant Biol 11:88–93
2. Steeves TA, Sussex IM (1989) Patterns in plant development. Shoot apical meristem mutants of Arabidopsis thaliana. New York: Cambridge University Press
3. Barton MK (2010) Twenty years on: the inner workings of the shoot apical meristem, developmental dynamo. Dev Biol 341:95–113
4. Satina S, Blakeslee AF, Avery AG (1940) Demonstration of the three germ layers in the shoot apex of Datura by means of induced polyploidy in periclinal chimeras. Am J Bot 27:895–905
5. Tilney-Bassett RAE (1986) Plant Chimeras. Edward Arnold, London
6. Baurle I, Laux T (2003) Apical meristems: the plant's fountain of youth. Bioessays 25:961–970
7. Birnbaum K, Shasha DE, Wang JY, Jung JW, Lambert GM, Galbraith DW, Benfey PN (2003) A gene expression map of the Arabidopsis root. Science 302:1956–1960
8. Brady SM, Orlando DA, Lee JY, Wang JY, Koch J, Dinneny JR, Mace D, Ohler U, Benfey PN (2007) A high-resolution root spatiotemporal

map reveals dominant expression patterns. Science 318:801–806
9. Yadav RK, Girke T, Pasala S, Xie M, Reddy GV (2009) Gene expression map of the Arabidopsis shoot apical meristem stem cell niche. Proc Natl Acad Sci USA 106:4941–4946
10. Guangyu C, Conner AJ, Christey MC, Fautrier AG, Field RJ (1997) Protoplast isolation from shoots of Asparagus cultures. Int J Plant Sci 158:537–542

Chapter 18

Translating Ribosome Affinity Purification (TRAP) for Cell-Specific Translation Profiling in Developing Flowers

Ying Wang and Yuling Jiao

Abstract

The development of a multicellular organism is accompanied by cell differentiation. In fact, many biological processes have cell specificity, such that distinct cell types respond differently to endogenous or environmental cues. To obtain cell-specific gene expression profiles, translating ribosome affinity purification (TRAP) has been developed to label polysomes containing translating mRNAs in genetically defined cell types. Here, we describe the immunopurification of epitope-labeled polysomes and associated RNAs from target cell types. TRAP has the additional advantage of obtaining only translating mRNAs, which are a better proxy to the proteome than a standard mRNA preparation.

Key words mRNA, Cell type, Posttranscriptional regulation, Transcriptome

1 Introduction

Widely used high-throughput transcriptome profiling approaches have been successful in dissecting gene regulatory networks. However, cell specificity in transcriptome profiling is often obscured by the existence of many cell types even within a simple organ. To obtain cell specificity while maintaining genome-wide coverage, several technologies have been developed [1, 2]. Laser Microdissection (LM) uses a laser beam to isolate cells of interest from fixed tissue sections under a microscope. This approach has been used in plants, and isolates cells based exclusively on their morphology and location within tissue sections [3]. Because of its labor-intensive nature, LM can obtain only a limited number of cells and has to be combined with RNA amplification, which usually introduces significant biases. Another approach to isolate specific cell types is Fluorescence Activated Cell Sorting (FACS), in which fluorescent proteins are expressed within cells of interest, and flow cytometry is used to isolate fluorescently labeled cells

José Luis Riechmann and Frank Wellmer (eds.), *Flower Development: Methods and Protocols*, Methods in Molecular Biology, vol. 1110, DOI 10.1007/978-1-4614-9408-9_18, © Springer Science+Business Media New York 2014

after protoplasting, i.e., cell wall digestion and cellular dissociation [4]. As a result of these treatments, the protoplasts that are purified through FACS undergo cellular stress during the procedure. In addition, the resulting sorted cell population can hardly be free of nonfluorescent cell contaminant. A similar approach is to label and isolate nuclei in cells of interest. Nuclei labeling and purification can be achieved either by expressing a fluorescent protein and flow cytometry [5], or by expressing an epitope-labeled nuclear envelope protein and subsequent affinity purification [6]. It should be noted that nuclear RNA preparations contain many unfinished mRNA molecules without further maturation, such as splicing, and may not reflect the relative abundance of translated mRNAs.

TRAP isolates translating mRNAs from cells of interest, and provides a better proxy to the proteome. Briefly, a ribosomal protein is epitope-labeled and expressed in target cells using cell type-specific promoters in transgenic plants or animals. In *Arabidopsis*, RPL18 protein can be labeled with His and FLAG tags (HF-RPL18) or with FLAG and GFP tags [7–10]. Both GFP-tagged RPL10a and HA-tagged RPL22 have been tested in mice [11, 12], and GFP-tagged RPL10a has been found to work efficiently in *Drosophila* [13]. The epitope-tagged ribosomal proteins are incorporated into ribosomes, and thus into polysomes, which can be affinity purified: total polysomes are isolated from transgenic plants or animals, and epitope tag-labeled polysomes are then purified using corresponding antibody-coated beads. TRAP does not need specialized equipment, and has the advantage of high RNA yield to readily proceed to RNA-seq without further RNA amplification [8]. The entire procedure, from polysome purification to RNA extraction, can be completed within 1 day.

2 Materials

1. Nuclease-free sterile water.
2. Nuclease-free sterile tubes.
3. Polysome extraction buffer: 100 mM Tris–HCl (pH 9.0), 200 mM KCl, 25 mM EGTA (pH 8.0), 36 mM $MgCl_2$, 1 % Brij L23 (Sigma-Aldrich, St. Louis, MO, USA), 1 % Triton X-100, 1 % IGEPAL CA-630 (Sigma-Aldrich, St. Louis, MO, USA), 1 % TWEEN 20, 2 % polyoxyethylene (10) tridecyl ether (Sigma-Aldrich, St. Louis, MO, USA), 1 % deoxycholic acid, 5 mM dithiothreitol, 1 mM phenylmethanesulfonyl fluoride, 50 μg/mL cycloheximide, 50 μg/mL chloramphenicol, 40 U/mL RNase inhibitor (such as RiboLock RNase Inhibitor, Fermentas, Burlington, ON, Canada), 1 mg/mL heparin (Sigma-Aldrich, St. Louis, MO, USA). Prepare the buffer with nuclease-free sterile water and store at 4 °C (*see* **Note 1**).

4. Wash buffer #1: 100 mM Tris–HCl (pH 8.5), 200 mM KCl, 25 mM EGTA (pH 8.0), 36 mM $MgCl_2$. Make up the buffer with nuclease-free sterile water and store at −20 °C.
5. Wash buffer #2: 100 mM Tris–HCl (pH 9.0), 200 mM KCl, 25 mM EGTA (pH 8.0), 36 mM $MgCl_2$, 5 mM dithiothreitol, 1 mM phenylmethanesulfonyl fluoride, 50 μg/mL cycloheximide, 50 μg/mL chloramphenicol, 40 U/mL RNase inhibitor. Make up with nuclease-free sterile water and store at 4 °C (*see* **Note 2**).
6. Monoclonal ANTI-FLAG M2 affinity gel (Sigma-Aldrich, St. Louis, MO, USA).
7. 3× FLAG peptide stock solution: dissolve 3× FLAG peptide lyophilized powder (Sigma-Aldrich, St. Louis, MO, USA) in 0.5 M Tris–HCl (pH 7.5) with 1 M NaCl at a concentration of 25 mg/mL. Dilute fivefold with water to prepare a 3× FLAG stock solution containing 5 mg/mL of 3× FLAG peptide.
8. RNeasy mini kit (Qiagen, Hilden, Germany) or an equivalent total RNA isolation kit or reagents.
9. *β*-mercaptoethanol.
10. 100 % ethanol.

3 Methods

Carry out all procedures on ice or at 4 °C in a cold room unless otherwise specified.

3.1 Polysome Extraction

1. Freeze in liquid nitrogen *Arabidopsis* tissue expressing HF-RPL18 immediately after collection and grind to a find powder under liquid nitrogen using a mortar and pestle (*see* **Note 3**). Transfer the suspension of tissue powder and liquid nitrogen into a liquid-nitrogen-cooled appropriately sized tube and allow the liquid nitrogen to evaporate without allowing the sample to thaw.
2. Add polysome extraction buffer as quickly as possible. Homogenize 1 Vol of pulverized *Arabidopsis* tissue powder with 2 Vol of polysome extraction buffer by gentle shaking. Incubate mixture on ice for 10 min.
3. Transfer mixture to centrifugation tubes and centrifuge at 16,000 × *g* for 10 min at 4 °C.
4. Transfer the supernatant to new centrifugation tubes and centrifuge at 16,000 × *g* for 10 min at 4 °C.
5. Transfer the supernatant to new centrifugation tubes and store at 4 °C for Subheading 3.3.

3.2 Preparation of the Anti-FLAG Agarose Beads

1. Thoroughly resuspend the anti-FLAG M2 affinity agarose gel in the vial to make a uniform suspension of the resin. Transfer 200 μL of suspension to a new centrifugation tube (*see* **Note 4**).
2. Centrifuge at 8,200 × *g* for 30 s at 4 °C.
3. Remove the supernatant with a pipette and add 2 mL wash buffer #1 to the resin.
4. Centrifuge at 8,200 × *g* for 30 s at 4 °C.
5. Remove the supernatant with a pipette and add 2 mL wash buffer #2 to the resin.
6. Centrifuge at 8,200 × *g* for 30 s at 4 °C.
7. Remove the supernatant with a pipette.

3.3 Immunoprecipitation of Polysome

1. Add polysome extract from Subheading 3.1 to prepared anti-FLAG agarose gel from Subheading 3.2.
2. Incubate for 2 h at 4 °C with gentle back-and-forth shaking (*see* **Note 5**).
3. Centrifuge at 8,200 × *g* for 30 s at 4 °C.
4. Transfer the supernatant to new centrifugation tubes, and keep beads at the bottom undisturbed (*see* **Note 6**).
5. Add 4 mL of wash buffer #2 to the beads, mix by gently inverting the tube, and centrifuge at 8,200 × *g* for 30 s at 4 °C.
6. Remove the wash buffer with a pipette and add 3 mL wash buffer #2 to the beads. Incubate tubes with gentle shaking for 10 min at 4 °C.
7. Centrifuge at 8,200 × *g* for 30 s at 4 °C.
8. Remove the wash buffer with a pipette and add 3 mL wash buffer #2 to the beads. Incubate tubes with gentle shaking for 10 min at 4 °C.
9. Centrifuge at 8,200 × *g* for 30 s at 4 °C.
10. Remove the wash buffer with a pipette and add 200 μL wash buffer #2 and 12 μL 3× FLAG peptide stock solution to the beads. Incubate with gentle shaking for 30–60 min at 4 °C (*see* **Note 7**).
11. Centrifuge at 8,200 × *g* for 30 s at 4 °C. Transfer the supernatant to a new tube.
12. Add another 100 μL wash buffer #2 and 6 μL 3× FLAG peptide stock solution to the beads. Incubate with gentle shaking for 30 min at 4 °C.
13. Centrifuge at 8,200 × *g* for 30 s at 4 °C. Combine both supernatant to the same tube.

3.4 RNA Extraction

1. Add 1.6 mL RLT buffer from RNeasy kit and 16 μL β-mercaptoethanol to the supernatant and mix well at room temperature.

2. Add 1.1 mL 100 % ethanol and mix by pipetting at room temperature.
3. Apply the sample into an RNeasy mini spin column. Centrifuge at 16,000 × *g* for 15 s at room temperature. Discard the flow through.
4. Add 700 μL RW1 buffer to the spin column and centrifuge at 16,000 × *g* for 15 s at room temperature. Discard the flow through.
5. Add 500 μL RPE buffer to the spin column and centrifuge at 16,000 × *g* for 15 s at room temperature. Discard the flow through. Repeat this step one more time.
6. Transfer the spin column to a new tube and centrifuge at 16,000 × *g* for 2 min at room temperature.
7. Transfer the spin column to a new 1.5 mL reaction tube and add 30 μL of nuclease-free sterile water at 50–55 °C. Incubate for 1 min.
8. Elute RNA by centrifuging at 16,000 × *g* for 1.5 min at room temperature.
9. Repeat the above elution step with another 30 μL nuclease-free sterile water. Combine eluted RNA solution (*see* **Note 8**).

4 Notes

1. Polysome extraction buffer needs to be prepared freshly and can be prepared from stock solutions. Stock solutions of Tris–HCl, KCl, EGTA, and $MgCl_2$ need to be autoclaved.
2. Wash buffer #2 needs to be prepared freshly and can be prepared from stock solutions. Stock solutions for Tris–HCl, KCl, EGTA, and $MgCl_2$ need to be autoclaved.
3. The amount of plant tissue required for TRAP depends on the number of cell that express the tagged construct within the tissue to be collected, and on the expression level of the HF-RPL18 protein in those cells. For flowers expressing *HF-RPL18* under the promoters of *APETALA1*, *APETALA3*, and *AGAMOUS* [14], 2.5–10 mL packed flower buds can be used.
4. The ratio of suspension to packed gel volume is around 2:1, and 200 μL of suspension contains about 100 μL packed gel.
5. Extended overnight incubation at 4 °C may increase the final RNA yield.
6. The supernatant from this step can be used to test immunoprecipitation efficiency with an anti-FLAG antibody by Western blotting.
7. Agarose beads bind RNA molecules nonspecifically. Polysomes must be eluted from agarose beads for RNA extraction.

8. RNA quality can be monitored using a bioanalyzer. Quantification of isolated by using a UV spectrometer is recommended. From the amount of starting floral tissue suggested above, we usually obtain 5–10 μg of total RNA. RNA from TRAP can be used directly for mRNA isolation and RNA-seq (*see* Chapter 23).

Acknowledgements

We thank J. Bailey-Serres for sharing protocols. This work was supported by the National Natural Science Foundation of China Grants 31222033 and 31171159, by the National Basic Research Program of China (973 Program) Grant 2012CB910902, and by the Hundred Talents Program of CAS.

References

1. Wang Y, Jiao Y (2011) Advances in plant cell type-specific genome-wide studies of gene expression. Front Biol 6:384–389
2. Rogers ED, Jackson T, Moussaieff A, Aharoni A, Benfey PN (2012) Cell type-specific transcriptional profiling: implications for metabolite profiling. Plant J 70:5–17
3. Kerk NM, Ceserani T, Tausta SL, Sussex IM, Nelson TM (2003) Laser capture microdissection of cells from plant tissues. Plant Physiol 132:27–35
4. Birnbaum K, Shasha DE, Wang JY, Jung JW, Lambert GM, Galbraith DW, Benfey PN (2003) A gene expression map of the Arabidopsis root. Science 302:1956–1960
5. Zhang C, Barthelson RA, Lambert GM, Galbraith DW (2008) Global characterization of cell-specific gene expression through fluorescence-activated sorting of nuclei. Plant Physiol 147:30–40
6. Deal RB, Henikoff S (2010) A simple method for gene expression and chromatin profiling of individual cell types within a tissue. Dev Cell 18:1030–1040
7. Mustroph A, Zanetti ME, Jang CJ, Holtan HE, Repetti PP, Galbraith DW, Girke T, Bailey-Serres J (2009) Profiling translatomes of discrete cell populations resolves altered cellular priorities during hypoxia in Arabidopsis. Proc Natl Acad Sci USA 106:18843–18848
8. Jiao Y, Meyerowitz EM (2010) Cell-type specific analysis of translating RNAs in developing flowers reveals new levels of control. Mol Syst Biol 6:419
9. Juntawong P, Bailey-Serres J (2012) Dynamic light regulation of translation status in Arabidopsis thaliana. Front Plant Sci 3:66
10. Ribeiro DM, Araujo WL, Fernie AR, Schippers JH, Mueller-Roeber B (2012) Translatome and metabolome effects triggered by gibberellins during rosette growth in Arabidopsis. J Exp Bot 63:2769–2786
11. Sanz E, Yang L, Su T, Morris DR, McKnight GS, Amieux PS (2009) Cell-type-specific isolation of ribosome-associated mRNA from complex tissues. Proc Natl Acad Sci USA 106:13939–13944
12. Heiman M, Schaefer A, Gong S, Peterson JD, Day M, Ramsey KE, Suarez-Farinas M, Schwarz C, Stephan DA, Surmeier DJ, Greengard P, Heintz N (2008) A translational profiling approach for the molecular characterization of CNS cell types. Cell 135: 738–748
13. Thomas A, Lee PJ, Dalton JE, Nomie KJ, Stoica L, Costa-Mattioli M, Chang P, Nuzhdin S, Arbeitman MN, Dierick HA (2012) A versatile method for cell-specific profiling of translated mRNAs in Drosophila. PLoS ONE 7:e40276
14. Krizek BA, Fletcher JC (2005) Molecular mechanisms of flower development: an armchair guide. Nat Rev Genet 6:688–698

Chapter 19

Laser-Assisted Microdissection Applied to Floral Tissues

Samuel E. Wuest and Ueli Grossniklaus

Abstract

Cellular context can be crucial when studying developmental processes as well as responses to environmental variation. Several different tools have been developed in recent years to isolate specific tissues or cell types. Laser-assisted microdissection (LAM) allows for the isolation of such specific tissue or single cell-types purely based on morphology and cytology. This has the advantage that (1) cell types that are rare can be isolated from heterogeneous tissue, (2) no marker line with cell type-specific expression needs to be established, and (3) the method can be applied to non-model species and species that are difficult to genetically transform. The rapid development of next-generation sequencing (NGS) approaches has greatly advanced the possibilities to perform molecular analyses in diverse organisms. However, there is a mismatch between currently available cell isolation tools and their applicability to non-model organisms. Therefore, LAM will become increasingly popular in the study of diverse agriculturally or ecologically relevant plant species. Here, we describe a protocol that has been successfully used for LAM to isolate either whole floral organs or even single cell types in plants, e.g., *Arabidopsis thaliana*, *Boechera* spp., or tomato.

Key words Laser-assisted microdissection, Single cell isolation, Microgenomics

1 Introduction

1.1 Advantages of Using Laser-Assisted Microdissection (LAM)

Cellular context can provide important insights into the molecular bases underlying developmental processes, as cell-specific processes are often masked in experiments that are performed at the organ or whole-plant level (reviewed in [1]). Over the last years, several methods have been developed that allow the isolation of specific cell types or tissues (*see* examples in [2]). However, several popular cell isolation methods, such as fluorescence-activated cell sorting [3] (*see* Chapter 17) or the isolation of nuclei from specific cell types by affinity purification [4, 5], rely on the use of transgenic lines and the availability of cell type-specific expression markers. At the same time, with the advances of NGS methods it has become possible to perform new types of experiments using non-model organisms. For example, RNA-Seq (*see* Chapter 23) offers the opportunity to study global gene expression even in organisms that lack reference sequences [6]. However, these non-model

José Luis Riechmann and Frank Wellmer (eds.), *Flower Development: Methods and Protocols*, Methods in Molecular Biology, vol. 1110, DOI 10.1007/978-1-4614-9408-9_19, © Springer Science+Business Media New York 2014

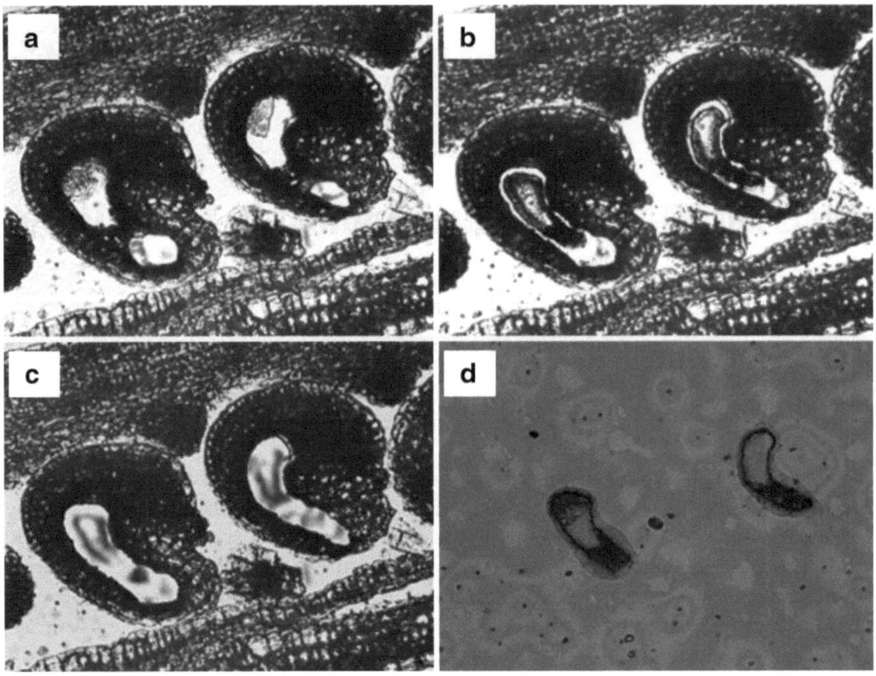

Fig. 1 Laser-assisted microdissection of single cell-types from flowers. The series shows LAM steps during the isolation of *Arabidopsis thaliana* central cells from 8 μm thick carpel sections. (**a**) Two ovules containing two intact central cells (large, vacuolated cells in the center). (**b**) Laser cutting performed with the MMI μCut system that uses a UV-laser beam to isolate tissue sections of interest. The laser beam has a diameter of approximately 1 μm. (**c**) Removal of central cells from the carpel sections by the use of an adhesive cap from a collection tube. (**d**) Isolated central cells attached to the adhesive cap, ready to be used for downstream applications (e.g., RNA extraction)

organisms might not be suitable for genetic transformation, meaning that there is a mismatch between the advances in NGS and the ability to understand cell type-specific expression. In these cases, laser-assisted microdissection (LAM) [7] can be considered the method of choice for the isolation specific cell or tissue types. No cell-specific expression marker line needs to be established, and the method can be applied to any species, including those that are difficult to genetically transform.

1.2 What Is LAM?

LAM allows for the isolation of tissue subtypes [8–15] or even single cell types [6, 16, 17], based purely on morphology or cytology (*see* Fig. 1). LAM has been a very popular tool in biomedical research for over 15 years, and also the plant research community has increasingly adopted it in recent years (e.g., in [9, 12, 16, 18–20]). In preparation for LAM the target tissue is chemically fixed, embedded into a resin or wax, and sectioned using a microtome. A LAM approach then allows the isolation of tissues or single cell types (or even subcellular domains [13]) from the sectioned specimens.

1.3 Existing Microdissection Devices

Several microdissection devices have been developed over the last 15 years. These can be roughly classified into laser-capture (use of infrared-lasers to melt the tissue area of interest to a thermoplastic polymer) and laser-cutting (use of UV-laser to photovolatize tissue and slide membrane); please refer to references [21] and [22] for an in-depth review of these approaches.

Our protocol describes the use of a Molecular Machines and Industries (MMI) laser-cutting device, such as the MMI CellCut Plus, which is equipped with a UV-laser that is well suited for the isolation of very small tissues or single cell types (e.g., egg cells with a diameter of 8–10 μm [16]). In principle, the MMI device is even suited for the isolation of certain subcellular domains (Marc Schmid and Ueli Grossniklaus, unpublished). Using this device, we have shown that it is also possible to dissect whole-mount tissues, whereby fine structures are attached directly to the membrane slide (Anupama Goyal, Samuel Wuest, and Ueli Grossniklaus, unpublished). Alternative devices for LAM include the Arcturus Pix Cell II, the PALM XZ, or the Leica AS LMD (*see* for example [21] and [23] for in-depth descriptions of the different devices). However, the fixation, embedding, and sectioning steps as described here can be applied for all of these microdissection devices.

1.4 A Simple Protocol for LAM Resembling In Situ Hybridization Procedures

If LAM and/or in situ hybridization are to be performed on a regular basis, it is advisable to automate certain steps of the protocols. For example, we generally perform the tissue embedding using the automated embedding system ASP200 from Leica. Such an embedding device allows for a streamlined, fast and reproducible workflow, cuts down on the time invested into sample collection, and provides results of higher reproducibility. However, such automated systems require a significant investment. Here, we describe a manual embedding method that takes several days, yet with little work required on most days. It is similar to a protocol typically used for in situ hybridization preparations (*see* Chapter 14), for which the expertise might already be available within a research group. Different methods, such as microwave-based embedding procedures that allow for shorter embedding periods, have been described elsewhere [23, 24].

1.5 Common Downstream Applications

Since the amount of isolated cellular material is often limited during LAM, downstream applications mostly focus on biomolecules that can be amplified before measurements. LAM is thus most commonly used in combination with gene expression profiling. Therefore, we consider here a protocol that aims at the extraction of RNA from the target cells for further downstream analyses. We advise on the use of precautions to preserve the integrity of the RNA in samples isolated by LAM, including carefully testing the RNA quality of the fixed, embedded, and laser dissected samples, respectively.

In addition to gene expression studies, LAM has also been used for genotyping tissue, for instance young embryos and endosperm in developing seeds [25]. Recent efforts have focused on generating protocols to study DNA methylation status in single cell types [26, 27]. These protocols are still being optimized and are not covered here.

1.6 Making the First Step

We have experienced that even though many researchers are interested in applying the technique to answer a specific scientific question, they hesitate to do so because LAM is considered a complicated, time-consuming and expensive technique. However, many steps of the LAM protocol are similar to those used in in situ hybridization methods, meaning that preexisting expertise and tools for this technique can be applied (e.g., tissue fixation and embedding, sectioning using a microtome). Furthermore, for local access options to a microdissection device, it is advisable to check the availability at other departments of the university campus, such as the biomedical department.

1.7 Experimental Design

Experimental design forms the base of any experiment in research, and in the case of LAM, it is advisable to consider the following points:

1. *Types of design and general design issues*: In any experiment that involves LAM, the need for the collection of specific cell types or tissues represents an additional sampling effort, which might include several rounds of optimization steps before the actual experiment can start. We have experienced in the past that once the workflow has been established, the sampling effort is often underestimated, and that sampling can extend over a substantial time period. For example, it is not rare that a molecular analysis fails due to limited input of biomolecules, especially when working at the technical limits. Also, samples from different treatments have to be processed in parallel and isolated at the same time, for otherwise confounding variables can be introduced.

   We therefore advise to (at least initially) perform experiments with relatively simple experimental designs (e.g., two categorical explanatory variables such as treatment 1/treatment 2, or cell type 1/cell type 2) and focus on appropriate replication and controls. It should also be kept in mind that confounding could also happen through comparison of cell types that are not equally amenable to isolation by LAM (e.g., in comparisons of large structures against small structures that can yield different RNA qualities after LAM).

   Furthermore, we argue that pooling of isolated tissue, which is often necessary to obtain sufficient amounts of biomolecules, does not substitute for proper experimental replication and randomization.

2. *Costs*: Even though the costs of consumables used during LAM are not excessively high, the downstream applications often involve optimized, commercially available kits (e.g., for RNA isolation and amplification, real-time RT-PCR). This makes intermediate quality control steps crucial and can quickly increase the overall costs. Again, this factor will promote the use of simple experimental design.
3. *Automatisation*: As described above, it can be beneficial to automate certain steps of the LAM protocol. For example, the use of automated embedding devices allows for more complex designs by improving reproducibility and speed of the experiments. In our laboratory, the embedding of material after fixation requires only an overnight step using the Leica ASP200 and reduces handling steps dramatically.

In summary, LAM has the potential to provide a detailed view of the abundance of biomolecules in specific plant tissues. It can provide important context to study developmental processes, but also cell type-specific responses to environmental signals.

2 Materials

2.1 Fixation

1. Absolute ethanol, analytical grade (*see* **Note 1**).
2. Glacial acetic acid, analytical grade (*see* **Note 2**).
3. Farmer's fixative: 3:1 (v/v) ethanol–acetic acid. Prepare fresh before fixation.

2.2 Dehydration and Embedding

1. RNAseZAP® (Ambion) or other cleaning agent for removing RNases (important for RNA-based downstream applications).
2. Xylene for analysis.
3. Paraplast Embedding Media (e.g., Sigma-Aldrich).
4. Glass vials (e.g., 20 mL) with screw tops.
5. Plastic weighing dishes.
6. Oven (e.g., hybridization oven).

2.3 Microtome Sectioning

1. Heating Table.
2. Light table (or any light source below a semitransparent surface).
3. Embedding cassettes (or any block that can be used to mount tissue for microtome sectioning).
4. Razorblades.
5. Microtome (e.g., RM2145 microtome, Leica).
6. RNase-free membrane slides ("MMI Molecular Machines & Industries AG, Glattbrugg, Switzerland").

2.4 Laser-Assisted Microdissection

1. Microdissection device (e.g., mmi CellCut Plus device, MMI Molecular Machines & Industries AG, Glattbrugg, Switzerland).
2. MMI collection tubes (without diffuser).
3. Microscope glass slides, pre-baked for 6 h at 180 °C.
4. For RNA-based work: PicoPure RNA extraction kit (Arcturus®).

3 Methods

3.1 Fixation

1. Prepare 40 mL of Farmer's fixative. Chill the fixative in a −20 °C freezer for at least 20 min (*see* **Note 4**).
2. Put 20 mL glass vials on ice and fill them with 5–10 mL of ice-cold Farmer's fixative per vial (*see* **Note 5**).
3. Collect target tissue using fine forceps (*see* **Note 6**) and submerge them into the fixative in the glass vials.
4. Apply a soft vacuum for 10 min and then release the vacuum slowly. Repeat the step once.
5. Change Farmer's fixative solution and keep tissue overnight on a shaker (shaking gently) at 4 °C.

Optional: after the fixation and overnight incubation, the tissue can be washed twice with 70 % ethanol prepared with DEPC-treated water, placed into ice-cold 70 % ethanol, and stored for a several days at 4 °C if necessary.

3.2 Dehydration and Embedding

Day 1:

1. Prepare aliquots of each 70, 80, 90 % (v/v) ethanol in DEPC-treated or RNase-free water, as well as pure ethanol and keep solutions at 4 °C. Keep samples on ice (or in the cold room) and on a shaker (shaking gently) for incubations.
2. Incubate tissue in 70 % ethanol for 1 h.
3. Incubate tissue in 80 % ethanol for 1 h.
4. Incubate tissue in 90 % ethanol for 1 h.
5. Incubate tissue in 100 % ethanol overnight.

Optional: use 0.1 % (w/v) Eosin Y in ethanol to stain tissues during the overnight step (advisable for small or delicate tissues).

Day 2:

1. Prepare aliquots of each 25, 50, and 75 % (v/v) xylene in ethanol. All steps can be performed at room temperature. Perform all steps involving xylene in a fume hood.
2. Incubate tissue in 100 % ethanol for 1 h.

3. Incubate tissue in 100 % ethanol for 1 h.
4. Incubate tissue in 25 % xylene in ethanol for 1 h.
5. Incubate tissue in 50 % xylene in ethanol for 1 h.
6. Incubate tissue in 75 % xylene in ethanol for 1 h.
7. Incubate tissue in 100 % xylene for 1 h.
8. Incubate tissue in 100 % xylene for 1 h.
9. Incubate in 100 % xylene with 1 Paraplast chip per 1 mL of xylene overnight.

Day 3:

1. Fill a 50 mL Falcon tube with Paraplast chips, heat to 54 °C, and let it melt (needed in **step 4**, below).
2. Add every half hour one Paraplast chip per 1 mL of xylene used, until the Paraplast chips no longer dissolve. During this step, keep samples at room temperature.
3. Put samples to 42 °C and slowly add Paraplast chips (one by one per 1 mL of xylene every half hour).
4. After 2–3 h, carefully decant half of the xylene–Paraplast mixture in the vials and replace with liquid Paraplast from point 1 at 54 °C.
5. Transfer samples into 54 °C.
6. After 2–3 h replace remaining xylene–Paraplast mixture in the vials with liquid Paraplast at 54 °C.
7. Refill the Falcon tube containing melted Paraplast with Paraplast chips.

Day 4-6:

1. Change Paraplast twice a day and keep samples at 54 °C.

Day 7:

1. Pre-warm weighing dishes (plastic "boats") at 54 °C on a heating table.
2. Pour Paraplast containing the target tissue into the dishes and then place the tissue samples in a desired position using pre-warmed tweezers or needles. Keep enough space between different samples.
3. Slowly cool down the dishes and store blocks at 4 °C until use (*see* **Note 7**).

3.3 Sectioning and Slide Preparation

1. Clean working area with RNaseZAP.
2. Dissect paraffin blocks containing the target tissue into small blocks (e.g., each containing one inflorescence) using a sharp

razor blade and mount them onto embedding cassettes using a hot spatula (*see* **Note 8**).

3. Chill the mounted blocks at 4 °C for 10–20 min (*see* **Note 9**).
4. Cut the paraffin blocks on a microtome to 6–15 μm thickness (*see* **Note 10**) after cleaning the microtome surfaces with RNaseZAP.
5. Transfer ribbons onto a black surface (e.g., a black sheet of paper in a plastic box) and examine them under a binocular microscope. If possible, remove parts of the ribbon that do not contain any tissue of interest.
6. Mount the ribbons onto membrane-coated microdissection slides (*see* **Note 11**). Mounting can be performed by "fishing" the ribbons floating in a small water bath (RNase-free water) in a pre-backed glass container (*see* **Note 12**). Alternatively, drop a few milliliters of water onto the slide surface, place the ribbons onto the water and then pour the water off over one edge (*see* **Note 13**).
7. Dry the slides overnight on a heating table at 42 °C.

3.4 Laser-Assisted Microdissection (LAM)

1. Deparaffinize slides in 100 % xylene at room temperature for 2 × 10 min before processing. Perform this step under a fume hood.
2. Dry the slides under the fume hood for at least 20 min before proceeding to LAM.
3. Turn on the microscope, laser and computer that are part of the LAM system.
4. Sandwich the tissue on the membrane-slide onto a normal glass slide (pre-baked at 180 °C for >6 h, *see* Subheading 2).
5. Open the MMI software tool that controls laser, stage motor, and cap lift. Set the laser speed, laser focus, and laser power, and calibrate the stage movement and laser position (*see* **Note 14**).
6. Use the software to define the slide edges and scan the slide at 4× magnification to obtain a roadmap of the slide.
7. Insert the isolation cap into the cap holder, open the tube, and attach the cap holder to the cap lift.
8. Jump from tissue section to tissue section (use the roadmap to locate tissue sections) and identify the cells of interest (*see* **Note 15**).
9. Once an area of interest has been identified, lower the cap lift onto the membrane slide using the cap-lift tool.
10. Use a selection tool (e.g., "hand pen"-tool) to outline the cells of interest on the computer screen and then press "cut".
11. Raise the cap lift to see whether the section of interest is stuck to the surface of the isolation cap lid (*see* **Note 16**).

12. Several sections can now be bulked on the same isolation cap. When finished, remove the isolation cap from the cap holder and close the tube.
13. Proceed to the extraction of biomolecules directly, or freeze the collection tubes until further use (e.g., at −80 °C for RNA-related work).

3.5 Extraction of RNA for Downstream Analyses and Quality Controls

1. *RNA-extraction methods and pooling*: Several methods for extracting RNA from laser-dissected material have been compared by Kerk and colleagues [12]. They found that the Arcturus® PicoPure RNA extraction kit (including a DNAse-treatment) results in highest yields, but other methods would also be suitable. We have used the PicoPure kit repeatedly and found it well suited to extract RNA from dissected material. The extraction method involves very simple procedures, and the kit manual provides easy-to-follow instructions.

   Pooling: It is often advisable to pool several tissue fragments together to achieve higher yields (*see* Fig. 2a). This can be done by (a) pooling different fragments on the same isolation cap, or (b) pooling several caps for RNA extraction. The latter can be achieved by incubating several caps independently with RNA extraction buffer, pool the resulting extracts in a single tube, adding an equivalent volume of the 70 % ethanol provided in the kit (e.g., 20 μL to 2 × 10 μL extracts), and applying the pooled extracts to a single column.
2. *Expected yields*: It is difficult to estimate the yields obtained from laser-dissected cells or tissue, since concentrations after extraction often lie outside the dynamic range of current assays for estimating RNA quantities (*see* **Note 17** for a short review of expected yields). Yields can be estimated, for example, by using the Qubit® Fluorometer (Life Technologies).
3. *RNA-quality*: Due to limited yields during LAM, the quality of the collected samples often cannot be directly assessed. Therefore, we recommend isolating large tissue fragments from non-target tissues from the processed slides to obtain an alternative quality estimate. RNA integrity can be best checked by electrophoresis, e.g., by the use of a microcapillary-based system such as the Bioanalyzer Pico RNA kit (Agilent; *see* **Note 18**).
4. *Downstream RNA-based applications*: RNA extracted from LAM samples can be used directly in combination with semi-quantitative or real-time RT-PCR, which is possible when larger tissue fragments such as floral buds or inflorescence meristems are isolated (e.g., [19, 28] and *see* Fig. 2c, d). For this, however, it is crucial to test the dynamic range of the PCR assays used. This can be done by performing serial dilution

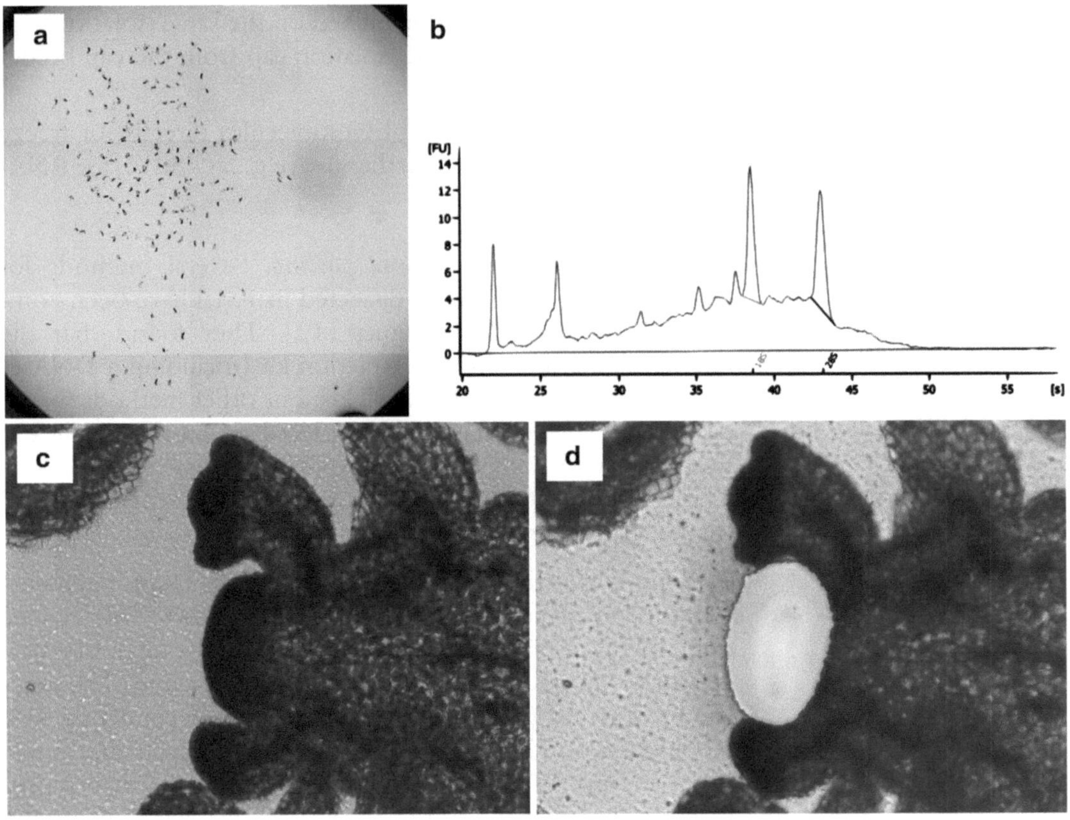

Fig. 2 Pooling tissue fragments and assessment of RNA quality for downstream application. (**a**) Pooling of several central cells on a single cap (approximately 150 cell fragments), in addition to on-column pooling of RNA extracts from 3 to 4 caps was necessary to perform a cell-specific expression profiling experiment [16]. The exact quantity of the extracted RNA could, however, not be determined from the pooled sample. Please note that pooling does not substitute for replication. (**b**) RNA electrophoresis profile (Bioanalyzer Pico Chip) of large floral structures isolated using LAM. In this example, the 28S and 18S rRNA peaks are well visible, indicative of high quality total RNA. If the RNA quality of samples cannot be assessed directly, neighboring non-target tissue can be isolated as a surrogate to estimate sample RNA quality. (**c**) and (**d**) Larger structures, such as *Arabidopsis* inflorescence meristems do not require as much pooling, and less than ten isolated fragments can be used directly in real-time RT-PCR experiments without the need of an additional RNA amplification step (Marko Sesartić and Samuel Wuest, unpublished)

curve experiments that include total RNA concentrations of less than 1 ng/μL. When working with few cells only, an amplification of the mRNA prior to the real-time RT-PCR experiment is usually necessary. For this, several protocols are available [29–31], in addition to a range of commercially available amplification kits (e.g., based on in vitro transcription amplification or PCR-based methods, *see* [21] for a review). Examples of commercially available kits are the NuGEN Ovation® Pico WTA systems (www.nugeninc.com and [32]), or mRNA amplification kits from Clontech (www.clontech.com).

The past few years have seen the rise of NGS-based transcriptional profiling, which is likely replacing hybridization-based methods such as microarray profiling in the future [6, 33]. Collecting rather large tissue fragments during LAM will enable direct use of the isolated mRNA for NGS approaches (e.g., sequencing with the Illumina HiSeq requires a little as 50–100 ng of total RNA for mRNA-Seq). Recently, the application of mRNA-Seq has also been adopted for small RNA inputs [34–37]. These methods either depend on PCR-based amplifications or in-vitro transcription RNA from cDNA templates; at the same time, new protocols are rapidly being developed and comparisons between methods are still rare (but *see* also [38] for a recent review of the methods). In addition, several kits have been developed very recently for applications such as mRNA-Seq. For example, an Illumina protocol recommends the use of the SMARTer ultralow input RNA kit (www.clontech.com, *see* also [35] for a use of the SMART technology), for which starting material ranging from 0.1 to 10 ng of total RNA can be amplified. Another recent release is the Ovation® RNA-Seq System V2 from NuGEN (www.nugeninc.com and [37]), where 0.5–100 ng of total RNA can be amplified and channeled into different existing sequencing library preparation protocols. Developments in the field are rapid, and optimized protocols and kits will soon become available for different NGS platforms, as well as different applications tailored for the different types of RNA molecules to be sequenced.

4 Notes

1. *CAUTION*: Ethanol liquid and vapor are highly flammable: keep away from heat sources and open flames and use appropriate ventilation.
2. *CAUTION*: Acetic acid causes severe skin burns and eye damage and liquid and vapor are flammable—wear protective gloves, protective clothing, and eye protection and would be best handled in a fume hood.
3. To prepare DEPC-treated water, add DEPC at a concentration of 0.1 % (v/v) to ultrapure water (e.g., 1 mL to 1 L). Stir and incubate overnight and then autoclave. Autoclaving will decompose DEPC and render it nontoxic. *CAUTION*: DEPC is toxic, and handling should be performed under a hood while wearing appropriate protection.
4. The choice of fixative generally determines (a) the histological preservation of sectioned tissues, and (b) the preservation and extractability of biomolecules from the dissected tissue [12, 39]. Farmer's fixative with subsequent embedding into Paraplast has been found to be a good choice for work with plant tissues [12, 23, 40]. Acetone- or phosphate buffer-based

fixation methods have also been applied in the plant sciences [19, 23]. While animal and biomedical researchers tend to snap-freeze and cryosection tissues for an optimal preservation of biomolecules [41], this procedure tends to yield poor histological preservation of plant tissues (especially in mature tissues containing large vacuoles) [21]. On the other hand, there is ample literature available that advises against the use of cross-linking fixatives such as paraformaldehyde (which generally yields better histological preservation)—these fixatives generally result in poor quality and low yields of biomolecules for downstream applications [12, 39].

5. The volume of fixative is dependent on the tissue type and size collected. 5–10 mL is well suited for the fixation of 10–20 *Arabidopsis* inflorescences. For larger tissues, the volume should be increased so that a large excess of fixative as compared to tissue is achieved.
6. When collecting inflorescences, it is advisable to remove as many large, undesired structures (such as maturing siliques) as possible. For collecting cells from young flowers or inflorescence meristems, remove all larger flowers on the inflorescence.
7. We have stored tissue embedded in paraffin blocks for several weeks in the fridge without considerable loss of RNA quality.
8. Mounting can be done be heating a spatula containing a small piece of Paraplast on a flame and pouring the liquid Paraplast onto the cassette. Immediately press the block onto the solidifying Paraplast. Re-heat the spatula and press it weakly against the lower side of the block to seal the remaining gaps between block and cassette.
9. Try to section samples on the microtome when the blocks are still cold. This will generally improve the tissue morphology.
10. The choice of the thickness of the microtome sections depends on the tissue of interest. It is advisable to cut thicker sections if larger tissue is harvested, since this will often lead to better integrity of the RNA. When dissecting single cells, sections should be thinner (e.g., 6–8 μm) to reduce contamination with non-target tissue and to render cytological features more prominent.
11. Mounting of the ribbons has to occur on the membrane-coated side of the MMI microdissection slide, so that the tissue can later be sandwiched between a pre-backed glass slide and the membrane.
12. All glassware used for RNA-based experiments should be baked before use at 180 °C for at least 6 h.
13. It has been suggested that RNA degradation can sometimes occur even when using RNAse-free water for mounting the

ribbons onto membrane-coated LAM slides [42]. Furthermore, when LAM is applied to very rare cell types for which only few cells can be harvested in a realistic time frame, high RNA quality of the target tissue is of central importance. A previously published method includes a tape-based paraffin section mounting system, which improved RNA quality considerably [42]. In our lab, the use of methanol instead of RNAse-free water for mounting the paraffin sections onto the slides yielded a more consistent quality of RNA (*see* also [17] and Anja Schmidt, personal communication; *CAUTION:* Methanol is toxic. Perform the mounting under a fume hood and wear lab coat, protective gloves, and safety glasses).

14. Laser and stage calibration have to be performed for each objective separately as described in the technical manual. Generally, these calibrations need only be performed once per computer user. For each of the objectives used during the actual laser dissection, the laser speed, focus and energy levels have to be adjusted separately. The laser focus is generally the most important parameter for optimal results. We recommend against the use of high power-settings, since scattered energy might result in damage and lower yields of biomolecules.

15. LAM is normally applied to fixed and dehydrated tissue sections. In these, the histological preservation might be different from normally mounted and rehydrated tissue sections. It is therefore advisable to assess in advance whether and how the cells of interest can be identified in dehydrated sections. Rehydration of the sections can be considered for some laser devices [40].

16. Sometimes, the dissected area of interest does not adhere to the isolation cap, or gets folded out of the focal plane when the cap lift is raised. This can have several reasons, for example: (a) the laser did not completely cut all the tissue. Try to optimize the laser settings, firstly by adjusting the laser focus, lowering the speed and, if necessary, increasing the power. Alternatively, set the cutting repeats to 2×. (b) The section of interest is too close to the border of the slide, so that lowering the isolation cap onto the membrane does not work anymore. The cap might be sitting on the metal frame of the slide. Lowering the cap onto the slide membrane should result in a change of the focal plane position of the tissue.

17. Yields also strongly depend on the organism, the cell type isolated, the fixation method used, RNA preservation during the isolation process, the thickness of microtome sections, and the size of laser-dissected fragments. For example, differentiated cells such as from the *Arabidopsis* hypocotyl are generally larger but contain large vacuoles and yield smaller RNA quantities

per area unit. On the other hand, meristematic tissues contain a large number of small, transcriptionally active cells with small vacuoles, and yield larger RNA quantities ([29] and personal observations). Thus, yield estimates for different organisms and cell types range widely between approximately 0.5–10 ng per 100 cells isolated [21]. Published reports of yields reflect this diversity. On one side, Kerk and colleagues obtained an average of approximately 12 ng and 17 ng per 100 isolated cells from maize bundle sheaths or radish hypocotyls, respectively. On the other hand side, Day and colleagues estimated RNA yields from *Arabidopsis* endosperm, seed coat, and embryo tissue, and obtained estimates of around 0.2–0.6 ng of RNA per isolated fragments (equivalent to sections of ~20–60 cells). They could isolate in the range of 20–60 ng of total RNA by pooling 100 isolated fragments. Cai and Lashbrook estimated an average yield of 1.0–1.5 ng of total RNA from 100 *Arabidopsis* petal or stamen abscission zone cells [42]. In our work, we have typically obtained yields similar to the latter two reports (e.g., when isolating *Arabidopsis* inflorescence meristems, data not shown). On the other hand, we were not able to accurately determine the yields isolated from pooled single gametophyte cells before two rounds of linear RNA amplification, but a microarray-based characterization of cellular transcriptomes required between 300 and 800 single cell fragments [16].

18. RNA integrity can be estimated by the RNA size distributions and the presence and size of ribosomal RNA peaks in total RNA extracts (*see* Fig. 2b). Embedded and laser-dissected plant material will generally result in some reduction of the RNA quality, but ribosomal peaks should still be visible in total RNA extracts. Even though RNA degradation could generally happen at multiple steps during the LAM protocol, it has been suggested that mounting the microtome-dissected ribbons onto the membrane-coated slides is a critical step during the process (*see* **Note 13**).

Acknowledgements

The authors would like to acknowledge Chloé Wuest and Anja Schmidt for reading of the manuscript, and Marc Schmid for providing Fig. 2b. S.E.W. is supported by a Marie Curie Intra European Fellowship within the 7th European Community Framework Programme, and work on cell-specific profiling methods in U.G.'s laboratory is supported by the University of Zürich and the Swiss National Science Foundation.

References

1. Rogers ED, Jackson T, Moussaieff A, Aharoni A, Benfey PN (2012) Cell type-specific transcriptional profiling: implications for metabolite profiling. Plant J 70(1):5–17
2. Taylor-Teeples M, Ron M, Brady SM (2011) Novel biological insights revealed from cell type-specific expression profiling. Curr Opin Plant Biol 14(5):601–607
3. Birnbaum K, Shasha DE, Wang JY, Jung JW, Lambert GM, Galbraith DW, Benfey PN (2003) A gene expression map of the Arabidopsis root. Science 302(5652):1956–1960
4. Henry GL, Davis FP, Picard S, Eddy SR (2012) Cell type-specific genomics of Drosophila neurons. Nucleic Acids Res 40(19):9691–9704
5. Deal RB, Henikoff S (2010) A simple method for gene expression and chromatin profiling of individual cell types within a tissue. Dev Cell 18(6):1030–1040
6. Schmid MW, Schmidt A, Klostermeier UC, Barann M, Rosenstiel P, Grossniklaus U (2012) A powerful method for transcriptional profiling of specific cell types in eukaryotes: laser-assisted microdissection and RNA sequencing. PLoS One 7(1):e29685
7. Emmert-Buck MR, Bonner RF, Smith PD, Chuaqui RF, Zhuang Z, Goldstein SR, Weiss RA, Liotta LA (1996) Laser capture microdissection. Science 274(5289):998–1001
8. Brooks L 3rd, Strable J, Zhang X, Ohtsu K, Zhou R, Sarkar A, Hargreaves S, Elshire RJ, Eudy D, Pawlowska T, Ware D, Janick-Buckner D, Buckner B, Timmermans MC, Schnable PS, Nettleton D, Scanlon MJ (2009) Microdissection of shoot meristem functional domains. PLoS Genet 5(5):e1000476
9. Ohtsu K, Smith MB, Emrich SJ, Borsuk LA, Zhou R, Chen T, Zhang X, Timmermans MC, Beck J, Buckner B, Janick-Buckner D, Nettleton D, Scanlon MJ, Schnable PS (2007) Global gene expression analysis of the shoot apical meristem of maize (Zea mays L.). Plant J 52(3):391–404
10. Spencer MW, Casson SA, Lindsey K (2007) Transcriptional profiling of the Arabidopsis embryo. Plant Physiol 143(2):924–940
11. Cai S, Lashbrook CC (2008) Stamen abscission zone transcriptome profiling reveals new candidates for abscission control: enhanced retention of floral organs in transgenic plants overexpressing Arabidopsis ZINC FINGER PROTEIN2. Plant Physiol 146(3):1305–1321
12. Kerk NM, Ceserani T, Tausta SL, Sussex IM, Nelson TM (2003) Laser capture microdissection of cells from plant tissues. Plant Physiol 132(1):27–35
13. Le BH, Cheng C, Bui AQ, Wagmaister JA, Henry KF, Pelletier J, Kwong L, Belmonte M, Kirkbride R, Horvath S, Drews GN, Fischer RL, Okamuro JK, Harada JJ, Goldberg RB (2010) Global analysis of gene activity during Arabidopsis seed development and identification of seed-specific transcription factors. Proc Natl Acad Sci USA 107(18):8063–8070
14. Walia H, Josefsson C, Dilkes B, Kirkbride R, Harada J, Comai L (2009) Dosage-dependent deregulation of an AGAMOUS-LIKE gene cluster contributes to interspecific incompatibility. Curr Biol 19(13):1128–1132
15. Deeken R, Ache P, Kajahn I, Klinkenberg J, Bringmann G, Hedrich R (2008) Identification of Arabidopsis thaliana phloem RNAs provides a search criterion for phloem-based transcripts hidden in complex datasets of microarray experiments. Plant J 55(5):746–759
16. Wuest SE, Vijverberg K, Schmidt A, Weiss M, Gheyselinck J, Lohr M, Wellmer F, Rahnenfuhrer J, von Mering C, Grossniklaus U (2010) Arabidopsis female gametophyte gene expression map reveals similarities between plant and animal gametes. Curr Biol 20(6):506–512
17. Schmidt A, Wuest SE, Vijverberg K, Baroux C, Kleen D, Grossniklaus U (2011) Transcriptome analysis of the Arabidopsis megaspore mother cell uncovers the importance of RNA helicases for plant germline development. PLoS Biol 9(9):e1001155
18. Casson S, Spencer M, Walker K, Lindsey K (2005) Laser capture microdissection for the analysis of gene expression during embryogenesis of Arabidopsis. Plant J 42(1):111–123
19. Inada N, Wildermuth MC (2005) Novel tissue preparation method and cell-specific marker for laser microdissection of Arabidopsis mature leaf. Planta 221(1):9–16
20. Chandran D, Inada N, Hather G, Kleindt CK, Wildermuth MC (2010) Laser microdissection of Arabidopsis cells at the powdery mildew infection site reveals site-specific processes and regulators. Proc Natl Acad Sci USA 107(1):460–465
21. Day RC, Grossniklaus U, Macknight RC (2005) Be more specific! Laser-assisted microdissection of plant cells. Trends Plant Sci 10(8):397–406
22. Espina V, Wulfkuhle JD, Calvert VS, VanMeter A, Zhou W, Coukos G, Geho DH, Petricoin EF 3rd, Liotta LA (2006) Laser-capture microdissection. Nat Protoc 1(2):586–603
23. Day RC (2010) Laser microdissection of paraffin-embedded plant tissues for transcript profiling. Methods Mol Biol 655:321–346

24. Chandran D, Inada N, Wildermuth MC (2011) Laser microdissection of plant-fungus interaction sites and isolation of RNA for downstream expression profiling. Methods Mol Biol 712:241–262
25. Nowack MK, Shirzadi R, Dissmeyer N, Dolf A, Endl E, Grini PE, Schnittger A (2007) Bypassing genomic imprinting allows seed development. Nature 447(7142):312–315
26. You W, Tyczewska A, Spencer M, Daxinger L, Schmid MW, Grossniklaus U, Simon SA, Meyers BC, Matzke AJ, Matzke M (2012) Atypical DNA methylation of genes encoding cysteine-rich peptides in Arabidopsis thaliana. BMC Plant Biol 12:51
27. Wöhrmann HJ, Gagliardini V, Raissig MT, Wehrle W, Arand J, Schmidt A, Tierling S, Page DR, Schob H, Walter J, Grossniklaus U (2012) Identification of a DNA methylation-independent imprinting control region at the Arabidopsis MEDEA locus. Genes Dev 26(16):1837–1850
28. Liu X, Kim YJ, Muller R, Yumul RE, Liu C, Pan Y, Cao X, Goodrich J, Chen X (2011) AGAMOUS terminates floral stem cell maintenance in Arabidopsis by directly repressing WUSCHEL through recruitment of Polycomb Group proteins. Plant Cell 23(10):3654–3670
29. Scanlon MJ, Ohtsu K, Timmermans MC, Schnable PS (2009) Laser microdissection-mediated isolation and in vitro transcriptional amplification of plant RNA. Curr Protoc Mol Biol Chapter 25:Unit 25A 23
30. Van Gelder RN, von Zastrow ME, Yool A, Dement WC, Barchas JD, Eberwine JH (1990) Amplified RNA synthesized from limited quantities of heterogeneous cDNA. Proc Natl Acad Sci USA 87(5):1663–1667
31. Wang E (2005) RNA amplification for successful gene profiling analysis. J Transl Med 3:28
32. Vermeulen J, Derveaux S, Lefever S, De Smet E, De Preter K, Yigit N, De Paepe A, Pattyn F, Speleman F, Vandesompele J (2009) RNA pre-amplification enables large-scale RT-qPCR gene-expression studies on limiting sample amounts. BMC Res Notes 2:235
33. Wang Z, Gerstein M, Snyder M (2009) RNA-Seq: a revolutionary tool for transcriptomics. Nat Rev Genet 10(1):57–63
34. Tang F, Barbacioru C, Wang Y, Nordman E, Lee C, Xu N, Wang X, Bodeau J, Tuch BB, Siddiqui A, Lao K, Surani MA (2009) mRNA-Seq whole-transcriptome analysis of a single cell. Nat Methods 6(5):377–382
35. Ramskold D, Luo S, Wang YC, Li R, Deng Q, Faridani OR, Daniels GA, Khrebtukova I, Loring JF, Laurent LC, Schroth GP, Sandberg R (2012) Full-length mRNA-Seq from single-cell levels of RNA and individual circulating tumor cells. Nat Biotechnol 30(8):777–782
36. Hashimshony T, Wagner F, Sher N, Yanai I (2012) CEL-Seq: single-cell RNA-Seq by multiplexed linear amplification. Cell Rep 2(3):666–673
37. Tariq MA, Kim HJ, Jejelowo O, Pourmand N (2011) Whole-transcriptome RNAseq analysis from minute amount of total RNA. Nucleic Acids Res 39(18):e120
38. Hebenstreit D (2012) Methods, challenges and potentials of single cell RNA-seq. Biology 1(3):658–667
39. Gillespie JW, Best CJ, Bichsel VE, Cole KA, Greenhut SF, Hewitt SM, Ahram M, Gathright YB, Merino MJ, Strausberg RL, Epstein JI, Hamilton SR, Gannot G, Baibakova GV, Calvert VS, Flaig MJ, Chuaqui RF, Herring JC, Pfeifer J, Petricoin EF, Linehan WM, Duray PH, Bova GS, Emmert-Buck MR (2002) Evaluation of non-formalin tissue fixation for molecular profiling studies. Am J Pathol 160(2):449–457
40. Day RC, McNoe LA, Macknight RC (2007) Transcript analysis of laser microdissected plant cells. Physiol Plantarum 129(2):267–282
41. Goldsworthy SM, Stockton PS, Trempus CS, Foley JF, Maronpot RR (1999) Effects of fixation on RNA extraction and amplification from laser capture microdissected tissue. Mol Carcinog 25(2):86–91
42. Cai S, Lashbrook CC (2006) Laser capture microdissection of plant cells from tape-transferred paraffin sections promotes recovery of structurally intact RNA for global gene profiling. Plant J 48(4):628–637

Part V

Molecular Biology, Genomics, and Systems Biology

Chapter 20

Identification of Arabidopsis Knockout Lines for Genes of Interest

José Tomás Matus, Thilia Ferrier, and José Luis Riechmann

Abstract

Determining gene function through reverse genetics has been an important experimental approach in the field of flower development. The method largely relies on the availability of knockout lines for the gene of interest. Insertional mutagenesis can be performed using either T-DNA or transposable elements, but the former has been more frequently employed in Arabidopsis. A primary concern for working with insertional mutant lines is whether the respective insertion results in a complete or rather a partial loss of gene function. The effect of the insertion largely depends on its position with respect to the structure of the gene. In order to quickly identify and obtain knockout lines for genes of interest in Arabidopsis, more than 325,000 mapped insertion lines have been catalogued on indexed libraries and made publicly available to researchers. Online accessible databases provide information regarding the site of insertion, whether a mutant line is available in a homozygous or hemizygous state, and outline technical aspects for plant identification, such as primer design tools used for genotyping. In this chapter, we describe the procedure for isolating knockout lines for genes of interest in Arabidopsis.

Key words *Arabidopsis thaliana*, T-DNA insertion, Mutant, Seed collection, Genotyping, Transposon

1 Introduction

The basic framework for flower development studies in Arabidopsis was established more than 20 years ago, through analyses of floral homeotic mutants that led to the formulation of the ABC genetic model of floral organ identity determination (*see* Chapter 1). The respective organ identity genes were identified and found to belong to large transcription factor gene families, and from early on it was apparent that other members of those gene families (initially, and in particular, of the MADS-box gene family) were also preferentially expressed in flowers [1]. This finding suggested the participation of additional gene family members in the process of flower development, and raised the possibility of functional redundancy among related genes. Since then, reverse genetics studies, in which genes with specific floral expression patterns and/or with high

José Luis Riechmann and Frank Wellmer (eds.), *Flower Development: Methods and Protocols*, Methods in Molecular Biology, vol. 1110, DOI 10.1007/978-1-4614-9408-9_20,

sequence similarity to know floral regulators are targeted, have been crucial to, for instance, refine and expand the ABC model through the addition to it of the *SEPALLATA* genes [2, 3], or to characterize carpel developmental pathways [e.g., 4, 5], among many other findings.

Among the different methods that can be used to disrupt a gene, the insertion of foreign DNA has proven to be a widely effective strategy in plant reverse genetics [6]. In Arabidopsis, insertional mutagenesis has been performed using both T-DNA [7] and transposable elements [8]. While T-DNA insertions are stable through multiple generations, transposons may translocate in the genome after integration. However, insertional mutagenesis using controlled transposable elements contained within the T-DNA sequence has also been described [9].

More than 325,000 mapped insertion lines are now publicly available for Arabidopsis researchers on indexed libraries (*see* Tables 1 and 2) [10]. A variety of vectors have been used to produce these collections, which may therefore differ in the positive selection marker(s) that the lines carry, or in the T-DNA border sequences, which are used to generate flanking sequence tags (FST). In addition, the laboratories that generated the collections may have used different strategies for the mapping and indexing of the insertional lines. This information is relevant for subsequent confirmatory sequencing, or for identifying plants carrying multiple mutations, obtained from crosses between lines from the same or from different collections. Activating-tagging elements, such as ubiquitous or constitutive promoters (e.g., the cauliflower mosaic virus 35S promoter, or 35SCaMV), have in some cases been included in the T-DNA, potentially resulting in gene over-expression when the insertion occurs in the respective upstream region [11]. The vast majority of insertional lines can be searched for on integrated Web sites such as the Salk Institute Genomic Analysis Laboratory (SIGnAL) T-DNA Express, The Arabidopsis Information Resource (TAIR), GABI-KAT, FLAG or directly from the institutes and laboratories that generated them.

For an insertion to have the possibility of disrupting a gene, it is generally assumed that the insertion point should be located between 500 nucleotides upstream of the annotated transcriptional start site and the stop codon. Using these criteria, potential insertional mutants can be obtained for most, but not all, of Arabidopsis genes: they have yet to be identified for nearly 12 % of them (considering TAIR9 annotation), despite the high number of T-DNA and transposon mutant lines that are available [10]. This shortcoming could result from biases in T-DNA insertion, which is reported to favor intergenic regions [12–14]. Moreover, an additional 8 % of genes are represented in these collections only in a hemizygous state (in which just one of the two copies of the gene is disrupted) [10], which could be a consequence of homozygous

Table 1
Available T-DNA collections and their features (adapted from [10]). A full list of reverse genetics projects and seed resources may be found at (http://www.arabidopsis.org/portals/mutants/worldwide.jsp)

Collection	Accession	Selection marker	Additional features in T-DNA	Seed stock center	Reference
Salk	Col-0	Kanamycin		ABRC NASC	[12]
SAIL	Col-0, Col-3	BASTA®	pBluescript SKII+, pLat52:GUSB[a]	ABRC NASC	[14]
GABI-KAT	Col-0	Sulfadiazine	35SCaMV SUL, 35SCaMV adjacent to RB (activation tagging), UBI4-2[b]	ABRC NASC	[13]
WISC-DS Lox insertion	Col-0	BASTA®	35SCaMV, Loxp site, Dissociation element (Ds)	ABRC NASC	[9]
FLAG—promoter trap	WS	Kanamycin BASTA®	Promoterless GUS fused to the right border	INRA[c]	[21]
Saskatoon (SK)	Col-4	BASTA®	35SCamV, pBstKS+	ABRC	[22]

[1]SAIL lines have been transformed with either one of two different binary vector systems. Lat52 promoter:GUSB is only found in SAIL lines transformed with the pCSA110 plasmid
[2]Four different plasmids have been used: pAC106, pAC161, pGABI1, and pADIS1, which differ in the use of 35SCaMV or of UBI4-2 promoters
[3]L'institut National de la Recherche Agronomique

Table 2
Available transposon collections and their features

Collection	System	Seed stock center	Reference
CSHL	Ds transposition Gene trap and enhancer trap	ABRC NASC	[23]
RIKEN	Ds transposition	RIKEN bioresource center	[24]
IMA	Ds transposition	ABRC NASC	[8]
SLAT JIC-SM	Maize enhancer/suppressor mutator element	NASC	[25]

mutations causing lethality or diminished fertility. When selecting insertional lines for phenotypic analyses, a primary concern is whether the respective insertions represent knockout mutations that completely disrupt the activity of the gene (for instance, by abolishing its expression, or by interrupting the coding sequence and precluding the synthesis of a functional protein) or rather result in a partial loss of gene function, in alterations in gene expression, or even in the production of aberrant truncated proteins whose synthesis may have phenotypic effects [15, 16]. The probability of an insertion to result in a knockout (or knockdown) alteration is related to the position of the insertion with respect to the structure of the gene. In general, the more distant the insertion is within the upstream region, the less likely it is that the gene is disrupted [6].

Seed stocks from the relevant lines can be obtained through the Arabidopsis Biological Resource Center (ABRC) or the Nottingham Arabidopsis Stock Center (NASC). Many of the seed batches distributed by these public stock centers correspond to T2 or T3 generation plants, and may contain both hemizygous and homozygous seeds. Thus, it is usually necessary to identify individual plants that are homozygous for the insertion.

The presence of a T-DNA insertion in a mutant line can be tested by analyzing selection marker resistance, and by PCR. However, only the later strategy provides confirmation of the presence of the insertion in the gene of interest. Multiple individuals can be genotyped by PCR to identify homozygous plants. Genomic DNA is extracted from a single leaf and PCR amplified. The PCR is performed using a gene-specific primer in combination with either a T-DNA-specific primer or a second gene-specific primer (*see* Fig. 1). These two combinations of primers may be used together in a single multiplexed PCR, or in two independent amplification reactions (PCR1 and PCR2 in Fig. 1). If the three primers are combined in a single reaction, artifact bands may result if the PCR conditions are suboptimal. With the first primer combination, a PCR product will only be formed if the T-DNA has been integrated in the gene of interest. On the other hand, the combination of the two gene-specific primers will only result in a PCR product if the T-DNA is absent, because in the mutagenized allele the primer annealing sites would be too far apart (separated by the insertion) (Fig. 1). In lines in which the insertion resulted in a lethal mutation, these analyses would lead to the identification of multiple individual hemizygous plants, but no homozygous plants would be detected.

To assess if a given insertion results in the disruption of the gene of interest, i.e., to determine if transcripts are produced, homozygous mutant plants can be analyzed by reverse transcription-polymerase chain reaction (RT-PCR) or by RNA-gel (Northern) blot hybridization. In addition, if an antibody against the target protein is available, the presence of protein product(s) (full-length or truncated) can also be tested. If phenotypic differences are detected between homozygous mutant plants and the corresponding

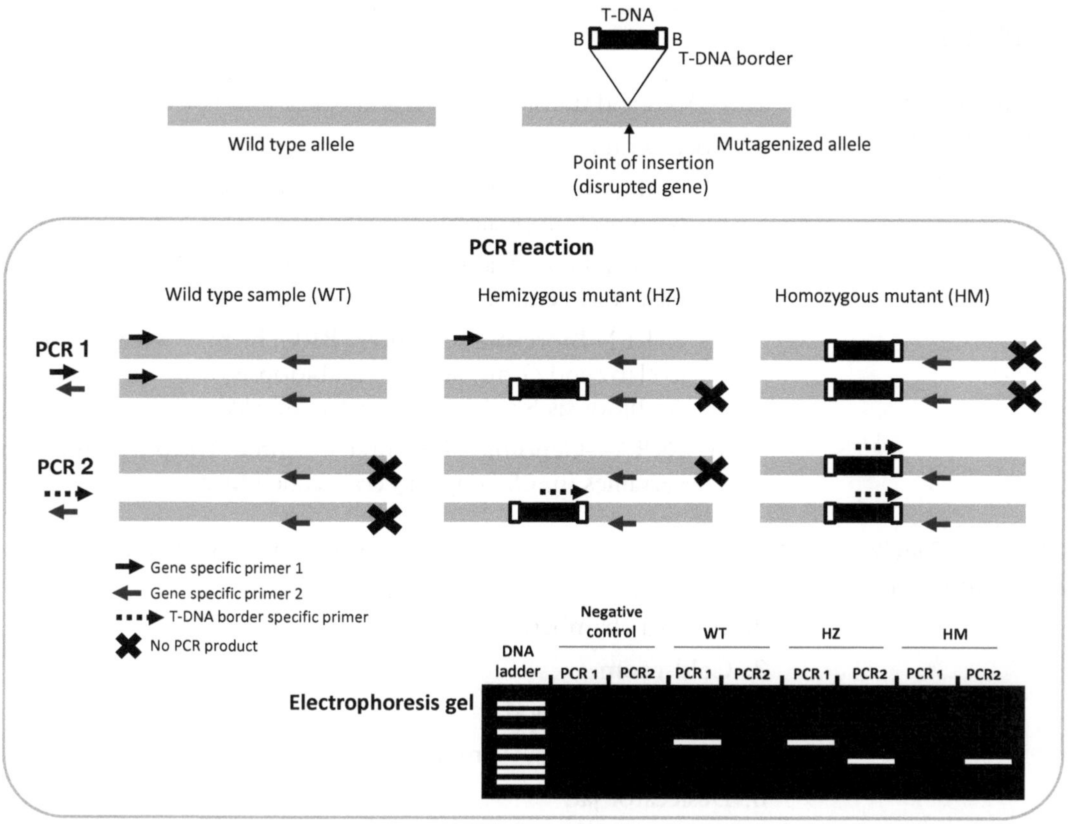

Fig. 1 Genotyping of T-DNA insertional mutant plants. T-DNA may integrate into the plant genome in either direction. For that reason, the T-DNA right and left borders are generically referred to in the diagram as "B". To facilitate analysis of results, primers are generally designed such that the PCR product derived from the gene-specific and the T-DNA-specific primer combination is smaller than that produced by the two gene-specific primers

control plants, it is important to obtain additional evidence to causally link the observed phenotype to the disrupted gene (*see* **Note 1**). To this end, additional independent insertional lines should be characterized, if available. Alternatively, the mutant allele(s) could be complemented by introducing a wild-type copy of the gene as a transgene.

Knockout mutations may not result in a discernable phenotypic alteration, for instance because the mutated gene is functionally redundant with another (often closely related) gene [4, 17]. A subsequent step would be to generate plants in which homologous gene(s) are also disrupted, through crosses with the corresponding insertional lines. The isolation of individuals that carry multiple insertions is facilitated if the original lines bear different selection markers.

2 Materials

2.1 Resources for Mutant Line Identification

1. T-DNA and transposon insertion collections (*see* Tables 1 and 2).
2. Stock centers:
   (a) Arabidopsis Biological Resource Center (ABRC, http://www.arabidopsis.org/abrc/).
   (b) Nottingham Arabidopsis Stock Center (NASC, http://nasc.nott.ac.uk).
   (c) RIKEN Bioresource Center (BRC, http://www.brc.riken.jp/lab/epd/Eng/species/arabidopsis)/ SENDAI Arabidopsis Seed Stock Center (SASSC).
   (d) INRA—Genomic Resource Center (http://www.ijpb.versailles.inra.fr/en/cra/cra_accueil.htm).

2.2 Plant Growth and In Vitro Culture

1. Seeds from *Arabidopsis* insertional lines and from the corresponding wild-type accession (*see* Table 1).
2. Growth chamber.
3. Cold-room.
4. Laminar flow hood.
5. Petri dishes (e.g., 9 cm diameter).
6. Desiccator jar.
7. Micropore 3M film.
8. Commercial bleach.
9. 37 % HCl.
10. Growth media: 0.5× Murashige and Skoog (MS) basal salt mixture including vitamins, 0.25 g/L MES (4-morpholine ethanesulfonic acid), pH 5.8, 0.8 % plant agar. Optional: depending on the seed collection, filtered herbicide or antibiotic (BASTA®-glufosinate-, Kanamycin or Sulfadiazine) may be added (*see* **Note 2**).
11. Soil mixture (e.g., 3:1:1 soil, perlite, vermiculite).
12. Plastic pots.
13. Plastic trays.
14. A pair of pointed-end forceps.
15. Labeling tape or printable labels.
16. Transparent plastic wrap.

2.3 Genotyping of Insertional Lines by PCR

2.3.1 Preparation of Genomic DNA

1. TissueLyser (QIAGEN).
2. Glass beads (4 mm diameter).
3. Microcentrifuge tubes (1.5 and 2 mL).
4. Tweezers (e.g., Dupont size #5).
5. Liquid nitrogen container.

6. Vortex.
7. Microcentrifuge.
8. Sterile deionized water.
9. Extraction buffer: 200 mM Tris–HCl, pH 7.5, 250 mM NaCl, 25 mM EDTA, 0.5 % SDS (w/v) (*see* **Note 3**).
10. 70 % ethanol.
11. Isopropanol.
12. 1–1.5 % (w/v) agarose gel in TAE buffer (40 mM Tris, 20 mM acetic acid, 1 mM EDTA, made from a 50× stock solution) containing ethidium bromide (5 μg/mL).

2.3.2 PCR Analysis

1. PCR thermocycler.
2. PCR tubes (0.2 mL).
3. GoTaq® Master mix (Promega) (*see* **Note 4**).
4. Gene-specific right and left side primers (*see* **Notes 5** and **6**).
5. T-DNA- or transposon-specific primers (*see* **Note** 7).
6. Sterile water.
7. Plant genomic DNA (freshly prepared).
8. 1–1.5 % (w/v) agarose gel in TAE buffer.
9. Electrophoresis apparatus and power supply.
10. Standard Loading buffer (30 % glycerol, 0.25 % bromophenol blue).
11. DNA ladder (1 kb).
12. Ultraviolet transilluminator.

2.4 Characterization of Insertion Lines by RT-PCR

2.4.1 Preparation of Total RNA

1. RNase-free microcentrifuge tubes (1.5 mL).
2. Plastic pellet pestles for 1.5 mL microcentrifuge tubes (optional: a mixer motor or an electric drill).
3. Forceps (e.g., Dupont size #5).
4. Liquid nitrogen.
5. Vortex.
6. Microcentrifuge.
7. Spectrum Plant Total RNA Kit (SIGMA) or an equivalent total RNA isolation kit or reagents (*see* **Note 8**).
8. Spectrophotometer (such as NanoDrop).
9. Loading dye.
10. 2 % (w/v) agarose gel in TAE buffer.

2.4.2 RT-PCR

1. RNase-free PCR tubes (0.5 mL).
2. Transcriptor First Strand cDNA Synthesis Kit (Roche) (*see* **Note 9**).

3. Thermocycler.
4. PCR tubes (0.2 mL).
5. GoTaq® Master mix (Promega).
6. Gene-specific primers (*see* **Note 10**).
7. Sterile water.
8. 1.2 % (w/v) agarose gel in TAE buffer.
9. Gel apparatus and power supply.
10. Loading dye.
11. DNA ladder (1 kb).
12. Ultraviolet transilluminator.

3 Methods

3.1 Seed Sterilization and MS Plate Preparation

1. Place approximately 50 of the seeds received from the stock center and up to 50 seeds of the corresponding wild-type control in 2 mL tubes. Label the tubes with a chlorine-resistant pen.
2. Prepare a sterilizing solution by placing 100 mL of commercial bleach in a beaker and carefully adding 3 mL of 37 % concentrated HCl (*see* **Note 11**).
3. Open the tubes containing the seeds and put them in the desiccator together with the beaker containing the sterilizing solution. Seal the desiccator jar, apply vacuum and allow sterilization to proceed for a period of 4 h.
4. Leave each tube open for 1 h under a sterile laminar flow hood.
5. Distribute seeds on MS growth media plates. The use of antibiotics is optional (*see* **Note 2**). You may saw wild-type seeds on MS plates with and without antibiotic as selection and growth controls.
6. Seal the plates with micropore 3M film and store in the dark at 4 °C for 3 or 4 days to synchronize germination.
7. Transfer plates to a growth chamber with the following long day conditions: photoperiod 16 h day (100 μmol m^{-2} s^{-1})/ 8 h night; temperature 22–24 °C day/20 °C night.
8. After 10–14 days (depending on the selection reagent used), a clear difference between resistant (with T-DNA insertion) and sensitive plants should be visible (*see* **Note 12**).

3.2 Transplantation of Seedlings from Media Plates to Soil

1. Fill plastic pots with well-watered soil mixture and place them on a plastic tray. Transfer at least ten plants from each line.
2. Label each pot (with labeling tape or printable labels) according to gene name, insertion line, plant identification number and date of transfer to growth chamber.

3. Use tweezers to remove plants from the MS plates by holding the hypocotyl very gently with the tweezer tips. Place the root into a small hole in the soil and cover it with soil.
4. Cover the tray with clear plastic wrap.
5. After 3 or 4 days, pierce the wrap with scissors or blade to decrease humidity inside the tray, and after 1 week or less remove it completely.

3.3 Confirmation of the Insertion in the Gene of Interest and Identification of Homozygous Lines

3.3.1 Isolation of Leaf Genomic DNA

1. Prepare as many 1.5 mL microcentrifuge tubes as individuals to analyze (including wild type), with two glass beads in each tube.
2. Remove one or two small rosette leaves from each plant and place them in the corresponding tube (*see* **Note 13**).
3. Put the tubes in liquid nitrogen to freeze the material (*see* **Note 14**). Once the tissue is frozen, place the tubes in the TissueLyser adaptor set and grind the leaves to a fine powder (*see* **Note 15**).
4. Add 400 μL of lysis buffer and mix the solution thoroughly and quickly by vortexing 30 s (*see* **Note 16**).
5. Centrifuge the extracts for 10 min at 14,000 rpm (20,000 × *g*) in a microcentrifuge.
6. Remove 300 μL of the supernatant and transfer it to a fresh 1.5 mL microcentrifuge tube.
7. Add 1 Vol of isopropanol and mix by inverting the tubes twice.
8. Incubate the mixture at room temperature for 2 min to precipitate nucleic acids.
9. Pellet the DNA by centrifuging 10 min at 14,000 rpm (20,000 × *g*) in a microcentrifuge.
10. Carefully remove the supernatant using a pipette, without disturbing the pellet at the bottom of the tube (*see* **Note 17**).
11. Wash the pellet with 500 μL of cold (−20 °C) 70 % ethanol. Try to detach the pellet from the bottom of the tube by shaking or pipetting.
12. Centrifuge for 5 min at 14,000 rpm (20,000 × *g*) in a microcentrifuge. Carefully remove the supernatant using a pipette, without disturbing the pellet at the bottom of the tube.
13. Dry the DNA pellet to remove any of the remaining liquid and resuspend in 50 μL of sterile deionized water (*see* **Note 18**).
14. Store the DNA solution at −20 °C until used (*see* **Note 19**).
15. Mix 2–5 μL of the DNA sample with the corresponding amount of loading buffer and run samples on a 1 % agarose gel, to analyze the integrity of the isolated DNA (*see* **Note 20**).

3.3.2 PCR for Genotyping

1. Two master mix solutions should be prepared, one for each of the primer pair combinations to be used (the volume to prepare depends on the number of plants to genotype). To genotype 10 plants, prepare mix for 12 reactions: 10 from the plants to genotype, 1 for the wild-type DNA control, and 1 for the negative (no template) control. The table below indicates reagent volumes for one reaction.

	Mix for PCR 1 (μL)	Mix for PCR 2 (μL)
Sterile water	8.5	8.5
GoTaq® Green Master Mix, 2×	12.5	12.5
10 μM LB T-DNA specific primer	–	1
10 μM gene-specific right side primer	1	1
10 μM gene-specific left side primer	1	–
Total volume	23	23

2. Mix the contents by vortexing for 5 s and spin the tube in a microcentrifuge for 10 s. Put the tubes on ice.
3. Pipet 23 μL of mix 1 into each of 12 PCR tubes.
4. Add 2 μL of the corresponding genomic DNA to the wild-type and sample tubes (or 2 μL of water to the negative—no template—control tube). Vortex for 5 s to mix the content.
5. Spin the PCR tubes in the microcentrifuge for 5 s and put the PCR tubes back on ice.
6. Repeat **steps 3–5** for Mix 2.
7. Perform PCR using the following profile:

1 cycle	Hot start 95 °C, 5 min
30 cycles	Denaturation 95 °C, 30 s Annealing 50–60 °C, 30 s Extension 72 °C, 1.5 min
1 cycle	Final extension 72 °C, 5 min

8. Prepare a 1–1.5 % agarose gel.
9. Load 8–10 μL of each PCR and the DNA ladder solution. Use the sample loading order indicated in Fig 1.
10. Run the gel at 100 V for 40–60 min.
11. Take a picture of the gel and determine the genotype (*see* Fig. 1).

3.4 Characterization of Insertion Lines by RT-PCR

3.4.1 RNA Extraction

1. Grow at least three seedlings, or one plant, of the wild type and of the selected homozygous mutant line (*see* **Note 21**).
2. Harvest around 100 mg of plant material in 1.5 mL RNase-free microcentrifuge tubes and freeze in liquid nitrogen.
3. Grind the tissue to a fine powder with the pellet pestles (and mixer motor) (*see* **Note 22**).
4. Follow the manufacturer's instructions for the RNA extraction kit.
5. Analyze the integrity of the isolated RNA by gel electrophoresis (2 % agarose gel) or in a Bioanalyzer, and determine the concentration by absorption at 260 nm with a NanoDrop spectrophotometer (*see* **Note 23**).

3.4.2 RT-PCR

1. Add 0.5–1 μg of total RNA into a 0.5 mL RNase-free PCR tube and follow the manufacturer's instructions of the reverse transcription-polymerase chain reaction kit (*see* **Note 24**).
2. Perform a PCR with 1–2 μL of first-strand cDNA (25–100 ng).
3. Run the PCR reaction and separate PCR products on a 2 % agarose gel.
4. Compare the amplification products belonging to the wild-type and homozygous mutagenized individual samples (*see* **Note 25**).

4 Notes

1. Both single and multiple independent insertions may occur on a transformation event. For the SALK, SAIL, and GABI-KAT seed collections, an average of 1.5 inserts per line has been reported [12–14]. However, the methods that are used for detecting insertions and obtaining flanking sequences allow indexing only one of the insertion events in lines that may carry more than one insertion. In addition, the process of DNA-transformation may lead to mutations that are not linked to the transferred DNA. As a result, a line may show a detectable phenotypic alteration that may not co-segregate with the indexed insertion.
2. Cases of gene silencing induced by T-DNA inserts have been reported, possibly associated to the presence of multiple independent insertions in the genome or of several tandem duplications of the T-DNA insert. Because of this, it may occur that the selection marker gene gets silenced. In fact, it has been reported that approximately 20% of lines within the SALK population are sensitive to kanamycin after two or

more generations [18]. To circumvent this problem, plants may be grown without antibiotic selection, and the presence of the insertion be confirmed only by PCR analysis. Recommended concentrations of selection reagents to be used for the different collections are as follows:

Collection	Selection	Final concentration
SALK	Kanamycin	50–100 μg/mL
SAIL	BASTA®	10–200 μg/mL
WISC-DS Lox	BASTA®	10–200 μg/mL
FLAG	BASTA®	10–200 μg/mL
SK	BASTA®	10–200 μg/mL
GABI-KAT	Sulfadiazine	7.5 mg/mL

For SALK lines, NASC recommends using a Kanamycin concentration of 10–40 μg/mL if gene silencing is occurring, (http://arabidopsis.info/CollectionInfo?id=19).

3. SDS precipitates at 4 °C, so the buffer should be kept at room temperature.
4. The reaction could be done with any other commercial Taq polymerase.
5. For the T-DNA mutant collections, a primer design tool (iSect Primer, http://signal.salk.edu/tdnaprimers.2.html) that selects the best gene-specific primers, is available. Primers can be designed not only in the coding or transcribed region of the gene, but also in the surrounding sequences. In addition, the software distinguishes the direction of the T-DNA insertion, and therefore determines the correct combination of primers.
6. The process of insertion of T-DNA in the genome may cause sequences flanking the insertion point to be deleted or duplicated. Therefore, primer annealing sites should not be designed too close to the predicted insertion site, but rather be positioned at least 300 nt apart [19].
7. For each collection, several different T-DNA or transposon-specific primers are commonly used. It has been noted that many tandem insertions found in some of the T-DNA collections occur on the right border side of the insert. Thus, for initial genotyping, primers designed for the Left Border (LB) are sufficient and preferred. But for a more detailed analysis (such as complete mapping of the insertion site, and determining if T-DNA tandem insertions are present), the Right Border (RB) primer is also needed. SAIL and GABI-KAT collections maintain the same left border sequence. Insertion-specific primers that are commonly used are listed below.

T-DNA specific left border (LB) and right border (RB) primers	Primer sequence (5′-3′)
Gabi-o8474 LB	ATAATAACGCTGCGGACATCTACATTTT
Gabi-o8409 LB	ATATTGACCATCATACTCATTGC
Gabi-o3144/35St RB	GTGGATTGATGTGATATCTCC
SALK LBb1	GCGTGGACCGCTTGCTGCAACT
SALK LBb1.3	ATTTTGCCGATTTCGGAAC
SALK LBa1	TGGTTCACGTAGTGGGCCATCG
SALK RB	CGCAATAATGGTTTCTGACGTA
SAIL LB1	GCCTTTTCAGAAATGGATAAATAGCCTTGCTTCC
SAIL LB2	GCTTCCTATTATATCTTCCCAAATTACCAATACA
SAIL LB3	TAGCATCTGAATTTCATAACCAATCTCGATACAC
SAIL RB	TAACAATTTCACACAGGAAACAGCTATGAC
FLAGdb LB	CTGCAAATTGCCTTTTCTTATCGA
FLAGdb RB	CTGATACCAGACGTTGCCCGCATAA
WISC-DS Lox p745 LB	AACGTCCGCAATGTGTTATTAAGTTGTC
SK-SKI015 RB	AGATCCGAAACTATCAGTG
SK-IK054 RB	ATGTGATATCTAGATCCGAAA C

8. The RNA preparation should be free of contaminating genomic DNA, so we recommend using a previously tested commercial kit for RNA isolation.
9. The reaction could be carried out with other commercial reverse transcriptase kits. In addition, gene expression may be tested by real-time, or quantitative, reverse transcription polymerase chain reaction (qRT-PCR) (*see* Chapter 21).
10. The relative position of the gene-specific primers is determined by the position of the insertion within the gene. If the insertion is located in the middle of the gene, upstream and downstream primers should be used [6]. The insertion of a T-DNA inside an exon may result in the synthesis of shorter and altered transcripts. The presence of truncated transcripts in the mutant plants should be assessed, as translation of these aberrant transcripts may produce protein products that could interfere with cellular processes, and thus result in phenotypic alterations [15, 16]. In addition, an insertion in an intron may sometimes have no effect on a gene's function if it is removed together with the intron during splicing [20].
11. Vapor phase seed sterilization with chlorine gas is an easy way to sterilize the seeds. The volume of sterilization solution that

is used and the time the seeds are subjected to chlorine gas may be changed. For example, the procedure may be carried out overnight using 33 mL of bleach and 1 mL of HCl.

12. Antibiotic resistance responses may vary according to the antibiotic's mode of action and concentration. Because of this, it is recommended to grow wild-type plants in parallel as control. For Sulfadiazine, yellow seedlings that do not form real leaves should be considered as sensitive, and thus negative for T-DNA insertion. At high concentrations, Kanamycin is a relatively toxic compound even for resistant plants. Plants may grow much slower and present anthocyanin accumulation on their leaves.
13. For plant DNA extraction, using younger tissues results in DNA preparations with lower levels of contamination by polysaccharides and polyphenol compounds.
14. Both fresh or frozen leaf tissue can be processed. When processing fresh tissue, add lysis buffer before tissue disruption. However, adding the buffer to frozen tissue before grinding results in low yields and degraded DNA.
15. Maximum DNA yields are obtained when tissue is subjected to grinding with the TissueLyser twice, for 1 min at 30 Hz each time.
16. At this stage, the DNA preparation can be left at room temperature.
17. Be very careful if pouring off the supernatant because the pellet may be loose and not stick well to the wall of the microcentrifuge tube.
18. Dry pellet in a SpeedVac at room temperature, or air dry by leaving the tube open on the bench. After adding the water, immediately freezing and thawing the tube may help with subsequent pellet resuspension.
19. To prevent DNA degradation, maintain DNA samples in the cold. These are crude preparations of genomic DNA, and may contain residual endonuclease activity.
20. A qualitative analysis is sufficient at this stage. DNA should not appear degraded (as a smear) on a gel electrophoresis. DNA concentration may also be determined using a spectrophotometer or fluorometer.
21. The expression pattern of the gene of interest should be first determined either through in silico analyses (using tools such as Genevestigator; https://www.genevestigator.com/gv/) or experimentally, by RT-PCR using wild-type plants, in order to identify the tissue or tissues in which the gene is expressed at higher levels. Then, this tissue should be selected to perform the RT-PCR analyses in the mutant individuals. For instance,

many genes that participate in flower development are specifically expressed in floral tissues, and Arabidopsis inflorescences are a good source for extracting high-quality RNA.

22. The presence of liquid nitrogen inside the microcentrifuge tubes during tissue grinding should be avoided, to prevent potential loss of tissue by nitrogen spill, or by the popping of the tube if closed with liquid nitrogen inside. Tubes can be pre-chilled in liquid nitrogen.
23. Pure RNA has an A260/A280 ratio between 1.8 and 2.1.
24. For the reaction, an anchored-oligo(dT)18 primer is used and the reaction is incubated 30 min at 55 °C. Perform a negative RT control reaction without transcriptase (using water instead of the enzyme) in parallel to the RT reaction of each sample. At the end of the reaction, the first-strand cDNAs can be diluted between five and tenfold.
25. Insertions in the coding region have the highest probability of knocking out a gene. Insertions in introns, 5′ and 3′ untranslated regions, or in the promoter, may instead simply alter gene expression and/or function (i.e., knockdown). In some rare cases, a hypermorph, or up-regulation of expression, can be observed when the insertion is localized in the promoter or the 3′UTR [10].

Acknowledgments

Work in our laboratory is supported by grants from Spanish Ministerio de Economía y Competividad (BFU2011-22734; and Programa Consolider-Ingenio, CSD2007-00036), and the Agència de Gestió d'Ajuts Universitaris i de Recerca (AGAUR) (grant SGR2009-GRC476). T. M. was supported by a fellowship from the European Molecular Biology Organization (EMBO), and T. F. by a fellowship from the Center for Research in Agricultural Genomics (CRAG).

References

1. Ma H, Yanofsky MF, Meyerowitz EM (1991) AGL1-AGL6, an Arabidopsis gene family with similarity to floral homeotic and transcription factor genes. Genes Dev 5(3):484–495
2. Ditta G, Pinyopich A, Robles P, Pelaz S, Yanofsky MF (2004) The *SEP4* gene of Arabidopsis thaliana functions in floral organ and meristem identity. Curr Biol 14(21): 1935–1940
3. Pelaz S, Ditta GS, Baumann E, Wisman E, Yanofsky MF (2000) B and C floral organ identity functions require *SEPALLATA* MADS-box genes. Nature 405(6783): 200–203
4. Liljegren SJ, Ditta GS, Eshed Y, Savidge B, Bowman JL, Yanofsky MF (2000) *SHATTERPROOF* MADS-box genes control seed dispersal in *Arabidopsis*. Nature 404(6779):766–770
5. Pinyopich A, Ditta GS, Savidge B, Liljegren SJ, Baumann E, Wisman E, Yanofsky MF (2003) Assessing the redundancy of MADS-box

genes during carpel and ovule development. Nature 424(6944):85–88

6. Wang YH (2008) How effective is T-DNA insertional mutagenesis in Arabidopsis? J Biochem Tech 1(1):11–20
7. Krysan PJ, Young JC, Sussman MR (1999) T-DNA as an insertional mutagen in Arabidopsis. Plant Cell 11(12):2283–2290
8. Parinov S, Sevugan M, De Y, Yang WC, Kumaran M, Sundaresan V (1999) Analysis of flanking sequences from dissociation insertion lines: a database for reverse genetics in Arabidopsis. Plant Cell 11(12):2263–2270
9. Woody ST, Austin-Phillips S, Amasino RM, Krysan PJ (2007) The WiscDsLox T-DNA collection: an arabidopsis community resource generated by using an improved high-throughput T-DNA sequencing pipeline. J Plant Res 120(1):157–165
10. O'Malley RC, Ecker JR (2010) Linking genotype to phenotype using the Arabidopsis unimutant collection. Plant J 61(6):928–940
11. Weigel D, Ahn JH, Blázquez MA, Borevitz JO, Christensen SK, Fankhauser C, Ferrándiz C, Kardailsky I, Malancharuvil EJ, Neff MM, Nguyen JT, Sato S, Wang ZY, Xia Y, Dixon RA, Harrison MJ, Lamb CJ, Yanofsky MF, Chory J (2000) Activation tagging in Arabidopsis. Plant Physiol 122(4):1003–1013
12. Alonso JM, Stepanova AN, Leisse TJ, Kim CJ, Chen H, Shinn P, Stevenson DK, Zimmerman J, Barajas P, Cheuk R, Gadrinab C, Heller C, Jeske A, Koesema E, Meyers CC, Parker H, Prednis L, Ansari Y, Choy N, Deen H, Geralt M, Hazari N, Hom E, Karnes M, Mulholland C, Ndubaku R, Schmidt I, Guzman P, Aguilar-Henonin L, Schmid M, Weigel D, Carter DE, Marchand T, Risseeuw E, Brogden D, Zeko A, Crosby WL, Berry CC, Ecker JR (2003) Genome-wide insertional mutagenesis of Arabidopsis thaliana. Science 301(5633):653–657
13. Rosso MG, Li Y, Strizhov N, Reiss B, Dekker K, Weisshaar B (2003) An Arabidopsis thaliana T-DNA mutagenized population (GABI-Kat) for flanking sequence tag-based reverse genetics. Plant Mol Biol 53(1–2):247–259
14. Sessions A, Burke E, Presting G, Aux G, McElver J, Patton D, Dietrich B, Ho P, Bacwaden J, Ko C, Clarke JD, Cotton D, Bullis D, Snell J, Miguel T, Hutchison D, Kimmerly B, Mitzel T, Katagiri F, Glazebrook J, Law M, Goff SA (2002) A high-throughput Arabidopsis reverse genetics system. Plant Cell 14(12):2985–2994
15. Gusmaroli G, Feng S, Deng XW (2004) The Arabidopsis CSN5A and CSN5B subunits are present in distinct COP9 signalosome complexes, and mutations in their JAMM domains exhibit differential dominant negative effects on development. Plant Cell 16(11):2984–3001
16. Okushima Y, Mitina I, Quach HL, Theologis A (2005) AUXIN RESPONSE FACTOR 2 (ARF2): a pleiotropic developmental regulator. Plant J 43(1):29–46
17. Bouche N, Bouchez D (2001) Arabidopsis gene knockout: phenotypes wanted. Curr Opin Plant Biol 4(2):111–117
18. Daxinger L, Hunter B, Sheikh M, Jauvion V, Gasciolli V, Vaucheret H, Matzke M, Furner I (2008) Unexpected silencing effects from T-DNA tags in Arabidopsis. Trends Plant Sci 13(1):4–6
19. Ulker B, Peiter E, Dixon DP, Moffat C, Capper R, Bouche N, Edwards R, Sanders D, Knight H, Knight MR (2008) Getting the most out of publicly available T-DNA insertion lines. Plant J 56(4):665–677
20. Carles CC, Choffnes-Inada D, Reville K, Lertpiriyapong K, Fletcher JC (2005) *ULTRAPETALA1* encodes a SAND domain putative transcriptional regulator that controls shoot and floral meristem activity in Arabidopsis. Development 132(5):897–911
21. Samson F, Brunaud V, Balzergue S, Dubreucq B, Lepiniec L, Pelletier G, Caboche M, Lecharny A (2002) FLAGdb/FST: a database of mapped flanking insertion sites (FSTs) of Arabidopsis thaliana T-DNA transformants. Nucleic Acids Res 30(1):94–97
22. Robinson SJ, Tang LH, Mooney BA, McKay SJ, Clarke WE, Links MG, Karcz S, Regan S, Wu YY, Gruber MY, Cui D, Yu M, Parkin IA (2009) An archived activation tagged population of Arabidopsis thaliana to facilitate forward genetics approaches. BMC Plant Biol 9:101
23. Sundaresan V, Springer P, Volpe T, Haward S, Jones JD, Dean C, Ma H, Martienssen R (1995) Patterns of gene action in plant development revealed by enhancer trap and gene trap transposable elements. Genes Dev 9(14): 1797–1810
24. Ito T, Motohashi R, Kuromori T, Mizukado S, Sakurai T, Kanahara H, Seki M, Shinozaki K (2002) A new resource of locally transposed dissociation elements for screening gene-knockout lines in silico on the Arabidopsis genome. Plant Physiol 129(4):1695–1699
25. Tissier AF, Marillonnet S, Klimyuk V, Patel K, Torres MA, Murphy G, Jones JD (1999) Multiple independent defective suppressor-mutator transposon insertions in Arabidopsis: a tool for functional genomics. Plant Cell 11(10):1841–1852

Chapter 21

Gene Expression Analysis by Quantitative Real-time PCR for Floral Tissues

Mariana Bustamante, Jian Jin, Oriol Casagran, Tania Nolan, and José Luis Riechmann

Abstract

Real-time, or quantitative, reverse transcription polymerase chain reaction (qRT-PCR), is a powerful method for rapid and reliable quantification of mRNA abundance. Although it has not featured prominently in flower development research in the past, the availability of novel techniques for the synchronized induction of flower development, or for the isolation of cell-specific mRNA populations, suggests that detailed quantitative analyses of gene expression over time and in specific tissues and cell types by qRT-PCR will become more widely used. In this chapter, we discuss specific considerations for studying gene expression by using qRT-PCR, such as the identification of suitable reference genes for the experimental setup used. In addition, we provide protocols for performing qRT-PCR experiments in a multiwell plate format (with the LightCycler® 480 system, Roche) and with nanofluidic arrays (BioMark™ system, Fluidigm), which allow the automatic combination of sets of samples with sets of assays, and significantly reduce reaction volume and the number of liquid-handling steps performed during the experiment.

Key words Real-time PCR, qRT-PCR, Quantitative PCR (qPCR), SYBR Green I dye

1 Introduction

Differential gene expression, over time or among different cell and tissue types, is central to the developmental processes of all organisms. In flower development studies, this aspect of gene function has usually been approached by using methods to characterize spatial patterns, or domains, of gene expression, such as in situ hybridization (*see* Chapter 14) and promoter–reporter gene fusions (*see* Chapter 15). In contrast, real-time, or quantitative, reverse transcription polymerase chain reaction (qRT-PCR), which is a powerful method for rapid and reliable quantification of mRNA abundance, has not featured prominently in flower development research. However, the development of techniques for the synchronized induction of flower development (*see* Chapter 16), or for the isolation of cell-specific mRNA populations (*see* Chapters 17–19),

José Luis Riechmann and Frank Wellmer (eds.), *Flower Development: Methods and Protocols*, Methods in Molecular Biology, vol. 1110, DOI 10.1007/978-1-4614-9408-9_21, © Springer Science+Business Media New York 2014

suggests that detailed quantitative analyses of gene expression over time and in specific tissues and cells will become more broadly used (not the least because such data will be needed for the characterization and ultimate modeling of the molecular gene regulatory networks that control the development of the flower).

The qRT-PCR method involves three processes: The conversion of mRNA into cDNA via reverse-transcription; the amplification of the resulting cDNA by PCR; and the detection and quantification in real time of the synthesized PCR amplification products [1–3]. The reliability of the data obtained in qRT-PCR experiments can be affected by several factors that impact on those processes, including template quality (RNA integrity [3, 4]), purity [3, 5] and quantity, efficiency of the RT reaction, PCR primer design, and efficiency of the PCR amplification [3]. To compensate for between-sample variations in the amount of starting material and in the efficiency of the qRT-PCR process, expression levels of the genes of interest are reported relative to one or more reference genes that are presumed to be uniformly and stably expressed across the tissues or conditions tested in the experiment, and whose abundance reflects the total amount of mRNA present in each sample. Thus, the reliability of qRT-PCR analyses is largely affected by the suitability of the gene (or genes) that is selected as a reference, i.e., by whether or not such a gene really fulfills the requirements of a normalization control [6, 7].

Housekeeping genes, which function in basic cellular processes and are expressed in all cells of an organism, have often been used as reference genes to normalize the data in qRT-PCR experiments (e.g., genes such as glyceraldehyde-3-phosphate dehydrogenase (*GAPDH*), elongation factor-1α (*EF-1α*), actin (*ACT*), or tubulin (*TUB*)). However, the initial evidence indicating that housekeeping genes are stably expressed was obtained using methods that are mostly qualitative (for instance, RNA gel-blots and end-point RT-PCR), and subsequent studies have demonstrated that in some circumstances their expression may be regulated, or be unstable, and thus show changes in transcript levels throughout development or among different conditions or tissues. In addition, housekeeping genes are usually expressed at higher levels than the typical genes of interest. For these reasons, using them as reference genes may introduce biases in the results obtained by qRT-PCR [6, 7]. For example, in a series of experiments designed to assess traditional *Arabidopsis* reference genes (including *ACT2*, *ACT7*, *ACT8*, *ADENINE PHOSPHORIBOSYLTRANSFERASE 1* (*APT1*), *EF1α*, *EUKARYOTIC TRANSLATION INITIATION FACTOR 4A1* (*eIF4A*), *TUB2*, *TUB6*, *TUB9*, *UBIQUITIN 4* (*UBQ4*), *UBQ5*, *UBQ10*, and *UBQ11*), it was found that *eIF4A* would appear to be stably expressed over the course of silique development when *APT1*, *UBQ5*, or *eF1α* were used to normalize the data, whereas its expression would appear quite variable when *TUB6* was used as reference gene [6].

In summary, the validity of "housekeeping" reference genes is not universal, and is highly dependent on the experimental conditions [7]. Thus, the selection of appropriate reference genes for the normalization of qRT-PCR data has emerged as a crucial component for successful expression studies carried out with this technology, and statistical algorithms like *geNorm* [8] or *BestKeeper* [9] have been developed for that purpose (*see* **Note 1**).

Concomitantly, the use of genome-wide technologies (i.e., DNA microarrays) to characterize gene expression changes across many different tissues and developmental stages, environmental conditions, or in response to biotic and abiotic stresses or perturbations, has resulted in very rich datasets (for instance, [10]) that can be mined to identify novel, better suited reference genes for the desired experimental setup. For instance, Czechowski et al. [11] analyzed a very large set of Arabidopsis data obtained with Affymetrix ATH1 GeneChip arrays to identify several hundred genes that outperform traditional reference genes in terms of expression stability throughout development and under a range of environmental conditions. Subsequent qRT-PCR experiments performed with a subset of those novel reference genes confirmed that they showed superior expression stability and lower absolute expression levels [11] (*see* Table 1) (*see* **Note 2**). The results obtained in *Arabidopsis* have informed the selection of reference genes in other plant species, as the corresponding orthologous genes may also show stable expression (for an example in Leafy spurge, *see* ref. 12). If candidate reference genes are selected based on orthology, however, their suitability needs to be confirmed experimentally, as such character is not always maintained across all experimental conditions in all organisms [3] (for instance, *see* ref. 13).

The approach of using genome-wide data to select reference genes has been further expanded and refined with *RefGenes*, an online tool that allows easy identification of condition-specific reference genes [14]. *RefGenes* is based on the Genevestigator database of normalized and well-annotated microarray experiments, and is accessible through the Genevestigator Web page (www.genevestigator.com). The appropriateness of using condition-specific reference genes is based on the observation that for each biological context a subset of stable genes exists that has a smaller variance than either commonly used reference genes or genes that were selected for their stability across all conditions [14]. In other words, there is no gene that is universally stable, and for each biological context and specific experimental condition, the most appropriate set of reference genes does vary.

Through *RefGenes*, users are able to select the microarray experiments that are most similar to their chosen experimental conditions (including tissue, developmental stage, treatment, etc.). Afterwards, the user indicates the set of target genes of interest (up to ten genes can be tested at once). A search is then triggered to identify those genes that have the lowest variance within the

Table 1
Traditional and novel Arabidopsis general reference genes, from Czechowski et al. [11]

Gene	Annotation	Forward primer (5′–3′)	Reverse primer (5′–3′)
Traditional reference genes			
AT1G13440	GAPDH	TTGGTGACAACAGGTCAAGCA	AAACTTGTCGCTCAATGCAATC
AT3G18780	ACT2	CTTGCACCAAGCAGCATGAA	CCGATCCAGACACTGTACTTCCTT
AT4G05320	UBQ10	GGCCTTGTATAATCCCTGATGAATAAG	AAAGAGATAACAGGAACGGAAACATAGT
AT5G25760	UBC	CTGCGACTCAGGGAATCTTCTAA	TTGTGCCATTGAATTGAACCC
AT5G60390	EF-1α	TGAGCACGCTCTTCTTGCTTTCA	GGTGGTGGCATCCATCTTGTTACA
Novel reference genes			
AT1G13320	PP2A subunit *PDF2*	TAACGTGGCCAAAATGATGC	GTTCTCCACAACCGCTTGGT
AT1G47770	Hypothetical protein	GTTCAATAAAATGGCGCATCTTG	GAAAAGGTGCAAAACGATCTCAC
AT1G58050	Helicase	CCATTCTACTTTTTGGCGGCT	TCAATGGTAACTGATCCACTCTGATG
AT1G62930	PPR gene	GAGTTGCGGGTTTGTTGGAG	CAAGACAGCATTTCCAGATAGCAT
AT2G07190	Hypothetical protein	CCGTCCAATCCAACAGATCAG	CGTCATCTAAAGACATTAGGTCCTGTAC
AT2G28390	SAND family	AACTCTATGCAGCATTTGATCCACT	TGATTGCATATCTTTATCGCCATC
AT2G32170	Expressed protein	ATCGAGCTAAGTTTGGAGGATGTAA	TCTCGATCACAAACCCAAAATG
AT3G01150	Polypyrimidine-tract-binding protein	GATCTGAATGTTAAGGCTTTTAGCG	GGCTTAGATCAGGAAGTGTATAGTCTCTG
AT3G32260	Hypothetical protein	CTGTTTGCCGAAGTTCCAGAGT	TTAAATCAGCAAGAACGTCGGATA
AT3G53090	Ubiquitin-transferase	TTCAAATACTTGCAGCCAACCTT	CCCAAAGAGAGGTATCACAAGAGACT
AT4G26410	Expressed protein	GAGCTGAAGTGGCTTCCATGAC	GGTCCGACATACCCATGATCC
AT4G27960	UBC9	TCACAATTTCCAAGGTGCTGC	TCATCTGGGTTTGGATCCGT
AT4G33380	Expressed protein	TTGAAAATTGGAGTACCGTACCAA	TCCCTCGTATACATCTGGCCA
AT4G34270	TIP41-like	GTGAAAACTGTTGGAGAGAAGCAA	TCAACTGGATACCCTTTCGCA
AT4G38070	bHLH	GAAAGCAAAGGCGGTGAGAG	CAAGGCACACTTGGTTCTTCC
AT5G08290	Mitosis protein	TTACTGTTTCGGTTGTTCTCCATTT	CACTGAATCATGTTCGAAGCAAGT
AT5G12240	Expressed protein	AGCGGCTGCTGAGAAGAA^GT	TCTCGAAAGCCTTGCAAAATCT
AT5G15710	F-box protein	TTTCGGCTGAGAGGTTCGAGT	GATTCCAAGACGTAAAGCAGATCAA
AT5G46630	Clathrin adaptor complex subunit	TCGATTGCTTGGTTTGGAAGAT	GCACTTAGCGTGGACTCTGTTTGATC
AT5G55840	PPR gene	AAGACAGTGAAGGTGCAACCTTACT	AGTTTTTGAGTTGTATTTGTCAGAGAAAG

selected set of microarray experiments and a range of expression that is similar to that of the target gene set. The result of the search is graphically displayed, showing the top 25 best candidate reference genes for the selected conditions. The behavior of these candidate genes in the chosen (or in additional) tissues or experimental conditions can then be explored using the *Conditions* tool of Genevestigator. In addition, the novel candidate reference genes that are identified using *RefGenes* should be validated for the specific biological conditions (tissue type, treatment, etc.) of the experiments to be performed, by using one of the aforementioned algorithms (*geNorm* or *BestKeeper*) and preferably together with commonly used reference genes.

The use of *RefGenes* to select reference genes for flower development studies is illustrated in Figs. 1 and 2 and in Table 2. A set of ten genes that participate in and/or are expressed at early stages of Arabidopsis flower development was used as the target set (including *SUPERMAN -SUP*, At3g23130-, *LEAFY -LFY*, At5g61850-, *AGL24*-At4g24540-, *YABBY3*-*YAB3*, At4g00180-, *APETALA2 -AP2*, AT4g36920-, *AGL42* -At5g62165-, *SHATTERPROOF2 -SHP2*, At2g42830-, *AGAMOUS -AG*, AT4g18960-, *SEPALLATA3*-*SEP3*, At1g24260-, and *APETALA3 -AP3*, At3g54340-, *see* ref. 15) to search for reference genes using a subset of the genome-wide expression profiling data available in Genevestigator (in particular, experiments under the Anatomy-Inflorescence category). *RefGenes* returns a list of candidate novel reference genes (Fig. 1, Table 2), which are then compared to traditional reference genes (*see* Fig. 2).

The detection of product formation in real time during the amplification reaction of qRT-PCR experiments is carried out by measuring the emission signal from either fluorescent double-stranded DNA-binding dyes (such as SYBR® Green I and EvaGreen®, see below), or template-specific fluorescent probes (such as the TaqMan® probe technology). A general protocol for using SYBR Green I dye in a qRT-PCR experiment performed in a LightCycler® 480 Real-Time PCR system (Roche) is provided below (equally suited real-time PCR machines are available from various manufacturers). In addition to standard real-time PCR systems, in which reactions are performed either in thin-wall PCR tubes or in multiwell plates, new systems for high-throughput analyses have been developed, which are based on nanofluidic arrays (such as the BioMark™ system, Fluidigm). These arrays contain nanofluidic networks that allow the automatic combination of sets of samples with sets of assays, and significantly reduce reaction volume (and thus the amount of material needed to perform an assay) and the number of liquid-handling steps performed during the experiment. A protocol for a qRT-PCR experiment using EvaGreen® and the BioMark™ system is also provided.

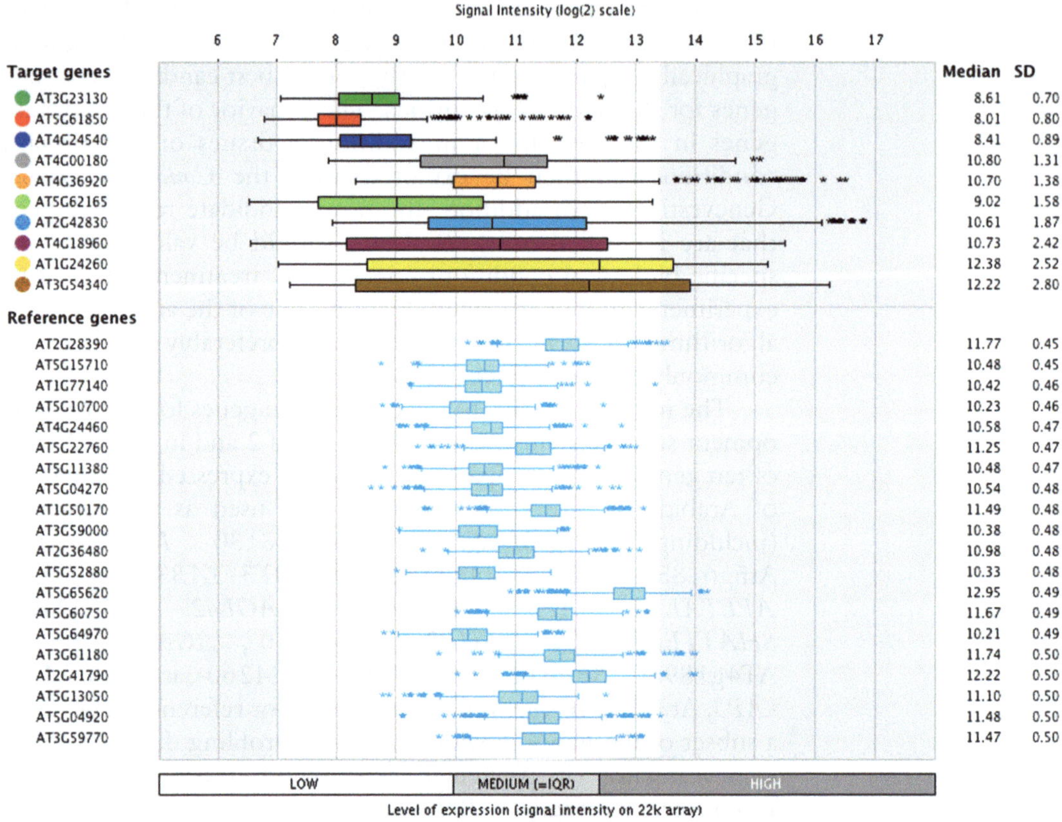

Fig. 1 Example of output results obtained when using the *RefGenes* tool with the indicated set of floral regulatory genes (*SUP* (AT3G23130), *LFY* (AT5G61850), *AGL24* (AT4G24540), *YAB3* (AT4G00180), *AP2* (AT4G36920), *AGL42* (AT5G62165), *SHP2* (AT2G42830), *AG* (AT4G18960), *SEP3* (AT1G24260), and *AP3* (AT3G54340)) and microarray experiments under the Anatomy-Inflorescence category in Genevestigator

2 Materials

2.1 Tissue Collection and RNA Extraction

1. RNase-free microcentrifuge tubes (1.5 mL).
2. Plastic pellet pestles for 1.5 mL microcentrifuge tubes (optional: a mixer motor or an electric drill).
3. Forceps (e.g., Dupont size #5).
4. Liquid nitrogen.
5. Vortex.
6. Microcentrifuge.
7. Spectrum Plant Total RNA Kit (Sigma-Aldrich) or an equivalent total RNA isolation kit or reagents (*see* **Note 3**).
8. Spectrophotometer (such as a NanoDrop).
9. Agilent Bioanalyzer and associated reagents (Agilent RNA 6000 Nano kit).

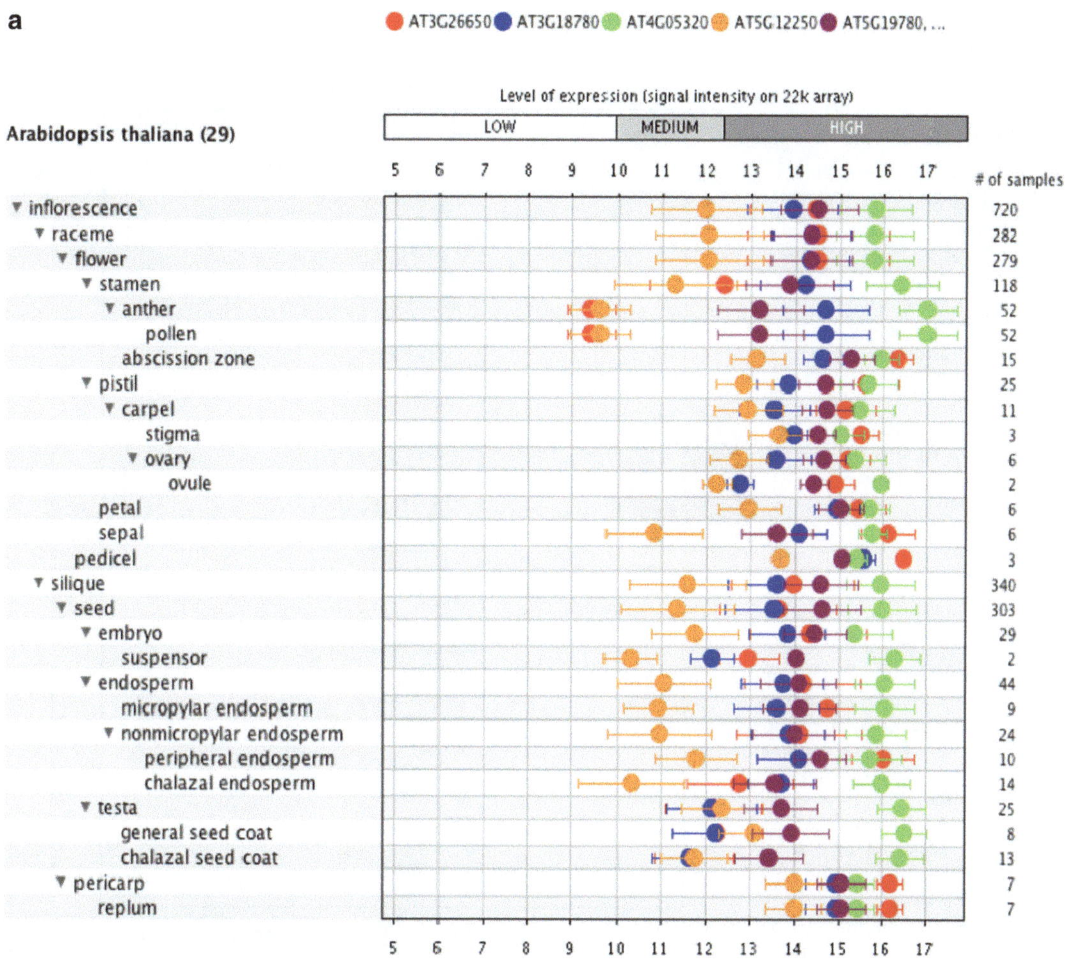

Fig. 2 Expression characteristics of some commonly used and novel reference genes in Arabidopsis inflorescences (Genevestigator; Inflorescence category microarray experiments). (**a**) Traditional reference genes: *GAPDH* (AT3G26650), *ACT2* (AT3G18780), *UBQ10* (AT4G05320), *TUB6* (AT5G12250), *TUA5* (AT5G19780). (**b**) Novel reference genes: AT2G28390, AT5G15710, AT1G77140, AT5G10700, and AT4G24460. The novel reference genes are more stably expressed throughout the experimental series, and their mean expression level is generally lower than that of traditional reference genes, and thus closer to that of the typical genes of interest

2.2 Reverse Transcription Reaction

1. High Capacity cDNA Reverse Transcription Kit (e.g., Applied Biosystems; other commercial kits are available, but the protocols provided below are based on this kit) containing dNTPs (100 mM), MultiScribe reverse transcriptase (50 U/mL), reverse transcription Random Primers, reverse transcription buffer (10×), RNase inhibitor (20 U/mL).
2. RNase-free PCR-tubes.
3. Nuclease-free water.

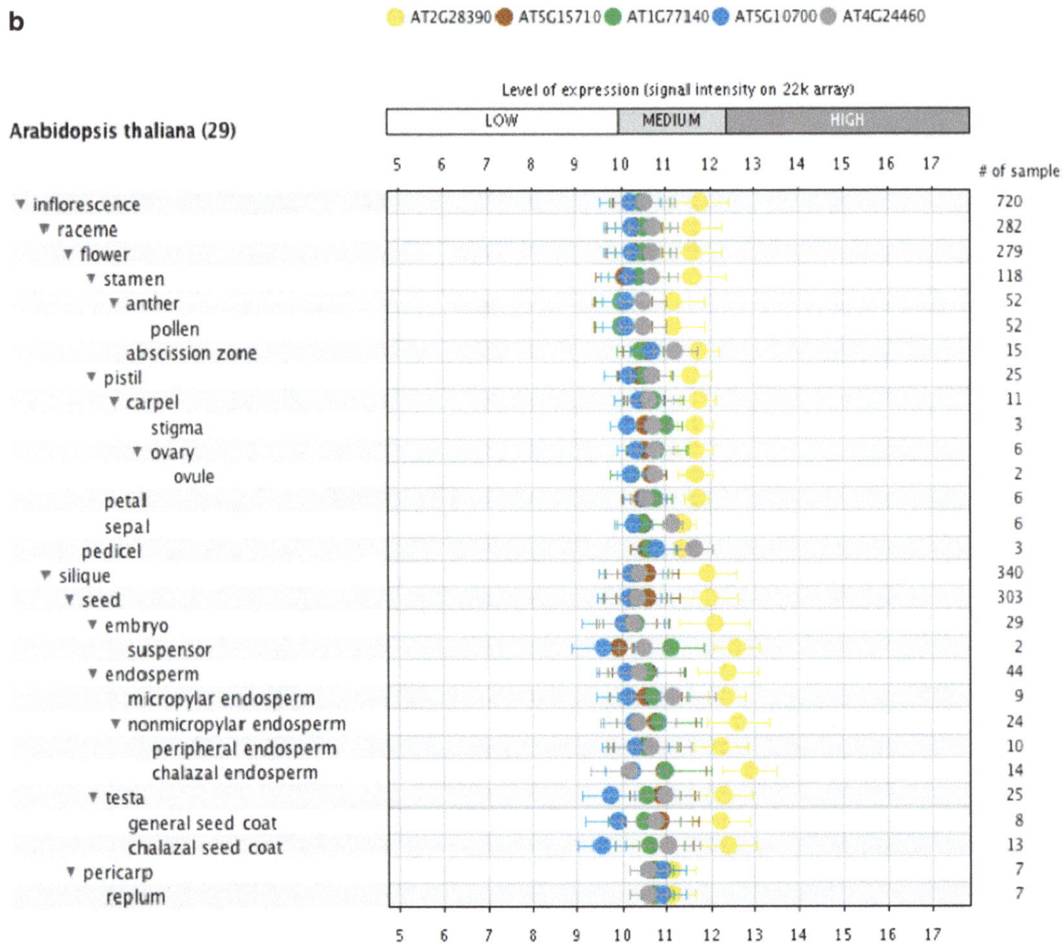

Fig. 2 (continued)

2.3 Quantitative Real-time PCR: LightCycler® 480 System

1. LightCycler® 480 SYBR Green I Master (Roche Diagnostics; other commercial kits are available, but the protocols provided below are based on this kit): ready-to-use hot-start PCR mix containing FastStart Taq DNA Polymerase, reaction buffer, dNTP mix (with dUTP, instead of dTTP), SYBR Green I dye, and $MgCl_2$.
2. LC 480 Multiwell Plate 96 (Roche Diagnostics) (*see* **Note 4**).
3. Forward and reverse PCR primers at 100 μM each.
4. Nuclease-free water.

2.4 Quantitative Real-time PCR: BioMark™ System

1. TaqMan PreAmp Master Mix 2× (Applied Biosystems).
2. SsoFast EvaGreen SuperMix with Low ROX (Bio-Rad): 2× real-time PCR mix, containing dNTPs, Sso7d fusion polymerase, $MgCl_2$, ROX passive reference dye and stabilizers.
3. 2× Assay Loading Reagent (Fluidigm).

Table 2
Candidate novel reference genes for Arabidopsis floral tissues identified using *RefGenes*

Gene	Annotation
AT2G28390	SAND family protein
AT5G15710	Galactose oxidase/kelch repeat superfamily protein
AT1G77140	Vacuolar protein sorting 45
AT5G10700	Peptidyl-tRNA hydrolase II (PTH2) family protein
AT4G24460	CRT (chloroquine-resistance transporter)-like transporter 2
AT5G22760	PHD finger family protein
AT5G11380	1-deoxy-d-xylulose 5-phosphate synthase 3
AT5G04270	DHHC-type zinc finger family protein
AT1G50170	Sirohydrochlorin ferrochelatase B
AT3G59000	F-box/RNI-like superfamily protein
AT2G36480	ENTH/VHS family protein
AT5G52880	F-box family protein
AT5G65620	Zincin-like metalloproteases family protein
AT5G60750	CAAX amino terminal protease family protein
AT5G64970	Mitochondrial substrate carrier family protein
AT3G61180	RING/U-box superfamily protein
AT2G41790	Insulinase (Peptidase family M16) family protein
AT5G13050	5-formyltetrahydrofolate cycloligase
AT5G04920	EAP30/Vps36 family protein
AT3G59770	sacI homology domain-containing protein/WW domain-containing protein

4. 20× DNA Binding Dye Sample Loading Reagent (Fluidigm).
5. Exonuclease I (*E. coli*) (20,000 U/mL; New England Biolabs).
6. Exonuclease I Reaction Buffer 10× (New England Biolabs).
7. Forward and reverse PCR primers at 100 μM each.
8. Nuclease-free water.
9. TE Buffer: 10 mM Tris–HCl, pH 8.0, 1.0 mM EDTA (TEKnova).
10. DNA Suspension Buffer; 10 mM Tris–HCl, pH 8.0, 0.1 mM EDTA (TEKnova).
11. 48.48 Dynamic Array IFC (Fluidigm).

3 Methods

The performance of the primers that are used in a qRT-PCR experiment is crucial for obtaining high-quality results, and there are several aspects that have to be considered for successful primer design (*see* **Note 5**). There are public, searchable databases of primer and probe sequences that have been used and validated in real-time PCR assays, and can thus be an alternative to the time-consuming primer design and experimental optimization steps (e.g., RTPrimerDB; http://medgen.ugent.be/rtprimerdb/). In addition, there are many online resources for primer design, some of which also provide access to a consultative design service, such as:

- Oligoarchitect: www.sigma.com/oligoarchitect
- RealTimeDesign: https://www.biosearchtech.com/display.aspx?pageid=54
- QuantPrime: http://www.quantprime.de/
- IDT-qPCR: http://eu.idtdna.com/scitools/Applications/RealTimePCR/
- Primer3: http://primer3.sourceforge.net/
- Primer-BLAST: http://www.ncbi.nlm.nih.gov/tools/primer-blast/

3.1 Tissue Collection and RNA Extraction

RNA quality (integrity and purity) is a critical factor for qRT-PCR experiments.

1. Harvest at least 100 mg of the desired plant tissue (e.g., inflorescences), into a 1.5 mL RNase-free microcentrifuge tube containing liquid nitrogen.
2. Grind the tissue to a fine powder with the pellet pestles (and a mixer motor), keeping the bottom of the tube immersed in liquid nitrogen throughout the grinding process in order to avoid RNA degradation (*see* **Notes 6** and **7**).
3. Follow the manufacturer's instructions for the RNA extraction kit.
4. Analyze the integrity of the isolated RNA using a Bioanalyzer (or by using the 3′/5′ integrity assay, *see* ref. 3) and determine the concentration by absorption at 260 nm (e.g., with a NanoDrop spectrophotometer).

3.2 Reverse Transcription Reaction

The reverse transcription reaction to synthesize cDNA from the starting RNA material can be performed with various priming strategies, enzymes, and experimental conditions [2, 3]. However, in order to compare gene expression data across different experiments or laboratories, these variables should be kept constant, particularly ensuring that the same amount of RNA is added to each reaction (or that the enzyme/protocol used results in a proportional cDNA yield).

1. Prepare an RT master mix in a 1.5 mL tube according to the following table:

Component	Volume (per reaction) (μL)
Water	4.2
10× RT Buffer (1×)	2
25× dNTP Mix (100 mM)	0.8
10× RT Random Primers	2
MultiScribe Reverse Transcriptase	1

2. Add 10 μL of mastermix to each individual PCR-tube. Then add 100–1,000 ng of each RNA sample, in a volume of 10 μL. The final reaction volume is 20 μL. No-RT control reaction(s) should be included in the experiment.
3. Briefly centrifuge the tubes to collect the contents and to eliminate any air bubbles.
4. Place the tubes in a thermal cycler using the following conditions:

	Step 1	Step 2	Step 3	Step 4
Temperature (°C)	25	37	85	4
Time	10 min	120 min	5 min	∞

5. Store cDNA samples at 4 °C (short term) or at −20 °C (for up to 6 months).

3.3 Quantitative Real-time PCR: LightCycler® 480 System

1. Set up your samples:
   (a) Every gene–primer-pair combination used in a qPCR should be tested to calculate primer efficiency (*see* **Note 8**).
   (b) The cDNA samples resulting from the RT reaction may be diluted in water, to obtain a final estimated concentration between 5 and 10 ng/μL (estimation based on the initial amount of RNA used in the RT reaction). This concentration range is ideal for the qRT-PCR. All amplification reactions should have a similar concentration of cDNA.
2. Before loading the PCR plate, and in order to minimize pipetting errors, it is important to prepare master mixes for each primer pair used. The accuracy of qPCR is highly dependent on accurate pipetting and thorough mixing of solutions. To prepare the qPCR Master Mix, add components in the order indicated in the table below. The protocol provided here

uses SYBR® Green I chemistry, but other PCR-product detection chemistries could be used (*see* **Note 9**).

Component	Volume (per reaction) (μL) for 96-well plate
LC480 SYBR® Green I Master (2×) (Roche Diagnostics)	10
Water	6.4
Primer Forward (10 μM)	0.8
Primer Reverse (10 μM)	0.8

3. Loading the plate: Once all master mixes for each pair of primers are prepared, start loading the plate by adding first the Master Mix (18 μL) and then the cDNA samples (2 μL). Avoid producing bubbles. The final reaction volume in each well is 20 μL. Then add the No Template Control (NTC) and no-Reverse Transcription control (no-RT, or RT-) reactions (*see* **Note 10**). Seal the plate with LightCycler® 480 Sealing Foil by pressing it firmly to the plate surface, using your hand or a scraper. Sealing the plate properly is crucial to eliminate evaporation at high temperatures.
4. Place the multiwell plate in a standard swing-bucket centrifuge equipped with a rotor for multiwell plates with suitable adaptors. Balance it with a suitable counterweight (e.g., another multiwell plate). Centrifuge the plate at 1,500 × *g* for 2 min.
5. Load the multiwell plate into the LightCycler® 480 Instrument and set up the qPCR program (annealing temperature in the PCR is primer-dependent):

	Temperature (°C)	Time	Acquisition
Activation	95	10 min	None
PCR (45 cycles)	95 60 72	10 s 30 s 30 s	None None Single
Melting	95 65 95	2 s 15 s –	None None Continuous
Cooling	40	30 s	None

3.4 Quantitative Real-time PCR: BioMark™ System

BioMark System arrays allow for the automatic combination of sets of samples with sets of assays, significantly reducing reaction volume and the number of liquid-handling steps performed during the experiment. For instance, using the 48 × 48 array (as described in this protocol), 48 different samples (e.g., time-points in a time-course experiment) can be tested with up to 48 different assays (e.g., genes).

1. Specific Target Amplification (STA): This step is recommended to increase the number of copies of target DNA.
   (a) STA Primer Mix (500 nM):
   - Pool together 1 μL aliquots of all of the 100 μM primer sets to be included in the STA reaction (up to 100 different assays).
   - Add DNA Suspension Buffer to make the final volume 200 μL.
   - Vortex to mix and briefly spin reaction tube.

   (b) STA Pre-Mix:
   - In a DNA-free hood, prepare a Pre-Mix for the STA reaction according to the following table:

Component	Volume (per reaction) (μL)
TaqMan PreAmp Master Mix	2.5
500 nM pooled STA Primer Mix	0.5
Water	0.75

   - Add 3.75 μL of STA Pre-Mix for each sample in a 96-well plate.
   - Add 1.25 μL of cDNA (at 5–25 ng/μL) to each reaction well, making a final volume of 5 μL. Include a no-PreAmplification control: add water instead of cDNA.
   - Seal the plate properly. Then, vortex and briefly spin the plate.

   (c) STA thermal cycle reaction:
   - Place the plate into the thermal cycler and run the following program (annealing temperature in the PCR is primer-dependent):

	Activation	16 cycles		Hold
Temperature (°C)	95	95	60	4
Time	10 min	15 s	4 min	∞

   - Eliminate the unincorporated primers from the STA amplification reaction. Prepare Exonuclease Mix as follows:

Component	Per 5 μL sample
Water	1.4 μL
Exonuclease I reaction buffer	0.2 μL
Exonuclease I (20 U/μL)	0.4 μL

- Add 2 μL of Exonuclease Mix to each 5 μL STA reaction. Vortex, centrifuge, and place in a thermal cycler.

	Digest	Inactivate	Hold
Temperature (°C)	37	80	4
Time	30 min	15 min	∞

- Dilute the STA reaction to an appropriate final product concentration, as shown below. A minimum dilution of fivefold should be used.

Volume of water or TE Buffer			
Volume of STA Rx	5-fold dilution	10-fold dilution	20-fold dilution
7 μL	18 μL	43 μL	93 μL

- Store diluted STA products at −20 °C or use immediately for on-chip PCR.

2. Sample and Assay Mix preparation:
   (a) Prepare Sample mix as shown below:

Component	Volume per inlet with overage (μL)
2× SsoFast EvaGreen Supermix with low ROX	3.0
20× DNA binding dye sample loading reagent	0.3

   (b) In a new 96-well plate aliquot 3.3 μL of Sample mix and add 2.7 μL of each STA and Exo I-treated sample.
   (c) Seal the plate properly. Then, vortex and spin plate. Keep on ice.
   (d) Prior to preparing the Assay mix, combine the two primers of each primer pair making a final concentration of 20 μM.
   (e) Prepare Assay mix as shown below:

Component	Volume per inlet with overage (μL)
2× Assay loading reagent	3.0
1× DNA suspension buffer	1.5

   (f) In a new 96-well plate, aliquot 4.5 μL of Assay mix and add in 1.5 μL of the primer pair mix at 20 μM.
   (g) Seal the plate properly. Then, vortex and spin the plate. Keep on ice.

3. Priming the 48 × 48 Dynamic Array™ IFC
   (a) Inject control line fluid into each accumulator on the chip. Load the chip within 60 min of priming (refer to instrument manufacturer's instructions for details).
   (b) Remove and discard the blue protective film from the bottom of the chip.
   (c) Place the chip into the IFC controller for the 48 × 48 Dynamic Array IFC.
   (d) Run the Prime script for the 48 × 48 Dynamic Array IFC.
   (e) Pipette 5 μL of each assay and 5 μL of each sample into their respective inlets on the chip.
   (f) Place the chip to the IFC controller and run the Load Mix program.
   (g) After the program has run, take out the chip from the IFC controller and remove any dust particle from the chip surface.
   (h) Place the chip in the Biomark System and run the following program (annealing temperature in the PCR is primer-dependent):

	Activation	30 Cycles		Melting	
Temperature (°C)	95	96	60	60	95
Time	60 s	5 s	20 s	3 s	1 °C/3 s

3.5 Data Analysis

Different methodologies can be used for determination of the Quantification Cycle, Cq [16] (previously referred to as Ct/Cp/takeoff point):

- The threshold cycle method measures the Cq at a constant fluorescence level. These constant threshold methods assume that all samples have the same amplicon DNA concentration at the threshold fluorescence. The strength of this method is that it is extremely robust, but the threshold value needs to be adjusted for each experiment.
- The second derivative method calculates Cq as the second derivative maximum of the amplification curve. It is not user-dependent and is widely used.

Before performing the actual analysis, it is important to validate the data according to a variety of criteria (preferably following the Minimum Information for Publication of Real-time PCR Experiments: MIQE guidelines, *see* **Note 11**, ref. 16). In particular:

- Check amplification curves. A normal amplification plot has three distinct phases; linear baseline, exponential and plateau.

- Check controls (RT-, NTC).
- Check that the slope of the standard curve is between −3.2 and −3.5.
- Check technical replicates. They should be within 0.5 Cq of each other.
- Check melting peaks (when using a binding dye, or probes such as Molecular Beacons or Scorpions that are not hydrolyzed during the reaction) to verify that single, specific amplification products have been synthesized in the reaction.

3.5.1 Absolute Quantification

Absolute quantification relies on measurement to a standards curve constructed using the real-time PCR data obtained from amplification of these standards of known concentrations of template. Commonly, standards are derived from purified dsDNA plasmid, in vitro-transcribed RNA or in vitro-synthesized ssDNA. A standard curve (plot of Cq value against log of amount of standard) is generated using different dilutions of the standard. The Cq value of the target is compared with the standard curve, allowing calculation of the initial amount of the target. It is important to select an appropriate standard for the type of nucleic acid to be quantified. This method requires having the same efficiency of amplification in all reactions (reactions with experimental samples and reactions with the external standards). When using absolute quantification for determination of mRNA concentration, it is usual to correct absolute copy number of the specific target relative to absolute copy number of one or more reference genes

3.5.2 Relative Quantification

Relative quantification relies on comparing the expression level of a target gene relative to a reference gene between a control sample and the test samples. Normalization to reference genes is the most common method for controlling for variation in qRT-PCR experiments. It is used to measure the relative change in mRNA expression levels. Many mathematical models are available. Most common relative quantification methods are:

(a) Pfaffl model [17]: combines gene quantification and normalization into a single calculation (Eq. 1). This model adjusts the amplification efficiencies (E) from target and reference genes in order to correct differences between the two assays.

$$Ratio = \frac{\left(E_{target}\right)^{\Delta Cq\ target\ (control-sample)}}{\left(E_{reference}\right)^{\Delta Cq\ reference\ (control-sample)}} \quad (1)$$

(b) $2^{-\Delta\Delta Cq}$ method [18]: This is a simpler version of the first model. Target and control amplification efficiency (E_{target} and $E_{\mathrm{reference}}$) are assumed to be maximum (100 %, i.e., a value of 2, indicating amplicon doubling during each cycle) (Eq. 2). In addition the relative expression of the target in all test samples is compared to that in a control or calibrator sample

$$Ratio = 2^{-[\Delta Cq\,Sample - \Delta Cq\,control]} \qquad (2)$$

4 Notes

1. *geNorm* is a widely used algorithm to determine the most stable reference from a given set of candidate genes on the basis of the *M* value (the *M* value is the internal control gene-stability measure, defined as the average pair-wise variation of a particular gene with all other control genes; genes with the lowest *M* values have the most stable expression) [8]. *geNorm* calculates and compares the *M* value of each pair of genes, and eliminates the gene with the highest *M* value, and then repeats this process with the remaining genes until the pair of genes with the lowest *M* value is identified. Thus, the genes forming this pair are considered as optimal reference genes among the initial candidate set.
2. The genome-wide analyses performed by Czechowski et al. led to the identification of many novel reference gene candidates, with purportedly better expression characteristics than traditional reference genes [11]. In these analyses the SD/MV ratio (SD/mean expression value, i.e., the coefficient of variation, or CV) for each gene in all the given experimental conditions (developmental series, abiotic stress series, hormone series, nutrient starvation and re-addition series, diurnal series, light series, and biotic stress series) is calculated. The gene that has the lowest CV value is considered as the gene with the most stable expression, and therefore a potential reference gene. Through these analyses, 25 reference genes, including 20 novel and 5 traditional ones, were recommended [11]. These genes were then validated by qRT-PCR and their expression stability ranked using the *geNorm* algorithm (*see* Table 1).
3. There are specific plates and films for the LC480 system that have been designed to ensure the best heat transfer from the thermal block and minimal autofluorescence, which is important in order to achieve a good signal-to-noise ratio in the detection of amplification products. In this protocol, we suggest using the LC 480 Multiwell Plate 96 from Roche.
4. The RNA preparation should be free of contaminating genomic DNA, so we recommend using a previously tested commercial kit for RNA isolation (*see* **Note 10**).
5. For primer design, it is important to consider the following points: (1) PCR products should be short (the ideal length is

from 70 to 250 bp); (2) The gene-specific forward and reverse primers should have similar melting temperatures (Tm) and length; (3) Primers should be between 15 and 25 nucleotides long and with a G/C content of around 50 %. (4) Primers should have low or no self-complementarity, in order to avoid the formation of primer dimers; (5) For the same reason, avoid pairs of primers that show sequence complementarity at their 3′ ends; (6) Primers that span introns or cross intron/exon boundaries are advantageous because they allow to distinguish amplification from cDNA or from contaminant genomic DNA. Primers should be ordered with desalt purification. Primer stock solutions should be prepared with DNase/RNase-free water. Make aliquots to avoid contamination and repeated freezing/thawing. Original stock of PCR primers should be stored at −20 °C, and working dilutions at 4 °C for up to 2 weeks.

6. The presence of liquid nitrogen inside the microcentrifuge tubes during tissue grinding should be avoided, to prevent potential loss of tissue by nitrogen spill, or by the popping of the tube if closed with liquid nitrogen inside. Tubes can be pre-chilled in liquid nitrogen. As an alternative for grinding the tissue, mortar and pestle could be used instead of pellet pestles and an electric drill.

7. Both fresh and frozen (−80 °C) tissue can be used as starting material, and ground plant material can be stored at −80 °C before RNA purification. However, do not allow the frozen material to thaw before grinding or before the first solution of the RNA purification procedure is added.

8. Make a four-step dilution series (1:4 dilutions) from cDNA samples. In order to evaluate the efficiency of the PCR reaction, it is important to generate at least one standard curve for each primer pair. A standard curve graph is made by plotting the Ct/Cp values on the *y*-axis and the logarithm of the input amounts on the *x*-axis. The slope of the line of this plot will give the efficiency of the reaction according to the equation $E = [10^{(-1/\text{slope})}] - 1$; slope should be between −3.2 and −3.5 and $R^2 > 0.98$.

9. SYBR® Green I and EvaGreen® are the most used dye chemistries, due to cost and simple optimization process. However, these dyes bind to any double-stranded DNA formed in the reaction, including primer-dimers and other nonspecific reaction products, which may result in an overestimation of the target concentration. Other methods, such as hydrolysis probes, may also be used. Probe-based qRT-PCR relies on the sequence-specific detection of a desired PCR product. It utilizes a fluorescently labeled target-specific probe, which results in increased specificity and sensitivity.

10. No template controls (NTC) should be included for each pair of primers tested to ensure that there is no reagent contamination. In these control reactions, water is added instead of sample, so no amplification is expected. In case the NTC reaction shows the synthesis of amplification products (i.e., the presence of a contaminant), measures such as pipette decontamination, using new primers aliquots, or thorough bench cleaning might be necessary. No reverse transcription controls (no-RT, or RT-) are used to detect the presence of contaminant genomic DNA in the RNA samples. If the RT- reaction shows the synthesis of amplification products, the corresponding RNA samples should be treated with DNase prior to their use in the reverse transcription reaction. If the primers were designed to span an intron or an intron/exon boundary, it is not necessary to perform a no-RT control.
11. MIQE Guidelines [16]. The MIQE guidelines were published in response to the recognition that several publications contain little information that describes the qPCR or that gives the reader the opportunity to determine the quality of the experiment. The result of these omissions is that several publications contain misleading conclusions based on inadequate quality control of the technical process. The MIQE guidelines contain a step-by-step guide and checklist, which leads the experimenter through the process of experiment validation. This has the additional function of providing a framework for publication analysis by peer reviewers and journal editors. Several publishing houses are now requiring that MIQE guidelines are followed for papers containing qPCR data.

Acknowledgements

Work in the authors' laboratory is funded by grants from Spanish Ministerio de Economía y Competividad (BFU2011-22734; and Programa Consolider-Ingenio, CSD2007-00036), and the Agència de Gestió d'Ajuts Universitaris i de Recerca (AGAUR) (grant SGR2009-GRC476). J. J. was supported by a fellowship from the European Molecular Biology Organization (EMBO).

References

1. Bustin SA, Benes V, Nolan T, Pfaffl MW (2005) Quantitative real-time RT-PCR—a perspective. J Mol Endocrinol 34(3):597–601
2. Kubista M, Andrade JM, Bengtsson M, Forootan A, Jonak J, Lind K, Sindelka R, Sjoback R, Sjogreen B, Strombom L, Stahlberg A, Zoric N (2006) The real-time polymerase chain reaction. Mol Aspects Med 27(2–3):95–125
3. Nolan T, Hands RE, Bustin SA (2006) Quantification of mRNA using real-time RT-PCR. Nat Protoc 1(3):1559–1582
4. Fleige S, Pfaffl MW (2006) RNA integrity and the effect on the real-time qRT-PCR performance. Mol Aspects Med 27(2–3):126–139
5. Nolan T, Hands RE, Ogunkolade W, Bustin SA (2006) SPUD: a quantitative PCR assay for

the detection of inhibitors in nucleic acid preparations. Anal Biochem 351(2):308–310

6. Gutierrez L, Mauriat M, Guenin S, Pelloux J, Lefebvre JF, Louvet R, Rusterucci C, Moritz T, Guerineau F, Bellini C, Van Wuytswinkel O (2008) The lack of a systematic validation of reference genes: a serious pitfall undervalued in reverse transcription-polymerase chain reaction (RT-PCR) analysis in plants. Plant Biotechnol J 6(6):609–618
7. Gutierrez L, Mauriat M, Pelloux J, Bellini C, Van Wuytswinkel O (2008) Towards a systematic validation of references in real-time RT-PCR. Plant Cell 20(7):1734–1735
8. Vandesompele J, De Preter K, Pattyn F, Poppe B, Van Roy N, De Paepe A, Speleman F (2002) Accurate normalization of real-time quantitative RT-PCR data by geometric averaging of multiple internal control genes. Genome Biol 3(7):RESEARCH0034
9. Pfaffl MW, Tichopad A, Prgomet C, Neuvians TP (2004) Determination of stable housekeeping genes, differentially regulated target genes and sample integrity: BestKeeper—Excel-based tool using pair-wise correlations. Biotechnol Lett 26(6):509–515
10. Schmid M, Davison TS, Henz SR, Pape UJ, Demar M, Vingron M, Scholkopf B, Weigel D, Lohmann JU (2005) A gene expression map of *Arabidopsis thaliana* development. Nat Genet 37:501–506
11. Czechowski T, Stitt M, Altmann T, Udvardi MK, Scheible WR (2005) Genome-wide identification and testing of superior reference genes for transcript normalization in Arabidopsis. Plant Physiol 139(1):5–17
12. Chao WS, Dogramaci M, Foley ME, Horvath DP, Anderson JV (2012) Selection and validation of endogenous reference genes for qRT-PCR analysis in leafy spurge (Euphorbia esula). PLoS One 7(8):e42839
13. Caldana C, Scheible WR, Mueller-Roeber B, Ruzicic S (2007) A quantitative RT-PCR platform for high-throughput expression profiling of 2500 rice transcription factors. Plant Methods 3:7
14. Hruz T, Wyss M, Docquier M, Pfaffl MW, Masanetz S, Borghi L, Verbrugghe P, Kalaydjieva L, Bleuler S, Laule O, Descombes P, Gruissem W, Zimmermann P (2011) RefGenes: identification of reliable and condition specific reference genes for RT-qPCR data normalization. BMC Genomics 12:156
15. Wellmer F, Alves-Ferreira M, Dubois A, Riechmann JL, Meyerowitz EM (2006) Genome-wide analysis of gene expression during early Arabidopsis flower development. PLoS Genet 2(7):e117
16. Bustin SA, Benes V, Garson JA, Hellemans J, Huggett J, Kubista M, Mueller R, Nolan T, Pfaffl MW, Shipley GL, Vandesompele J, Wittwer CT (2009) The MIQE guidelines: minimum information for publication of quantitative real-time PCR experiments. Clin Chem 55(4):611–622
17. Pfaffl MW (2001) A new mathematical model for relative quantification in real-time RT-PCR. Nucleic Acids Res 29(9):e45
18. Livak KJ, Schmittgen TD (2001) Analysis of relative gene expression data using real-time quantitative PCR and the 2(-Delta Delta C(T)) Method. Methods 25(4):402–408

Chapter 22

Misexpression Approaches for the Manipulation of Flower Development

Yifeng Xu, Eng-Seng Gan, and Toshiro Ito

Abstract

The generation of dominant gain-of-function mutants through activation tagging is a forward genetic approach that complements the screening of loss-of-function mutants and that has been successfully applied to studying the mechanisms of flower development. In addition, the functions of genes of interest can be further analyzed through reverse genetics. A commonly used method is gene overexpression, where strong, often ectopic expression can result in an opposite phenotype to that caused by a loss-of-function mutation. When overexpression is detrimental, the misexpression of a gene using tissue-specific promoters can be useful to study spatial-specific function. As flower development is a multistep process, it can be advantageous to control gene expression, or its protein product activity, in a temporal and/or spatial manner. This has been made possible through several inducible promoter systems, as well as by constructing chimeric fusions between the ligand binding domain of the glucocorticoid receptor (GR) and the protein of interest. Upon treatment with a steroid hormone at a specific time point, the fusion protein can enter the nucleus and activate downstream target genes. All these methods allow us to genetically manipulate gene expression during flower development. In this chapter, we describe methods to produce the expression constructs, method of screening, and more general applications of the techniques.

Key words Overexpression, Misexpression, Promoter, Protein tagging, Activation tagging, Glucocorticoid receptor

1 Introduction

Over the last few decades, loss-of-function mutants produced by chemical, irradiation, and insertional mutagenesis have contributed greatly to our understanding of flower development. However, the screening for loss-of-function mutations (*see* Chapter 6), or the uncovering of gene function through the analysis of knockout mutants (*see* Chapter 20), is often hampered by genetic redundancy, i.e., where the activity of one gene can compensate for the loss of another. To circumvent this problem, plant transformation vectors containing multimerized transcriptional enhancers of the constitutive Cauliflower Mosaic Virus (CaMV) *35S* promoter have been used to generate dominant mutants, which typically result

José Luis Riechmann and Frank Wellmer (eds.), *Flower Development: Methods and Protocols*, Methods in Molecular Biology, vol. 1110, DOI 10.1007/978-1-4614-9408-9_22, © Springer Science+Business Media New York 2014

from the transcriptional activation, or over-expression, of genes in proximity of the resulting T-DNA insertions [1–3]. Thus, unlike the screening of loss-of-function mutants, such activation tagging lines can be screened in the T1 generation for phenotypic alterations resulting from enhanced gene function.

Once a floral regulator has been identified by forward or reverse genetics, its function can be further characterized by manipulating its expression in a number of ways. In addition to knocking out and reducing gene expression, the simplest and most commonly used approach is to overexpress the gene by placing it under the control of an ectopic and constitutive promoter, such as the CaMV *35S* promoter. Ectopic misexpression of a gene may result in a phenotype that helps to elucidate its functions at the cellular and organismal level. Whereas the analysis of a loss-of-function mutant might allow inferring that the affected gene is essential for a certain developmental function, overexpression studies can reveal whether such a gene is sufficient to carry out that function in a certain development context. For example, the ectopic expression of the floral organ identity gene *AGAMOUS* (*AG*) causes the homeotic transformation of perianth to reproductive organs, suggesting that it is sufficient to determine reproductive organ identity in the context of floral primordia [4].

Although useful and simple to apply, constitutive overexpression approaches have some limitations. Firstly, the CaMV *35S* promoter is often not completely constitutive, but rather exhibits both developmental and tissue specificity. One example is pollen, in which the promoter shows no or only very low activity [5]. Secondly, the high levels of transcript abundance that result from constitutive overexpression might mask temporal and spatial-specific effects, and/or may lead to lethality or sterility (which would make it impossible to obtain stable transformants). In other cases, the constitutive and ectopic expression of genes may lead to artificial phenotypes, by activating or repressing signalling cascades that are not normally linked with the gene's function. Thus, the molecular analyses of downstream cascades and phenotypic interpretation need to be carefully conducted and correlated with those of the corresponding loss-of-function mutants. Furthermore, to overcome the limitations of constitutive overexpression, several methods can also be employed to control gene expression in a temporal and/or spatial-dependent fashion.

Tissue-specific promoters and inducible promoter systems can be applied in different situations. Tissue-specific promoters restrict the activity of the transgenes to certain tissues, which allows assessing the function of a gene in specific cell types. In addition to directly driving a gene of interest under a specific promoter, two-component systems are also available, which utilize a trans-activator and the binding sites conjugated with a minimal promoter, such as the LhG4/*pOp* system [6, 7]. The LhG4 activator component carries the DNA-binding domain of a mutant *lac* repressor and a transcriptional activation domain of the yeast GAL4 protein. The DNA-binding domain of a mutant *lac* repressor binds to the *lac* repressor-binding sites (*lac* operators or Op) with particularly high-affinity. Expression of the gene of interest, which has been placed under the control of the *lac* repressor binding sites, is thus determined by the expression pattern of the LhG4 activator, which is controlled by the chosen (tissue-specific) promoter. An important advantage of two-component systems is that they facilitate combinatorial experimental designs, where a number of genes of interest can be studied in different developmental contexts by generating and crossing the appropriate driver and reporter lines [8, 9].

In addition to tissue-specific expression, the ability to temporally modulate gene activity is essential to study downstream cascades and to identify immediate or direct effects of a gene in developmental processes. Chemical inducible systems are one of the solutions to temporally control transgene expression and/or activity, with those based on the glucocorticoid receptor (GR) being the most commonly used [10]. A one-component GR inducible strategy is suitable for nuclear-localized proteins, such as transcription factors and chromatin remodeling factors. In this case, the respective protein-coding sequence is translationally fused (at its N- or C-terminus) with the sequence coding for the GR ligand-binding domain. In transgenic plants, these chimeric transcription factors are retained in the cytoplasm by heat shock proteins in the absence of the ligand. Upon treatment of the tissue or cells with a glucocorticoid such as dexamethasone, the GR fusion protein is released from its cytoplasmic retention and imported into the nucleus, where it may be functionally active and thus activated. The addition of epitope tags, such as c-Myc or HA, at the N- or C-terminal end of the fusion protein is helpful for biochemical assays such as Chromatin Immunoprecipitation (ChIP) to examine the binding of a transcriptional regulator to specific target sites [11, 12]. Functionality of the fusion protein needs to be confirmed in the respective mutant background by phenotypic rescue upon dexamethasone induction. Furthermore, transgenic lines with an optimal inducibility need to be isolated, because the expression of transgenes can depend on, or be affected by, their place of insertion in the genome.

Two-component inducible promoter systems utilize GR-fused trans-activators and the corresponding binding sites, such as GR-LhG4/*pOp* to temporally activate transgenes [6, 13]. In this case, the GR domain is fused to the N-terminus of the synthetic transcription factor LhG4, whose expression is under the control of the constitutive CaMV *35S* promoter or of a tissue-specific promoter. Other chemically inducible two-component systems have also been successfully used. For instance, the GAG system utilizes a fusion protein between the Gal4 DNA binding domain, which binds to Gal4 UAS, and the VP16 activation domain, followed by GR [14]. The estradiol-inducible XVE system is based on a fusion between the DNA binding domain of LexA, VP16 and an estradiol receptor ligand binding domain (ER) [15]. The ethanol-inducible AlcR/AlcA system is based on the transcriptional activator AlcR from *Aspergillus nidulans* and an AlcR-responsive promoter element, *alcA* [16].

This chapter describes our current protocols for activation tagging, as well as for the use of ectopic over-expression and tissue-specific expression of a gene of interest in flower development. In addition, the use of two different types of inducible systems, direct GR fusions to transcription factors (induction of protein activity) and the two-component LhG4-GR/*pOP* system (transcriptional induction), is described.

2 Materials

2.1 Construction of Expression Cassettes

1. Vectors (*see* Fig. 1).
2. cDNA prepared from tissue(s) expressing the gene of interest, or genomic DNA for tissue-specific promoters.
3. Competent cells of appropriate *E. coli* strains.
4. Luria–Bertani (LB) media (broth and agar plate).
5. Antibiotics for selection of plasmids in *E. coli*.
6. Oligonucleotide primers.
7. Restriction enzymes, *Taq* DNA polymerase, high fidelity and proofreading DNA polymerase, and T4 DNA ligase.

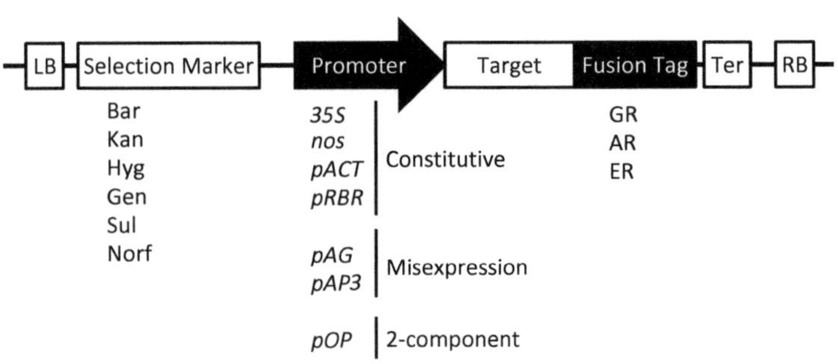

Fig. 1 Schematic representation of expression cassettes used for the misexpression of genes during flower development. Selection markers, promoters, gene fusions, transcriptional terminators (Ter), as well as Left Border (LB) and Right Border (RB) sequences are indicated. Selection markers: *Bar* Basta (or glufosinate), *Kan* Kanamycin, *Hyg* Hygromycin, *Gen* Gentamycin, *Sul* Sulfadiazine, *Norf* Norflurazon. Promoters: *35S* CaMV *35S*, *nos* Nopaline synthase promoter, *pACT ACTIN* promoter, *pRBR RETINOBLASTOMA-RELATED* promoter, *pAG AGAMOUS* promoter, *pAP3 APETALA3* promoter, *pOP lac* operators positioned upstream of a minimal promoter. Protein fusions: *GR* glucocorticoid receptor, *AR* androgen receptor, *ER* estrogen receptor

8. Agarose.
9. Gel extraction kit.
10. PCR purification kit.
11. Plasmid Miniprep kit.

2.2 *Agrobacterium*-Mediated Transformation of *Arabidopsis*

1. *Agrobacterium tumefaciens* strains.
2. Luria–Bertani (LB) media (broth and agar plate).
3. Antibiotics for selection of plasmids in *Agrobacteria*.
4. Infiltration medium: 5 % (w/v) (50 g/L) sucrose, 1× (4.4 g/L) Murashige & Skoog (MS) salt, 0.03 % (v/v) (300 μL/L) Silwet L-77.

2.3 Screening of Transgenic Plants

1. MS solid medium: 1× MS salts, 3 % sucrose, 0.8 % Phytagel™, pH 5.7 (adjusted with KOH).
2. Petri plates (100-mm square), around 40 mL of MS solid medium per plate.
3. Seed sterilization solution: freshly diluted bleach (10 % bleach) containing 0.01 % (v/v) Triton X-100.
4. 100 % ethanol and 95 % ethanol prepared with sterilized water.
5. Antibiotics for selection of transgenic plants.

2.4 Dexamethasone Induction and Protein Synthesis Inhibition Treatment

1. Plastic container.
2. Dexamethasone stock solution: 30 mM dexamethasone. Resuspend 23.55 mg of dexamethasone in 2 mL of ethanol. Store the stock solution at −20 °C, where it will last for at least 1 month.
3. Dexamethasone working solution: 0.5–25 μM (10 μM is often used as starting concentration) dexamethasone in deionized water with 0.015 % (v/v) Silwet L-77.
4. Cycloheximide stock solution: 10 mM cycloheximide in water. Solution can be stored in aliquots at −20 °C for several months.

2.5 Screening and Identification of Activation-Tagged Genes

1. Plant DNA extraction kit.
2. Restriction enzyme (*see* **Note 1**).
3. PCR purification kit.
4. T4 DNA ligase.
5. Competent *E. coli* cells.
6. Luria–Bertani (LB) media (broth and agar plate).

2.6 Observation of Phenotypes

1. Stereomicroscope.

3 Methods

3.1 Construction of Expression Cassettes

All cloning procedures are done in accordance with standard protocols [17]. A typical expression cassette flanked by the Left and Right Border is shown in Fig. 1.

3.1.1 Choice of Promoters

A good promoter of choice for overexpression is the Cauliflower Mosaic Virus (CaMV) *35S* promoter [18–20]. There are different versions of the promoter. We frequently use the double 35S promoter with success for strong ectopic expression [18]. Other constitutive promoters are also available, such as nopaline synthase (*nos*), octopine synthase (*ocs*), and mannopine synthase (*mas*) promoters [21–23], or promoters of housekeeping genes, such as actin (p*ACT*) or ubiquitin (p*UBQ*) [24, 25]. The *RETINOBLASTOMA-RELATED* promoter (*pRBR*) can be used to drive expression in the male and female gametophyte [26, 27], in which the *35S* promoter is only weakly active.

The phenotypes that result from constitutive overexpression of a gene may not represent its normal function in a native environment. Moreover, gene overexpression may cause plant lethality and sterility, making it impossible to obtain stable transformants. In these situations, using tissue-specific expression or misexpression might be a better choice. Typically, a clone containing 2–3 kb upstream of the transcriptional start site would be able to capture the endogenous expression of the gene. In some cases, regulatory elements are located in intragenic regions (as in the case of *AGAMOUS* (*AG*) and *SUPERMAN* (*SUP*) [28–30]) or in the 3′ untranslated or downstream regions of genes [31, 32]. Therefore, the use of the corresponding promoters to drive the expression of other genes would require including those regions in the constructs, to ensure proper expression patterns. *See* Table 1 for a list promoters used for studies in flower/meristem development.

For the two-component system, a gene of interest can be positioned downstream of the *pOP* promoter sequence. Depending on the need of the experiment, expression of the synthetic transcription factors LhG4 or LhG4-GR can be driven by the constitutive *35S* promoter or by tissue-specific promoters [7, 33]. Several different strategies can be used to combine the *pOP:GENE* construct and the *promoter::LhG4* construct *in planta*: the two constructs can be inserted into the same transformation vector; the two constructs could be used for sequential plant transformation (if they carry different selection markers); or doubly transgenic plants may be generated by crossing single transformants [8, 9].

3.1.2 Choices of Inducible Systems

Although dexamethasone-inducible GR fusions are the focus of this protocol, there are other chemical inducible systems that can be used, such as the dihydrotestosterone (DHT)-inducible

Table 1
Promoters used for misexpression studies in flower/meristem development

Gene	Expression pattern	References
ACTIN 4 (*ACT4*)	Ubiquitously expressed including developing pollen	[43]
RETINOBLASTOMA RELATED (*RBR*)	Ubiquitously expressed including developing pollen	[26, 27]
CLAVATA3 (*CLV3*)	In vegetative and floral stem cells	[44, 45]
UNUSUAL FLORAL ORGANS (*UFO*)	In shoot apical meristems and floral primordia at stages 1–3, restricted later to the junction between whorls 1 and 2	[45, 46]
CUP-SHAPED COTYLEDON2 (*CUC2*)	In meristems and organ boundaries, also in leaf margins	[47]
LEAFY (*LFY*)	Gradually increased in vegetative meristems and highly expressed in inflorescence and floral meristems	[45, 48]
AINTEGUMENTA (*ANT*)	Throughout emerging lateral organ primordia of all above-ground organs	[49, 50]
ASYMMETRIC LEAVES1 (*AS1*)	Throughout emerging lateral organ primordia of all above-ground organs	[51, 52]
FILAMENTOUS FLOWER (*FIL*)	In the peripheral region of vegetative and inflorescence meristems, later restricted to abaxial side of the vegetative and floral organ primordia	[7, 53, 54]
APETALA1 (*AP1*)	In floral meristems, later restricted to whorls 1 and 2 from stage 3 in developing flowers	[55, 56]
APETALA3 (*AP3*)	In whorls 2 and 3 from stage 3 in developing flowers	[52, 57]
AGAMOUS (*AG*)	In whorls 3 and 4 from stage 3 in developing flowers	[28]
CRABS CLAW (*CRC*)	In developing carpels and nectaries from stage 6	[58, 59]

Names of promoters, their expression patterns, and relevant references are given

Androgen Receptor (AR) [34] and the estradiol-inducible Estrogen Receptor (ER) [15]. These receptor tags can be fused with the protein-coding sequence directly in a one-component system, or with the activator in a two-component system such as LhG4-GR in the *pOP*/LhGR system [6]. Besides these chemical inducers, there are other inducible systems utilizing heat, cold, or alcohol (*see* Table 2), each with their own advantages and disadvantages. Although they are not the focus of this chapter, they are listed here for comparison.

The GR-based system is the most commonly used inducible system as glucocorticoid can easily permeate plant tissues and rapidly induces gene activation [14]. It can also be systematically transported, thus dexamethasone treatment can be done via spraying or dipping into a dexamethasone solution or planting on MS plates

Table 2
Inducible systems used to temporally control gene expression

Inducible system	Vector used	Mode of activation	References
GR fusion	pBI-ΔGR	Dexamethasone	[10, 60]
GVG/UAS	pBI101-GVG-UAS-Luc	Dexamethasone	[14]
pOP/LhGR	pBIN-LhGR-N, pOP6-GUS pBIN-LR-LhGR, pOP6-GUS pOPOff1	Dexamethasone	[6, 13]
AR fusion	pGREEN-35S-AR	Dihydrotestosterone	[34]
ER fusion	pSKM36-ESR2-ER	Estradiol	[43, 61]
XVE/OlexA	pER8	Estradiol	[15]
AlcR/AlcA	*alcR alcA::reporter*	Ethanol	[16]
pHSP18.2	pTT101 pGA482-pHSP18.2	Heat treatment (37 °C)	[62] [63]
CBF3/*pRD29A*	pMDC-CBF3 pFAJ-PAP1	Cold treatment (4 °C)	[40]

Names of vector used, their mode of activation, and relevant references are shown

containing dexamethasone [35]. On the other hand, estradiol is not easily transported in adult plants and hence, for local induction, an ER-based inducible system may be a better choice [36]. Another advantage of using these chemical inducers is that, at the concentration used, they are generally nontoxic to plants and show no observable pleiotropic physiological effect.

The ethanol-inducible AlcR/AlcA system is another system that can be easily applied to drive gene expression [16]. The inducer, ethanol, is nontoxic at the concentrations typically used. As ethanol evaporates quickly, ethanol vapor can be used for pulsed and local induction [37]. However, the volatility of ethanol means that it may also activate nearby plants during treatment. Extra care should also be taken when handling seedlings in an ethanol-sterilized environment.

Utilizing the heat-shock promoter *pHSP18.2*, genes can be induced simply by placing the plants in 37 °C condition for 1–2 h [38]. Compared to chemical treatment, heat treatment can be applied uniformly to plants at a large scale. Nevertheless, repeated heat-shock treatment may produce pleiotropic effects, especially affecting pollen development [39]. Unlike heat treatment, cold treatment can be used for a longer period in the CBF3/*pRD29A* system [40]. However, it should be noted that prolonged cold treatment impairs cell cycle progression and also causes pollen sterility [39].

3.1.3 Cloning Step

1. Amplify the coding region of the gene of interest using high fidelity DNA polymerase with cDNA (target gene) or gDNA (promoter) as the template (*see* **Note 2**).
2. Analyze the PCR amplification product by agarose gel electrophoresis, to check whether the size of the amplicon is correct, followed by extraction of the DNA fragment of interest from the gel with any gel extraction kit.
3. Digest amplicon and vector with appropriate restriction enzymes (*see* **Note 3**).
4. After the digestion reaction, cleanup the amplicon and the vector with a PCR purification kit if no big fragment is cut out. Otherwise, run the digestion products in a standard agarose gel and extract the DNA fragment(s) of interest with a gel extraction kit.
5. Ligate the vector with the insert (*see* **Note 4**).
6. Transform the ligation reaction using heat shock competent cells or by electroporation (*see* **Note 5**).
7. Add LB broth without antibiotic up to 1 mL and incubate for 1 h at 37 °C in a shaker at around 250 rpm. Spin down cells with a microcentrifuge at maximum speed. Pour off the supernatant. Resuspend the cell pellet with the residual LB Broth and spread on LB agar plates with the appropriate antibiotic. Incubate plates overnight at 37 °C.
8. Screen for positive clones by colony PCR, and transfer the clone into a culture tube with 3 mL of LB broth with the appropriate antibiotic. Incubate overnight at 37 °C in a shaker at around 250 rpm.
9. Extract the plasmid from the cells with a Miniprep kit according to the manual provided by the supplier.
10. Verify the plasmid sequence by Sanger DNA sequencing.

3.1.4 Activation Tagging

Many activation tagging vectors have been described in the literature and can be requested from the groups that generated them (*see* Table 3).

3.2 Transformation of Arabidopsis (Agrobacterium-Mediated Transformation)

1. Grow *Arabidopsis* on soil in trays or pots under a 16-h photoperiod at 22 °C. The plants are ready for transformation when their inflorescence stems are at least 5 cm in length. This occurs typically between 4 and 6 weeks after germination.
2. Transform an appropriate *Agrobacterium*-strain with the desired plant transformation vector, using either standard electroporation techniques or freeze-thaw transformation (*see* **Note 6**). After a recovery period of continuous shaking at 28 °C for 2–3 h, spread or streak the *Agrobacterium* cells onto LB agar plates containing the appropriate antibiotics for selection. Incubate at 28–30 °C for 2 days.

Table 3
Vectors used for activation tagging

Plasmid	Plant species	Plant selection	Reference
pPCVICEn4HPT	*Arabidopsis*, Petunia, Tobacco	Hygromycin	[1]
pSKI015	*Arabidopsis*	Basta	[2]
pSKI074	*Arabidopsis*	Kanamycin	[2]
pTag2B4A1	*Catharanthus*	Hygromycin	[64]
pGA2715	Rice	Hygromycin	[65]
pAG3202	Tomato	Kanamycin	[66]
pER16 (inducible)	*Arabidopsis*	Kanamycin	[67]

Names of plasmids, the plant species in which the vectors have been used, the selection markers for the isolation of transgenic plants, and relevant references are shown

3. Inoculate three individual colonies into 3 mL of LB Broth with the appropriate antibiotics. Incubate overnight at 28–30 °C with continuous shaking at around 250 rpm. in a standard orbital shaker.
4. Inoculate 500 mL of LB broth containing the appropriate antibiotics with 3 mL of the overnight culture. Grow overnight at 28–30 °C with continuous shaking until the culture reaches an OD_{600} of 1.0–1.5.
5. Harvest *Agrobacterium* cells by centrifugation at 8,000 × *g* for 15 min at room temperature. Resuspend the pellet with infiltration medium to an OD_{600} of 0.5–1.0.
6. Transfer 500 mL of the *Agrobacterium*-infiltration solution into a suitable plastic container. Trays or pots of *Arabidopsis* plants are placed upside down with the inflorescence submerged in the infiltration solution. Shake at 80 rpm on an orbital shaker for 3 min. Be careful not to let the soil contact the solution (*see* **Note 7**).
7. Cover the plants with plastic wrap to maintain high humidity and place transformed plants under low light intensity overnight.
8. Remove the plastic wrap and transfer the plants to growth room or chamber. Water the plants until they set seed. Allow the plants to dry gradually. Harvest dried seeds and store in a desiccator cabinet for at least 1 week.

3.3 Screening of Transformants

1. Sterilize T1 seeds for plating. Put T1 seeds in an appropriately sized tube and add approximately 10× volume of freshly prepared seed sterilization solution. Shake at high speed for 15 min.

Aspirate off liquid and rinse with approximately 10× volume of 95 % ethanol for three times.

2. Add around 10× volume of pure ethanol and pipette seeds onto filter paper (Whatman) to dry (*see* **Note 8**). Disperse dried seeds evenly onto MS medium plates with appropriate antibiotics.
3. Stratify the seeds by placing the plates at 4 °C in the dark for 2–5 days to synchronize germination, followed by placing the plates in a growth chamber at 22 °C with 16-h photoperiod.
4. Identify putative transgenic plants. Perform genotyping PCR, and assay for transgene expression by RT-PCR for verification. In the case of dexamethasone-inducible lines, select lines without any leaky phenotype (*see* **Note 9**).
5. Optional I: to screen the T1 transgenic plants for resistance to the herbicide Basta, the seeds also can be sown on soil and sprayed with 0.2 % (v/v) Basta solution after germination.
6. Optional II: perform Western blot analysis to identify lines that produce high levels of the fusion protein, using an antibody that recognizes the epitope tag introduced in the fusion protein (such as c-Myc or HA).

3.4 Dexamethasone Induction and Protein Synthesis Inhibition Treatment

Option 1: Grow seedlings on dexamethasone-containing MS plates for up to 21 days after germination. This approach leads to a sustained activation of the GR fusion protein and is thus not suitable for transient activation experiments.

1. Sterilize seeds and grow them on MS plates as described in Subheading 3.3. One set of MS plates contains dexamethasone at concentrations between 0.1 and 25 μM to test (*see* **Note 10**). Prepare another set of MS plates, which serve as controls, and replace the dexamethasone solution with an equal volume of ethanol.

Compare the phenotypes of the dexamethasone and mock-treated plants to identify phenotypic abnormalities likely caused by the activation of the GR fusion protein.

Option 2: Treat plants by submersion in dexamethasone-containing water. This approach is advisable for both transient and sustained inductions.

1. Freshly prepare a dexamethasone work solution and a mock-solution in which the dexamethasone stock solution is replaced with an equal volume of ethanol.
2. Submerge plants with an inflorescence shoot of around 5 cm into a plastic container with dexamethasone-containing or mock solutions. Shake at low speed (around 70 rpm) for 3 min to remove bubbles and allow the solution to penetrate thoroughly.

3. For sustained activation, treatments can be repeated daily (*see* **Note 11**). However, because a prolonged treatment with Silwet L-77 might have a toxic effect on plants, reduce its concentration after the initial treatment, or replace it with Triton X-100 (used at a concentration of 0.1 % (v/v)).
4. In transient activation experiments, 10 μM cycloheximide in water can be applied prior to or at the same time as the dexamethasone treatment, to inhibit de novo protein synthesis. This treatment is useful for testing whether gene expression changes are the result of a direct or indirect regulation of the transcription factor under study.

3.5 Screening and Determination of the Activation Tag

Screening of T1 seeds is done as described in Subheading 3.3. After selecting the lines with the desired phenotypes, it is important to determine the site of T-DNA insertion. This can be done through the method of Plasmid Rescue:

1. Extract genomic DNA from the transgenic lines (*see* **Note 12**).
2. Digest about 100 ng of purified genomic DNA with restriction enzyme (*see* **Note 1**).
3. Clean-up the digested genomic DNA with a PCR purification kit.
4. Use T4 DNA ligase to self-ligate the digested DNA at 16 °C at low concentration.
5. Clean-up the ligated plasmid with a PCR purification kit.
6. Transform the purified plasmid into competent *E. coli* cells.
7. Amplify the plasmid and analyze the flanking sequence by sequencing using the T-DNA specific primer.
8. Examine the genes located nearby the T-DNA insertion site (*see* **Note 13**).
9. Generate independent transgenic lines that overexpress the candidate gene(s) identified through activation tagging, to confirm its capability to cause the desired phenotype. The rescued plasmid can be transformed into plants if the full coding region of the gene is contained in it.

3.6 Observation of Phenotypes

After obtaining the desired lines, a phenotypic characterization is frequently done in plants of the T2 or T3 generation (*see* **Note 14**). *Arabidopsis* flower development can be divided into 12 stages during early development (Fig. 2) and another eight stages during late stage development after the bud opens [41]. Depending on the stage of development, observation of phenotypes may be done under a scanning electron microscope (*see* Chapters 7 and 13) or a stereomicroscope. Look for plants with floral phenotypes such as alteration of meristem structure or size, floral homeotic conversion, early or delayed termination of floral meristem, abnormalities in floral organ number, size, or shape, etc. It is also advisable to grow

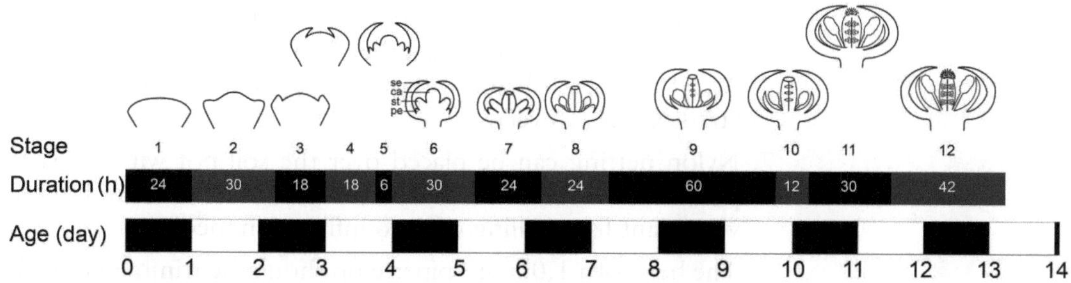

Fig. 2 Early *Arabidopsis* flower development. The duration of each stage is estimated to the nearest 6 h interval when wild-type Arabidopsis plants are grown at 25 °C under constant illumination [41]. Sizes of flower buds shown are not to scale

wild-type plants side by side to the transgenic lines as controls to account for potential environmental effects on plant growth and development.

4 Notes

1. The restriction enzyme to use depends on the activation tagging vector employed. For example, in pSKI015 and pSKI074, the restriction enzymes *Kpn*I, *Eco*RI, and *Hin*dIII can be used for the isolation of sequences next to the right border of the T-DNA [2].
2. For direct digestion of PCR products, the primers can be designed with a protective extension of two to four nucleotides at the 5′ end, depending on the restriction enzyme and the corresponding recognition site. Alternatively, to increase the efficiency of the restriction digestion, the amplicon can be inserted into TA cloning vectors (e.g., from Invitrogen or Promega) first.
3. To reduce the occurrence of self-ligation of the vector, Calf Intestinal Alkaline Phosphatase can be added to the restriction digestion mix to remove terminal phosphates from the DNA backbone.
4. Ligation reactions can be carried out with rapid ligation mixture at room temperature for a few minutes, or by standard conditions at 16 °C overnight. The total reaction volume is usually 5–10 μL. The amount of vector commonly used is 50 ng and the molar ratio of vector to insert is 1:3.
5. Prior to transformation into *E. coli* by electroporation, the DNA from the ligation reaction needs to be purified and desalted by ethanol precipitation and then resuspended in water.
6. Compared with the electroporation technique, freeze-thaw transformation is a much cheaper method. Thaw competent *Agrobacterium* cells on ice and add about 100 ng of plasmid to 250 μL of cell suspension. Incubate the mixture on ice for 5 min.

Transfer mixture to liquid nitrogen and incubate for 5 min. Incubate the mixture for another 5 min in a 37 °C water bath. Add 1 mL of LB Broth without antibiotics and shake for 2–4 h at 28 °C for recovery.

7. Nylon netting can be placed over the soil pot with seeds, to prevent the contact of soil with infiltration medium and to prevent plant from falling off into infiltration medium.
8. The head of a 1,000 μL pipette tip should be trimmed to make the size of the opening much bigger to pipette seeds easily. Filter tips are advised to prevent contaminations.
9. While dexamethasone-inducible regulation is tight, we sometimes observe leaky effects of fusion proteins in the absence of hormone. At least 20 independent lines should be characterized for optimal inducibility.
10. 10 μM dexamethasone is the standard concentration, although concentrations ranging from 0.1 to 25 μM have been used successfully. If treatments of 10 μM dexamethasone results in toxic effects to the plants, reduced dexamethasone concentrations should be tested.
11. In the case of *ag-1* 35S::AG-GR, inflorescences that were subjected to a single treatment with 10 μM dexamethasone solution, it was determined that the AG-GR protein was retained in the nucleus for up to 2–3 days [42]. For full rescue of the *ag* mutant phenotypes, at least four times of daily dexamethasone treatments was necessary.
12. High quality genomic DNA is important for successful cloning of the flanking sequence. Hence, we recommend using a suitable extraction kit following the manufacturer's instructions.
13. Four tandemly connected 35S enhancers of the *pPCVICEn4HPT* vector are active up to around 10 kb from the insertion site. If a gene showing enhanced expression cannot be found in the vicinity of the activation construct, consider that there may be an unannotated gene present, such as a gene coding for a noncoding small regulatory RNA or a gene coding for a small peptide.
14. Molecular analysis should be carried out in homozygous lines to prevent sample variation.

Acknowledgement

This work was supported by research grants to T.I. from Temasek Life Sciences Laboratory (TLL), the National Research Foundation Singapore under its Competitive Research Programme (CRP Award No. NRF-CRP001-108) and PRESTO, Japan Science and Technology Agency, 4-1-8 Honcho Kawaguchi, Saitama, Japan.

References

1. Walden R, Fritze K, Hayashi H, Miklashevichs E, Harling H, Schell J (1994) Activation tagging: a means of isolating genes implicated as playing a role in plant growth and development. Plant Mol Biol 26(5):1521–1528
2. Weigel D, Ahn JH, Blazquez MA, Borevitz JO, Christensen SK, Fankhauser C, Ferrandiz C, Kardailsky I, Malancharuvil EJ, Neff MM, Nguyen JT, Sato S, Wang ZY, Xia Y, Dixon RA, Harrison MJ, Lamb CJ, Yanofsky MF, Chory J (2000) Activation tagging in Arabidopsis. Plant Physiol 122(4):1003–1013
3. Ito T, Meyerowitz EM (2000) Overexpression of a gene encoding a cytochrome P450, CYP78A9, induces large and seedless fruit in Arabidopsis. Plant Cell 12(9):1541–1550
4. Mizukami Y, Ma H (1992) Ectopic expression of the floral homeotic gene AGAMOUS in transgenic Arabidopsis plants alters floral organ identity. Cell 71(1):119–131
5. Wilkinson JE, Twell D, Lindsey K (1997) Activities of CaMV 35S and nos promoters in pollen: implications for field release of transgenic plants. J Exp Bot 48(2):265–275
6. Moore I, Samalova M, Kurup S (2006) Transactivated and chemically inducible gene expression in plants. Plant J 45(4):651–683
7. Goldshmidt A, Alvarez JP, Bowman JL, Eshed Y (2008) Signals derived from YABBY gene activities in organ primordia regulate growth and partitioning of Arabidopsis shoot apical meristems. Plant Cell 20(5):1217–1230
8. Moore I, Galweiler L, Grosskopf D, Schell J, Palme K (1998) A transcription activation system for regulated gene expression in transgenic plants. Proc Natl Acad Sci U S A 95(1): 376–381
9. Baroux C, Blanvillain R, Betts H, Batoko H, Craft J, Martinez A, Gallois P, Moore I (2005) Predictable activation of tissue-specific expression from a single gene locus using the pOp/LhG4 transactivation system in Arabidopsis. Plant Biotechnol J 3(1):91–101
10. Lloyd AM, Schena M, Walbot V, Davis RW (1994) Epidermal cell fate determination in Arabidopsis: patterns defined by a steroid-inducible regulator. Science 266(5184): 436–439
11. Earley KW, Haag JR, Pontes O, Opper K, Juehne T, Song K, Pikaard CS (2006) Gateway-compatible vectors for plant functional genomics and proteomics. Plant J 45(4):616–629
12. Knop M, Siegers K, Pereira G, Zachariae W, Winsor B, Nasmyth K, Schiebel E (1999) Epitope tagging of yeast genes using a PCR-based strategy: more tags and improved practical routines. Yeast 15(10B):963–972
13. Craft J, Samalova M, Baroux C, Townley H, Martinez A, Jepson I, Tsiantis M, Moore I (2005) New pOp/LhG4 vectors for stringent glucocorticoid-dependent transgene expression in Arabidopsis. Plant J 41(6):899–918
14. Aoyama T, Chua NH (1997) A glucocorticoid-mediated transcriptional induction system in transgenic plants. Plant J 11(3):605–612
15. Zuo J, Niu QW, Chua NH (2000) Technical advance: an estrogen receptor-based transactivator XVE mediates highly inducible gene expression in transgenic plants. Plant J 24(2):265–273
16. Roslan HA, Salter MG, Wood CD, White MR, Croft KP, Robson F, Coupland G, Doonan J, Laufs P, Tomsett AB, Caddick MX (2001) Characterization of the ethanol-inducible alc gene-expression system in Arabidopsis thaliana. Plant J 28(2):225–235
17. Green MR, Sambrook J (2012) Molecular cloning: a laboratory manual, 4th edn. Cold Spring Harbor Laboratory Press, New York
18. Fang RX, Nagy F, Sivasubramaniam S, Chua NH (1989) Multiple cis regulatory elements for maximal expression of the cauliflower mosaic virus 35S promoter in transgenic plants. Plant Cell 1(1):141–150
19. Benfey PN, Chua NH (1990) The cauliflower mosaic virus 35S promoter: combinatorial regulation of transcription in plants. Science 250(4983):959–966
20. Benfey PN, Ren L, Chua NH (1990) Tissue-specific expression from CaMV 35S enhancer subdomains in early stages of plant development. EMBO J 9(6):1677–1684
21. Leisner SM, Gelvin SB (1988) Structure of the octopine synthase upstream activator sequence. Proc Natl Acad Sci U S A 85(8):2553–2557
22. DiRita VJ, Gelvin SB (1987) Deletion analysis of the mannopine synthase gene promoter in sunflower crown gall tumors and Agrobacterium tumefaciens. Mol Gen Genet 207(2–3): 233–241
23. An G (1986) Development of plant promoter expression vectors and their use for analysis of differential activity of nopaline synthase promoter in transformed tobacco cells. Plant Physiol 81(1):86–91
24. An YQ, McDowell JM, Huang S, McKinney EC, Chambliss S, Meagher RB (1996) Strong, constitutive expression of the Arabidopsis ACT2/ACT8 actin subclass in vegetative tissues. Plant J 10(1):107–121

25. Callis J, Raasch JA, Vierstra RD (1990) Ubiquitin extension proteins of Arabidopsis thaliana. Structure, localization, and expression of their promoters in transgenic tobacco. J Biol Chem 265(21):12486–12493
26. Johnston AJ, Kirioukhova O, Barrell PJ, Rutten T, Moore JM, Baskar R, Grossniklaus U, Gruissem W (2010) Dosage-sensitive function of retinoblastoma related and convergent epigenetic control are required during the Arabidopsis life cycle. PLoS Genet 6(6):e1000988
27. Borghi L, Gutzat R, Futterer J, Laizet Y, Hennig L, Gruissem W (2010) Arabidopsis RETINOBLASTOMA-RELATED is required for stem cell maintenance, cell differentiation, and lateral organ production. Plant Cell 22(6):1792–1811
28. Sieburth LE, Meyerowitz EM (1997) Molecular dissection of the AGAMOUS control region shows that cis elements for spatial regulation are located intragenically. Plant Cell 9(3):355–365
29. Busch MA, Bomblies K, Weigel D (1999) Activation of a floral homeotic gene in Arabidopsis. Science 285(5427):585–587
30. Ito T, Sakai H, Meyerowitz EM (2003) Whorl-specific expression of the SUPERMAN gene of Arabidopsis is mediated by cis elements in the transcribed region. Curr Biol 13(17): 1524–1530
31. Kaufmann K, Wellmer F, Muino JM, Ferrier T, Wuest SE, Kumar V, Serrano-Mislata A, Madueno F, Krajewski P, Meyerowitz EM, Angenent GC, Riechmann JL (2010) Orchestration of floral initiation by APETALA1. Science 328(5974):85–89
32. Ito T, Wellmer F, Yu H, Das P, Ito N, Alves-Ferreira M, Riechmann JL, Meyerowitz EM (2004) The homeotic protein AGAMOUS controls microsporogenesis by regulation of SPOROCYTELESS. Nature 430(6997): 356–360
33. Lenhard M, Bohnert A, Jurgens G, Laux T (2001) Termination of stem cell maintenance in Arabidopsis floral meristems by interactions between WUSCHEL and AGAMOUS. Cell 105(6):805–814
34. Sun B, Xu Y, Ng KH, Ito T (2009) A timing mechanism for stem cell maintenance and differentiation in the Arabidopsis floral meristem. Genes Dev 23(15):1791–1804
35. Geng X, Mackey D (2011) Dose-response to and systemic movement of dexamethasone in the GVG-inducible transgene system in Arabidopsis. Methods Mol Biol 712:59–68
36. Tornero P, Chao RA, Luthin WN, Goff SA, Dangl JL (2002) Large-scale structure-function analysis of the Arabidopsis RPM1 disease resistance protein. Plant Cell 14(2): 435–450
37. Maizel A, Weigel D (2004) Temporally and spatially controlled induction of gene expression in Arabidopsis thaliana. Plant J 38(1):164–171
38. Masclaux F, Charpenteau M, Takahashi T, Pont-Lezica R, Galaud JP (2004) Gene silencing using a heat-inducible RNAi system in Arabidopsis. Biochem Biophys Res Commun 321(2):364–369
39. Zinn KE, Tunc-Ozdemir M, Harper JF (2010) Temperature stress and plant sexual reproduction: uncovering the weakest links. J Exp Bot 61(7):1959–1968
40. Feng Y, Cao CM, Vikram M, Park S, Kim HJ, Hong JC, Cisneros-Zevallos L, Koiwa H (2011) A three-component gene expression system and its application for inducible flavonoid overproduction in transgenic Arabidopsis thaliana. PLoS One 6(3):e17603
41. Smyth DR, Bowman JL, Meyerowitz EM (1990) Early flower development in Arabidopsis. Plant Cell 2(8):755–767
42. Ito T, Ng KH, Lim TS, Yu H, Meyerowitz EM (2007) The homeotic protein AGAMOUS controls late stamen development by regulating a jasmonate biosynthetic gene in Arabidopsis. Plant Cell 19:3516–3529
43. Ikeda Y, Banno H, Niu QW, Howell SH, Chua NH (2006) The ENHANCER OF SHOOT REGENERATION 2 gene in Arabidopsis regulates CUP-SHAPED COTYLEDON 1 at the transcriptional level and controls cotyledon development. Plant Cell Physiol 47(11): 1443–1456
44. Fletcher JC, Brand U, Running MP, Simon R, Meyerowitz EM (1999) Signaling of cell fate decisions by CLAVATA3 in Arabidopsis shoot meristems. Science 283(5409):1911–1914
45. Deveaux Y, Peaucelle A, Roberts GR, Coen E, Simon R, Mizukami Y, Traas J, Murray JA, Doonan JH, Laufs P (2003) The ethanol switch: a tool for tissue-specific gene induction during plant development. Plant J 36(6):918–930
46. Durfee T, Roe JL, Sessions RA, Inouye C, Serikawa K, Feldmann KA, Weigel D, Zambryski PC (2003) The F-box-containing protein UFO and AGAMOUS participate in antagonistic pathways governing early petal development in Arabidopsis. Proc Natl Acad Sci U S A 100(14):8571–8576
47. Steiner E, Efroni I, Gopalraj M, Saathoff K, Tseng TS, Kieffer M, Eshed Y, Olszewski N, Weiss D (2012) The Arabidopsis O-linked N-acetylglucosamine transferase SPINDLY interacts with class I TCPs to facilitate cytokinin responses in leaves and flowers. Plant Cell 24(1):96–108

48. Blazquez MA, Soowal LN, Lee I, Weigel D (1997) LEAFY expression and flower initiation in Arabidopsis. Development 124(19):3835–3844
49. Elliott RC, Betzner AS, Huttner E, Oakes MP, Tucker WQ, Gerentes D, Perez P, Smyth DR (1996) AINTEGUMENTA, an APETALA2-like gene of Arabidopsis with pleiotropic roles in ovule development and floral organ growth. Plant Cell 8(2):155–168
50. Long J, Barton MK (2000) Initiation of axillary and floral meristems in Arabidopsis. Dev Biol 218(2):341–353
51. Eshed Y, Baum SF, Perea JV, Bowman JL (2001) Establishment of polarity in lateral organs of plants. Curr Biol 11(16):1251–1260
52. Tretter EM, Alvarez JP, Eshed Y, Bowman JL (2008) Activity range of Arabidopsis small RNAs derived from different biogenesis pathways. Plant Physiol 147(1):58–62
53. Sawa S, Watanabe K, Goto K, Liu YG, Shibata D, Kanaya E, Morita EH, Okada K (1999) FILAMENTOUS FLOWER, a meristem and organ identity gene of Arabidopsis, encodes a protein with a zinc finger and HMG-related domains. Genes Dev 13(9):1079–1088
54. Siegfried KR, Eshed Y, Baum SF, Otsuga D, Drews GN, Bowman JL (1999) Members of the YABBY gene family specify abaxial cell fate in Arabidopsis. Development 126(18): 4117–4128
55. Mandel MA, Gustafson-Brown C, Savidge B, Yanofsky MF (1992) Molecular characterization of the Arabidopsis floral homeotic gene APETALA1. Nature 360(6401):273–277
56. Ohno CK, Reddy GV, Heisler MG, Meyerowitz EM (2004) The Arabidopsis JAGGED gene encodes a zinc finger protein that promotes leaf tissue development. Development 131(5): 1111–1122
57. Tilly JJ, Allen DW, Jack T (1998) The CArG boxes in the promoter of the Arabidopsis floral organ identity gene APETALA3 mediate diverse regulatory effects. Development 125(9):1647–1657
58. Bowman JL, Smyth DR (1999) CRABS CLAW, a gene that regulates carpel and nectary development in Arabidopsis, encodes a novel protein with zinc finger and helix-loop-helix domains. Development 126(11):2387–2396
59. Alvarez JP, Goldshmidt A, Efroni I, Bowman JL, Eshed Y (2009) The NGATHA distal organ development genes are essential for style specification in Arabidopsis. Plant Cell 21(5):1373–1393
60. Simon R, Igeno MI, Coupland G (1996) Activation of floral meristem identity genes in Arabidopsis. Nature 384(6604):59–62
61. Che P, Lall S, Howell SH (2008) Acquiring competence for shoot development in Arabidopsis: ARR2 directly targets A-type ARR genes that are differentially activated by CIM preincubation. Plant Signal Behav 3(2):99–101
62. Matsuhara S, Jingu F, Takahashi T, Komeda Y (2000) Heat-shock tagging: a simple method for expression and isolation of plant genome DNA flanked by T-DNA insertions. Plant J 22(1):79–86
63. Yoshida K, Kasai T, Garcia MR, Sawada S, Shoji T, Shimizu S, Yamazaki K, Komeda Y, Shinmyo A (1995) Heat-inducible expression system for a foreign gene in cultured tobacco cells using the HSP18.2 promoter of Arabidopsis thaliana. Appl Microbiol Biotechnol 44(3–4):466–472
64. van der Fits L, Hilliou F, Memelink J (2001) T-DNA activation tagging as a tool to isolate regulators of a metabolic pathway from a genetically non-tractable plant species. Transgenic Res 10(6):513–521
65. Jeong DH, An S, Kang HG, Moon S, Han JJ, Park S, Lee HS, An K, An G (2002) T-DNA insertional mutagenesis for activation tagging in rice. Plant Physiol 130(4):1636–1644
66. Mathews H, Clendennen SK, Caldwell CG, Liu XL, Connors K, Matheis N, Schuster DK, Menasco DJ, Wagoner W, Lightner J, Wagner DR (2003) Activation tagging in tomato identifies a transcriptional regulator of anthocyanin biosynthesis, modification, and transport. Plant Cell 15(8):1689–1703
67. Zuo J, Niu QW, Frugis G, Chua NH (2002) The WUSCHEL gene promotes vegetative-to-embryonic transition in Arabidopsis. Plant J 30(3):349–359

Chapter 23

Next-Generation Sequencing Applied to Flower Development: RNA-Seq

Jun He and Yuling Jiao

Abstract

Genome-wide study of gene expression, or transcriptome profiling, is critical for our understanding of biological functions, including developmental processes. Recent breakthroughs in high-throughput sequencing technologies have revolutionized gene expression profiling to study the transcriptome at the nucleotide level, which is known as RNA-seq. RNA-seq, also called "whole transcriptome shotgun sequencing," uses next-generation sequencing technologies to sequence cDNA in order to infer a sample's RNA content. Here we describe a detailed bench-ready protocol to generate RNA-seq libraries for high-throughput single-end or pair-end sequencing compatible with the Illumina sequencing platform.

Key words Next-generation sequencing, RNA-seq, Transcriptome

1 Introduction

High-throughput gene expression study is central to many biological processes including flower development, which is under the control of complicated transcriptional regulatory networks [1]. Evolved from Southern blotting, DNA microarrays containing tens of thousands of DNA probe spots attached to a solid surface expanded gene expression profiling to the genome scale [2]. However, hybridization-based microarray methods have several limitations, including reliance on genome sequence information, high background levels owing to cross-hybridization, and a limited dynamic range of detection owing to both background and saturation of signals [3]. Moreover, it is hard to compare expression levels among different experiments [3]. Recent advances in next-generation sequencing technologies have made it possible to directly sequence the transcriptome [4–12]. Deep-sequencing based RNA-seq has overcome the limitations encountered by microarray technologies [13]. In addition, RNA-seq has the single-base resolution that is needed to study detailed features of each transcript, such as splicing isoforms, transcription start site(s), and

José Luis Riechmann and Frank Wellmer (eds.), *Flower Development: Methods and Protocols*, Methods in Molecular Biology, vol. 1110, DOI 10.1007/978-1-4614-9408-9_23,

polyadenylation site(s). In fact, RNA-seq has been rapidly adopted for many biological studies, including flower development [14–16]. However, the analysis of the data generated in RNA-Seq experiments can be complex, and the methods for data processing are still evolving. Whereas RNA-Seq data analysis is not described in this chapter, the reader is referred to several recent reviews addressing that topic [17–22].

2 Materials

Prepare all solutions with diethyl pyrocarbonate (DEPC)-treated water to avoid RNA degradation.

2.1 Purification of Total RNA

1. Nuclease-free mortar and pestle.
2. Liquid nitrogen.
3. Total RNA extraction kit or reagent (such as the RNeasy plant mini kit, Qiagen, Hilden, Germany).
4. DEPC-treated water: add 1 mL of DEPC to 1 L of distilled water. Mix thoroughly with magnetic stirrer overnight at room temperature. Autoclave and then cool to room temperature before use.

2.2 mRNA Purification from Total RNA

1. Dynabeads mRNA purification kit (Life Technologies, Carlsbad, CA, USA).
2. Magnetic stand (Promega, Madison, WI, USA).
3. Washing buffer: 10 mM Tris–HCl, 0.15 M LiCl, 1 mM EDTA, pH 7.5 at 25 °C.
4. Beads binding buffer: 20 mM Tris–HCl, 1 M LiCl, 2 mM EDTA, pH 7.5 at 25 °C.
5. 10 mM Tris buffer: 10 mM Tris–HCl, pH 7.5 at 25 °C.

2.3 Fragmentation of mRNA

1. 5× Fragmentation buffer: 200 mM Tris-acetate, 500 mM potassium acetate, 150 mM magnesium acetate.
2. Stop buffer: 200 mM EDTA, pH 8.0 at 25 °C.
3. Glycogen (20 μg/μL, Life Technologies, Carlsbad, CA, USA).
4. 3 M NaOAC, pH 5.2 at 25 °C.
5. 70 % ethanol, dilute with DEPC-treated water.

2.4 cDNA Synthesis

1. Superscript II reverse transcriptase (200 U/μL, Life Technologies, Carlsbad, CA, USA).
2. 5× First-strand buffer (Life Technologies, Carlsbad, CA, USA).
3. 0.1 M dithiothreitol (DTT, Life Technologies, Carlsbad, CA, USA).

4. Random primers (3 μg/μL, Life Technologies, Carlsbad, CA, USA).
5. RNase inhibitor (40 U/μL, such as RiboLock RNase inhibitor, Fermentas, Burlington, ON, Canada).
6. 10× Second-strand buffer: 500 mM Tris–HCl, 50 mM $MgCl_2$, 10 mM DTT, pH 7.8 at 25 °C.
7. 25 mM dNTP mix.
8. RNase H (5 U/μL, Fermentas, Burlington, ON, Canada).
9. DNA polymerase I (10 U/μL, Enzymatics, Beverly, MA, USA).
10. PCR purification kit (such as Zymoclean DNA clean & concentrator-5, Zymo, Orange, CA, USA).

2.5 End Repair

1. 10× End repair buffer: 50 mM Tris–HCl, 10 mM $MgCl_2$, 1 mM ATP, 10 mM DTT, pH 7.5 at 25 °C.
2. T4 DNA polymerase (3 U/μL, New England, Biolabs, Ipswich, MA, USA).
3. Klenow DNA polymerase (5 U/μL, New England, Biolabs, Ipswich, MA, USA).
4. T4 polynucleotide kinase (10 U/μL, New England, Biolabs, Ipswich, MA, USA).

2.6 dA Addition

1. 10× A-Tailing buffer (i.e., NEB Buffer 2, New England, Biolabs, Ipswich, MA, USA).
2. 1 mM dATP.
3. Klenow (3′–5′exo-) (5 U/μL, Enzymatics, Beverly, MA, USA).

2.7 Adaptor Ligation

1. 2× Rapid T4 DNA ligase buffer (Enzymatics, Beverly, MA, USA).
2. T4 DNA ligase (600 U/μL, Enzymatics, Beverly, MA, USA).
3. NEBNext adapter for Illumina (15 μM, New England, Biolabs, Ipswich, MA, USA).
4. USER enzyme (1 U/μL, New England, Biolabs, Ipswich, MA, USA).

2.8 Gel Purification and Size Selection

1. Low-Range ultra agarose.
2. TAE buffer.
3. 6× DNA Orange loading dye: 10 mM Tris–HCl, 0.15 % Orange G, 0.03 % xylene cyanol FF, 60 % glycerol, 60 mM EDTA, pH 7.6 at 25 °C.
4. 100 bp DNA ladder.

5. Ethidium bromide solution, 2 μg/mL.
6. Gel purification kit (such as the QIAquick gel extraction kit, Qiagen, Hilden, Germany).

2.9 PCR Amplification

1. NEBNext universal PCR primer (25 μM) (New England, Biolabs, Ipswich, MA, USA).
2. NEBNext index primers (25 μM) (New England, Biolabs, Ipswich, MA, USA).
3. 5× Phusion high fidelity PCR buffer (New England, Biolabs, Ipswich, MA, USA).
4. Phusion high fidelity DNA polymerase (2 U/μL, New England, Biolabs, Ipswich, MA, USA).
5. 25 mM dNTPs.

3 Methods

Keep samples on ice to minimize RNA degradation unless otherwise specified. Perform all centrifugation steps at 4 °C, unless indicated otherwise. Wear gloves and use sterile techniques when working with RNA. All glassware and plasticware should be RNase-free.

3.1 Purification of Total RNA

Total RNA can be purified from floral tissue by standard methods (*see* **Note 1**). The starting total RNA should be of high quality as determined by gel electrophoresis or by using a Bioanalyzer, and quantified by using a UV spectrometer such as NanoDrop 2000c (*see* **Note 2**).

3.2 mRNA Purification from Total RNA

1. Dilute a minimum of 5 ng total RNA with RNase-free water to 50 μL in a non-sticky tube, heat the sample at 65 °C for 5 min to disrupt RNA secondary structure and then place the sample on ice.
2. Mix well the Dynabeads oligo(dT)$_{25}$ suspension with a vortex at 4 °C, aliquot 15 μL of Dynabeads oligo(dT)$_{25}$ into a 1.5 mL non-sticky tube.
3. Wash the beads twice with 100 μL beads binding buffer each time and remove the supernatant by using a magnetic stand (*see* **Note 3**).
4. Resuspend the beads in 50 μL beads binding buffer, add the 50 μL total RNA sample from **step 1**, and rotate the tube at room temperature for 5 min. Remove the supernatant by using a magnetic stand.
5. Wash the beads twice with 200 μL washing buffer each time and remove the supernatant by using a magnetic stand.

6. Add 50 μL of 10 mM Tris–HCl to the beads and mix well gently by finger tapping.
7. Heat the sample at 80 °C for 2 min, then immediately put the tube on a magnetic stand. Transfer the supernatant containing mRNA to a new RNase-free non-sticky Eppendorf tube and add 50 μL beads binding buffer.
8. Heat the sample at 65 °C for 5 min to disrupt RNA secondary structure then place the sample on ice.
9. Wash the beads from **step 7** twice with 200 μL washing buffer each and remove the supernatant.
10. Add 100 μL sample from **step 8** to the beads from **step 9**, mix well by finger tapping, and then rotate for 5 min at room temperature. Remove the supernatant by using a magnetic stand.
11. Wash the beads twice with 200 μL washing buffer each and remove the supernatant by using a magnetic stand.
12. Add 17 μL of 10 mM Tris–HCl to the beads, mix well gently by finger tapping, heat the sample at 80 °C for 2 min to elute the mRNA from the beads, and then immediately put the tube on the magnetic stand. Transfer the supernatant containing mRNA to an RNase-free 200 μL thin-wall PCR tube. The resulting volume of mRNA should be approximately 16 μL.

3.3 Fragmentation of mRNA

1. Add 5× fragmentation buffer (4 μL) to the mRNA sample (16 μL).
2. Incubate the tube in a PCR thermal cycler at 94 °C for 3–4 min (*see* **Note 4**).
3. Add 2 μL stop buffer and place the tube on ice.
4. Transfer the solution to a 1.5 mL RNase-free non-sticky Eppendorf tube and add 2 μL NaOAc (3 M, pH 5.2), 2 μL glycogen, and 60 μL ethanol (100 %) to the tube. Incubate at −80 °C for 30 min.
5. Centrifuge the tube at 17,000 × *g* for 25 min at 4 °C in a microcentrifuge.
6. Pipette out the supernatant carefully without dislodging the RNA pellet, then wash the pellet with 300 μL of 70 % ethanol.
7. Centrifuge the tube at 17,000 × *g* for 20 min at 4 °C and carefully pipette out the supernatant.
8. Dry the pellet for 5 min in a SpeedVac or 10 min in a laboratory hood.
9. Resuspend the mRNA pellet in 11.1 μL of RNase-free water.

3.4 cDNA Synthesis

1. Add 1 μL random primers (3 μg/μL) to the mRNA (11.1 μL) and incubate the sample in a PCR thermal cycler at 65 °C for 5 min, then place the tube on ice.

2. Assemble the first strand cDNA synthesis mix in a PCR tube at RT:

mRNA w/random primers	12.1 μL
5× First-strand buffer	4 μL
0.1 M DTT	2 μL
25 mM dNTP mix	0. 4 μL
RNase inhibitor (40 U/μL)	0.5 μL

The final volume should be 19 μL.

3. Incubate the sample at 25 °C in a thermal cycler for 2 min.
4. Add 1 μL Superscript II reverse transcriptase (200 U/μL) to the sample and run the following PCR program:
   (a) 25 °C for 10 min.
   (b) 42 °C for 50 min.
   (c) 70 °C for 15 min.
   (d) Hold at 4 °C.
5. Place the tube on ice, and add 62.8 μL of water to the first strand cDNA synthesis mix.
6. Add 10 μL of 10× second-strand buffer and 1.2 μL of dNTP mix (25 mM), mix well and incubate on ice for 5 min.
7. Add 1 μL of RNase H (2 U/μL) and 5 μL of DNA polymerase I (10 U/μL), mix well and incubate at 16 °C in a thermal cycler for 2.5 h.
8. Purify the sample using a DNA clean and concentrator kit and elute in 50 μL of water.

3.5 End Repair

1. Assemble the end repair mix in a 1.5 mL reaction tube on ice:

Eluted DNA	50 μL
10× End repair buffer	10 μL
25 mM dNTP mix	1.6 μL
T4 DNA polymerase (3 U/μL)	5 μL
Klenow DNA polymerase (5 U/μL)	1 μL
T4 polynucleotide kinase (10 U/μL)	5 μL
Nuclease-free water	27.4 μL

The final volume should be 100 μL.

2. Incubate the sample in a thermal cycler at 20 °C for 30 min.
3. Purify the sample using a DNA clean and concentrator kit and elute in 32 μL water.

3.6 dA Addition

1. Assemble the dA adding mix in a 1.5 mL reaction tube on ice:

Eluted DNA	32 μL
A-tailing buffer	5 μL
1 mM dATP	10 μL
Klenow (3′–5′exo-) (5 U/μL)	3 μL

   The final volume should be 50 μL (*see* **Note 5**).
2. Incubate the sample in a thermal cycler at 37 °C for 30 min.
3. Purify the sample using a DNA clean and concentrator kit and elute in 23 μL water.

3.7 Adaptor Ligation

1. Assemble the adaptor ligation mix in a PCR reaction tube on ice:

Eluted DNA	23 μL
2× Rapid T4 DNA ligase buffer	25 μL
Adapter (15 μM)	1 μL
T4 DNA ligase (600 U/μL)	1 μL

   The final volume should be 50 μL.
2. Incubate the sample in a thermal cycler at 20 °C for 15 min.
3. Add 3 μL USER enzyme (1 U/μL), mix well by pipetting, and then incubate at 37 °C for 15 min.
4. Purify the sample using a DNA clean and concentrator kit and elute in 10 μL water.

3.8 Gel Purification and Size Selection

1. Prepare a 2 % low-range agarose gel with 1× TAE.
2. Load the samples as follows (*see* **Note 6**):
   (a) 1 μL DNA ladder in the first well.
   (b) 10 μL DNA elute from 3.7 mixed with 2 μL of 6× DNA loading dye (with Orange G) in the second well.
   (c) 1 μL DNA ladder in the third well.
3. Run the gel at 100 V for 35–40 min or until Orange G reaches the bottom of the gel.
4. Post-stain in ethidium bromide solution (2 μg/mL) for 20 min, then destain in distilled water for 20 min.
5. Place the gel on top of a plastic wrap on a UV box. Be sure to minimize exposure time, and cut the gel at 200 bp (±25 bp) (*see* **Note 7**).
6. Purify the sample using a QIAquick gel extraction kit and elute in 37 μL of elution buffer.

3.9 PCR Amplification

1. Assemble the PCR reaction mix in a 200 μL thin wall PCR tube:

Eluted DNA	37 μL
Universal PCR primer (25 μM)	1 μL
Index primer (X) (25 μM) (*see* **Note 8**)	1 μL
5× Phusion high fidelity PCR buffer	10 μL
25 mM dNTP mix	0.5 μL
Phusion high fidelity DNA polymerase (2 U/μL)	0.5 μL

   The final volume should be 50 μL.

2. Run the PCR with the following program (*see* **Note 9**):
   (a) 98 °C for 10 s.
   (b) 98 °C for 10 s.
   65 °C for 30 s.
   72 °C for 30 s.
   Repeat (b) for 15 cycles.
   (c) 72 °C for 5 min.
3. Purify the PCR product using a DNA Clean Kit and elute in 10 μL elution buffer. This product is the sample library, ready for sequencing (*see* **Note 10**).

4 Notes

1. We recommend the use of RNeasy plant mini kit, TRIzol reagent, or similar RNA extraction kits, and to follow manufacturer's protocols for plant materials. Ensure that the working areas, reaction tubes, and tips are free of RNase contaminants. For floral buds, we routinely start from a 0.1 to 0.5 mL volume of packed tissues, and obtain 50–500 μg total RNA.
2. If a Bioanalyzer is used to check RNA integrity, it is recommended that only samples with an RNA Integrity Number (RIN) value greater than 8 are used for library construction. For a typical RNA-Seq library construction experiment, the use of 5 ng to 10 μg total RNA is recommended. Within that range, using a larger amount of starting total RNA is recommended to reduce the number of PCR amplification cycles during library preparation, which will result in more even distribution of mapped reads within target genes. If the starting amount of total RNA is less than 5 ng, exponential or linear RNA amplification can be used [15, 23].

3. Mix well the beads with buffer by gentle shaking using finger tapping. When removing the supernatant, put the tube on a magnetic stand for about 2 min and remove the buffer without taking the tube out from the magnetic stand.
4. Fragment length distribution depends on the incubation time. A 3-min incubation is recommended if a 400 bp band will be excised in Subheading 3.7, and a 4 min incubation for a 200 bp band.
5. Please note that dATP is very sensitive to freeze-thaw cycles. To avoid freeze-thawing of dATP stocks, store in small aliquots.
6. Using ladders on both sides of the sample lanes helps to locate the gel area to be excised as the DNA sample may not be visible. When running two samples, leave one empty lane between samples and ladders to prevent cross-contamination. Do not run more than two samples on the same gel to avoid cross contamination.
7. The target DNA size may range from 200 to 600 bp depending on your experimental need. Longer target DNA size (such as 400 bp) is recommended for paired-end sequencing.
8. Indexed PCR primers are needed if multiple samples will be combined in the same sequencing lane. The NEBNext multiplex oligos for Illumina contains 12 index primers, each with a different index. For each reaction, only one of the 12 primer indices is used for one library. Make sure to combine samples amplified using different indexed primers into the same sequencing lane.
9. PCR cycle number can be adjusted from 4 to 25 depending on the amount of starting material. For example, use 4 cycles of amplification if starting with 5 μg or more total RNA, 18 cycles for 100 ng, and 25 cycles for 5 ng of starting total RNA.
10. A sample library can be quality controlled by using an Agilent 2100 Bioanalyzer with the Agilent DNA 1000 chip kit to check the size, purity, and concentration of the sample. The concentration of a sample library can be quantified using a Life Technologies Qubit fluorometer. The identity of a sample library can be confirmed by cloning 1 μL of the library into a blunt-end cloning vector, and sequence using Sanger sequencing. For *Arabidopsis*, at least 10 M mapped 50 bp single-end reads are needed to obtain a reasonable coverage of the transcriptome for gene expression quantification. Longer reads, such as 100 bp, can increase mapping accuracy by reducing multi-mapping reads. To study alternative splicing, paired-end sequencing with deeper coverage is recommended [24, 25].

Acknowledgements

We thank B.A. Williams and B.J. Wold for sharing protocols. This work was supported by the Ministry of Agriculture of China (Grant 2011ZX08010-002), by the Knowledge Innovation Program of CAS Grant KYQY-150, and by the Hundred Talents Program of CAS.

References

1. Krizek BA, Fletcher JC (2005) Molecular mechanisms of flower development: an armchair guide. Nat Rev Genet 6:688–698
2. Schena M, Shalon D, Davis RW, Brown PO (1995) Quantitative monitoring of gene expression patterns with a complementary DNA microarray. Science 270:467–470
3. Wang Z, Gerstein M, Snyder M (2009) RNA-Seq: a revolutionary tool for transcriptomics. Nat Rev Genet 10:57–63
4. Marioni JC, Mason CE, Mane SM, Stephens M, Gilad Y (2008) RNA-seq: an assessment of technical reproducibility and comparison with gene expression arrays. Genome Res 18:1509–1517
5. Morin R, Bainbridge M, Fejes A, Hirst M, Krzywinski M, Pugh T, McDonald H, Varhol R, Jones S, Marra M (2008) Profiling the HeLa S3 transcriptome using randomly primed cDNA and massively parallel short-read sequencing. Biotechniques 45:81–94
6. Mortazavi A, Williams BA, McCue K, Schaeffer L, Wold B (2008) Mapping and quantifying mammalian transcriptomes by RNA-Seq. Nat Methods 5:621–628
7. Nagalakshmi U, Wang Z, Waern K, Shou C, Raha D, Gerstein M, Snyder M (2008) The transcriptional landscape of the yeast genome defined by RNA sequencing. Science 320:1344–1349
8. Cloonan N, Forrest AR, Kolle G, Gardiner BB, Faulkner GJ, Brown MK, Taylor DF, Steptoe AL, Wani S, Bethel G, Robertson AJ, Perkins AC, Bruce SJ, Lee CC, Ranade SS, Peckham HE, Manning JM, McKernan KJ, Grimmond SM (2008) Stem cell transcriptome profiling via massive-scale mRNA sequencing. Nat Methods 5:613–619
9. Lister R, O'Malley RC, Tonti-Filippini J, Gregory BD, Berry CC, Millar AH, Ecker JR (2008) Highly integrated single-base resolution maps of the epigenome in Arabidopsis. Cell 133:523–536
10. Wilhelm BT, Marguerat S, Watt S, Schubert F, Wood V, Goodhead I, Penkett CJ, Rogers J, Bahler J (2008) Dynamic repertoire of a eukaryotic transcriptome surveyed at single-nucleotide resolution. Nature 453:1239–1243
11. Ozsolak F, Milos PM (2011) RNA sequencing: advances, challenges and opportunities. Nat Rev Genet 12:87–98
12. Martin JA, Wang Z (2011) Next-generation transcriptome assembly. Nat Rev Genet 12: 671–682
13. Malone JH, Oliver B (2011) Microarrays, deep sequencing and the true measure of the transcriptome. BMC Biol 9:34
14. Jiao Y, Meyerowitz EM (2010) Cell-type specific analysis of translating RNAs in developing flowers reveals new levels of control. Mol Syst Biol 6:419
15. Schmid MW, Schmidt A, Klostermeier UC, Barann M, Rosenstiel P, Grossniklaus U (2012) A powerful method for transcriptional profiling of specific cell types in eukaryotes: laser-assisted microdissection and RNA sequencing. PLoS One 7:e29685
16. Yang H, Lu P, Wang Y, Ma H (2011) The transcriptome landscape of Arabidopsis male meiocytes from high-throughput sequencing: the complexity and evolution of the meiotic process. Plant J 65:503–516
17. Oshlack A, Robinson MD, Young MD (2010) From RNA-seq reads to differential expression results. Genome Biol 11:220
18. Kvam VM, Liu P, Si Y (2012) A comparison of statistical methods for detecting differentially expressed genes from RNA-seq data. Am J Bot 99:248–256
19. Fang Z, Martin JA, Wang Z (2012) Statistical methods for identifying differentially expressed genes in RNA-Seq experiments. Cell Biosci 2:26
20. Garber M, Grabherr MG, Guttman M, Trapnell C (2011) Computational methods for transcriptome annotation and quantification using RNA-seq. Nat Methods 8:469–477
21. Gao D, Kim J, Kim H, Phang TL, Selby H, Tan AC, Tong T (2010) A survey of statistical software for analysing RNA-seq data. Hum Genomics 5:56–60

22. Cumbie JS, Kimbrel JA, Di Y, Schafer DW, Wilhelm LJ, Fox SE, Sullivan CM, Curzon AD, Carrington JC, Mockler TC, Chang JH (2011) GENE-counter: a computational pipeline for the analysis of RNA-Seq data for gene expression differences. PLoS One 6:e25279
23. Tang F, Barbacioru C, Nordman E, Li B, Xu N, Bashkirov VI, Lao K, Surani MA (2010) RNA-Seq analysis to capture the transcriptome landscape of a single cell. Nat Protoc 5:516–535
24. Marquez Y, Brown JW, Simpson C, Barta A, Kalyna M (2012) Transcriptome survey reveals increased complexity of the alternative splicing landscape in Arabidopsis. Genome Res 22:1184–1195
25. Filichkin SA, Priest HD, Givan SA, Shen R, Bryant DW, Fox SE, Wong WK, Mockler TC (2010) Genome-wide mapping of alternative splicing in Arabidopsis thaliana. Genome Res 20:45–58

Chapter 24

Next-Generation Sequencing Applied to Flower Development: ChIP-Seq

Emmanuelle Graciet, Diarmuid Seosamh Ó'Maoiléidigh, and Frank Wellmer

Abstract

Over the past 20 years, classic genetic approaches have shown that the developmental program underlying flower formation involves a large number of transcriptional regulators. However, the target genes of these transcription factors, as well as the gene regulatory networks they control, remain largely unknown. Chromatin immunoprecipitation coupled to next-generation sequencing (ChIP-Seq), which allows the identification of transcription factor binding sites on a genome-wide scale, has been successfully applied to a number of transcription factors in *Arabidopsis*. The ChIP-Seq procedure involves chemical cross-linking of proteins to DNA, followed by chromatin fragmentation and immunoprecipitation of specific protein–DNA complexes. The regions of the genome bound by a specific transcription factor can then be identified after next-generation sequencing.

Key words Arabidopsis, Flower development, ChIP-Seq, Transcription factors

1 Introduction

The different developmental programs required to form specific organs are largely controlled at the transcriptional level, through the specific regulation of gene expression by transcription factors [1]. In plants, genetic analyses have identified a large number of transcriptional regulators that are involved in the specification of the different floral organs and in the regulation of flower development ([2], reviewed in ref. 3). Although these studies have led to an understanding of how these transcription factors act, the molecular basis for their function, as well as the topology of the underlying gene regulatory networks, has remained largely unknown.

The development of chromatin immunoprecipitation (ChIP) in the 1980s by Varshavsky and colleagues [4] was an important step forward in the identification of regions of the genome that were bound by specific proteins. However, this method alone was

José Luis Riechmann and Frank Wellmer (eds.), *Flower Development: Methods and Protocols*, Methods in Molecular Biology, vol. 1110, DOI 10.1007/978-1-4614-9408-9_24, © Springer Science+Business Media New York 2014

not sufficient to identify, at a genome-wide scale, the binding sites of transcription factors. Recent advances in sequencing technologies, in particular the development of next-generation or ultra high-throughput sequencing (reviewed in ref. 5), and their coupling to ChIP (termed ChIP-Seq) have opened new possibilities to investigate, at a genome-wide scale, how transcription factors act to regulate complex developmental programs [6–10].

The use of whole-genome tiling arrays to identify the regions of the genome bound by a protein (termed ChIP-chip) has also been used for the genome-wide identification of transcription factors binding sites [11–13]. However, this approach leads to lower resolution around the binding sites [12] and has been superseded by ChIP-Seq. This chapter therefore focuses on the use of next-generation sequencing, with an emphasis on the use of the Illumina sequencing platform.

2 Materials

2.1 Tissue Collection and Fixation

1. Liquid nitrogen.
2. MC buffer: 10 mM sodium phosphate buffer pH 7.0, 50 mM NaCl, 0.1 M sucrose.
3. Fixation buffer: 1 % (w/v) formaldehyde in MC buffer (*see* **Note 1**).
4. Solution to stop fixation: 2.5 M glycine.
5. Vacuum pump and vacuum desiccator.

2.2 Chromatin Preparation

1. M1 buffer: 10 mM sodium phosphate buffer pH 7.0, 100 mM NaCl, 1 M hexylene glycol, 10 mM 2-mercaptoethanol, 1× Complete Protease Inhibitor cocktail (Roche) (*see* **Note 2**).
2. M2 buffer: 10 mM sodium phosphate buffer pH 7.0, 100 mM NaCl, 10 mM $MgCl_2$, 0.5 % (v/v) Triton X-100, 1 M hexylene glycol, 10 mM 2-mercaptoethanol, 1× Complete Protease Inhibitor cocktail (Roche) (*see* **Note 2**).
3. M3 buffer: 10 mM sodium phosphate buffer pH 7.0, 100 mM NaCl, 10 mM 2-mercaptoethanol, 1× Complete Protease Inhibitor cocktail (Roche) (*see* **Note 2**).
4. Lysis buffer: 50 mM Tris–HCl pH 8.1, 10 mM EDTA, 1 % (w/v) SDS, 1× Complete Protease Inhibitor cocktail (Roche) (*see* **Notes 2** and **3**).
5. ChIP dilution buffer: 16.7 mM Tris–HCl pH 8.1, 167 mM NaCl, 1.2 mM EDTA, 1.1 % (v/v) Triton X-100, 0.01 % (w/v) SDS, 1× Complete Protease Inhibitor cocktail (Roche) (*see* **Note 2**).
6. 10 mL syringe.

7. Miracloth.
8. Bioruptor (Diagenode).
9. Qubit instrument and Quant-iT dsDNA HS assay kit (Invitrogen).

2.3 Testing Chromatin Size

1. RNase A (DNase-free).
2. Proteinase K.
3. 100 % ice-cold ethanol.
4. 80 % ice-cold ethanol.
5. 3 M sodium acetate pH 5.2.
6. 2.5 M glycogen (from blue mussel).

2.4 Chromatin Pre-clearing

1. Protein A or Protein G sepharose (*see* **Note 4**).
2. 0.5 mg/mL lipid-free bovine serum albumin (BSA) prepared in IP buffer (see below).
3. IP buffer: 50 mM Hepes pH 7.5, 150 mM NaCl, 10 μM $ZnSO_4$, 5 mM $MgCl_2$, 1 % (v/v) Triton X-100, 0.05 % (w/v) SDS.
4. Rotator at 4 °C.

2.5 Chromatin Immunoprecipitation

1. Antibody against the protein of interest (*see* **Note 5**).
2. Protein A or Protein G Sepharose (*see* **Note 4**).
3. IP buffer (see above).
4. Elution buffer: 0.1 M glycine, 0.5 M NaCl, 0.05 % (v/v) Tween 20; adjust to pH 2.8 using HCl.
5. Neutralization solution: 1 M Tris base, pH 9.0.
6. Incubator shaker at 37 °C.

2.6 DNA Purification

1. Proteinase K.
2. RNase A (DNase-free).
3. Phenol–chloroform–isoamyl alcohol solution (25:24:1).
4. Phase-lock gel (heavy) microcentrifuge tubes (e.g., from Eppendorf).
5. 100 % ice-cold ethanol.
6. 80 % ice-cold ethanol.
7. 3 M sodium acetate pH 5.2.
8. 2.5 M glycogen (from blue mussel).
9. TE buffer.
10. Incubator or water bath at 65 °C.
11. Qubit instrument and Quant-iT dsDNA HS assay kit (Invitrogen).

2.7 ChIP Validation Using Quantitative PCR (qPCR)

1. SYBR Green.
2. Primers for negative loci (i.e., regions that are not bound by the protein of interest; *see* Table 1) and positive regions (i.e., loci that are known or suspected to be bound by the protein of interest).

2.8 Library Preparation for Illumina Sequencing

1. Illumina ChIP-Seq DNA Sample Prep Kit (*see* **Note 6**).
2. Certified Low Range Ultra Agarose (Bio-Rad).
3. QIAquick Gel Extraction Kit (Qiagen).
4. QIAquick PCR Purification Kit (Qiagen).
5. MinElute PCR Purification Kit (Qiagen).
6. Dark Reader non-UV transilluminator (*see* **Note 7**).
7. DNA High Sensitivity Kit and Bioanalyzer (Agilent).
8. Qubit instrument and Quant-iT dsDNA HS assay kit (Invitrogen).
9. SYBR Green.
10. Primers for negative loci (i.e., regions that are not bound by the protein of interest; *see* Table 1) and positive regions (i.e., loci that are known or suspected to be bound by the protein of interest).

3 Methods

3.1 General Considerations

Specific parameters need to be considered when designing a ChIP-Seq experiment (*see* Fig. 1). One of the most important factors is the quality of the antibody used to immunoprecipitate the DNA-binding protein of interest. In order to avoid artifacts due to ectopic expression, the use of the endogenous promoter to express the gene encoding the protein of interest is recommended. The use of protein-specific antibodies avoids the need to generate plants with complex genetic backgrounds for the ChIP-Seq experiment. Protein-specific antibodies can be generated by inoculating animals (typically rabbits) using a recombinant full-length protein, a recombinant fragment of the protein or a peptide corresponding to a specific sequence in the protein of interest. Although the use of peptides as antigens has advantages, such as having the possibility to choose a unique region to avoid cross-reactivity with closely related proteins, we found that using larger fragments to immunize rabbits tends to yield better quality antibodies for ChIP. When generating your own polyclonal antibodies, it is worth testing the serum without any additional purification step in a ChIP-qPCR experiment. If the strength of the signal obtained by quantitative PCR is too low, the antibody of interest can be affinity-purified using either the recombinant protein or the peptide used to immunize the animal.

Table 1 Genomic regions and primers used to calculate the average background [9]

Gene ID	Alias	Forward primer (5′ → 3′)	Reverse primer (5′ → 3′)
At4g26930	REF1	TCTCCGACCTTTCTTCACACCCATTCC	GTCTCCGCTTAGGAGCACGAAAGCTATC
At4g39400	BRI1	ACCCAGCACTAAACAGAAGATCAG	CCCAACCACCTATCTCTTGATTCTC
At5g09810	ACTIN	CGTTTCGCTTTCCTTAGTGTTAGCT	AGCGAACGGATCTAGAGACTCACCTTG
At4g03870	Mu	GATTTACAAGGAATCTGTTGGTGGT	CATAACATAGGTTTAGAGCATCTG C

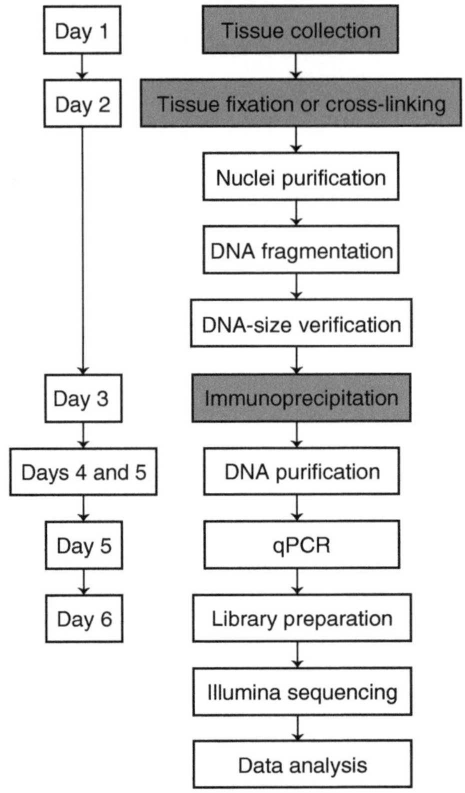

Fig. 1 Overview of the ChIP procedure. Different steps of the ChIP-Seq protocol are outlined, including a timeline for the procedure. Steps *shaded* in *grey* usually require optimization, depending on the protein of interest and the antibody used

If protein-specific antibodies of sufficient quality cannot be produced, the protein of interest may be expressed as a fusion with a tag (e.g., Green Fluorescent Protein (GFP) or other epitope tags such as hemagglutinin, myc, flag, etc.). When using this strategy, it is important to confirm that the protein of interest remains functional after addition of the tag (e.g., by testing if the expression of the tagged protein is sufficient to phenotypically rescue a plant that is mutant for the gene of interest). The choice of tag used, the sequence between the protein of interest and the tag (linker), as well as the position of the tag (at the N- or C-terminus of the protein) need to be considered carefully, as they can affect the function of the protein. Many companies sell ChIP-grade antibodies for a wide range of tags, but the quality of the antibodies can vary greatly from one batch to another, so it is important to test a batch before buying a large amount of antibody of the same batch for a ChIP-Seq experiment. Finally, polyclonal antibodies tend to perform better for ChIP-Seq than monoclonal antibodies, although the latter are usually more specific.

While the use of endogenous promoters is optimal, this can result in additional complications. For example, the gene of interest might only be expressed in a few cells, at very low levels, or at a specific stage of flower development, which makes ChIP-Seq more difficult, compared to genes that are expressed at higher levels throughout most of flower development. To overcome this problem, a constitutive promoter such as the Cauliflower Mosaic Virus 35S promoter has often been used. However, novel techniques such as the INTACT method allow the use of endogenous promoters for genes that are expressed at low levels or in a few cells only [14].

Another important factor to obtain good quality ChIP-Seq data is the choice of the negative control to filter background signal. The use of epitope-tagged proteins offers some advantages, as plants with a similar genotype as those used for ChIP-Seq, minus the tagged protein, can be used for background control. In this case, the background generated by binding of the antibody used to immunoprecipitate the tagged protein to other DNA-binding proteins can be assessed directly. When using antibodies raised against the protein of interest, the choice of the negative control is more difficult, especially if the mutant plants for the gene of interest have phenotypes that are very different from the wild type. In this case, the pre-immune serum from the animal that produced the ChIP antibody can be used to generate the background control. When performing a ChIP for the background control, the DNA yield can be a problem, but this can be overcome by pooling together numerous ChIP experiments.

3.2 Tissue Collection and Fixation (See **Note 8***)*

1. Collect 0.5 mL of floral tissue in a 1.5 mL tube and freeze in liquid nitrogen (*see* **Note 9**). The tissue can be stored at −80 °C until use. The protocol below is for 0.5 mL of tissue. All steps are performed at 4 °C unless otherwise stated.
2. Place the 1.5 mL tube containing 0.5 mL tissue on ice in a vacuum desiccator and add 1 mL of ice-cold Fixation buffer. Apply vacuum (800 mbar or until bubbles start to emerge from the tissue) for 15 min, then release vacuum and invert the tube three times. To ensure that the tissue is immersed in Fixation buffer the tissue may be centrifuged at 9,500 × *g* for 30 s. Repeat vacuum infiltration for 15 min and mixing three more times (total fixation time: ~60 min) (*see* **Note 10**).
3. Stop fixation by adding 50 μL of 2.5 M glycine (final glycine concentration: ~125 mM), mix well by inverting the tube three times and spin down at 9,500 × *g* for 30 s. Vacuum infiltrate on ice for 5 min.
4. Remove the Fixation buffer and wash the tissue by adding 1 mL MC buffer, inverting the tube three times, followed by spinning down the tissue at 9,500 × *g* for 30 s. Repeat the wash step with MC buffer twice more. After the last wash, remove as much MC buffer as possible and snap freeze the tissue in liquid nitrogen (*see* **Note 11**).

3.3 Chromatin Preparation

1. Add 100 μL M1 buffer and grind tissue on ice (*see* **Note 12**) until the tissue suspension appears homogenous. Add 150 μL M1 buffer and grind the tissue further. Finally, add 1 mL of M1 buffer.
2. Put a circular piece of Miracloth (diameter ~3 cm) into a 10-mL syringe. Pre-wet the miracloth with approximately 300 μL M1 buffer. Filter the solution through the miracloth. Squeeze the miracloth with the plunger. Repeat filtration once more with a fresh piece of miracloth.
3. Centrifuge the cell suspension at 7,700 × *g* for 1 min at 5 °C.
4. Remove the supernatant and resuspend the pellet in 0.9 mL M2 buffer. Centrifuge at 7,700 × *g* for 1 min at 5 °C. Repeat washes with M2 buffer twice more.
5. Wash the pellet once with 0.9 mL M3 buffer and centrifuge at 2,000 × *g*, 5 min, 5 °C (*see* **Note 13**).
6. Resuspend the pellet in 200 μL Lysis buffer by pipetting and incubate 10 min on ice (*see* **Note 14**).
7. Add 800 μL ChIP dilution buffer and resuspend chromatin by pipetting until the extract is homogenous again.
8. Split sample in 3 × 300 μL aliquots and sonicate the sample to solubilize the chromatin and shear DNA into fragments around 300 bp in size. Use the Bioruptor at high intensity and carry out 12 cycles of sonication (i.e., in 30 s pulses, with 30 s in between) (*see* **Note 15**).
9. Centrifuge the fragmented chromatin at 12,000 × *g* for 10 min at 5 °C to pellet insoluble material. Pool the supernatants and keep 100 μL to check chromatin size.
10. Measure the DNA concentration using the Qubit (*see* **Note 16**).

3.4 Checking Chromatin Size

1. Take the 100 μL aliquot of sonicated chromatin from **step 9**, Subheading 3.3 and add 150 μL TE buffer.
2. Add 1.5 μL RNase and incubate 30 min at 37 °C.
3. Add 1.5 μL proteinase K and incubate 30 min at 55 °C.
4. Precipitate the DNA by adding 1 μL glycogen, 0.1 vol. sodium acetate pH 5.4, and 2.5 vol. ice-cold 100 % ethanol. Mix by pipetting and incubate at −20 °C for at least 30 min.
5. Centrifuge 15 min at 18,000 × *g* and 5 °C.
6. Remove supernatant and add 1 mL ice-cold 80 % ethanol.
7. Centrifuge 5 min at 18,000 × *g*, 5 °C.
8. Remove supernatant and let the pellet dry.
9. Resuspend the pellet in 15 μL water.
10. Run the purified DNA on a 1.5 % agarose gel. The DNA should be around 300 bp on average (Fig. 2) (*see* **Note 15** if the DNA is not of the expected size range).

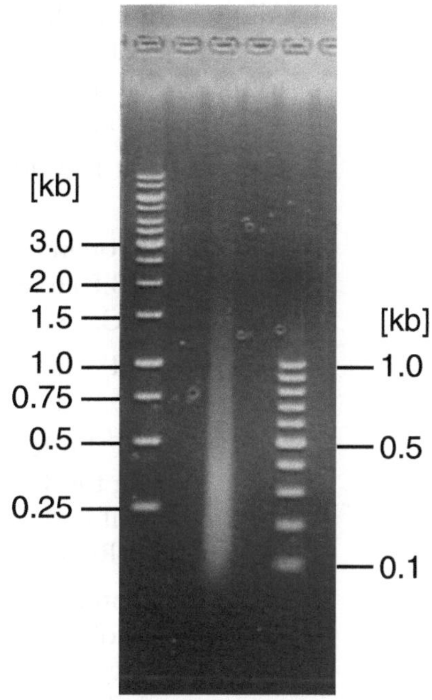

Fig. 2 Gel electrophoresis of fixed chromatin after fragmentation by sonication. A 1.5 % agarose gel was used to separate the sheared chromatin. The size of the DNA ladders is indicated on either side of the gel. The average size for the chromatin is around 300 bp

3.5 Chromatin Pre-clearing

1. To equilibrate the protein A sepharose, repeat five times the following procedure: add 1 mL IP buffer to 120–500 μL protein A sepharose slurry (120 μL of a 50 % slurry of protein A sepharose is sufficient for one IP). Invert the tube to resuspend the beads and centrifuge 1 min at 2,400 × *g*, 4 °C. Remove the supernatant and repeat.
2. Block the protein A sepharose by adding 1 mL of 0.5 mg/mL lipid-free BSA. Keep on a rotator at 4 °C for 20 min. Centrifuge 2,400 × *g*, 4 °C for 1 min and remove the supernatant. Estimate as accurately as possible the volume of beads and add an equal amount of IP buffer to obtain a 50 % slurry.
3. Add 1.1 mL IP buffer to 900 μL sheared chromatin and spin twice at 18,000 × *g* for 10 min at 5 °C (transfer the supernatant to a new tube in between the two centrifugation steps). Keep the chromatin on ice.
4. Add 60 μL of BSA-blocked protein A sepharose per 2 mL of chromatin. Incubate on a rotator for 90 min at 4 °C.
5. Centrifuge 1 min at 18,000 × *g*, 4 °C and keep the supernatant. If there were several tubes for pre-clearing, pool together the supernatants before taking the "input" aliquot. Keep the remaining resin prepared in **step 1** at 4 °C.
6. Remove 100 μL aliquot as "input" fraction.

3.6 Immuno-precipitation

1. Aliquot the pre-cleared chromatin in 2 mL tubes (~1.8 mL per tube). Add the antibody specific to the protein of interest to the pre-cleared chromatin and incubate 1 h to overnight at 4 °C on a rotator (*see* **Note 17**).
2. Centrifuge at 18,000 × *g*, 10 min, 4 °C to remove precipitated material.
3. Add 60 μL protein A beads (as a 50 % slurry; prepared in **step 1**; Subheading 3.5) to each supernatant and incubate on a rotator for 2 h at 4 °C.
4. Centrifuge 1 min at 2,400 × *g*, 4 °C and discard the supernatant.
5. Add 1 mL IP buffer and keep the tubes on their sides in ice while shaking for 8 min (a rotator in a cold room could also be used). Centrifuge 1 min at 2,400 × *g*, 4 °C and discard the supernatant. Repeat this step four more times.
6. Elute the protein–DNA complexes from the beads by adding 100 μL ice-cold elution buffer, incubate for 1 min at 37 °C while shaking vigorously, and centrifuge for 1 min at maximum speed. Transfer the eluate to a 2 mL tube and add 50 μL 1 M Tris base pH 9 to neutralize. Elute from the beads twice more, but for the last elution step incubate for 4 min at 37 °C (final combined volume of eluate will become ~450 μL).
7. Spin the eluate at maximum speed at 4 °C for 2 min. Transfer eluate (protein–DNA complexes) to a new 2 mL tube without touching any pellet that may have been formed (i.e., residual beads of protein A sepharose).
8. Add 4 μL proteinase K to the eluate and incubate overnight at 37 °C. Also add 330 μL TE to the "input" DNA sample (**step 6**, Subheading 3.5) and add 4 μL proteinase K. Treat the "input" sample similarly as the eluate obtained after ChIP.
9. Add 2 μL DNAse-free RNAse to the DNA samples and incubate 30 min at 37 °C.
10. Add another 4 μL proteinase K and incubate at 65 °C for at least 6 h to reverse cross-links.

3.7 DNA Purification

1. To the "IP sample" and "input" tubes from **step 10**, Subheading 3.6, add 500 μL of phenol–chloroform–isoamyl alcohol, vortex, and transfer to a pre-spun Phase-lock gel (Heavy) tube.
2. Centrifuge at 12,000 × *g* for 10 min at 4 °C.
3. Precipitate DNA with 1 μL glycogen, 0.1 vol. sodium acetate pH 5.4, and 2.5 vol. ice-cold 100 % ethanol.
4. Incubate overnight at −20 °C.
5. Spin at maximum speed for 15 min at 4 °C.
6. Wash pellet by adding 80 % ethanol and spin for 5 min at maximum speed and 4 °C.

7. Remove any ethanol and dry pellet.
8. Resuspend pellet in 50 μL 1× TE pH 8.0. Incubate 20 min at 37 °C. If multiple immunoprecipitation reactions need to be pooled, transfer the solution from the first tube into the next DNA pellet. Incubate again 20 min at 37 °C and repeat until all pellets have been resuspended.
9. Measure the DNA concentration using 3 μL DNA and the Qubit (*see* **Note 18**). This DNA is ready for library preparation without any additional purification step.

3.8 ChIP Validation Using Quantitative PCR (qPCR)

Before using the ChIP DNA for library preparation, it is better to test the quality of the ChIP by qPCR, using oligonucleotides for positive regions (e.g., known binding sites) and for negative regions (regions of the genome not expected to be bound). For the latter, we usually use four different sets of primers (*see* Table 1) and use the average of the signal obtained for these negative regions ("average background") to calculate the enrichment at a specific locus relative to the average background.

When designing primers for qPCR, aim at amplicons that are between 60 and 200 bases. We usually aim at a primer annealing temperature of 60 °C, although this is not always possible.

1. Prepare a master mix containing both primers at a final concentration of 5 μM each
2. Dilute the ChIP DNA to 50 pg/μL and the input DNA to 5 ng/μL (*see* **Note 19**).
3. Dilute the SYBR green (for one reaction, use 5 μL SYBR green and add 3 μL H_2O).
4. Combine the following reagents (for 1 reaction):

DNA	1 μL
Primer mix	1 μL
Diluted SYBR green	8 μL

5. Run the following qPCR program:

   Initial denaturation (1 cycle): 95 °C, 10 min

   Quantification (45 cycles): 95 °C, 10 s; 60 °C, 20 s; 72 °C, *X*s (where the time *X* is determined by the size of the largest amplicon)
6. Determine the Cp values.
7. Data analysis:

   For a specific region of the genome, the amount of DNA is normalized using the input sample, so that:

   enrichment at a specific locus = percentage of input DNA relative to total DNA used for ChIP (using this protocol: 5 %) × $2^{\Delta Cp}$, with $\Delta Cp = Cp_{input} - Cp_{ChIP}$

To calculate the enrichment relative to the average background:

(a) Calculate average background = (sum of enrichment values for all negative regions)/(number of negative regions tested).

(b) Calculate the relative enrichment = (enrichment at locus of interest)/(average background).

3.9 Library Preparation for Illumina Sequencing

Once a satisfactory ChIP sample has been obtained (*see* **Note 20**), a sequencing library can be prepared. The ultimate goal of the library preparation protocol is to attach adapter sequences to the ChIP DNA, which allows PCR amplification using adapter sequence-specific primers. The ChIP sample requires several steps of processing before amplification. First, single-stranded DNA overhangs that are present in the ChIP sample must be converted into phosphorylated blunt ends. This allows the addition of single base adenine overhangs to the 3′ end of the DNA fragments. The DNA adapter sequences include single base thymine overhangs, which allow their ligation to the modified ChIP DNA fragments. After PCR amplification of the adapter-ChIP sample, the sample can be size-selected. Finally, the quality of library is estimated by quantification of the DNA concentration, size verification of the amplified DNA and by qPCR. A detailed account of these steps is provided below.

1. To convert the single stranded DNA overhangs to blunt phosphorylated ends, combine the ChIP DNA (*see* **Note 21**) and the following reagents as outlined below and incubate for 30 min at 20 °C.

ChIP DNA	30 μL
Sterile Water	10 μL
T4 DNA ligase buffer with 10 mM ATP	5 μL
10 mM dNTP mix	2 μL
T4 DNA polymerase	1 μL
Klenow DNA polymerase	1 μL
T4 PNK	1 μL
Total reaction volume	50 μL

2. Purify the blunting reaction using a QIAquick PCR purification kit. Elute the purified DNA in 34 μL of EB.
3. To incorporate an adenine base to the 3′ end of the blunt phosphorylated DNA fragments, combine the DNA from **step 2** with the following reagents. Incubate the reaction for 30 min at 37 °C.

Blunt phosphorylated ChIP DNA	34 μL
Klenow Buffer	5 μL
dATP	10 μL
Klenow exo (3′–5′ minus)	1 μL
Total reaction volume	50 μL

4. Clean up **step 3** reaction using a MinElute PCR purification kit. Elute the purified DNA in 10 μL of EB.
5. To ligate the adapter DNA sequences to the ChIP DNA, combine the DNA from **step 4** to the following reagents and incubate for 15 min at room temperature (*see* **Note 22**).

"A" overhang ChIP DNA	10 μL
DNA ligase buffer	15 μL
Diluted adapter mix	1 μL
DNA ligase	4 μL
Total reaction volume	30 μL

6. Purify the DNA from **step 5** using a QIAquick PCR purification kit. Elute the purified DNA in 36 μL of EB.
7. For sequencing, suitable concentrations of DNA are required. Therefore, the ChIP sample must be amplified using a high fidelity enzyme and adapter-specific DNA oligonucleotides, as described below.

Adapter-DNA	36 μL
5× Phusion buffer	10 μL
10 mM dNTP mix	1.5 μL
PCR primer 1.1	1 μL
PCR primer 2.1	1 μL
Phusion polymerase	0.5 μL
Total reaction volume	50 μL

PCR amplification protocol: (a) 30 s at 98 °C, (b) (10 s at 98 °C, 30 s at 65 °C, 30 s at 72 °C) for 18 cycles, (c) 5 min at 72 °C. (d) Hold at 4 °C.

8. Make a 2 % agarose gel using Certified Low Range Agarose and ethidium bromide. After adding loading dye, load the PCR mix directly into a lane of the gel. Load a 100 bp and a 1 kb ladder to estimate the size of the DNA fragments that were amplified. Run a low current through the gel for at least 6 h.

9. Gel-extract the entire smear of DNA that has been amplified (*see* **Note 23**).
10. Purify the DNA from **step 9** using a QIAquick Gel extraction kit. Elute the purified DNA in 30 μL of EB.
11. Repeat **step 8** (*see* **Note 24**).
12. Repeat **step 9**.
13. Purify the DNA from **step 12** using a QIAquick Gel extraction kit (*see* **Note 25**). Elute the purified DNA in 15 μL of EB.
14. Measure the DNA concentration using 1 μL of DNA from **step 13** and the Qubit.
15. Estimate the size range that has been extracted by using the Agilent Bioanalyzer with the Agilent High Sensitivity DNA Kit (*see* **Note 26**).
16. Estimate the quality of the library by qPCR as described in Subheading 3.8 (*see* **Note 27**).
17. Dilute the prepared library to 10 nM (*see* **Note 28**). Samples are now ready to be submitted for sequencing.

4 Notes

1. Use stabilized formaldehyde (e.g., Sigma #252549) and do not use the stock solution if a precipitate has formed. Add the formaldehyde to the fixation buffer immediately before use.
2. Add hexylene glycol, 2-mercaptoethanol and the Complete Protease Inhibitor cocktail just before use.
3. Keep the Lysis buffer at room temperature, as the SDS tends to precipitate.
4. Protein A and protein G have different affinities for antibodies from different organisms (e.g., protein G has higher affinity for specific mouse IgGs, compared to protein A), so the type of resin needs to be chosen carefully depending on the antibody used to immunoprecipitate the protein of interest. We use protein A sepharose (GE Healthcare; GE#17-5280-01).
5. The quality of the antibody is essential for a successful ChIP assay. Many companies sell "ChIP-grade" antibodies, but they need to be tested carefully, keeping in mind that different batches can vary greatly. Antibodies against the protein of interest can also be custom-made using standard protocols. Their efficiency for ChIP needs to be tested carefully.
6. DNA sample preparation kits from other suppliers (such as NEB) have been successfully used to generate sequencing libraries.
7. Use of a dark reader is recommended but not essential.

8. All buffers should be ice-cold, except the Lysis buffer.
9. The amount of tissue indicated is for inflorescence tissue collected from the floral induction system described in Chapter 16. ChIP assays can also be carried out using whole inflorescences, but this usually results in a lower DNA yield after chromatin preparation. Hence, the volume of tissue collected should be adjusted to the type of tissue used for ChIP.
10. Be sure to resuspend any tissue that has been lodged at the bottom of the tube. If tissue remains lodged in the lid, the tubes can be spun for 30 s at 9,500 × *g* and 4 °C before applying vacuum again.
11. After this step, the tissue can be stored at −80 °C for long-term storage.
12. You might want to let everything defrost slightly after freezing in liquid nitrogen.
13. The pellet contains the purified nuclei.
14. If the tissue has not been fixed properly, when resuspending the nuclei pellet with the Lysis buffer, the solution will become very viscous.
15. We find the sonication step works better using tubes from Diagenode (#M-50001). Keep the Bioruptor in the cold room and check the temperature of the water bath before shearing the chromatin. The starting temperature should be at 4 °C or slightly below but will increase once sonication has begun. The time of sonication can be adjusted depending on the size of the DNA obtained (e.g., increase the number of cycles if the average size of the DNA fragments obtained is too high).
16. A chromatin concentration over 50 ng/μL of chromatin per 0.5 mL of inflorescence-like tissue is normal. At this stage, the chromatin can be frozen at −80 °C until further use.
17. The amount of antibody used, as well as the incubation time with the antibody, needs to be optimized for each antibody. If the DNA yield is low at the end of the procedure, a longer incubation with the antibody (up to overnight at 4 °C), and/or higher amounts of antibody per ChIP can be tested. In general, we find that the quality of the antibody is the most important parameter.
18. The yield after immunoprecipitation depends on several factors, such as protein abundance, as well as quality and timing of the tissue collection. Using GFP-tagged transcription factors such as APETALA3, PISTILLATA, or AGAMOUS, a total DNA yield over 2 ng is normal.
19. If the yield after ChIP is very low, the qPCR can be carried out with less DNA (down to 25 pg for the ChIP DNA and 1 ng for the input).

20. The fold-enrichments expected after ChIP can vary depending on the protein being investigated, the genomic regions that are being tested and the antibody being used. We found that average fold-enrichments of 10 (when comparing 16 different positively bound regions to the average of 4 negatively bound regions) were sufficient to produce high quality sequencing libraries. Additionally, we have successfully prepared and sequenced ChIP libraries from as little as 5 ng total of starting ChIP DNA. However, a higher concentration, if available, is desirable. We normally aim for a starting concentration of 10 ng.
21. The phenol–chloroform and ethanol precipitated ChIP DNA can be used directly to generate the sequencing library. An additional column purification prior to end repair can be performed if desired, however, this step is unnecessary.
22. Allowing the ligation reaction to run for longer may increase the number of adapter-DNA molecules formed, however, our results indicate that 15 min at room temperature is sufficient. If the ligation reaction time is increased, incubate the tubes at 16 °C. We have successfully used as little as a 1:40 dilution of the adapter mix to prepare the sequencing libraries; however, a 1:10 dilution has also been used successfully. The choice of dilution should be based on the starting concentration of DNA used and practice rounds using "input" DNA, if possible.
23. We found that size-selecting predefined ranges of DNA produced poorer quality libraries. Therefore, we advise that the entire DNA smear be extracted. The largest range of DNA we ultimately sequenced spanned from 150 to 500 bp. When visualizing the PCR-amplified DNA, oligonucleotide dimers and adapters are often visible at the bottom of the gel. If the oligonucleotide dimers or adapters are very close to the DNA you wish to extract, allow the DNA to migrate further.
24. A second gel extraction dramatically reduces the presence of oligonucleotide dimers and adapters.
25. Do not heat the gel in order to dissolve it. The column will need to be loaded several times with the dissolved gel. After the dissolved gel solution has been filtered through the binding column, add 500 μL of Buffer QG to the column and discard the flow-through. Resume the Qiagen gel extraction protocol.
26. Oligonucleotide dimers and adapters can be observed and their concentration relative to the rest of the sample can be estimated by using the Agilent Bioanalyzer with the DNA High Sensitivity Kit. After the second gel extraction, contaminations should be at a minimum.
27. We have observed a compression of fold-enrichment values once the sequencing library has been prepared, relative to the starting material used. If possible, a sequencing library should

also be prepared with a negative control ChIP or the INPUT DNA. This will normalize for any sequence-specific problems.

28. Use the average size of DNA indicated on the Bioanalyzer to estimate the concentration.

Acknowledgements

Work in our laboratory is funded by grants from Science Foundation Ireland to F.W. and E.G.

References

1. Kaufmann K et al (2010) Regulation of transcription in plants: mechanisms controlling developmental switches. Nat Rev Genet 11(12):830–842
2. Bowman JL et al (1991) Genetic interactions among floral homeotic genes of Arabidopsis. Development 112(1):1–20
3. Krizek BA, Fletcher JC (2005) Molecular mechanisms of flower development: an armchair guide. Nat Rev Genet 6(9):688–698
4. Solomon MJ, Varshavsky A (1985) Formaldehyde-mediated DNA-protein crosslinking: a probe for in vivo chromatin structures. Proc Natl Acad Sci U S A 82(19):6470–6474
5. Metzker ML (2010) Sequencing technologies—the next generation. Nat Rev Genet 11(1):31–46
6. Johnson DS et al (2007) Genome-wide mapping of in vivo protein-DNA interactions. Science 316(5830):1497–1502
7. Kaufmann K et al (2010) Orchestration of floral initiation by APETALA1. Science 328(5974): 85–89
8. Yant L et al (2010) Orchestration of the floral transition and floral development in Arabidopsis by the bifunctional transcription factor APETALA2. Plant Cell 22(7):2156–2170
9. Wuest SE et al (2012) Molecular basis for the specification of floral organs by APETALA3 and PISTILLATA. Proc Natl Acad Sci U S A 109(33):13452–13457
10. Gerstein MB et al (2012) Architecture of the human regulatory network derived from ENCODE data. Nature 489(7414):91–100
11. Morohashi K, Grotewold E (2009) A systems approach reveals regulatory circuitry for Arabidopsis trichome initiation by the GL3 and GL1 selectors. PLoS Genet 5(2): e1000396
12. Kaufmann K et al (2009) Target genes of the MADS transcription factor SEPALLATA3: integration of developmental and hormonal pathways in the Arabidopsis flower. PLoS Biol 7(4):e1000090
13. Mathieu J et al (2009) Repression of flowering by the miR172 target SMZ. PLoS Biol 7(7):e1000148
14. Deal RB, Henikoff S (2010) A simple method for gene expression and chromatin profiling of individual cell types within a tissue. Dev Cell 18(6):1030–1040

Chapter 25

Live-Imaging of the Arabidopsis Inflorescence Meristem

Marcus G. Heisler and Carolyn Ohno

Abstract

The aboveground tissues of higher plants are derived from a small population of stem cells located at the shoot apex within a structure called the shoot apical meristem (SAM). The SAM not only includes the stem cells but also incorporates a region from which lateral organs arise. The SAM is therefore of prime interest for understanding plant growth and development. In this chapter we outline methods for using confocal microscopy to image the Arabidopsis inflorescence SAM. This method enables detailed examination of cell division and growth patterns (Reddy et al., Development 131:4225–4237, 2004) as well as gene expression and protein localization patterns over time (Heisler et al. Curr Biol 15:1899–1911, 2005). When combined with perturbation approaches, the method offers an extremely powerful system for investigating SAM function in great detail.

Key words Arabidopsis, Confocal, Imaging, Shoot meristem, Flowers

1 Introduction

Development is a dynamic process involving the growth and proliferation of cells to form distinct tissues. In plants new tissues arise continuously from apical meristems located at the tips of the root and shoot. The shoot apical meristem (SAM) not only contains stem cells from which all aboveground tissues are derived but also contributes to the continuous formation of lateral organs on its flanks. To fully understand these developmental processes it is important to be able to examine gene expression and protein localization within the SAM with adequate spatial and temporal resolution.

During the reproductive phase of the model species *Arabidopsis thaliana*, the stem elongates and the SAM gives rise to flowers rather than leaves. Due to changes in the shape of the meristem and the different morphology of the lateral organs, the inflorescence shoot apex and early flower buds become visible from above and therefore accessible for microscopy. The most important limitations, due to the morphology of the apex, are that a cover slip

José Luis Riechmann and Frank Wellmer (eds.), *Flower Development: Methods and Protocols*, Methods in Molecular Biology, vol. 1110, DOI 10.1007/978-1-4614-9408-9_25, © Springer Science+Business Media New York 2014

cannot easily be positioned, and that a relatively large distance between the meristem surface and objective lens needs to be maintained due to the presence of the floral buds. These challenges are easily met, however, by a family of objective lenses initially designed for electrophysiology called dipping lenses. Dipping lenses are immersed in water but unlike typical water immersion lenses, they do not require a cover slip and feature large working distances.

Two approaches were initially developed to utilize dipping lenses and confocal microscopy to monitor SAM cell division dynamics and gene expression patterns. The difference between the approaches was the use of either an upright or inverted microscope. In the upright case the plant is mounted in a small plastic container filled with water and observed from above [1]. This is a relatively simple set up compared to imaging the SAM using an inverted microscope since in this case either the plant has to be mounted upside down or the stem needs to be bent such that the apex is inverted [3]. Another difference between those first two studies was the use of the auxin transport inhibitor NPA to temporarily inhibit organ formation so that better access to the meristem is possible. However, adequate access is also possible without such treatment and this avoids any potential NPA associated artifacts.

In this chapter, a detailed description of the method developed for imaging inflorescence SAMs using an upright confocal microscope [1] is provided. The chapter covers both preparation of the plants for mounting on the microscope, as well as important points and tips for imaging. The points on imaging are brief; however, as a general description of confocal imaging is covered elsewhere (e.g., *see* ref. 4), As to the choice of GFPs, it is best to learn from examples in the plant imaging literature.

2 Materials

1. Confocal microscope: As mentioned above, for the method described here an upright confocal is required. More recent models are generally more sensitive and able to distinguish a greater variety of fluorophores compared to older models. However, no specific manufacturer of microscopes is recommended.
2. Dipping lens: These lenses are available from most microscope manufacturers. They typically have magnifications of 20, 25, 40, and 63×. Most microscope manufacturers now make low-power dipping lenses that have numerical apertures (NAs) between 0.95 and 1.1. These lenses are preferable since they provide great flexibility for looking at samples of varying size at high resolution. However, some of them require specialized turrets associated with fixed-stage microscopes. The more typical 40× 0.8NA and 63× 0.9NA lenses are also very useful.

3. Boxes: For mounting plant apices on the microscope stage small transparent plastic boxes are required (e.g., Clear boxes 2 7/8″ length × 2″ width × 1 ¼″ height, Durphy Packaging Co. Ivyland, PA, USA).
4. GM medium: 1 % sucrose, 1× Murashige and Skoog basal salt mixture, MES 2-(MN-morpholino)-ethane sulfonic acid, 0.8 % Bacto Agar, 1 % MS vitamins, pH 5.7 with 1 M potassium hydroxide solution.
5. Plants for imaging. These can express fluorescent markers, or fluorescent stains can be used for cell membranes.
6. Sterile and desalted water.
7. Large clear plastic boxes to store and help keep sterile the small boxes containing the seedlings during their growth.
8. Ethanol.
9. Large beakers (2–5 L) for submerging the imaging boxes in ethanol for sterilization.
10. Tweezers (e.g., Dumont 5.5 INOX).
11. Sterile hood.
12. Air permeable tape (e.g., 3M Scotch Filter Tape, from Carolina Biologicals).

3 Methods

Our general approach to imaging plant inflorescence apices is to submerge them under water while utilizing dipping lenses to observe the meristems and young floral buds from above. Specimens are mounted in small plastic containers, usually boxes, which can be filled up with water. The apex is anchored within the box by having either the plant root system (in the case of time-lapse imaging) or just the stem dissected below the apex (for one-time snapshots) embedded in solid growth medium or agarose, respectively. Furthermore, plants can either be grown in the boxes from the seedling stage or transplated into the box (including roots) from soil. The procedure for all three cases is described in detail below.

3.1 One-Time Snapshots of Dissected Apices

1. Prepare boxes for sample mounting: pour 1 % molten agarose (in water) into each box to a depth of approximately 1 cm. Leave to cool and solidify (*see* **Note 1**).
2. Prepare an apex for mounting: dissect the inflorescence apex from the growing plant, leaving approximately 1 cm of stem under the apex. Place the apex within a dissecting dish filled with water (*see* **Note 2**).
3. Dissect away the larger flowers and any siliques attached to the stem (Fig. 1c, d) (*see* **Note 3**).

Fig. 1 Plant preparation for imaging. (**a**) Shows a view before dissection of a whole Arabidopsis plant that has been transplanted into a mounting box. (**b**) View of same plant as in (**a**) but after dissection of several leaves and flowers that either obstructed a view of the meristem or impeded the microscope objective lens from coming into close proximity with the SAM. (**c**) Close-up view of the un-dissected inflorescence meristem shown in (**a**). (**d**) Close-up view of the dissected inflorescence meristem shown in (**b**). Note the clear view of the meristem in the middle of the picture. (**e**) View of plant mounted within its box on the microscope stage. A 40× dipping lens is positioned above the apex, within the water

4. Mount the apex within the box: first use the fine point of the forceps to make vertical incisions or holes into the agar down to the bottom (*see* **Note 4**). Next gently transfer the apex from the dissecting dish into the box and drop it somewhere near

the newly created hole. Do not put water in the box yet since this makes it difficult to insert and mount the specimen. Although the hole may be difficult to see, with the help of a dissecting scope, position the apex such that the bottom of the stem is as close as possible to the hole. Then try to push the bottom of the stem gently, directly down, into the hole. Once it goes partway in, change the grip on the apex to more easily push the stem downward into the agar such that the apex is positioned rigidly above the agar surface.

5. After mounting the apex in the agarose, water can be added to the box to prevent dehydration. The apex should be completely submerged but the water should not be at a level such that it easily spills out of the box.
6. Next, use a 100 μL pipette to create jets of water aimed at the meristem and young flower buds to dislodge trapped air (*see* **Note 5**). Once most of the trapped air has been removed, use a dissecting microscope and forceps to make fine adjustments to the angle of the stem in the agar so that the meristem is maximally visible from directly above (repeat the procedure for the removal of trapped air whenever necessary). Lastly, again using fine forceps, dissect away any remaining flower buds that still obscure the inflorescence meristem by pinching where the pedicel meets the stem.
7. When there is a clear view of the shoot apical meristem from above (e.g., compare Fig. 1c, d), the box can be placed carefully (without spilling the water still in it) on the microscope stage (Fig. 1e) (*see* **Note 6**).
8. Once the plant apex within the box is roughly positioned under the objective lens, raise the microscope stage (or lower the objective) until the tip of the objective touches the water. Then, make sure there are no bubbles under the objective by looking from the side. If there are bubbles raise the objective out of the water and then repeat the immersion. If bubbles persist use a pasteur pipet to blow away any remaining bubbles with water.
9. Next, while using epifluorescence illumination (fluorescence filters work well), position the meristem within the microscope field of view by manually using the X–Y controls. Still looking from the side, position the plant apex at the focal point by moving the stage in all three dimensions until there is a maximum of deflected epifluorescence light coming from the apex (Fig. 1e).
10. Finally, look down the eye-pieces while controlling the X, Y, and Z controls until a bright signal is seen. In general the meristem is located centrally within the floral buds but it will take some practice to be able to recognize it reliably and quickly. A low power (e.g., 20–40×) objective will facilitate this. Proceed to Subheading 3.4 for tips on confocal imaging of the inflorescence meristem.

3.2 Growing in Boxes

In this approach, seedlings are first grown on plates and then transplanted into boxes where they grow until the transition to flowering. This approach seems to work well for keeping plants healthy during time-lapse imaging as the root system is fully developed within the box. The down side is that it can be difficult to maintain sterility and so typically many more plants need to be prepared than would actually be used for imaging to compensate for the loss due to contamination. However, the use of antibiotics and fungicides in the growth medium can reduce this problem.

1. Sow sterilized seed onto GM plates and allow to germinate.
2. Sterilize boxes by soaking, immersing them in 70 % ethanol briefly. Allow to dry within a flow hood with lids open. Expose to UV light from tissue culture flow hood for a few hours to further sterilize (*see* **Note** 7).
3. Fill each box with GM medium (containing 300 mg/L carbenicillin) to a depth of around 1 cm.
4. Using sterile conditions, transplant one seedling from the plates into each box. Position the plant centrally and push the root into them medium.
5. Close the box and seal with air permeable tape. Repeat for all plants/boxes.
6. To try to further prevent contamination, the boxes can be enclosed within a larger transparent box that has been surface sterilized with ethanol and also sealed with air permeable tape.
7. Place boxes in growth cabinet or growth room in a position that minimizes condensation.
8. Grow plants until flower buds are visible at the center of the rosette prior to bolting (Fig. 1a, c).
9. Next, proceed to **step 11** in Subheading 3.3 below.

3.3 Transplanting from Soil

This method is a convenient way to do short term (1–2 days) time-lapse imaging with a minimum of preparation. Plants are simply transferred from soil at the right stage directly into boxes without sterilization. It does help, however, to have antibiotics such as carbenicillin in the solid medium.

1. Grow plants in soil until flower buds are visible at the center of the rosette prior to bolting (Fig. 1a, c).
2. Prepare boxes as described in Subheading 3.2, **step 3**.
3. Remove the solid growth medium from a central area (1–2 cm^2) within the box for insertion of the plant root system.
4. Using forceps, dig into the soil surrounding the plant to be transplanted so that the plant (including much of the root system) can be lifted out of the pot (*see* **Note 8**).
5. Place plant into petri dish containing water.

6. Using forceps gently agitate the roots together with clumped soil under water so as to wash away much of the dirt.
7. Again using forceps, very gently try to dislodge away into the water any large clump of dirt or vermiculite that is still clinging to the roots. However, avoid breaking the roots. Removal of all soil or vermiculite particles will not be possible without great damage so the aim here is just to reduce the size of the root system plus soil such that it can fit into the hole previously created within the mounting box growth medium (*see* **step 3**, above).
8. Next pick up the plant and gently insert the root system into the hole within the mounting box growth medium. The plant should be oriented such that the inflorescence is facing directly upward to expose the meristem optimally later on.
9. Now pour some nutrient concentrate (e.g., Miracle grow) into the hole to partially cover the roots (*see* **Note 9**).
10. Finally, using a 1 mL pipette, squirt molten 1 % agarose gradually into the hole on top of the roots so as to solidify the root system. This should be done gradually so as to minimize heat damage.
11. To anchor and position the plant for subsequent imaging, more 1 % molten agarose is used. This molten agarose is applied to the leaves where they contact the solid agarose medium and around the petioles so as to rigidify the plant. Again, the agarose should be applied gradually and care should be taken not to immerse the meristem itself in agarose. While applying the agarose, the plant should be held in an orientation compatible with subsequent imaging. This can either mean trying to keep the plane of the rosette leaves horizontal, or if there are only a few leaves, trying to keep the inflorescence meristem facing upwards with the help of direct observation using a dissecting microscope. If necessary, after solidification the agarose can be broken up with tweezers to reposition. Ultimately, the orientation should be such that the inflorescence meristem or young buds are visible through a dissecting microscope looking down directly from above (Fig. 1a, c).
12. Fill the box with water to a level just above the inflorescence meristem (Fig. 1e).
13. Remove air bubbles as described in Subheading 3.1, **step 6**.
14. With a razor blade dissect away both leaves and flowers that obscure the meristem. Also remove leaves that will prevent the objective lens from approaching the meristem from above (Fig. 1b, d).
15. Again as described in Subheading 3.1, **step 6**, remove any air bubbles from the meristem region and proceed as described in **steps** 7 and **8** of that section.

3.4 Points for Consideration When Imaging the Inflorescence Apex

This section does not describe confocal imaging in detail since such information can be obtained elsewhere. Rather, it provides some general bullet points for live-imaging the inflorescence meristem.

1. Minimize photodamage by using the lowest laser power that still gives adequate signal. If necessary, the pinhole aperture can also be increased to collect more light, although Z resolution will suffer. For gaining a general impression of the morphology and cellular structure optical, sections can be up to 2 μm apart (we usually use 1 μm). Although a smaller spacing between slices may better match the optical slice thickness, smaller spacing also means more optical slices and therefore more photodamage for time-lapse experiments.
2. Realize that the stem below the apex may be growing rapidly. This means that as you image the tissue, the tissue may be moving upwards in the *Z* direction. Also, be aware that growth may have to be taken into account when setting the top section of a Z-stack. These complications can be avoided if the plant has not bolted yet or if a dissected apex has been left underwater for 20–30 min before imaging. Also, an intact plant can be immersed under ice-cold water within its mounting box for 30 min before imaging to stop growth temporarily.
3. For time-lapse imaging, the time intervals chosen should reflect both the biological question being asked as well as the risk of photodamage. The settings required to avoid exposure of the plant to excessive photodamage depends on many factors, including the sensitivity of the microscope, the brightness of the signal and the number of optical sections acquired at each time-interval. Typically, a feel for these parameters must be gained through trial and error; however, for bright signals, intervals of 3 h over 5 day is possible [1].
4. To gain an intuitive feel for the morphology, 3D animations of volume rendered Z stacks are most useful. Maximum intensity projections are also helpful.

4 Notes

1. Although one box can accommodate many specimens, it is usually best to prepare multiple boxes since any boxes not used immediately can be stored at 4 °C for future use. Boxes to be stored should be kept in plastic wrap to prevent dehydration of the agarose.
2. Submersion under water soon after separating the apex from the plant helps prevent dehydration of the tissue, especially

during the subsequent dissection of flowers (even though initially the apex will float).

3. Successful imaging of the inflorescence meristem depends on making sure that no flowers obstruct a view of the meristem from above. Also, to mount the apex the stem must be inserted into the agar. Therefore any siliques as well as older flowers need to be removed. However, care must be taken not to use forceps to hold the apex by the stem, thereby crushing it. Instead, one pair of forceps can be used to hold the flower or silique that is to be removed, while the other pair of forceps is used to pinch the corresponding pedicel at a point adjacent to the stem. Then, by simultaneously pinching and pulling the two points of contact apart the silique or flower can be removed without further damage. As these dissections are taking place, an effort should be made to keep the apex submerged as much as possible, especially the wounded regions. Roughly, all the flowers that can easily be removed in the dissecting dish should be removed. In practice, this usually means at least all flowers older than stage 12 [5]. Any smaller flower buds that obscure the apex can more easily be dissected once the apex is mounted in the agar within the box.
4. In general the incisions should be made towards the center of the box since if the specimen is placed too close to the wall of the box, the objective lens may not gain access to it.
5. Trapped air is visible under the dissecting microscope as a silvery region usually obscuring the smaller flower buds and meristem.
6. This can be a little tricky. Firstly the microscope stage is lowered to its extreme bottom position. Often the objective lens will also have to be moved out of the way by rotating the microscope turret. Also, any objective lens in positions adjacent to the dipping lens will have to be removed. Lastly, the box itself may need to be tilted temporarily to enable the dipping lens to move back into position above the sample.
7. Every effort must be made to keep the boxes sterile throughout this procedure, otherwise the plants will become contaminated and die.
8. This digging should occur at a distance from the plant in order to leave the soil directly contacting the bulk of the root system intact. Dig around and then under the plant and try to minimize damage to the roots themselves. Do not try to strip away dirt from the roots at this stage.
9. The aim here is to both provide some nutrients for growth as well as reduce the heat damage caused by the subsequent addition of molten agarose on top of the roots.

Acknowledgements

We acknowledge the Australian Research Council and European Research Council for support.

References

1. Reddy GV, Heisler MG, Ehrhardt DW, Meyerowitz EM (2004) Real-time lineage analysis reveals oriented cell divisions associated with morphogenesis at the shoot apex of *Arabidopsis thaliana*. Development 131:4225–4237
2. Heisler MG, Ohno C, Das P, Sieber P, Reddy GV, Long JA, Meyerowitz EM (2005) Patterns of auxin transport and gene expression during primordium development revealed by live imaging of the Arabidopsis inflorescence meristem. Curr Biol 15:1899–1911
3. Grandjean O, Vernoux T, Laufs P, Belcram K, Mizukami Y, Traas J (2004) In vivo analysis of cell division, cell growth, and differentiation at the shoot apical meristem in Arabidopsis. Plant Cell 16:74–87
4. Pawley J (ed) (2006) Handbook of biological confocal microscopy. Springer, New York
5. Smyth DR, Bowman JL, Meyerowitz EM (1990) Early flower development in Arabidopsis. Plant Cell 2:755–767

Chapter 26

Gene Regulatory Network Models for Floral Organ Determination

Eugenio Azpeitia, José Davila-Velderrain, Carlos Villarreal, and Elena R. Alvarez-Buylla

Abstract

Understanding how genotypes map unto phenotypes implies an integrative understanding of the processes regulating cell differentiation and morphogenesis, which comprise development. Such a task requires the use of theoretical and computational approaches to integrate and follow the concerted action of multiple genetic and nongenetic components that hold highly nonlinear interactions. Gene regulatory network (GRN) models have been proposed to approach such task. GRN models have become very useful to understand how such types of interactions restrict the multi-gene expression patterns that characterize different cell-fates. More recently, such temporal single-cell models have been extended to recover the temporal and spatial components of morphogenesis. Since the complete genomic GRN is still unknown and intractable for any organism, and some clear developmental modules have been identified, we focus here on the analysis of well-curated and experimentally grounded small GRN modules. One of the first experimentally grounded GRN that was proposed and validated corresponds to the regulatory module involved in floral organ determination. In this chapter we use this GRN as an example of the methodologies involved in: (1) formalizing and integrating molecular genetic data into the logical functions (Boolean functions) that rule gene interactions and dynamics in a Boolean GRN; (2) the algorithms and computational approaches used to recover the steady-states that correspond to each cell type, as well as the set of initial GRN configurations that lead to each one of such states (i.e., basins of attraction); (3) the approaches used to validate a GRN model using wild type and mutant or overexpression data, or to test the robustness of the GRN being proposed; (4) some of the methods that have been used to incorporate random fluctuations in the GRN Boolean functions and enable stochastic GRN models to address the temporal sequence with which gene configurations and cell fates are attained; (5) the methodologies used to approximate discrete Boolean GRN to continuous systems and their use in further dynamic analyses. The methodologies explained for the GRN of floral organ determination developed here in detail can be applied to any other functional developmental module.

Key words Gene regulatory networks, Functional module, Flower development, Cell differentiation, Attractors, Morphogenesis, Dynamics, Floral organ determination, Attractors, Basins of attraction, Stochastic networks, Mathematical models, Computational simulations, Robustness

---

Eugenio Azpeitia, José Davila-Velderrain, and Carlos Villarreal contributed equally to this work.

José Luis Riechmann and Frank Wellmer (eds.), *Flower Development: Methods and Protocols*, Methods in Molecular Biology, vol. 1110, DOI 10.1007/978-1-4614-9408-9_26,

1 Introduction

The mapping of the genotype unto the phenotypes implies the concerted action of multiple components during cell differentiation and morphogenesis that comprise development [1]. These components are part of regulatory motifs, which hold nonlinear interactions that produce complex behaviors [2, 3]. Such complexity cannot be understood in terms of individual components, and rather emerges as a result of the interactions among the components of the whole system. In order to integrate the action of multiple molecular components and follow their dynamics, it is indispensable to postulate mathematical and computational models. Gene regulatory network (GRN) models have appeared as one of the most powerful tools for the study of complex molecular systems. Small GRNs can sometimes be studied with analytical mathematical formulations, while medium or large size GRNs are amenable for dynamical analyses only with computer simulations [4]. As following the dynamics of the genomic interactomes is still intractable even with the most powerful computers, and given the fact that genomic networks are composed of multiple structural and functional modules, others and we have proposed to search for such modules for the study of biomolecular systems dynamics using GRN models (e.g., [5–7]).

Boolean models are probably the simplest type of formalism employed for the study of GRNs. Nonetheless, Boolean models provide meaningful information about the system. Importantly, Boolean GRNs can be approximated to continuous models that enable the use of additional mathematical tools [4, 8]. Given that: (a) the logic of GRNs is adequately formalized with Boolean models; (b) obtaining real biological parameters from biological molecular systems is still a complicated task; and (c) the use of realistic models can be computationally expensive, we believe that Boolean models and their continuous approximations are becoming a fundamental and practical tool to study GRN dynamics and to understand the complex behaviors observed in developmental processes (*see* refs. 9–11).

Based on the above rationale, the first step in building a GRN model is the identification of a developmental module and the integration of all the experimental data on the molecular components participating in it. The ABC genetic model of floral organ determination (*see* refs. 3, 12) (*see* Chapter 1) is part of a clearly circumscribed developmental module that underlies the sub-differentiation of the floral meristem in four concentric rings early on during flower development. From the outermost part of the floral meristem to its center, each ring comprises the primordial cells of sepals, petals, stamens, and carpels. Based on experimental evidence [13], it became obvious that although necessary, the ABC genes are not sufficient to specify floral organs. The ABC model has been instrumental to understanding flower development and evolution. However, it does not constitute a dynamic model able to recover the ABC combinatory code, as well as explain how the expression profiles of the set of molecular components included in the flower organ determination GRN, which includes the ABC genes, is established to promote the sepal, petal, stamen and carpel cell fates. Importantly, such a dynamic GRN model is the basis to understand how such cell types are determined in time and space, and thus, how the morphogenetic pattern that characterizes young floral meristems will form adult flowers [12, 14].

In order to uncover the necessary and sufficient set of interacting components involved in floral organ specification, the first step implies recovering the experimental evidence of ABC gene interacting components that include both regulated and regulator genes. In the case of Boolean models, the experimental data is formalized in the form of Boolean functions, which determine the dynamics of the GRN. In Boolean or any other type of discrete network, it is possible to fully explore the whole set of configurations or states of the system, and find the steady state configurations (attractors; see below). Kauffman postulated that the attractors to which GRNs converge, could correspond to the states characterizing differentiated cells [15]. More recently, Boolean GRNs have been grounded on experimental data ([5]; *see* review in ref. 3) showing that the attractors of developmental networks indeed correspond to the stable gene configuration observed in different types of cells, as long as a sufficient set of components involved in a given developmental module are incorporated.

In this Chapter we focus on the regulatory module underlying floral organ determination in *Arabidopsis thaliana* during early stages of flower development. Some of the methodologies explained here have been used in previous publications on such GRN [5, 7, 16–19]. In this chapter we will use examples extracted mainly from our own studies to explain how to develop and extend experimentally supported Boolean GRN models. Then, we explain how to incorporate stochastic properties in the model, which can allow us to explore the temporal sequence with which attractors or cell gene configurations and cell-fates are attained (e.g., [4]). Finally, we explain how we can approximate the Boolean model to a continuous one that can then be used in other types of models, for example, to explore spatial aspects of morphogenesis [14]. It is important to keep in mind that the tools presented in this Chapter can be applied to any GRN. Consequently, we begin with general explanations and afterwards we use examples from the literature to illustrate each methodological step.

2 Methods

2.1 Definitions

GRN nodes and edges: In GRNs, nodes represent genes, proteins or other types of molecular components such as miRNAs and hormones, while edges represent regulatory interactions among the components. Usually the interactions are positive (activations) or negative (inhibitions), but other type of interactions can be included (e.g., protein-protein interactions).

Variables: Variables are the elements that describe the system under study (usually the nodes) and which can take different values at each time.

Variable/Gene state: The value that a node takes at a certain time represents its state. The state can be a discrete or continuous value. In the case of Boolean networks the states can only be "0" when "OFF" and "1" when "ON."

Network State/Configuration: The vector composed by a set of values, where each value corresponds to the state of a specific gene of the network. In a Boolean network such vectors or network configurations are arrays of "0's" and "1's."

Attractors: Stationary network configurations are known as attractors. Single-state, stationary configurations are known as fixed-point attractors (Fig. 1a) and these are generally the ones that correspond to the arrays of gene activation states that characterize

a

Time	GEN1	GEN2	GEN3
1	1	0	0
2	1	0	0
3	1	0	0
.			
.			
.			
n-1	1	0	0
n	1	0	0

Fixed-point attractor

b

Time	GEN1	GEN2	GEN3
1	0	1	1
2	1	0	1
3	0	1	1
.			
.			
.			
n-1	1	0	1
n	0	1	1

Cyclic attractor

c

Time	GEN1	GEN2	GEN3
1	1	1	1
2	1	1	0
3	0	0	1
.			
.			
.			
n-1	0	0	0
n	0	1	0

Transitory states

Fig. 1 Fixed-point attractors, cyclic attractors, and transitory states. (**a**) An example of a fixed-point attractor. As observed, fixed-point attractors have one unique state where they stay indefinitely unless something perturbs them. (**b**) An example of a cyclic attractor. Cyclic attractors are composed of two or more network states that orderly repeat. In this case we observe a two state cyclic attractor. (**c**) Transitory states. Transitory states are states that lead to an attractor, but are not attractors themselves

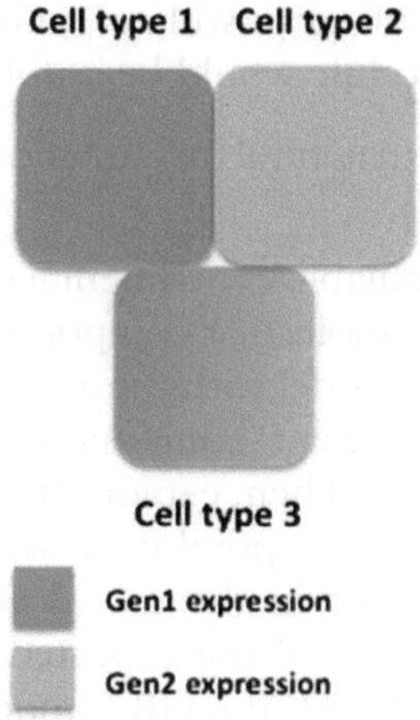

Expected attractors	GEN1	GEN2
Cell type 1	1	0
Cell type 2	0	1
Cell type 3	1	1

Fig. 2 The set of expected attractors. As explained in the main text, the set of expected attractors is obtained from the experimental information. In the case of cell types, the attractors correspond to the observed stable gene configuration of each cell type. Thus, if our system consists in three different cell types, one cell type with GEN1 expression, other with GEN2 expression, and a third one with both GEN1 and GEN2 expression, our set of expected attractors will be exactly this

different cell types. Whereas a set of network states that orderly repeat cyclically correspond to cyclic attractors (Fig. 1b).

Transitory states: All states that are not or do not form part of an attractor are transient or transitory states (Fig. 1c).

Basin of attraction: The set of all the initial configurations that eventually lead to a particular attractor constitute its basin of attraction.

Expected or observed attractors: Gene expression profiles or configurations that have been obtained from experimental assays and reported in the scientific literature for particular cell types are referred to here as the expected or observed attractors. Such attractors are expected to be recovered by the postulated GRN (Fig. 2).

Model Validation: The task of evaluating a model by means of contrasting its predictions with experimental results. For Boolean GRNs, model validation would imply, among others: recovering the observed gene configurations for the cells under study under *wt* and mutant or overexpression conditions, robustness analyses, etc. (see below).

Robustness: The ability of a system to maintain an output in the face of perturbations. For the case of a Boolean GRN model, it is evaluated, for example, by assessing if the system's attractors are still recovered under different transient and permanent mutations (alterations in the Boolean functions, nodes, or GRN topology).

2.2 General Protocol

A generic protocol to postulate a GRN model for a particular developmental module would be as follows:

(i) Identify a structural or functional developmental module (*see* **Note 1**).

(ii) Based on available experimental data, select the set of potential nodes or molecular components that will be incorporated in the GRN model with the aim of integrating the key necessary and sufficient components of the functional module under analysis. Then, explore the experimental data concerning the spatio-temporal expression patterns of the genes to be incorporated in the model and assemble a table with a Boolean format of the expected configurations that should be recovered with the GRN model (such configurations are the "expected attractors") (*see* **Note 2**).

(iii) Integrate and formalize the experimental data concerning the interactions among the selected nodes using Boolean logical functions that will rule the Boolean GRN dynamics.

(iv) The GRN is modeled as a dynamic system by exploring the states attained, given all possible initial configurations and the Boolean functions defined in (iii). The GRN is initialized in all possible configurations and followed until it reaches a fixed-point or cyclic attractor (*see* **Note 3**).

(v) Compare the simulated attractors to the ones observed experimentally (expected attractors; *see* **item (ii)** above). A perfect coincidence would suggest that a sufficient set of molecular components (nodes) and a fairly correct set of interactions have been considered in the postulated GRN model. If this is not the case, additional components and interactions can be incorporated or postulated, or the Boolean functions can be modified. This allows to refine interpretations of experimental data, or to postulate novel interactions to be tested experimentally in the future. In any case, the process can be repeated several times based on the dynamical behavior of the modified versions of the GRN under study until a regulatory module is postulated. Such module can include some novel hypothetical interactions or components, integrate available experimental data, and identify possible experimental contradictions or holes.

(vi) To validate the model, it is addressed if it recovers the *wt* and mutant (loss of function and gain of function) gene activation configurations that characterize the cells being considered. Perturbation analyses of the nodes and interactions, or the Boolean functions, can also be used for validating the model in order to test the robustness of the GRN under study. Eventually, novel predictions can be made and tested experimentally.

(vii) To recover the dynamics of the GRN and the temporal pattern of attractor attainment, the logical functions can be modeled as stochastic ones. Observed temporal patterns of cell-fate or gene configurations attainment can be used to validate the GRN model under consideration.

(viii) For further applications and also in cases that continuous functions are appropriate to describe the behavior of some of the components, the Boolean model can be approximated to a continuous one (*see* Subheading 2.5). Besides being useful for further modeling procedures, the continuous approximation is also a means of performing a robustness analysis of the GRN under study. Such a task hence implies as well a further validation of the model being postulated.

(ix) Equivalent approaches to the ones summarized in (vi) and (vii) for discrete systems can be used in continuous ones.

There are two types of materials needed when modeling dynamic GRNs. First, the expected results to be recovered by the model that are extracted from the literature and depend on the aims of the model and the nature of the developmental module being considered, but generally include stable gene configurations (attractors), mutant phenotypes, and developmental transitions, to name a few. The second set is the software required for the analyses of the GRN. Currently there are several available programs for GRN analyses (*see* **Note 8**). In the following sections, we explain with more detail and specific examples how this general protocol can be applied. We start by explaining the simplest Boolean approach for dynamical GRN modeling.

2.3 Deterministic Boolean GRN Model

In Boolean GRN models, nodes can only attain one of two possible values: "1" if the node is "ON," and "0" if the node is "OFF." A "0" node value usually represents that a gene is not being expressed, but can also represent the absence of a protein or hormone, while a "1" node value represents that a gene is expressed or another type of molecular component is present. As mentioned above, the first step in building a network is to extract the necessary experimental information to define the set of components to be considered in the GRN model, the set of expected attractors, and the Boolean functions that formally integrate the experimental data and define the dynamics of the GRN.

2.3.1 Expected Attractors

In Boolean GRNs, the network states (*see* Subheading 2.1) are defined by vectors of 0s and 1s. While a formal mathematical definition of attractors can be found on the chapter "Implicit Methods for Qualitative Modeling of Gene Regulatory Networks" of another Springer Protocols book [20], in Subheading 2.1 we give a more pragmatic definition of attractors, and we prefer to stick to it. In 1969, Kauffman proposed that the attractors of a GRN model could correspond to stable gene configurations characteristic of particular cell types or physiological states (*see* Subheading 2.1; Fig. 2). Consequently, the expected attractors are defined from gene expression patterns obtained from the literature, as well as from other data sources that clearly define the spatio-temporal gene configuration of the system. For example, Espinosa-Soto and collaborators [7] defined the expected attractors from the gene expression patterns reported in scientific publications. In another study, La Rota and collaborators [19] integrated experimental data into a gene expression map for the sepal primordium. Based on its expression map they defined zones with different combinations of gene expression, and each zone corresponded to an expected attractor. Defining the expected set of attractors is an indispensable step when building the GRN model, because they are used to validate the GRN (see below). Although it should be clear that the postulation of the Boolean functions is an independent task, and hence, it does not imply circularity.

2.3.2 Boolean Functions

In a Boolean GRN model the state of expression of each gene changes along time according to the dynamic equation

$$x_i(t+\tau) = f_i\left(x_1(t), x_2(t), \ldots, x_k(t)\right), \qquad (1)$$

in which the future state of gene i evolves temporally as a function of the current state of its k regulators. Boolean functions f_i can be formalized as logical statements or as truth tables. Logical statements use the logical operators "AND," "OR" and "NOT" to describe gene interactions, while in truth tables the state of the gene of interest is given for all possible state combinations of its k regulators (*see* **Note 4**). Logical operators can be combined in order to describe complex gene regulatory interactions, and can always be translated into an equivalent truth table. In Fig. 3, we provide examples of common gene regulatory interactions formalized as logical statements with their equivalent truth table. Consequently, in general, Boolean functions are generated from experimental evidence (but *see* **Note 5**). For example, if TGEN (a target gene) is ectopically expressed in a GEN1 loss-of-function background, it is inferred that GEN1 is a negative regulator of TGEN, and we use the "NOT" logical operator to describe GEN1 regulation over TGEN or its equivalent truth table (Fig. 4). In this Boolean function, the state of TGEN at time $t+\tau$ is 1 if GEN1 value is 0 at time t, and TGEN value at time $t+\tau$ is 0 if GEN1 value is 1 at time t (*see* **Note 6**).

The Boolean functions of the GRN developmental module being used here as an example, were grounded on available experimental information [5, 7, 17–19]. As with expected attractors, Boolean functions can be grounded on different types of experimental data, as long as they clearly state how genes interact (*see* **Note 7**). We now will provide an example of how the

GEN1(t)	GEN2(t)	TGEN(t+τ)
0	0	0
0	1	0
1	0	0
1	1	1

TGEN(t+τ) = GEN1(t) & GEN2(t)

GEN1(t)	GEN2(t)	TGEN(t+τ)
0	0	0
0	1	1
1	0	1
1	1	1

TGEN(t+τ) = GEN1(t) | GEN2(t)

GEN1(t)	GEN2(t)	TGEN(t+τ)
0	0	0
0	1	0
1	0	1
1	1	0

TGEN(t+τ) = GEN1(t) & ! GEN2(t)

GEN1(t)	GEN2(t)	TGEN(t+τ)
0	0	0
0	1	1
1	0	0
1	1	0

TGEN(t+τ) = ! GEN1(t) & GEN2(t)

Fig. 3 Examples of common Boolean functions. Here we present four examples of common Boolean functions for a target gene, in this case TGEN, with two regulators, namely, GEN1 and GEN2

experimental information was integrated and formalized as a Boolean function. During the transition from inflorescence to flower meristem, the expression of *TERMINAL FLOWER 1* (*TFL1*) needs to be repressed [21, 22], because *TFL1* is a promoter of inflorescence development [23]. *TFL1* is transcribed in the center of the meristem and from there it moves to peripheral cells [24]. *EMF1* is assumed to be a positive regulator of *TFL1* because the *emf1* mutant is epistatic to *tfl1* loss-of-function mutant, and both, *tfl1* and *emf1* mutants have similar phenotypes in terms of inflorescence meristem identity [25]. The over expression phenotype of *AP1* is similar to the loss-of-function of *TFL1*, and in the *ap1* mutant *TFL1* is ectopically expressed, suggesting that *AP1* is a negative regulator of *TFL1* [26]. Similarly, *TFL1* expression is not observed in *LFY* over expression and is ectopically expressed in *LFY* loss-of-function mutants [27]. According to these results, *EMF1* is a positive regulator of *TFL1*, while *AP1* and *LFY* are

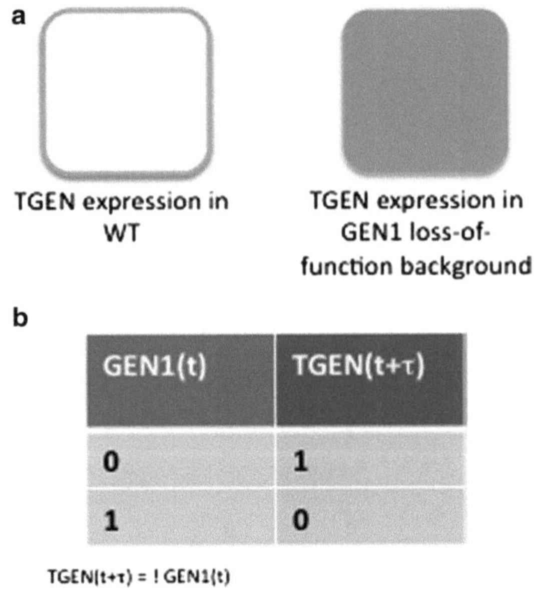

Fig. 4 Truth table and logical statement of the example explained in the main text. (**a**) TGEN expression is not observed in the GEN1 loss-of-function background. Hence, we can assume that GEN1 is a negative regulator of TGEN. This Boolean function can be represented with a (**b**) truth tab. or a (**c**) logical statement

negative regulators of *TFL1*. These results were formalized as a logical statement [18] as follows:

$$TFL1 = EMF1\ AND\ NOT\ AP1\ AND\ NOT\ LFY$$

A complete list of the Boolean functions and the experimental evidence for this model can be found in refs. 7, 18; *note some typographical errors corrected in* refs. 1, 12.

2.3.3 Validating the GRN: Simulated Attractors vs. Expected Attractors

Once the Boolean functions and the set of expected attractors of the GRN are obtained, we can proceed to make a first, necessary validation of the GRN. The first step is to use numerical simulations to recover the attractors that our set of Boolean functions generates (*see* **Note 8**). The attractors recovered in the simulations must coincide with the expected attractors, based on experimental data. In Espinosa-Soto and collaborators [7] ten attractors were recovered. Four out of the ten attractors corresponded to gene activation configurations that characterize meristematic cells of inflorescence meristems, while the rest corresponded to the gene configurations observed in sepal, petal, stamen and carpel primordial cells (Fig. 5). In the GRN for sepal development formulated by La Rota and collaborators [19], at least two attractors were recovered; one corresponding to the abaxial and the other one to the adaxial cells of the floral organ.

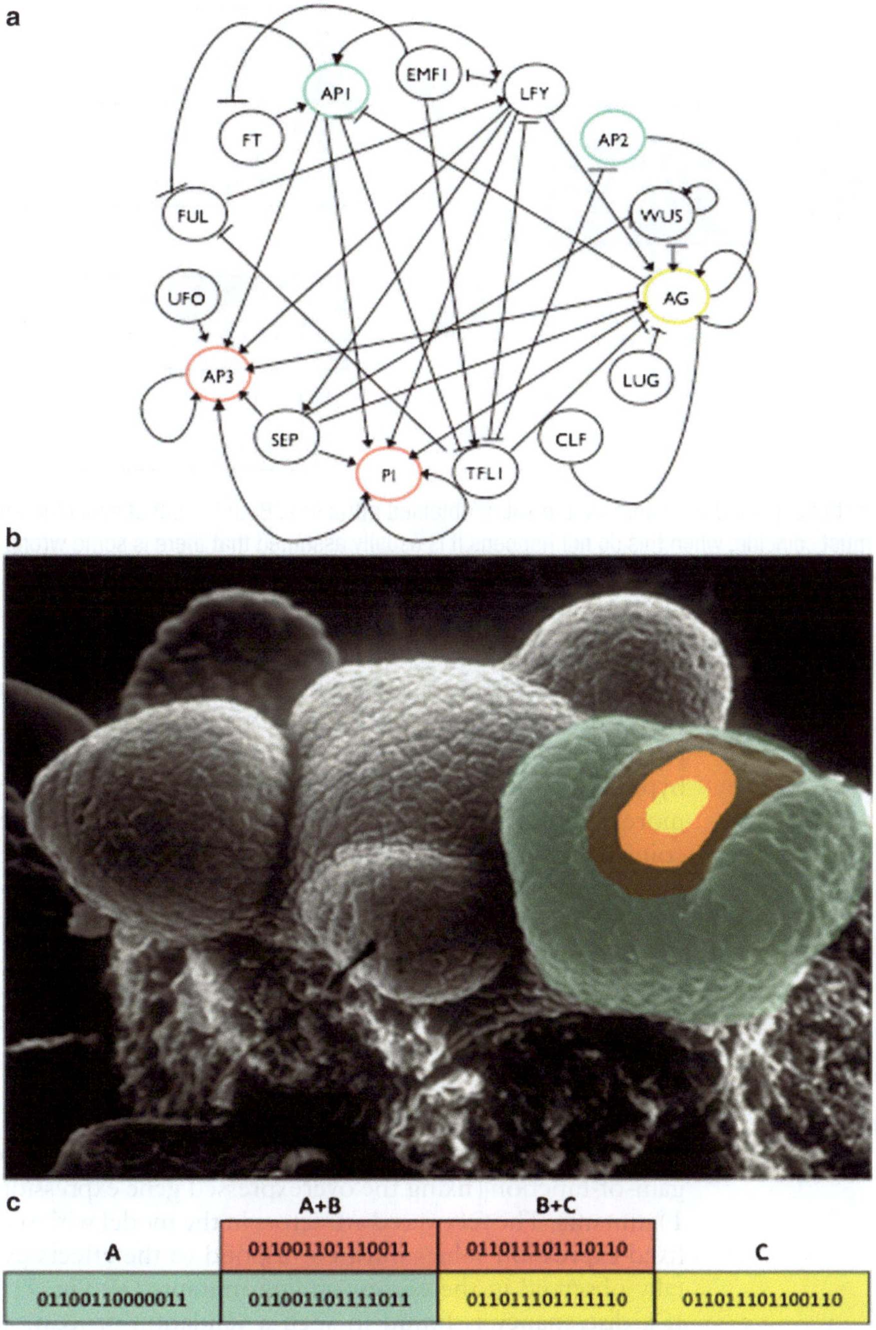

Fig. 5 Obtained attractors of the flower organ specification GRN. In (**a**) we present the graph of the flower organ specification GRN proposed by Espinosa-Soto and collaborators [7]. The GRN recovered 10 fixed-point attractors. Six of the attractors corresponded to the observed gene configuration in the primordial cell of sepals (one attractor), petals (two attractors), statements (two attractors), and carpels (one attractor). (**b**) A flower meristem in which the primordial sepal cells are colored in *green*, primordial petal cells in *brown grey*, primordial stamens in *orange*, and primordial carpel cells in *yellow*. In (**c**), the ABC model and the floral organ determination GRN attractors that correspond to A, A + B, B + C, and C gene combinations, which specify sepal, petal, stamen, and carpel primordial cells, respectively. The activation states correspond to each of the GRN nodes starting on the left with "EMF1" and consecutively progressing clockwise the rest of the genes in the GRN shown in (**a**)

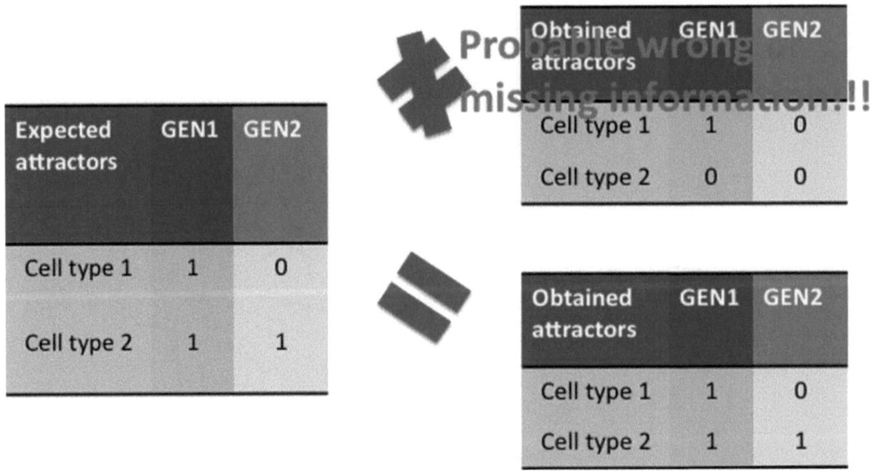

Fig. 6 The set of expected attractors vs. the set of obtained attractors. Both the set of expected and obtained attractors must coincide, when this do not happens it is usually assumed that there is some wrong or missing information

In cases in which the attractors recovered by the simulated GRN under study and those observed experimentally do not coincide, additional nodes or interactions can be considered, or the postulated Boolean functions can be modified (Fig. 6). Such novel hypotheses can be tested by running the GRN dynamics once more, and if the simulated and observed (expected) attractors now coincide, the model can be used to postulate novel interactions, missing data, or contradictions among those that had been proposed previously. For example, in Espinosa-Soto and collaborators [7] four missing interactions were predicted. Importantly, some of these predictions have been experimentally validated by independent and posterior research, demonstrating the predictive capacity and usefulness of this approach.

2.3.4 Mutant Analysis

An additional means to validate a GRN model is to simulate loss-of-function (fixing the mutated gene expression value to 0) and gain-of-function (fixing the overexpressed gene expression value to 1) mutants. The recovered attractors in the model with such altered fixed expression values must correspond to the effects experimentally observed in the corresponding mutants (*see* Fig. 7; **Note 9**). If a discrepancy is found in such a validation process, additional hypotheses concerning new nodes or interactions can be postulated. For the postulated GRN module underlying floral organ determination, most of the recovered attractors in the simulated mutants corresponded to the genetic configurations that have been observed experimentally [7, 17, 18]. In some cases, the simulated and observed (expected) attractors did not coincide and new interactions were postulated. For example, in Espinosa-Soto and collaborators [7] a positive feedback loop was predicted for the

WT GEN1 simulation

GEN2(t)	GEN3(t)	GEN1(t+τ)
0	0	0
0	1	0
1	0	0
1	1	1

TGEN expression

If GEN1 = 1 TGEN = 1
If GEN1 = 0 TGEN = 0

lof GEN1 simulation

GEN2 (t)	GEN3(t)	GEN1(t+τ)
0	0	0
0	1	0
1	0	0
1	1	0

TGEN = 0

gof GEN1 simulation

GEN2(t)	GEN3(t)	GEN1(t+τ)
0	0	1
0	1	1
1	0	1
1	1	1

TGEN = 1

Fig. 7 Loss-of-function and gain of function mutant simulations. Loss-of-function and gain-of-function mutant simulations are done by fixing the state of the desired gene to 0 and 1, respectively. In (**a**) the Boolean function of a non-mutated GEN1. In (**b**) and (**c**) the Boolean function of the same gene in a loss-of-function and a gain-of-function simulation, respectively. The Boolean functions are presented as truth tables and as logical statements. lof = loss-of-function, gof = gain-of-function

gene *AGAMOUS* (*AG*), even though this seemed unlikely because in the *ag-1* loss-of-function mutant plants, the *AG* expression pattern is the same as in wild-type plants [28]. In a posterior study in an independent laboratory, the prediction was verified experimentally [29].

Simulations of mutants are also useful when trying to predict the effects of multiple mutants, which are complicated to generate in the laboratory. Moreover, even when the GRN involved in flower determination in *Arabidopsis* and *Petunia* seems to be conserved, the mutant phenotypes are not identical. Espinosa-Soto and collaborators [7] used mutant analyses to test the effect of a duplication in B genes that has been reported in *Petunia*, and recovered the single mutant that had been described, and at the same time predicted the expected phenotype for the double mutant of the two duplicates.

2.3.5 Robustness Analyses

Experimental and theoretical work has demonstrated that living organisms are robust against perturbations. Moreover, at the molecular level the processes involved in different biological behaviors are also robust against internal and external variations. Such robustness implies that the overall functionality of the system remains when perturbed [30, 31]. In the case of GRNs, attractors should be robust when the Boolean functions are altered. In Espinosa-Soto and collaborators [7] the output value of every line of the truth tables was changed one by one. Interestingly, we found that the original attractors did not change for more than 95 % of the logical table alterations, indicating that the functionality of the postulated developmental module is robust to this type of perturbation. There are other types of perturbation analyses. For example, we could change with a certain probability the value of a line of the truth table, or the state of the network. Similarly, if we perturb the GRN with these other types of perturbations, the systems' attractors are expected to be maintained.

2.4 Stochastic Boolean GRN Model: Temporal Sequence of Cell-Fate Attainment

In deterministic GRN models, as the Boolean model exposed above, the system under study always converges to a single attractor if initialized from the same configuration, and once it attains such steady-state, it remains there indefinitely. However, during a developmental process, cells change from one stable cell configuration to another one in particular temporal and spatial or morphogenetic patterns. In order to explore questions such as how differentiating cells decide between one of the available attractors, or the order in which the system converges to the different attractors, given an initial condition, and to make statistical predictions of such possible behaviors, a stochastic formalism is needed.

In this section we develop a discrete stochastic model as an extension of the deterministic Boolean GRN. We then show how this approach can be used to explore the patterns of cell-fate attainment. Specifically, the model formalism explained here allows the investigation of the temporal sequence with which attractors are visited in the GRN when noise or random perturbations drive the system from one attractor to any other one.

2.4.1 From Deterministic to Stochastic Models

In a Boolean GRN model the dynamics given by Eq. 1 is deterministic: for a given set of Boolean functions f_i (*see* Subheading 2.3.2), the configuration of the network at time t completely determines the configuration of the network at the next time step $t+1$ (conventionally $\tau=1$). If Eq. 1 is iterated starting from a given initial configuration (defined by an array of n entries with 0s and 1s representing the activation states of the n genes), the network will eventually converge to an attractor. This deterministic version implies that once the system reaches an attractor, it remains there for all subsequent iterations. However, if noise is introduced into either the Boolean functions, or the gene states, there is a finite probability for the system to "jump" from one basin of attraction to another one (for definitions, *see* Subheading 2.1) and consequently, from one attractor to another one. Such a stochastic Boolean model of the GRN enables the study of transitions among attractors.

Noise can be implemented in a Boolean GRN model in several ways (*see* **Note 10**). Here we implement noise by introducing a constant probability of error ξ for the deterministic Boolean functions. In other words, at each time step, each gene "disobeys" its Boolean function with probability ξ, such that in the stochastic version, Eq. 1 is extended to

$$x_i(t+\tau) = \begin{cases} f_i(t), & \textit{with prob.} 1-\xi \\ 1-f_i(t) & \textit{with prob.} \xi \end{cases} \quad (2)$$

Note that the stochastic version (e.g., Eq. 2) reduces to a deterministic one (Eq. 1)) when $\xi = 0$. In the model, the stochastic perturbations are applied independently and individually to each gene at each iteration. This implementation of noise for stochastic Boolean modeling of GRNs has been referred to as the stochasticity in nodes (SIN) model with the assumption of a single fault at a time [20, 32].

2.4.2 The Transition Probability Matrix

When Eq. 2 is iterated, both the set of Boolean functions f_i and the error probability ξ determine the configuration of the network at the next time step. Under this stochastic dynamics, a given initial configuration will no longer converge to the same attractor each time. This situation allows us to estimate a probability of transition from one network state to another state as the frequency with which this transition occurs in a large number of repetitions of the same iteration (see below). The estimated transition probabilities can then be used to study the behavior of the system and to make statistical predictions.

As we want the model to be useful in the exploration of the patterns of temporal cell-fate attainment, the network states that we are interested in are the fixed-point attractor states that represent the cell types. Thus, we need to estimate the probability p_{ij} of transition from the attractor i to the attractor j. From the deterministic Boolean model, we already know to which attractors the network converges. In the following we use the term attractor to refer to both, the attractor and its basin. Thus, we can define a scalar (single-valued) variable X_t to describe the state of the network in terms of the specific attractor in which the network is in at time t. Then, X_t will take at time t any value from the ordered set $(1,2,i,\ldots,K)$ where each i represents one specific attractor from the available k attractors. The configuration of the network at time t is then related to the configuration at time $t+1$ through what is known as the transition probabilities. If the network is in attractor i at time t, at the next time step $t+1$, it will either stay in attractor i or move to another attractor j.

Formally, p_{ij} denotes a one-step transition probability that is defined as the following conditional probability:

$$p_{ij} = Prob\{X_{t+1} = j / X_t = i\}, \tag{3}$$

the probability that the network at time $t+1$ is in the attractor j given that it was in the attractor i at the previous time t, where $i,j = 1,2,\ldots,K$ for K attractors. The set of probabilities p_{ij} can be expressed in matrix form:

$$P = \begin{pmatrix} p_{11} & \cdots & p_{1k} \\ \vdots & \ddots & \vdots \\ p_{k1} & \cdots & p_{kk} \end{pmatrix}.$$

As the number of attractors K is finite, P is a $K \times K$ transition matrix. Operationally, under the current model, one can estimate the probabilities of the i-th row by first iterating Eq. 2 one time step starting from a given initial configuration corresponding to the basin of attraction of attractor i. If, after the iteration, the system remains in the same attractor, or the same basin of attraction, one count is added to the diagonal entry that corresponds to P_{ii}. If the configuration ends up in a different basin j, the count is added to the column j that corresponds to p_{ij}. This process is repeated a large number of times (e.g., 10,000) for each of the possible $\Omega = 2^n$ initial conditions. For each state (attractor), the one-step transition probabilities should satisfy $\sum_{j=i}^{K} p_{ij} = 1$ and $p_{ij} \geq 0$. This means that in the transition matrix P, the rows must sum to 1. This is achieved by dividing the number of counts in each matrix entry by the total number of configurations that started in the corresponding matrix row (e.g., basin i). As the dynamics in Eq. 2 are driven by both the Boolean functions f_i and the error probability ξ, given a fixed set of Boolean functions, different values of ξ will result in different values of the transition probabilities p_{ij} (*see* **Note 11**).

2.4.3 The Probabilistic Dynamics of Cell-Fate Attainment

Once the transition matrix P is calculated, it can be used in a dynamic model to describe how the probability of being in a particular attractor changes in time. In other words, we are now in position to derive a probabilistic dynamic model to simulate the dynamics of temporal cell-fate attainment.

In the previous subsection, the dynamics of transition between attractor states were defined in terms of transition probabilities.

When this is the case, the state of the network at any given time X_t can only be represented by its associated discrete probability distribution. We denote this distribution by the vector $p_X(t)=(p_1(t), p_2(t),\ldots,p_K(t))$, where $p_i(t)$ represents the probability of the network being in attractor i at time t, and $\sum_{i=i}^{K} p_i(t)=1$.

Given $p_X(t)$, the probability distribution associated with X_{t+1} can be found by multiplying the transition matrix P by $p_X(t)$. We obtain the following dynamic equation

$$p_X(t+1)=p_X(t)P, \tag{4}$$

this latter equation projects the process forward in time, and it allows us to follow the dynamics of the probabilities of cell-fate attainment by means of straightforward iteration.

In order to do so, it is necessary to specify an initial vector $p_X(t=0)$ which represents the probability distribution of the network state at time $t=0$. In biological terms, this initial vector can be interpreted as the representation of how a large population of cells is distributed over the available attractors. In other words, how many cells of each type are in the population at the initial time $t=0$. As the probabilities p_i sum to one, an underlying assumption is that the number of cells in the population remains constant. In the next subsection we show how this initial distribution can be chosen based on a biological motivation in order to explore a specific question regarding the dynamics of cell-fate attainment during floral organ formation. When the matrix P and the initial vector $p_X(0)$ are specified, Eq. 4 can be iterated (*see* **Note 12**); this process will generate a trajectory for the temporal evolution of the probability of each of the attractors. Every attractor will have a maximum in the probability of being reached at particular times. This maximum corresponds to the moment at which the corresponding cell-fate is most likely. Thus, the order in which the maximal probability of the different attractors is reached may serve as an intrinsic explanation for the emerging temporal order during early stages of development. Note that, as the transition probabilities of the matrix P depend on the value of ξ used in Eq. 2, the trajectories for the probability of attractor attainment will vary for different values of the error probability ξ.

2.4.4 Temporal Cell-Fate Pattern During Early Stages of Flower Development

In this subsection we show how the modeling formalism presented above can be applied to propose mechanistic explanations of observed patterns of temporal cell-fate attainment. In the modeling framework presented here, stochasticity may seem just as a modeling artifact that allows the study of transitions among attractors. However, a multitude of studies have demonstrated both theoretically and experimentally that stochasticity and the so-called biological noise are ubiquitously present in biological systems given the chemical nature of biological processes (for example *see* refs. 33–36).

Table 1
Example of a transition matrix *P* estimated from the GRN model for the floral organ determination of *A. thaliana*. The matrix elements are the transition probabilities among pairs of the six attractors (S, P1, P2, S1, S2, and C). Probabilities where calculated in Alvarez-Buylla et al. [4] using (ξ=0.01)

	sep	pe1	pe2	st1	st2	car
sep	0.939395	0.001943	0.009571	0.000083	0.000490	0.048517
pe1	0.036925	0.904162	0.009250	0.033900	0.000488	0.015275
pe2	0.009067	0.000464	0.941609	0.000024	0.048374	0.000461
st1	0.000084	0.001893	0.000020	0.936514	0.009960	0.051530
st2	0.000020	0.000001	0.002074	0.000356	0.987953	0.009597
car	0.002045	0.000034	0.000020	0.001951	0.010020	0.985930

Under the hypothesis that random fluctuations in a system may be important for cell behavior and pattern formation, Alvarez-Buylla and collaborators proposed a discrete stochastic model to address whether noisy perturbations of the GRN model for the floral organ determination of *A. thaliana* are sufficient to recover the stereotypical temporal pattern in gene expression during flower development [4]. As mentioned above, previous analysis of the deterministic Boolean GRN showed that the system converges only to ten fixed-point attractors, which correspond to the main cell types observed during early flower development [7]. Six of the attractors correspond to the four floral organ primordial cells within the flower meristem: sepals, petals, stamens, and carpels (S, P1, P2, S1, S2, and C).

Following Subheading 2.4.2, we can study the dynamics of cell-fate attainment of the floral organ primordial cells by defining a variable X_t which can take as a value any of the attractors (S, P1, P2, S1, S2, and C) at each time t. Then, given the six attractors of interest, we would like to estimate the transition matrix P, with the transition probabilities p_{ij} of transition from attractor i to attractor j as components. This matrix can be estimated by iterating Eq. 2 and following the algorithm described in Subheading 2.4.2. Alvarez-Buylla and collaborators [4] followed a similar approach, and estimated the matrix P shown in Table 1. This matrix was estimated using a value of 0.01 for the probability of error ξ in Eq. 2.

We follow the temporal evolution of the probability of reaching each attractor by iterating Eq. 4 using as P the matrix just estimated (*see* Table 1). However, as mentioned in Subheading 2.4.3, it is necessary to specify an initial distribution $p_X(0)$, which defines what fraction of the whole cell population corresponds to each of the cell-types (S, P1, P2, S1, S2, and C) at the initial time of the

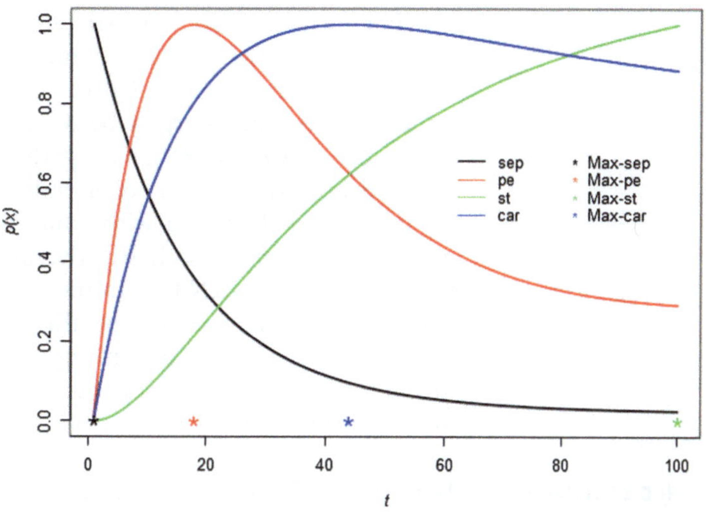

Fig. 8 Temporal sequence of cell-fate attainment pattern under the stochastic Boolean GRN model. Maximum relative probability *p* of attaining each attractor, as a function of time (in iteration steps). The value of the error probability used was $\xi = 0.01$. *Stars* mark the time when maximal probability of each attractor occurs. The most probable sequence of cell attainment: sepals, petals, carpels, and stamens

simulation. Since sepal primordial cells are the first to attain their fate in flower development, we use as an initial distribution a vector in which the value corresponding to the fraction of sepal cells is set to 1 and all the other values are set to zero; this is $p_X(0) = (1,0,0,0,0,0)$, where the order of the values is (S, P1, P2, S1, S2, and C). Thus, initially, all of the population of cells within a floral primordium is in the sepal attractor. Then, Eq. 4 can be iterated to follow the changes in the probability of reaching each one of the other attractors over time, given that the entire system started in the sepal configuration. The resulting normalized trajectories for the case in point are shown in Fig. 8 (*see* **Note 13**). The graph clearly shows how the trajectory for each of the attractor's probability reaches its maximum at a given time. One star for each of the attractors was drawn in the graph just above the x-axis at the time when its maximal probability occurs. In accordance with biological observations, the results show that the most probable sequence of cell attainment is: sepals, petals, and the stamens and carpels almost concomitantly.

The results presented here were calculated using just one value for the probability of error ($\xi = 0.001$). In the work of Alvarez-Buylla and collaborators [4], it was shown that the system exhibited a sequence of transitions among attractors that mimics the sequence of gene activation configurations observed in real flowers for a level of noise (value of ξ) of around 0.5–10 % (*see* **Note 11**).

The nonintuitive, constructive role of moderated noise perturbing the dynamics of nonlinear systems is a well-known phenomenon in physics [37]. Currently, there is a growing interest in understanding the interplay between noise and the nonlinearity of biological networks [38]. Using the model formalism presented here, Alvarez-Buylla and collaborators concluded that the stereotypical temporal pattern with which floral organs are determined may result from a stochastic dynamic system associated with a highly nonlinear GRN [4]. In the light of these findings, the modeling framework exposed in this section constitutes a simple approach to understanding morphogenesis, providing predictions on the population dynamics of cells with different genetic configurations during development.

2.5 Approximation to a Continuous GRN Model

2.5.1 Deterministic Approach

Boolean GRNs have been useful to study the complex logic of transcriptional regulation involved in cell differentiation because it seems that the qualitative topology of such networks, rather than the detailed form of the kinetic functions of gene interactions, rule the attractors reached. However, for some further mathematical developments and also for studies of the detailed behavior of GRN dynamics, the differences in genetic expression decay rates, threshold expression values, saturation rates, and other quantitative aspects of GRNs can become very relevant. These aspects of GRNs cannot be contemplated by a discrete approach. Hence, it becomes necessary to investigate also continuous representations of GRN dynamics. Several studies reviewed here show that such continuous approximations of the discrete GRNs lead to novel predictions, but at the same time recover consistent results with those arising in the Boolean framework.

Several approaches have been used to describe the Boolean GRN as a continuous system. A well-known scheme is the piece-wise linear Glass dynamics of the network [39]. This model is based on a set of differential equations in which each continuous variable x_i, representing the level of expression of a given gene, has an associated discrete variable that represents the state of expression of that gene. This is accomplished by introducing the discrete variables $\hat{x}_i$ defined as $\hat{x}_i = H(x_i - \theta_i)$, where θ_i represents a threshold, and $H(x)$ is the Heaviside step function: $H(x) = 1$ if $x > 1$, and $H(x) = 0$ if $x < 1$. This definition implies that gene n displays a dichotomic expression driven by a more gradual continuous dynamics. The piece-wise continuous Glass dynamics of the GRN is described by

$$\frac{\mathrm{d}x_i(t)}{\mathrm{d}t} = \mu\left[f_i\left(\hat{x}_1(t),\ldots,\hat{x}_k(t)\right) - x_i(t)\right] \quad (5)$$

where f_i are the input functions of the discrete Boolean model, and $\mu = 1/\tau$ is the relaxation rate of the gene expression profile. Within this description, the microscopic configuration of the GRN at a given time is described by the set of continuous values $\{x_1(t),\ldots,x_k(t)\}$; this set induces in turn the set of corresponding discrete values $\{\breve{x}_1(t),\ldots,\breve{x}_k(t)\}$ as the Boolean configuration of the network. The equilibrium states of the GRN that determine a given phenotype may be obtained from the condition $dx_i/dt=0$, which leads to

$$x_i^S = f_i\left(\breve{x}_1^S(t),\ldots,\breve{x}_k^S(t)\right) \tag{6}$$

independently of the value of the relaxation rate. Even when the Boolean input functions f_i are the same in the discrete and continuous approaches, there are infinitely many microscopic configurations compatible with the same Boolean configuration, and the discrete model of the GRN and the corresponding continuous piece-wise linear model are not necessarily equivalent, since the attractors of the two models can be different. However, numerical simulations to study the GRN for floral organ differentiation in *A. thaliana*, show that the Glass dynamics generate exactly the same ten fixed-point attractors obtained in the Boolean model, although the size of the corresponding attraction basins may display some variation [4].

An alternative approach consists in considering that the input functions display a saturation behavior characterized by a logistic or a Hill function, usually employed in biochemistry to describe ligand saturation as a function of its concentration. In the first case, the input associated to node i may be included in the form

$$\Theta\left[f_i(x_1,\ldots,x_k)\right]=\frac{1}{1+\exp\left[-b_i\left[f_i(x_1,\ldots,x_k)-\epsilon_i\right]\right]} \tag{7}$$

where ϵ_i is a threshold level (usually $\epsilon_i=1/2$), and b_i the input saturation rate. It may

be easily seen that for $b_i>>1$, the input function becomes a Heaviside step function:

$$\Theta[f_i-\epsilon_i]\rightarrow H[f_i-\epsilon_i], \tag{8}$$

and thus displays a dichotomic behavior (in practice this may be achieved for, e.g., $b_i>10$). This approach has been employed, for example, in the modeling of the GRN for differentiation of Th cells of the immune response by Mendoza and Xenarios [40], or in the study of floral organ specification in *A. thaliana* [1].

On the other hand, Hill-type inputs of GRNs have been employed in a number of investigations on biological development and differentiation (*see* the review in ref. 41). They have the following structure:

$$\Xi^{(n)}[f_i]=\frac{A_i(f_i)^n}{(\epsilon_i)^n+(f_i)^n}, \tag{9}$$

with the parameter n, an integer number, and A_i the maximum asymptotic value attained by the input. The latter approach was used by Zhou et al. [42], to model pancreatic cell fates; and by Wang and coworkers [43] to study myeloid and erythroid cell fates. The approximation to be used depends on the nature of the problem under study. In fact, the GRN inputs could be described also by any set of polynomial functions that reflect the biological interactions of the network.

Another approach that can be used to translate the logical into continuous functions involves the use of "fuzzy logics" proposed by L. A. Zadeh [44] to study systems that do not follow strictly 1 or 0 truth-values. This is achieved by using the following rules

$$\begin{aligned} &x_i(t)\; and\; x_j(t) \rightarrow \min\left[x_i(t), x_j(t)\right] \\ &x_i(t)\; or\; x_j(t) \rightarrow \max\left[x_i(t), x_j(t)\right]\;. \\ &not\; x_i(t) \rightarrow 1 - x_i(t) \end{aligned} \tag{10}$$

Here, the operators, min and max mean to choose between the minimum and maximum values of the functions x_i and x_j at a given time t. It can be shown that these rules lead to a Boolean algebra [1]. One possible disadvantage of this proposition is that it involves only piece-wise differential functions. Another possibility is to consider the following algorithm:

$$\begin{aligned} &x_i(t)\; and\; x_j(t) \rightarrow x_i(t) \cdot x_j(t) \\ &x_i(t)\; or\; x_j(t) \rightarrow x_i(t) + x_j(t) - x_i(t) \cdot x_j(t). \\ &not\; x_i(t) \rightarrow 1 - x_i(t) \end{aligned} \tag{11}$$

The structure of the expressions associated to the logical connectors "and" and "not" is obvious, while the expression for "or" is derived by substituting such expressions into De Morgan's law: not(x_i or x_j) = (not x_i) and (not x_j). As before, it may be straightforwardly checked that these rules define a Boolean algebra. For example, a logic input like

$$f_1 = (x_1\; or\; x_2)\; and\; not(x_3)$$

would read:

$$f_1 = (x_1 + x_2 - x_1 \cdot x_2)(1 - x_3).$$

We now proceed to write the equation for the GRN continuous dynamics. By assuming that the source of gene activation can be characterized, for example, by a logistic-type behavior, we may introduce the following set of differential equations:

$$\frac{\mathrm{d}x_i}{\mathrm{d}t} = \Theta\left[f_i(x_1, \ldots, x_k)\right] - \mu_i x_i \tag{12}$$

where $\mu_i = 1/\tau_i$ represents the expression decay rate of node i of the GRN. Notice that within this approach we consider that, in general, each gene may have its own characteristic decay rate. This assumption introduces further richness into the description, as a hierarchy of times of genetic expression may define alternative routes to cell fates. In particular, notice that the steady states of the GRN, given by the condition $dx_i/dt = 0$, lead to the expression

$$x_i^S = \frac{1}{\mu_i}\Theta\left[f_i\left(x_1^S,\ldots,x_k^S\right)\right]. \tag{13}$$

Taking into account that the node inputs are defined by logical sentences with a Boolean architecture, then the attractor set obtained in this case is equivalent by construction to the set derived in the discrete Boolean approach. Thus, if a given attractor arising in the discrete Boolean approach has an expression pattern like {1,0,0,1,1,...}, the corresponding pattern in the continuous approach would have the structure $\{1/\mu_i, 0, 0, 1/\mu_4, 1/\mu_5, \ldots\}$, so that they become identical when $\mu_i = 1$ (with the possible exception of some isolated attractors). The consideration of the several relaxation rates for gene expression dynamics introduces an important difference with respect to Glass dynamics. For example, in the case that a gene has a large decay rate, corresponding to $\mu_i >> 1$, then $x_i^S \rightarrow 0$, and the expression pattern would differ with that arising when $\mu_i = 1$. Then, the dynamic behavior of a gene with a large decay rate (short expression time) would be equivalent to an effective mutation associated to lack of functionality. Similarly, the case $\mu_i << 1$ would correspond to an over-expression of that gene. We conclude that the gene expression dynamics is not only regulated by the GRN interactions topology, but also by the hierarchy of relative expression times of its components.

On the other hand, the system also may acquire very different behaviors depending on the value of the saturation rate. As mentioned before, for $b_i >> 1$, the input function becomes a Heaviside step function. In the case, $b_i = 1$, the input function would show a softer behavior. It turns out that in this latter case the attractor set may change drastically with respect to that obtained in the Boolean-like case. This plasticity could be employed to study regulatory systems with a hybrid functionality consisting of transcriptional regulatory logics that are well described with Boolean GRN, and external or coupled signaling transduction pathways that have continuous behaviors and which can impact the dynamics of some of the GRN components.

3 Notes

1. A developmental module incorporates a set of necessary and sufficient molecular components for a particular cell differentiation or morphogenetic process. It is considered a module because it is largely robust to initial conditions and it attains

GEN1(t)	GEN2(t)	GEN3(t)	TGEN(t+τ)
0	0	0	0
0	0	1	0
0	1	0	0
0	1	1	0
1	0	0	0
1	0	1	1
1	1	0	1
1	1	1	1

=

TGEN(t+τ) = GEN1(t) & (GEN2(t) | GEN3(t))

TGEN(t+τ) = (GEN1(t) & GEN3(t)) | (GEN1(t) & GEN2(t))

TGEN(t+τ) = (GEN1(t) & ! GEN2(t) & GEN3(t)) | (GEN1(t) & GEN2(t) & ! GEN3(t)) | (GEN1(t) & GEN2(t) & GEN3(t))

TGEN(t+τ) = ! (! GEN1(t) & (! GEN2(t) | ! GEN3(t)))

Fig. 9 Equivalence between truth tables and logical statements. As observed each truth table have many equivalent logical statements while each logical statement is represented by a unique truth table

certain attractors robustly. The uncovered GRN underlying the ABC patterns of gene activation and the early subdifferentiation of the flower meristem into four concentric regions or primordial floral organ cells, thus constitutes a developmental model. Other developmental modules involved in flower development could be those involved in: the cellular subdifferentiation of each one of the floral organ primordia during organ maturation, determining floral organ number and spatial disposition, in the dorso-ventrality or shape of floral organs, ovule maturation, etc.

2. In the table that formalizes the experimental data, if the gene or protein is expressed register a "1," and if not a "0." If some components have expression patterns with cyclic behavior, they could be part of cyclic attractors. In some cases, a discrete network with more than two activation states can be postulated if deemed necessary. Quantitative variation in expression levels can be also incorporated later in a continuous model approximated from the discrete one.

3. Several other algorithms exist to numerically find the attractors of a Boolean Network in an efficient way. For examples, *see* ref. 20.

4. It is important to keep in mind that the "AND" and "OR" logical operators can be interconverted. For instance, the logical statement "GEN1 AND GEN2" is equivalent to the logical statement "NOT (NOT GEN1 OR NOT GEN2)." Because of this, most truth tables (except the simplest ones, like the constants) have many equivalent logical statements. Consequently, each Boolean function can be formalized as a unique truth table, but can be described with one or many equivalent logical statements (Fig. 9).

5. Sometimes, the experimental information is not enough to completely define the Boolean functions. For example, in La

Complete characterized Boolean function

GEN1(t)	GEN2(t)	TGEN(t+τ)
0	0	0
0	1	0
1	0	0
1	1	1

Incomplete characterized Boolean function

GEN1(t)	GEN2(t)	TGEN(t+τ)
0	0	*
0	1	0
1	0	0
1	1	1

Fig. 10 Complete and incomplete characterized Boolean functions. While in complete characterized Boolean functions the value of TGEN in all row of the truth tables is specified, in incomplete characterized Boolean functions in one or more rows of the truth table is not specified. Incomplete characterized Boolean function can be the result of missing information data, asynchrony or environmental perturbations and can be resolved with different approaches as explained in the main text

Rota and collaborators [19] Boolean functions were first generated considering only confirmed direct molecular interactions. However, gaps in the experimental information precluded the generation of a unique set of Boolean function determining the GRN. Consequently, they predicted possible interactions by looking for consensus binding sites in the promoters of the included nodes and introducing some speculative hypothesis of molecular interactions.

For example, imagine that TGEN expression disappears when you generate single loss-of-function alleles of GEN1 and GEN2, while TGEN expression is promoted if we over-express both GEN1 and GEN2. Consequently, we conclude that GEN1 and GEN2 are both positive and necessary regulators for TGEN expression. However, this experimental data do not say anything about what happens to TGEN expression in the simultaneous absence of GEN1 and GEN2. In such a case we would have an incompletely characterized Boolean Function (Fig. 10). Such incompletely characterized Boolean functions can also appear due to asynchrony and interactions with the environment [45]. The inclusion of asynchrony in the model provides a more realistic description of our system, while environmental inputs influence is pervasive in biological systems. Hence, the incorporation of incomplete Boolean functions in a model is an instrumental tool. There are many ways to approach this problem: we could test all possible Boolean functions (as in ref. 19), introduce asynchrony in our model, give a probability to each possible Boolean function, or even directly work with incomplete Boolean functions. Several free software programs are capable of considering asynchrony, probabilities for different logical functions or can work with incomplete Boolean functions, such as ANTELOPE [45] and BoolNet [46].

6. Sometimes we cannot represent the available experimental data with a Boolean formalism because we need more values to represent our nodes' activity. For example, imagine that GEN1 differentially affect TGEN in the loss-of function, when normally expressed and when over expressed. This can be resolved replacing the Boolean formalism with a multivalued or a continuous approach. In a multivalued approach, the nodes can take as many values as necessary. In the last example, we could allow GEN1 to have three values, namely, 0 when is OFF, 1 when is normally expressed and 2 when is over expressed. It is important to note that a Boolean formalism can be approximated to a continuous one as was explained in the last section of this paper. For example, Espinosa-Soto and collaborators [7] initially followed a multivalued modeling approach, which was later shown to yield the same qualitative results when transformed into a Boolean system [17]. Similar situations have been documented when transforming a continuous into a Boolean model (e.g., [6, 47]). Currently some software applications allow the analysis of discrete multivalued networks (e.g., GINSIM) [48].
7. As mentioned above, sometimes the experimental information is not enough to generate the Boolean function. We can also find contradictory information linked to particular gene interactions. For example, one author may report that GEN1 positively regulates TGEN, while another one may report that GEN1 is a negative regulator of TGEN. In cases like this, models are extremely helpful, even when they could be considered incomplete. With models we can test both suggestions in a fast and cheap way. The result that better reproduces the experimentally observed system's behavior should be considered the most likely hypothesis. For example, in La Rota and collaborators [19] GRN model of sepal primordium they generated multiple sets of Boolean functions describing their GRN and selected those that recovered the expected attractors and mutant phenotypes. At other times GRN models can be also used to explain apparent contradictions or disputes concerning the interpretation of experimental data.
8. There are several free software packages to recover the attractors and basins of attraction of Boolean GRN, including ANTELOPE [45], GINSIM [48], BoolNet [46], Atalia [12], GNbox [49], GNA [50], and BioCham [51].
9. It is important to note that recovering the expected attractors when the mutants are simulated does not guarantee that the model is correct, because networks with different topologies can sometimes reach the same attractors [52]. However, we can assure that a GRN model that is unable to reproduce all mutants is incorrect.

10. Although stochasticity in Boolean models of GRNs is commonly modeled using the SIN model (*see* Subheading 2.4.1), another method called the stochasticity in functions (SIF) has been introduced recently. The objective of this method is to model stochasticity at the level of biological functions (i.e., Boolean functions in the GRN), and not just by flipping the state of a gene as in the SIN model (for details *see* refs. 20, 32).

11. It could be the case that interesting, nontrivial behaviors may be uncovered just at certain levels of the error probability ξ (e.g., noise). Thus, as customary in numerical explorations, it is necessary to test different values of ξ. However, one expects generic, robust behavior to be observed under a relatively wide range of noise levels. Moreover, the stochastic modeling of GRN can thus be useful to make inferences concerning the range of noise levels that are experienced in particular developmental systems under study.

12. When trying to iterate Eq. 4, make sure that the order in which the position corresponding to each attractor state in the initial vector $p_X(0)$ is the same as the one for the columns in the transition the matrix *P*. In other words, if the fraction of cells in attractor *A* is specified in the position *i* of the initial vector, the row *i* of the transition matrix should correspond to the probabilities of transition from attractor *A* to the other attractors.

13. It can be the case that the heights of the trajectories, which correspond to the temporal evolution of the probability of being in each attractor, differ considerably. This is to be expected; given that the basins of the different attractors vary in size, and so do their absolute probabilities. One way to transform the data in order to obtain a graph where the heights of the trajectories are of comparable size is to normalize each probability value with respect to the maximum of each attractor's curve (e.g., dividing the probability value by the maximum value). We followed this approach to obtain the graph in Fig. 8, where also the trajectories corresponding to attractors *se1* and *se2*; and *st1* and *st2* where respectively added to obtain only one trajectory for the attractor *se* and one for *st*. However, it is important to note that, as we are interested in the temporal order in which the attractors reach its maximum probability, this normalization process is not necessary. The order of appearance of the maximum value of the probability of each attractor in the original simulated trajectories would be the same as the one observed in the normalized trajectories. The normalization step just allows us to obtain a clearer graph. In the graph in Fig. 7, we draw one star for each of the attractors just above the x-axis at the time when its maximal probability occurs. The observed pattern is exactly the same in the simulated trajectories before the normalization.

References

1. Villarreal C, Padilla-Longoria P, Alvarez-Buylla ER (2012) General theory of genotype to phenotype mapping: derivation of epigenetic landscapes from N-node complex gene regulatory networks. Phys Rev Lett 109(118102):1–5
2. Alvarez-Buylla ER, Balleza E, Benítez M, Espinosa-Soto C, Padilla-Longoria P (2008) Gene regulatory network models: a dynamic and integrative approach to development. SEB Exp Biol Ser 61:113–139
3. Alvarez-Buylla ER, Azpeitia E, Barrio R, Benítez M, Padilla-Longoria P (2010) From ABC genes to regulatory networks, epigenetic landscapes and flower morphogenesis: making biological sense of theoretical approaches. Semin Cell Dev Biol 21(1):108–117
4. Alvarez-Buylla ER, Chaos A, Aldana M, Benítez M, Cortes-Poza Y, Espinosa-Soto C, Hartasánchez DA, Lotto RB, Malkin D, Escalera Santos GJ, Padilla-Longoria P (2008) Floral morphogenesis: stochastic explorations of a gene network epigenetic landscape. PLoS One 3(11):e3626
5. Mendoza L, Alvarez-Buylla ER (1998) Dynamics of the genetic regulatory network for *Arabidopsis thaliana* flower morphogenesis. J Theor Biol 193(2):307–319
6. Albert R, Othmer HG (2003) The topology of the regulatory interactions predicts the expression pattern of the segment polarity genes in *Drosophila melanogaster*. J Theor Biol 223(1): 1–18
7. Espinosa-Soto C, Padilla-Longoria P, Alvarez-Buylla ER (2004) A gene regulatory network model for cell-fate determination during *Arabidopsis thaliana* flower development that is robust and recovers experimental gene expression profiles. Plant Cell 16:2923–2939
8. Azpeitia E, Benítez M, Vega I, Villarreal C, Alvarez-Buylla ER (2010) Single-cell and coupled GRN models of cell patterning in the *Arabidopsis thaliana* root stem cell niche. BMC Syst Biol 4:134
9. Albert I, Thakar J, Li S, Zhang R, Albert R (2008) Boolean network simulations for life scientists. Source Code Biol Med 3:16
10. Albert R, Wang RS (2009) Discrete dynamic modeling of cellular signaling networks. Methods Enzymol 467:281–306
11. Assmann SM, Albert R (2009) Discrete dynamic modeling with asynchronous update, or how to model complex systems in the absence of quantitative information. Methods Mol Biol 553:207–225
12. Alvarez-Buylla ER, Benítez M, Corvera-Poiré A, Chaos CA, de Folter S, Gamboa de Buen A, Garay-Arroyo A, García-Ponce B, Jaimes- MF, Pérez-Ruiz RV, Piñeyro-Nelson A, Sánchez-Corrales YE (2010) Flower development. Arabidopsis Book 8:e0127
13. Pelaz S, Tapia-López R, Alvarez-Buylla ER, Yanofsky MF (2001) Conversion of leaves into petals in Arabidopsis. Curr Biol 11(3): 182–184
14. Barrio RÁ, Hernández-Machado A, Varea C, Romero-Arias JR, Alvarez-Buylla E (2010) Flower development as an interplay between dynamical physical fields and genetic networks. PLoS One 5(10):e13523
15. Kauffman S (1969) Homeostasis and differentiation in random genetic control networks. Nature 224:177–178
16. Mendoza L, Thieffry D, Alvarez-Buylla ER (1999) Genetic control of flower morphogenesis in *Arabidopsis thaliana*: a logical analysis. Bioinformatics 15(7–8):593–606
17. Chaos Á, Aldana M, Espinosa-Soto C et al (2006) From genes to flower patterns and evolution: dynamic models of gene regulatory networks. J Plant Growth Regul 25(4):278–289
18. Sanchez-Corrales YE, Alvarez-Buylla ER, Mendoza L (2010) The *Arabidopsis thaliana* flower organ specification gene regulatory network determines a robust differentiation process. J Theor Biol 264:971–983
19. La Rota C, Chopard J, Das P, Paindavoine S, Rozier F, Farcot E, Godin C, Traas J, Monéger F (2011) A data-driven integrative model of sepal primordium polarity in Arabidopsis. Plant Cell 23(12):4318–4333
20. Garg A, Mohanram K, De Micheli G, Xenarios I (2012) Implicit methods for qualitative modeling of gene regulatory networks. Methods Mol Biol 786:397–443
21. Alvarez J, Guli CL, Yu XH, Smyth DR (1992) terminal flower: a gene affecting inflorescence development in *Arabidopsis thaliana*. Plant J 2(1):103–116
22. Shannon S, Meeks-Wagner DR (1991) A mutation in the Arabidopsis TFL1 gene affects inflorescence meristem development. Plant Cell 3(9):877–892
23. Parcy F, Bomblies K, Weigel D (2002) Interaction of LEAFY, AGAMOUS and TERMINAL FLOWER1 in maintaining floral meristem identity in Arabidopsis. Development 129(10):2519–2527

24. Conti L, Bradley D (2007) TERMINAL FLOWER1 is a mobile signal controlling Arabidopsis architecture. Plant Cell 19(3): 767–778
25. Chen L, Cheng JC, Castle L, Sung ZR (1997) EMF genes regulate Arabidopsis inflorescence development. Plant Cell 9(11):2011–2024
26. Liljegren SJ, Gustafson-Brown C, Pinyopich A (1999) Interactions among APETALA1, LEAFY, and TERMINAL FLOWER1 specify meristem fate. Plant Cell 11(6):1007–1018
27. Ratcliffe OJ, Bradley DJ, Coen ES (1999) Separation of shoot and floral identity in Arabidopsis. Development 126(6):1109–1120
28. Gustafson-Brown C, Savidge B, Yanofsky MF (1994) Regulation of the Arabidopsis floral homeotic gene APETALA1. Cell 76(1): 131–143
29. Gómez-Mena C, de Folter S, Costa MMR, Angenent GC, Sablowski R (2005) Transcriptional program controlled by the floral homeotic gene agamous during early organogenesis. Development 132(3):429–438
30. Kitano H (2007) Towards a theory of biological robustness. Mol Syst Biol 3:137
31. Whitacre JM (2012) Biological robustness: paradigms, mechanisms, and systems principles. Front Genet 3:67
32. Garg A, Mohanram K, Di Cara A, De Micheli G, Xenarios I (2009) Modeling stochasticity and robustness in gene regulatory networks. Bioinformatics 25:i101–i109
33. Samoilov MS, Price G, Arkin AP (2006) From fluctuations to phenotypes: the physiology of noise. Sci STKE 2006:re17
34. Hoffmann M, Chang HH, Huang S, Ingber DE, Loeffler M, Galle J (2008) Noise-driven stem cell and progenitor population dynamics. PLoS One 3(8):e2922
35. Eldar A, Elowitz MB (2010) Functional roles for noise in genetic circuits. Nature 467(7312):167–173
36. Balázsi G, van Oudenaarden A, Collins JJ (2011) Cellular decision making and biological noise: from microbes to mammals. Cell 144(6):910–925
37. Horsthemke W, Lefever R (1984) Noise-induced transitions: theory and applications in physics, chemistry, and biology. Springer, Berlin
38. Chalancon G, Ravarani CNJ, Balaji S, Martinez-Arias A, Aravind L, Jothi R, Babu MM (2012) Interplay between gene expression noise and regulatory network architecture. Trends Genet 28(5):221–232
39. Glass L (1975) Classification of biological networks by their qualitative dynamics. J Theor Biol 54:85–107
40. Mendoza L, Xenarios I (2006) A method for the generation of standardized qualitative dynamical systems of regulatory networks. Theor Biol Med Model 3:13
41. Ferrell JE Jr (2012) Bistability, bifurcations, and Waddington's epigenetic landscape. Curr Biol 22:R458–R466
42. Zhou JX, Brusch L, Huang S (2011) Predicting pancreas cell fate decisions and reprogramming with a hierarchical multi-attractor model. PLoS One 6(3):e14752
43. Wang J, Zhang K, Xua L, Wang E (2011) Quantifying the Waddington landscape and biological paths for development and differentiation. Proc Natl Acad Sci 108:8257–8262
44. Zadeh LA (1965) Fuzzy sets. Inf Control 8:338–353
45. Arellano G, Argil J, Azpeitia E, Benítez M, Carrillo M, Góngora P, Rosenblueth DA, Alvarez-Buylla ER (2011) "Antelope": a hybrid-logic model checker for branching-time Boolean GRN analysis. BMC Bioinformatics 12:490
46. Müssel C, Hopfensitz M, Kestler HA (2010) BoolNet—an R package for generation, reconstruction and analysis of Boolean networks. Bioinformatics 26(10):1378–1380
47. von Dassow G, Meir E, Munro EM, Odell GM (2000) The segment polarity network is a robust developmental module. Nature 406(6792):188–192. doi:10.1038/35018085
48. Naldi A, Berenguier D, Fauré A, Lopez F, Chaouiya C (2009) Logical modelling of regulatory networks with GINsim 2.3. Biosystems 97(2):134–139
49. Corblin F, Fanchon E, Trilling L (2010) Applications of a formal approach to decipher discrete genetic networks. BMC Bioinformatics 11(1):385
50. de Jong H, Geiselmann J, Hernandez C, Page M (2003) Genetic network analyzer: qualitative simulation of genetic regulatory networks. Bioinformatics 19(3):336–344
51. Calzone L, Fages F, Soliman S (2006) Biocham: an environment for modeling biological systems and formalizing experimental knowledge. Bioinformatics 22(14):1805–1807
52. Azpeitia E, Benítez M, Padilla-Longoria P, Espinosa-Soto C, Alvarez-Buylla ER (2011) Dynamic network-based epistasis analysis: boolean examples. Front Plant Sci 2:92

INDEX

A

B

C

José Luis Riechmann and Frank Wellmer (eds.), *Flower Development: Methods and Protocols*, Methods in Molecular Biology, vol. 1110, DOI 10.1007/978-1-4614-9408-9, © Springer Science+Business Media New York 2014

MIX
Papier aus verantwortungsvollen Quellen
Paper from responsible sources
FSC® C105338

If you have any concerns about our products,
you can contact us on
ProductSafety@springernature.com

In case Publisher is established outside the EU,
the EU authorized representative is:
Springer Nature Customer Service Center GmbH
Europaplatz 3, 69115 Heidelberg, Germany

Printed by Libri Plureos GmbH
in Hamburg, Germany